THE LIBRARY
ST. MARY'S COLLEGE OF MARYLAND
ST. MARY'S CITY, MARYLAND 20686

S0-CIF-188

TRANSFER RNA:
Biological Aspects

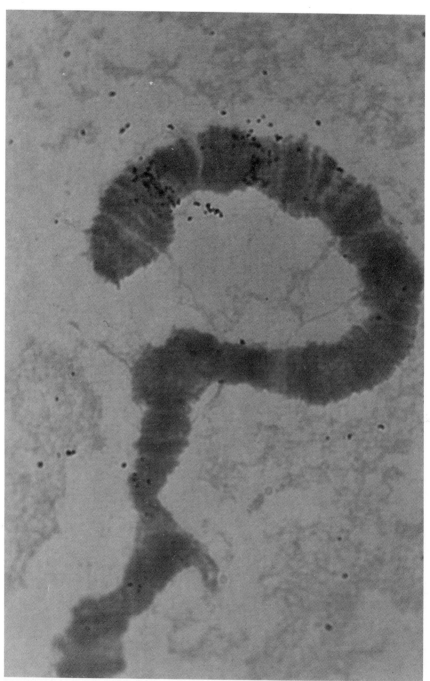

Hybridization of purified tRNA$_2^{Lys}$ to *Drosophila* salivary gland polytene chromosomes. (Photo courtesy of Shizu Hayashi and Gordon M. Tener, University of British Columbia, Vancouver.)

TRANSFER RNA:
Biological Aspects

Edited by

Dieter Söll
Yale University

John N. Abelson
University of California, San Diego

Paul R. Schimmel
Massachusetts Institute of Technology

Cold Spring Harbor Laboratory
1980

**COLD SPRING HARBOR
MONOGRAPH SERIES**
The Lactose Operon [1]
The Bacteriophage Lambda [2]
The Molecular Biology of Tumour Viruses [3]
Ribosomes [4]
RNA Phages [5]
RNA Polymerase [6]
The Operon [7]
The Single-Stranded DNA Phages [8]
Transfer RNA (*in 2 parts*):
  Structure, Properties, and Recognition, 9A
  Biological Aspects, 9B

TRANSFER RNA: Biological Aspects
© 1980 by Cold Spring Harbor Laboratory
All rights reserved
Printed in the United States of America
Book design by Emily Harste

Library of Congress Cataloging in Publication Data
Main entry under title:

Transfer RNA: biological aspects.

(Cold Spring Harbor monograph series ; 9B)
1. Ribonucleic acid, Transfer. I. Söll, Dieter.
II. Abelson, John. III. Schimmel, Paul R.
IV. Series.
QP623.T715   574.8'734   79-17578
ISBN 0-87969-129-8

Contents

Preface, xi

Biosynthesis

tRNA Synthesis, 3
G. P. Mazzara and W. H. McClain

In Vitro Synthesis of tRNA: Identification and Processing of Primary Transcripts, 29
V. Daniel, M. Zeevi, and A. Goldfarb

tRNA Precursors in RNase P Mutants, 43
Y. Shimura, H. Sakano, S. Kubokawa, F. Nagawa, and H. Ozeki

The Purification of 3' Processing Nucleases Using Synthetic tRNA Precursors, 59
R. K. Ghosh and M. P. Deutscher

Aspects of RNase P Structure and Function, 71
S. Altman, E. J. Bowman, R. L. Garber, R. Kole, R. A. Koski, and B. C. Stark

Biochemical Characterization of RNase P: A tRNA Processing Activity with Protein and RNA Components, 83
C. Guthrie and R. Atchison

rRNA and tRNA Processing Signals in the rRNA Operons of *Escherichia coli*, 99
R. A. Young, R. J. Bram, and J. A. Steitz

RNA Processing in an *Escherichia coli* Strain Deficient in Both RNase P and RNase III, 107
H. D. Robertson, E. G. Pelle, and W. H. McClain

Processing of Spacer tRNAs from rRNA Transcripts of *Escherichia coli*, 123
E. Lund, J. E. Dahlberg, and C. Guthrie

Processing of rRNA and tRNA in *Escherichia coli*: Cooperation between Processing Enzymes, 139
D. Apirion, B. K. Ghora, G. Plautz, T. K. Misra, and P. Gegenheimer

The Interaction of RNase M5 with a 5S rRNA Precursor, 155
N. R. Pace, B. Meyhack, B. Pace, and M. L. Sogin

Enzymatic Removal of Intervening Sequences in the Synthesis of Yeast tRNAs, 173
R. C. Ogden, G. Knapp, C. L. Peebles, H. S. Kang, J. S. Beckmann, P. F. Johnson, S. H. Fuhrman, and J. N. Abelson

Yeast tRNA Precursors: Structure and Removal of Intervening Sequences by an Excision-Ligase Activity, 191
P. Valenzuela, P. Z. O'Farrell, B. Cordell, T. Maynard, H. M. Goodman, and W. J. Rutter

Gene Arrangement

The Organization of tRNA Genes, 211
J. Abelson

Escherichia coli tRNA^{Tyr} Gene Clusters: Organization and Structure, 221
J. Rossi, J. Egan, M. L. Berman, and A. Landy

Cloning of Two Chemically Synthesized Genes for a Precursor to the su^+3 Suppressor tRNA^{Tyr}, 245
M. J. Ryan, E. L. Brown, R. Belagaje, H. G. Khorana, and H.-J. Fritz

tRNA Genes in rRNA Operons of *Escherichia coli*, 259
E. A. Morgan, T. Ikemura, L. E. Post, and M. Nomura

Yeast Suppressor tRNA Genes, 267
M. V. Olson, G. S. Page, A. Sentenac, K. Loughney, J. Kurjan, J. Benditt, and B. D. Hall

Mapping of tRNA Genes on the Circular DNA Molecule of *Spinacia oleracea* Chloroplasts, 281
A. Steinmetz, M. Mubumbila, M. Keller, G. Burkard, J. H. Weil, A. J. Driesel, E. J. Crouse, K. Gordon, H.-J. Bohnert, and R. G. Herrmann

Transcription and Processing of Nematode tRNA Genes Microinjected into the Frog Oocyte, 287
R. Cortese, D. Melton, T. Tranquilla, and J. D. Smith

tRNA Genes of *Drosophila melanogaster*, 295
G. M. Tener, S. Hayashi, R. Dunn, A. Delaney, I. C. Gillam, T. A. Grigliatti, T. C. Kaufman, and D. T. Suzuki

The Localization of the Genes for $tRNA_4^{Glu}$ and $tRNA_2^{Asp}$ in *Drosophila melanogaster* by In Situ Hybridization, 309
E. Kubli, T. Schmidt, and A. H. Egg

4S RNA Gene Organization in *Drosophila melanogaster*, 317
R. T. Elder, O. C. Uhlenbeck, and P. Szabo

Arrangement and Transcription of *Drosophila* tRNA Genes, 325
B. Hovemann, O. Schmidt, H. Yamada, S. Silverman, J. Mao, D. DeFranco, and D. Söll

Suppression and Coding

Genetics of Nonsense Suppressor tRNAs in *Escherichia coli*, 341
H. Ozeki, H. Inokuchi, F. Yamao, M. Kodaira, H. Sakano, T. Ikemura, and Y. Shimura

Applications of Temperature-sensitive Suppressors to the Study of Cellular Biochemistry and Physiology, 363
M. P. Oeschger

Characterization of Nonsense Suppressor tRNAs from *Saccharomyces cerevisiae*: Identification of the Mutational Alterations That Give Rise to the Suppressor Function, 379
P. W. Piper

i⁶A-deficient tRNA from an Antisuppressor Mutant of *Saccharomyces cerevisiae*, 395
H. Laten, J. Gorman, and R. M. Bock

Nonsense Suppressor tRNAs in *Schizosaccharomyces pombe*, 407
J. Kohli, F. Altruda, T. Kwong, A. Rafalski, R. Wetzel, D. Söll, G. Wahl, and U. Leupold

Interactions between UGA-suppressor tRNA^{Trp} and the Ribosome: Mechanisms of tRNA Selection, 421
R. H. Buckingham and C. G. Kurland

Use of Protein Synthesis In Vitro to Study Codon Recognition by *Escherichia coli* tRNA^{Leu} Isoaccepting Species, 427
E. Goldman and G. W. Hatfield

Nontriplet tRNA-mRNA Interactions, 439
J. F. Atkins

Other Roles of tRNA

Comments on the Role of Aminoacyl-tRNA in the Regulation of Amino Acid Biosynthesis, 453
H. E. Umbarger

Role of tRNA^{Trp} and Leader RNA: Secondary Structure in Attenuation of the *trp* Operon, 469
S. P. Eisenberg, L. Soll, and M. Yarus

Role of tRNA^{Leu} in Branched-chain Amino Acid Transport, 481
S. C. Quay and D. L. Oxender

Biochemistry and Biology of Aminoacyl-tRNA-Protein Transferases, 493
R. L. Soffer

tRNAs as Primers for Reverse Transcriptases, 507
J. E. Dahlberg

In Vitro Selective Binding of tRNAs to rRNAs of Vertebrates, 517
W. K. Yang and D. L. R. Hwang

tRNA-like Structures in Viral RNA Genomes, 539
A.-L. Haenni and F. Chapeville

Appendices

Appendix I   Localization of tRNA Genes on *Drosophila melanogaster* Polytene Salivary Gland Chromosome, 559
E. Kubli

Appendix II   Known Locations of tRNA Genes on the *Escherichia coli* Map, 563
A. Y. P. Cheung

Appendix III   Codon Usage in Several Organisms, 565
H. Grosjean

Author Index, 571

Subject Index, 573

Preface

During recent years, the advent of recombinant DNA technology, the rapid advances in DNA sequencing methods, and the establishment of efficient cell-free protein synthesizing systems from eukaryotes have opened new horizons in tRNA research and provided new tools for studies in established areas. *Transfer RNA: Biological Aspects* summarizes current knowledge in the areas of tRNA biosynthesis, tRNA gene organization and structure, genetic suppression and coding, and the role of tRNA in regulatory processes. The volume gives an overall perspective of current research in these rapidly expanding fields.

Transfer RNA: Structure, Properties, and Recognition the first portion of the Transfer RNA set, treated the structure and properties of tRNA and of aminoacyl-tRNA synthetases and the interaction of tRNAs with proteins, including ribosomes. These areas are the foundations of the tRNA field.

The chapters for both texts were solicited by the editors from individuals who delivered invited lectures at the August 1978 Cold Spring Harbor Laboratory meeting on tRNA. Many of the chapters have been updated to mid-1979. We wish to thank our colleagues for their help and suggestions.

In the organization of the meeting, we were helped enormously by Gladys Kist and Winifred Modzeleski of the Cold Spring Harbor Laboratory Meetings Office. This volume was made possible only through the conscientious and skillful editorial help of Mary-Teresa Halpin, Annette Zaninovic, and Nancy Ford, Director of Publications. We are also indebted to Jim Watson for his enthusiastic advice and counsel.

Dieter Söll
John Abelson
Paul Schimmel

Biosynthesis

tRNA Synthesis

Gail P. Mazzara and William H. McClain
Department of Bacteriology
University of Wisconsin
Madison, Wisconsin 53706

The involvement of precursor RNAs in the biosynthesis of tRNA was first observed in mammalian cells. Results of these early studies were subsequently confirmed and extended with prokaryotic cells. For example, with bacteriophage T4 the terminal steps leading to the production of T4 tRNA^{Pro} and tRNA^{Ser} have been defined in detail (McClain 1977). More recently, analysis of tRNA biosynthesis with eukaryotic cells has made substantial progress as a result of the technological developments in cloning and DNA sequencing. The study of tRNA biosynthesis in yeast has revealed the presence within certain tRNA genes of intervening sequences that are transcribed and subsequently excised from the precursor RNAs in a processing reaction that apparently has no counterpart in prokaryotic tRNA synthesis. In this paper we shall describe how it has been possible to elucidate the details of precursor RNA processing reactions to reveal the various steps in the pathway of tRNA biosynthesis.

The initial product of transcription of a tRNA gene is a longer RNA molecule (precursor RNA) that must be processed to yield mature tRNA. Because the precursor RNA contains extra nucleotide residues not present in the mature molecule, such processing includes one or more enzymatic cleavages of the precursor RNA to excise the functional tRNA. Additional processing events include enzymatic modification of nucleotide residues and addition of nucleotides to the terminal regions of precursor RNA molecules.

The discovery that tRNAs are synthesized via larger precursor RNAs has raised several questions related to the process by which these molecules are converted into tRNAs. Of special interest are the identities of the participating enzymes (processing enzymes) and the nucleotide sequences of the initial precursor RNA and of all precursor RNA intermediates generated in the production of the tRNAs. An appreciation of these details should provide a framework for asking how a precursor RNA is enzymatically recognized and handled so that its accurate conversion into tRNA is guaranteed. Perhaps the most intriguing question, but also the one we have the least expectation of answering in the near future, relates to the biological significance of the entire process: Why are precursor RNAs utilized in the production of tRNAs?

What progress has been made toward answering the questions related to tRNA biosynthesis? Much of our knowledge about this process has come from studies with microorganisms, where RNA sequence determination and mutant isolation are more readily carried out. Analysis of RNA processing in prokaryotic organisms has revealed the structure of several precursor RNA intermediates, as well as the identities of a few of the processing enzymes. From such analyses has emerged some appreciation of the interactions between these enzymes and precursor RNAs. However, many questions remain to be answered. Definition of a complete pathway for tRNA biosynthesis will require isolation and characterization of the initial tRNA gene transcript and of subsequent precursor RNA intermediates, as well as identification of every enzyme that participates in precursor RNA processing. Still to be achieved is an understanding of the molecular features underlying the specificity of interactions between the processing enzymes and precursor RNAs. Attempts to answer these questions have relied upon two major approaches: mutant methodology, which makes available mutants altered either in processing enzymes or in the tRNA genes, and in vitro reconstruction of processing reactions from partially purified precursor RNA intermediates and enzymes. Although these approaches have yielded much information about the process of tRNA biosynthesis, each is subject to limitation. With mutants in which a processing enzyme is inactivated by mutation, alternate pathways for the processing of precursor RNAs may continue to operate; the effects of such a mutation may also be obscured by residual activity of the mutant enzyme, and determination of the level of enzyme activity may represent only a gross qualitative estimate of enzyme levels within the cell. For example, enzymes that perform quite similar reactions in vitro may have entirely different functions in vivo as a result of compartmentalization or localization of these enzymes in the cell. Furthermore, if processing in the cell is carried out by a specific enzyme complex, such a complex might be difficult to reconstruct in vitro, especially from soluble enzyme fractions. In vitro studies must be approached with these limitations in mind.

ORGANIZATION AND TRANSCRIPTION OF tRNA GENES

A basic question is how the tRNA genes are arranged. One may ask whether these genes are clustered together or distributed randomly on the chromosome. Such information can yield clues to the expected products of transcription of tRNA genes as well as allow speculation about the control of tRNA biosynthesis.

Escherichia coli

In *E. coli*, tRNA genes were initially mapped by means of suppressor tRNA mutants. These genetic mapping experiments have shown that tRNA genes are distributed at various positions in the chromosome (Bachmann et al. 1976), although many of these genes are organized within small clusters that contain several closely linked tRNA sequences. These mapping data have been supported by biochemical experiments in which tRNAs have been hybridized to fragments of *E. coli* DNA (Brenner et al. 1970). More recently, Ikemura and Ozeki (1977) have determined the chromosomal locations of *E. coli* genes specifying more than 20 different tRNA species; their results confirm earlier observations that tRNA genes are clustered at several locations in the chromosome (see Appendix I). These results give some indication of the maximum size of tRNA gene transcripts; however, it has not been possible to reach any general conclusions about which kinds of tRNA are grouped together. Those clusters that contain more than one copy of a tRNA gene quite possibly arose by gene duplication; however, tRNA genes with similar sequences may occupy quite different positions on the genetic map, and genes within the same cluster may possess entirely different sequences.

Certain tRNA genes are located within some or all of the seven rRNA gene clusters in *E. coli* (Nomura et al. 1977). These tRNA genes were mapped by hybridizing tRNA to plasmids or transducing phages that carry complete or partial rRNA gene clusters.

Among the rRNA gene clusters, three encode $tRNA_{1B}^{Ala}$ and $tRNA_1^{Ile}$ in the spacer region between 16S and 23S sequences; the other four contain the genes for $tRNA_2^{Glu}$ in the spacer regions. Additional tRNAs, $tRNA^{Asp}$, $tRNA^{Trp}$, and a $tRNA^{Thr}$ species, are encoded by genes at the promoter-distal ends of rRNA gene clusters, although not all of the gene clusters contain distal tRNA genes. There is no apparent correlation between kinds of spacer tRNA genes and kinds of distal tRNA genes contained within an rRNA gene cluster. Should an entire gene cluster be transcribed as a unit, different processing pathways might operate to yield only rRNAs, only tRNAs, or both the rRNAs and their associated tRNAs. Thus, posttranscriptional processing of these tRNAs may overlap with that of rRNAs, and synthesis of these tRNAs might be coordinately controlled with the synthesis of rRNA.

Bacteriophage T4

In bacteriophage T4, which codes for eight tRNA species as well as two larger stable RNAs of unknown function (McClain et al. 1972), all ten low-molecular-weight RNAs map together in a region of less than 2500 bp (McClain et al. 1972; Wilson et al. 1972). The three precursor RNAs that have been thus far identified each contain the sequences of two tRNAs (McClain et al. 1972; Guthrie et al. 1975). These precursor RNAs are not primary transcripts, since they arise by cleavage of a larger primary transcription product. The results of in vitro transcription of the T4 tRNA genes (Goldfarb et al. 1978) suggest that the T4 tRNAs and at least some of the associated stable RNAs are transcribed as a single unit; however, attempts to isolate a single polycistronic transcript from infected cells have thus far been unsuccessful, possibly because processing of this putative precursor RNA begins before transcription of the entire region is complete. Studies currently underway should eventually yield the DNA sequence of the region coding for all eight T4 tRNAs and the associated stable RNAs (Mazzara and McClain, in prep.). This information will help to define the first step in the pathway for biosynthesis of tRNAs by revealing information about the primary transcript of these genes.

Eukaryotes

E. coli has about 60 tRNA genes and possesses one or a few copies of a particular gene for each tRNA sequence (Schweizer et al. 1969); by contrast, eukaryotes have a much higher number of tRNA genes: in yeast, 320–400 (Schweizer et al. 1969); in *Drosophila*, 750 (Weber and Berger 1976); in HeLa cells, 1300 per haploid genome (Hatlen and Attardi 1971); and in *Xenopus*, about 8000 (Clarkson et al. 1973b). Despite the large number of tRNA genes in eukaryotes, the chromatographic profiles of the tRNAs from these organisms show little more complexity than *E. coli* tRNA. One interpretation of these results is that relatively few individual tRNA genes are present in multiple copies.

The location and arrangement of eukaryotic tRNA genes have been studied in several organisms. Molecular hybridization experiments with purified tRNA and DNA from animal cells confirmed the idea that, with the exception of mitochondrial tRNAs, tRNA molecules arise as a result of transcription from nuclear DNA (Burdon 1975). In HeLa cells (Aloni et al. 1971) and in *Drosophila* (Grigliatti et al. 1973), the tRNA genes are distributed among chromosomes of all size ranges; in yeast, the eight tRNA^{Tyr} genes are unlinked (Hawthorne and Mortimer 1968; Olson et al. 1977). Recently, however, in vitro transcription and processing of the yeast genes for tRNA$_3^{Arg}$ and tRNA^{Asp}, which are cloned into the plasmid vector pBR322, have demonstrated that these tRNAs are cotranscribed in vitro to yield a dimeric precursor RNA that is subsequently processed to yield the mature tRNAs. The presence of this dimeric precursor RNA has not yet been demonstrated in vivo (Ogden et al. 1979).

In *Xenopus*, the majority of tRNA genes are clustered within DNA segments that are about ten times the size of one tRNA sequence (Clarkson et al. 1973a). It is not known whether each tRNA gene is separated from its neighbors by spacer DNA sequences or whether small clusters of adjacent tRNA genes exist. It is also not yet known how these tRNA sequences are transcribed. The only tRNA precursor RNAs isolated from eukaryotes in sufficient purity and amount for sequencing contain only one tRNA sequence, but these are probably not the primary transcripts of tRNA genes.

PRECURSOR RNA STRUCTURE

E. coli

In *E. coli*, genetic manipulation has resulted in the production of strains that accumulate precursor RNAs as a result of mutations blocking or delaying their precursor RNA processing, namely, mutations altering the processing enzymes or changing individual tRNA genes. This strategy was used for the isolation of the first precursor RNA that was sequenced, the precursor to su^+3 tRNA^{Tyr} (Altman and Smith 1971). The precursor RNA was isolated from mutants altered in the structural gene for su^+3 tRNA^{Tyr}; such mutants were identified by their altered suppressor properties. The biosynthesis of the mutant tRNA was followed in *E. coli* infected with the transducing phage $\phi 80psu^+3$, part of whose genome is replaced by a segment of the bacterial chromosome carrying the su^+3 tRNA$_1^{Tyr}$ gene. In this system, the expression of the tRNA gene is greatly amplified as a result of replication of the phage DNA.

Several mutants of the su^+3 tRNA$_1^{Tyr}$ gene were isolated and these were shown to contain single-base substitutions in the su^+3 tRNA. Most of these mutants did not synthesize the mature tRNA in normal amounts and transiently accumulated a precursor RNA. Nucleotide sequence analysis of precursor RNAs from several mutants showed that they consist of a 41-nucleotide leader sequence beginning with pppG, followed by the tRNA sequence including the CCA terminal with two or three additional residues attached to the 3' end. The precursor RNA contained few modified nucleotides.

Other *E. coli* precursor RNAs have been isolated from a temperature-sensitive mutant for the activity of the endoribonuclease RNase P (Schedl et al. 1975). When the mutant strain is grown at the non-permissive temperature, a variety of precursor RNAs accumulate; these range in size from 75 to about 600 nucleotides. The larger ones contain several tRNA sequences separated by inter-tRNA spacer nucleotides; in addition, the 3' and 5' ends of the precursor RNAs contain additional nucleotides not present in the mature tRNA sequences. However, most of the precursor RNAs that accumulate are small and contain only one tRNA sequence; sometimes, the same tRNA sequence is found in one or more larger precursor RNAs. One explanation for the presence of multiple precursor RNAs containing the same tRNA sequence is that the smaller molecules are derived from the larger precursor RNAs by endonucleolytic cleavage in the spacer regions separating tRNA sequences. An endonuclease activity with the appropriate specificity, called RNase P2 (Schedl et al. 1975) or RNase O (Sakano and Shimura 1975), has been identified and partially purified.

Some of the precursor RNAs that accumulate in RNase-P-defective cells have been subjected to nucleotide sequence analysis. Among these are monomeric precursors to $tRNA^{Phe}$, $tRNA_2^{Glu}$, and $tRNA^{Asp}$ (Vögeli et al. 1975, 1977). These precursor RNAs contained extra sequences at the 5' terminals, but not at the 3' terminals. A dimeric precursor RNA containing $tRNA_2^{Gly}$ and $tRNA^{Thr}$ has also been sequenced (Chang and Carbon 1975). Like the $tRNA_1^{Tyr}$ precursor RNA, it contains extra residues at both the 5' and 3' terminals, in addition to a 6-nucleotide spacer region between the two tRNA sequences. All of the *E. coli* precursor RNAs thus far examined contain the 3'-terminal CCA sequence of the tRNA; as will be discussed, precursor RNA chains of other organisms do not always contain the 3'-terminal CCA residues.

Bacteriophage T4

The tRNAs of bacteriophage T4 have provided a useful system for the study of tRNA biosynthesis. Following bacteriophage infection, T4 DNA is transcribed to yield several precursor RNAs that are then processed by the host enzyme system into eight tRNAs. Since *E. coli* DNA is not transcribed during infection, one can selectively label with ^{32}P bacteriophage-specific precursor RNAs and tRNAs. The labeled RNAs can be purified by polyacrylamide gel electrophoresis and subjected to nucleotide sequence analysis. These procedures permitted the identification of eight T4 tRNA species, two large stable RNAs, and three precursor RNAs, each of which contained the sequences of two tRNA species (McClain et al. 1972; Guthrie et al. 1975).

The processing of one of the T4 precursor RNAs, which contains the nucleotide sequences of $tRNA^{Pro}$ and $tRNA^{Ser}$ (proline-serine precursor RNA), has been studied in some detail. As a first step in this analysis, the nucleotide sequences of the mature tRNAs and of the precursor RNA were determined; these are shown schematically in Figure 1. One feature particularly distinguishes the proline-serine precursor RNA from *E. coli* precursor RNAs thus far examined: It lacks the CCA sequence corresponding to the 3' terminals of the mature tRNA molecules.

Inspection of the proline-serine precursor RNA sequence reveals the alterations required to convert it to mature $tRNA^{Pro}$ and $tRNA^{Ser}$: (1) a pair of endonucleolytic cleavages to generate two smaller precursor RNAs that contain the 5' terminals of the tRNAs; (2) nucleolytic removal of residues from the 3' terminals of these smaller precursor RNAs; (3) synthesis of the

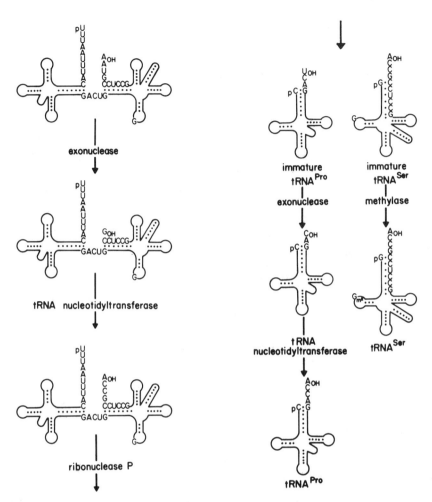

Figure 1 Seven terminal steps in the biosynthetic pathway leading from proline-serine precursor RNA to tRNA^{Pro} and tRNA^{Ser}. Nucleotide residues not associated with the reactions are represented by a continuous line. Dots represent the hydrogen-bonded base pairs of the cloverleaf forms. Note that two changes occur in the third reaction; RNase P cleaves the precursor RNA twice, but the order of these cleavages has not yet been determined (McClain et al. 1975; Seidman et al. 1975a).

3′-terminal CCA sequence of the newly generated precursor RNAs; and (4) formation of Gm in tRNA^{Ser}.

E. coli mutant strains defective in specific enzymes associated with RNA metabolism were used to identify the enzymes that catalyze these reactions and to elucidate the order of steps in the biosynthetic pathway. To determine whether a particular enzyme participates in T4 tRNA synthesis, the ability of T4 suppressor tRNAs to function in a mutant strain lacking the enzyme was examined. T4 strains carrying a suppressor form of tRNA^{Ser} together with a suppressible nonsense mutation were tested for growth on mutant and wild-type cells, both of which are nonpermissive for nonsense mutant strains. When formation of the suppressor tRNA is blocked in mutant cells, the bacteriophage will not grow. Using this technique it was shown that tRNA^{Ser} is not produced in cells lacking RNase P, tRNA nucleotidyl transferase, or an exonuclease called BN exonuclease (Deutscher et al. 1974; Seidman et al. 1975a,b). In these mutant strains, appearance of the mature tRNAs is blocked, and precursor RNAs, which are the substrates of the missing enzymes, accumulate. Nucleotide-sequence analysis of each precursor RNA intermediate indicates the kinds of enzymatic alterations to it that have occurred in previous steps; this analysis permitted the elucidation of a stepwise biosynthetic pathway leading from proline-serine precursor RNA to mature tRNAs. This pathway is shown in Figure 1.

Processing is initiated by the replacement of UAA$_{OH}$ by CCA$_{OH}$ at the 3' terminal of the proline-serine precursor RNA chain through the combined action of BN exonuclease and tRNA nucleotidyl transferase. The intermediate thus generated is then cleaved twice by RNase P to yield two tRNA-size products that contain the 5' terminals of the tRNA species; the order in which these two cleavages occur has not been established. The cleavage product derived from the 3' terminal of the precursor RNA is immature tRNA^{Ser}; addition of a 2'-O-methyl group to the molecule is the final step in tRNA^{Ser} biosynthesis. The other RNase P cleavage product is immature tRNA^{Pro}, which lacks part of the 3'-terminal CCA$_{OH}$ residues; the combined action of exonuclease and tRNA nucleotidyl transferase brings about the formation of mature tRNA^{Pro}. Reconstitution of these reactions in vitro, using purified enzymes and substrates, corroborated the results derived from in vivo analysis of the pathway (Schmidt 1975; Seidman et al. 1975b; Schmidt et al. 1976; Schmidt and McClain 1978).

A striking feature of the biosynthetic pathway that generates tRNA^{Pro} and tRNA^{Ser} is the highly specific nature of the enzymatic reactions involved. Although the molecular mechanisms underlying enzyme-RNA interactions that give rise to such specificity cannot be precisely defined yet, appreciation of the grosser features associated with this interaction has been achieved through the study of bacteriophage T4 strains that are blocked in the production of mature tRNA because of mutations in the tRNA genes (McClain et al. 1975). In general, these studies have supported the proposal that the three-dimensional conformation of precursor RNAs, which is defined by the conformations of its member tRNA species, is important for their successful interaction with processing enzymes.

Because the dimeric precursor RNAs isolated from bacteriophage T4 are not primary transcripts, other processing steps must precede the production of these precursor RNAs. One enzyme likely to be involved in this early processing is RNase III (McClain 1979).

Eukaryotes

Precursor RNAs have been identified in mammalian cells (Burdon and Clason 1969; Bernhart and Darnell 1969), silkworm silk glands (Garber et al. 1978), and yeast (Blatt and Feldman 1973; Hopper et al. 1978; Knapp et al. 1978; O'Farrell et al. 1978). They correspond to the small version of prokaryotic precursor RNAs, containing only one tRNA sequence per precursor RNA, and probably represent intermediate steps in the biosynthesis of tRNA. For the most part, the characteristics of eukaryotic precursor RNAs have not been examined in detail because of the difficulty in obtaining radiochemically pure preparations of individual species. However, in two systems, the silk glands of *Bombyx mori* and yeast, individual precursor RNAs have been isolated and characterized.

In the silk glands of *B. mori*, the synthesis of a few tRNA species, namely, tRNA^{Gly}, tRNA^{Ala}, and tRNA^{Ser}, is selective (Garber et al. 1978), a feature that facilitates isolation of individual precursor RNAs from this organism. Following two-dimensional polyacrylamide gel electrophoresis of ^{32}P-labeled RNAs from silk glands, radiochemically purified precursor RNAs to tRNA$^{Ala}_{2a}$, tRNA$^{Ala}_{2b}$, tRNA$^{Gly}_{1}$, and tRNA$^{Gly}_{2}$ were identified and characterized by fingerprint analysis. This analysis revealed that each precursor RNA contained a single tRNA sequence along with precursor-specific sequences at the 3' and 5' ends, and the tRNA sequences contained within the precursor RNAs were at least partially modified. Whether these particular precursor RNAs contained nucleoside-5'-triphosphates at the 5' terminals, thus representing primary transcripts, was not determined; however, unfractionated precursor RNA mixtures contained pppA and pppG, indicating that a fraction of these were primary transcription products. The precursor RNAs could be cleaved in vitro by cell extracts to yield tRNA molecules, but the identity of the participating enzymes and their relationship to those that function in vivo was not established.

In yeast, analysis of tRNA biosynthesis was facilitated by the determination of the DNA sequence of genes encoding tRNA^{Tyr} and tRNA^{Phe}. Yeast tRNA^{Tyr} is encoded by eight unlinked genes, each of which can mutate to a tyrosine-inserting nonsense suppressor (Hawthorne and Leupold 1974; Piper et al. 1977). Correspondingly, eight different restriction fragments produced by digestion of yeast DNA with *Eco*RI can hybridize to tRNA^{Tyr}. Since only one tRNA^{Tyr} sequence was detected in wild-type yeast cells, the eight genes are presumed to be identical or nearly identical in the region that encodes the mature tRNA sequence.

Goodman et al. (1977) identified the yeast EcoRI restriction fragment that encoded a particular tRNA^{Tyr} suppressor locus and cloned this fragment from wild-type and suppressor yeast strains. The DNA sequences of these genes and of two other genetically uncharacterized tRNA^{Tyr} genes that had also been cloned were then determined. Strikingly, the sequence analysis revealed that perfect sequence homology exists within the DNA encoding the mature tRNA^{Tyr} sequence, but flanking sequences are almost totally dissimilar. Thus, if 5' leader regions of yeast tRNAs are transcribed and then subsequently removed by a single enzyme, as they are for prokaryotic tRNAs, processing at the 5' terminals of the precursor RNAs would depend very little, if at all, on the nucleotide sequence of the 5' leader region.

None of the tRNA^{Tyr} genes contained the 3'-terminal CCA residues. This result demonstrates the necessity for posttranscriptional addition of these residues to precursor RNA.

An unusual feature of these DNA sequences is that they differ from that of tRNA^{Tyr} by virtue of a 14-bp tract that occurs just to the 3' side of the anticodon triplet. Such extra DNA sequences, called intervening sequences, have been found in a number of other eukaryotic genes, including genes coding for mouse and rabbit globin, ovalbumin, and immunoglobin (see Darnell 1978). Four alleles of yeast tRNA^{Phe} genes that have been sequenced also contain an intervening 18- or 19-bp sequence at a similar position in the tRNA sequence (Valenzuela et al. 1978). The presence of an intervening sequence within the tRNA^{Tyr} and tRNA^{Phe} genes indicates that the biosynthetic pathways for these tRNAs must involve elimination of the intervening sequence. Various possibilities for removal of this sequence include DNA splicing, RNA polymerase "jumping," or cleavage and ligation of precursor RNA. Recent investigations have provided evidence for the latter mechanism. Hopper et al. (1978) demonstrated that the yeast mutant *ts*136, which has a lesion in the *rna1* gene, accumulates a small number of precursors to tRNAs at the nonpermissive temperature. Some of these precursor RNAs contain modified nucleotides characteristic of mature tRNAs and several of them hybridized to yeast tRNA genes carried on *E. coli* plasmids. Subsequently, Knapp et al. (1978) and O'Farrell et al. (1978) characterized two of these precursor RNAs, the precursors to tRNA^{Tyr} and tRNA^{Phe}. They demonstrated, by nucleotide-sequence analysis, that these precursor RNAs contain the intervening sequences but do not possess additional nucleotides at the 5' ends. The 3' ends contain the mature CCA terminal, which is not encoded by the genes for these tRNAs and thus has been added posttranscriptionally. Modified nucleotides are also present in the precursor RNAs, although modification is incomplete.

An enzyme activity from wild-type yeast capable of excising the intervening sequence from these precursor RNAs in vitro was also identified.

These results demonstrated that the intervening sequence within the tRNA gene is transcribed and subsequently removed during processing of the precursor RNA. Removal of the intervening sequence is presumably a late step in the processing pathway, since the mature 3' and 5' terminals of the precursor RNAs are already present in the molecules analyzed by Knapp et al. (1978) and O'Farrell et al. (1978). Evidence indicates that the intervening sequences in at least some mRNAs are also transcribed and subsequently excised from the RNA molecule (Darnell 1978).

Analysis of the transcription and processing of cloned yeast tRNA genes by intact *Xenopus* oocytes (De Robertis and Olson 1979) or by extracts from *Xenopus* oocytes (Ogden et al. 1979) has confirmed and extended knowledge of the process by which these tRNAs are synthesized. In the case of tRNA^{Tyr}, for example, biosynthesis begins with the transcription of a precursor RNA that contains extra sequences at the 5' end, lacks the 3'-CCA sequence, and possesses an intervening sequence. In an ordered series of steps, this primary transcript is processed: the 5' leader sequence is removed in at least two steps, the 3'-CCA sequence is added, and several base modifications occur to yield a precursor RNA with mature 3' and 5' terminals that still contains the intervening sequence. The final major processing step is the removal of the intervening sequence and the religation of the RNA (De Robertis and Olson 1979).

O'Farrell et al. (1978) also demonstrated that the precursor RNA cannot be aminoacylated in vitro. This result may reflect the non-tRNA-like conformation of the precursor RNA. Analysis of the structure of the precursor RNA indicated that although it is similar to that of mature tRNA, the anticodon region in the precursor RNA is rearranged to accommodate residues of the intervening sequence in a base-paired structure. Thus far, no one has reported attempts to aminoacylate precursor RNAs from prokaryotic cells; these molecules might be substrates for aminoacyl-tRNA synthetases because they are similar in conformation to mature tRNA.

Interestingly, only a few precursor RNAs accumulate in $ts136$, and those precursor RNAs that have been analyzed all contain intervening sequences. By contrast, yeast tRNA$_3^{Arg}$ and tRNA^{Asp} genes do not contain intervening sequences, and their corresponding precursor RNAs do not accumulate in $ts136$ (Knapp et al. 1978). This information suggests that the *rna1*-gene function is required directly or indirectly in the removal of intervening sequences. It has been shown, however, that strain $ts136$ contains the enzyme(s) required for the removal of intervening sequences. The physiological reason that only a fraction of precursor RNAs contain intervening sequences is not known, nor is the function of the intervening sequence in the production of biologically active tRNA.

CONVERSION OF PRECURSOR RNA TO MATURE tRNA: PROCESSING ENZYMES

Conversion of the primary transcript to mature tRNA entails a number of changes in the precursor RNA. In *E. coli*, several of the enzymes responsible for catalyzing these reactions have been identified. Enzymes with similar activities in vitro have been partially purified from eukaryotic cells. A novel processing activity (or activities), which catalyzes removal of intervening sequences from yeast precursor RNAs, has been found in extracts of yeast cells (Knapp et al. 1978; O'Farrell et al. 1978) and of *Xenopus* oocytes (Ogden et al. 1979), but has not yet been substantially purified. Thus, it is not known whether one or two enzymes perform the two functions of excision of the intervening sequence and ligation of the precursor RNA chain.

Endoribonucleases

In *E. coli*, removal of nucleotides to generate the mature 5' terminal of tRNA is accomplished by the endoribonuclease RNase P. Robertson et al. (1972) purified this enzyme from *E. coli* and showed that it specifically cleaves the precursor to tRNA$_1^{Tyr}$. The enzyme has also been shown to cleave in vitro a variety of other precursor RNAs of known sequence; fingerprint analysis of the cleaved products has indicated that RNase P cleaves the precursor RNA to generate the correct 5' terminal of the tRNA contained therein (Robertson et al. 1972; Chang and Carbon 1975). Further, the properties of RNase-P-defective mutants, which accumulate a wide variety of precursor RNAs, have shown that the enzyme is involved in the processing of virtually all *E. coli* tRNAs (Schedl et al. 1975). It is also required for the biosynthesis of bacteriophage T4 tRNAs (Seidman et al. 1975a; Guthrie 1975) and the tRNAs specified by phage BF23, a relative of T5 (Sakano et al. 1974).

Examination of the nucleotide sequences in precursor RNAs that are cleaved by RNase P shows that enzyme specificity is not determined by the nucleotide sequence immediately surrounding the cleavage site. In addition, the correct nucleotide sequence of the cleavage site is not in itself sufficient for RNase P action. Schmidt et al. (1976) isolated the two oligonucleotides from the precursor to T4 tRNA^{Pro} and tRNA^{Ser} that span the RNase P cleavage sites, and tested these as possible substrates for RNase P. Under conditions where 40% of the precursor RNA was cleaved to mature tRNAs by RNase P, no cleavage of the oligonucleotides was detected. These results demonstrate that precursor RNA secondary and/or tertiary structure is required for RNase P cleavage. This conclusion is supported by the finding that mutations at numerous locations in precursor RNAs, distinct from the site of RNase P cleavage, affect processing by this enzyme (Altman and Smith 1971; McClain et al. 1975). Since the effect of these mutations is to alter the conformation of the precursor RNAs, often by disrupting the cloverleaf structure of the tRNAs contained therein, RNase P presumably recognizes the tRNA conformations of these precursor RNAs. A tRNA-like structure for precursor RNAs has, in fact, been supported by several types of experimental results (Chang and Smith 1973; Seidman and McClain 1975). The ability to recognize a common precursor RNA structure would enable RNase P to cleave a variety of precursor RNAs of different nucleotide sequences.

Thus, in interacting with precursor RNAs, processing enzymes are looking for the common features shared by these molecules, namely, their three-dimensional, tRNA-like structures. This property of processing enzymes offers an interesting contrast to the aminoacyl-tRNA synthetases with regard to their specificity. Enzymes of the latter group use as substrates only a small number of tRNA chains. In contrast, processing enzymes appear to act on a great number of precursor RNA chains. Thus, aminoacyl-tRNA synthetases recognize the differences between tRNA molecules, whereas processing enzymes recognize the similarities. Enzymes of the latter group nevertheless seem to achieve a high degree of specificity in catalyzing their reactions. But it is important to point out that we do not know if this specificity is comparable in magnitude to that of aminoacyl-tRNA synthetases, for the absence in normal cells of inaccurately processed precursor RNAs may in part reflect the action of degradative nucleases.

The tRNA-like conformation of precursor RNAs brings distant regions of the tRNA chain together to form the three-dimensional substrate structure recognized by the processing enzymes. Similarly, the secondary structure surrounding the 16S rRNA in the 30S ribosomal precursor RNA predicts that distantly spaced, complementary RNA sequences may come together to form a functional recognition site for RNase III, an enzyme involved in the processing of the 30S precursor RNA (Young and Steitz 1978).

The structural features associated with accurate RNase P cleavage have not been precisely defined. On the 5' side of the cleavage site, as few as 9 nucleotides are sufficient for cleavage of the $tRNA_1^{Tyr}$ precursor RNA (Altman et al. 1975). However, the rate at which precursor RNA containing only nine additional residues at the 5' end was cleaved by RNase P was not reported; knowledge of this rate compared to the rate of cleavage of intact precursor RNA is important for the evaluation of the relevance of in vitro data to reactions that actually occur in the cell.

The nucleotide sequence at the 3' terminal of precursor RNAs is involved in recognition of the substrate by RNase P, presumably because of its spatial proximity to the cleavage site. The state of the 3' terminal is important for the processing of bacteriophage T4 tRNAs by RNase P both in vivo and in vitro; these precursor RNAs must have a complete CCA sequence at the 3' terminal before they can be efficiently cleaved by RNase P (Seidman et al. 1975a,b; McClain and Seidman 1975). By contrast, removal of 3' extra residues from the precursor to *E. coli* tRNA$_1^{Tyr}$ decreases the ability of RNase P to cleave the precursor RNA in vitro. Altman et al. (1975) have reported that chemical or enzymatic removal of 3 or 4 nucleotides from the 3' end of tRNA$_1^{Tyr}$ precursor RNA decreases the rate of cleavage by RNase P to about 50% of that found with intact precursor RNA. These in vitro results indicate the importance of 3' terminal residues for RNase P cleavage, although, in contrast to the findings with T4 precursor RNAs, a 3'-CCA sequence in the tRNA$_1^{Tyr}$ precursor RNA is not required for RNase P action. The data also imply that in vivo RNase P cleavage occurs before trimming of the 3' terminal of the tRNA$_1^{Tyr}$ precursor RNA is completed.

How RNase P recognizes and handles a variety of precursor RNAs is not known. Robertson et al. (1972) suggested that RNase P may be associated with some nucleic acid. Guthrie and Atchison (this volume) have shown that RNase P is in association with RNA. Although preparations of RNA-free protein are devoid of RNase P activity, cleavage function can be restored in vitro by addition of RNA derived from partially purified RNase P preparations. Furthermore, Stark et al. (1978) and Altman et al. (this volume) have presented evidence that RNase P requires an associated RNA species for its function. They concluded that highly purified RNase P preparations contain a discrete RNA species; furthermore, treatment of this purified enzyme preparation with micrococcal nuclease or RNase A abolished its activity. Based on these findings, Stark and co-workers hypothesized that the interaction of RNase P with its substrates depends upon the presence of RNA in the enzyme complex. Why the RNA is required, e.g., for stabilization of an RNA-protein complex or involvement in substrate recognition, is not known. A genetic analysis of the putative RNase-P-associated RNA molecule would be invaluable. Should it be demonstrated that an RNA is essential for RNase P function, its role may be to align the enzyme on the precursor RNA by interacting with the invariant residues shared by all tRNAs.

A second endonucleolytic activity may be responsible for the partial cleavage of precursor RNAs that occurs in RNase P mutant cells. Many small precursor RNAs found in RNase P mutants are not primary transcripts but are instead cleavage products of larger, multimeric precursor RNAs (Schedl et al. 1975). Thus, another endonuclease may split within the spacer regions of these multimeric precursor RNAs to generate smaller precursor RNAs with extra sequences at the 3' and 5' terminals. An endonucleolytic activity that has these properties in vitro, called RNase P2 (Schedl et al. 1975) or RNase O (Sakano and Shimura 1975), has been partially purified. In vitro, this enzyme is capable of cleaving a number of dimeric and multimeric precursor RNAs to give molecules containing one or more tRNA sequences.

Schedl et al. (1975) have studied the actions of RNase P and RNase P2 in vitro, using as substrates the large precursor RNAs that accumulate in RNase P mutant cells. Although these precursor RNAs could be cleaved by both endonucleases, their conversion to products containing only one tRNA sequence was only achieved by a combination of RNase P and RNase P2. These results suggested that cleavage of precursor RNAs occurs sequentially, such that cleavage by one endonuclease exposes a site for the second enzyme. However, these data must be interpreted with caution, since in vitro reactions may not always be relevant to the biology of cells in vivo. For example, in vivo some processing may occur during transcription; in this case, any limitations on the action of enzymes imposed by the conformation of an intact precursor RNA in vitro may have no counterpart in vivo. A similar objection does not apply to the T4 proline-serine precursor RNA, because processing of this molecule begins at the 3' terminal, subsequent to the completion of transcription.

Studies of RNase P2 action in vitro are subject to additional limitations. The characterization of RNase P2 and RNase O is not complete; it is not known whether the two activities represent the same enzyme, or whether they differ from another *E. coli* ribonuclease, RNase III, which shares some properties with these enzymes (Dunn 1976). Their involvement in the processing of precursor RNAs can be rigorously demonstrated only by the isolation of appropriate mutants affecting these enzymes.

RNase III has been shown to participate in some instances of precursor RNA processing. In RNase-III-defective *E. coli* strains, T4 tRNA^{Gln} is not produced, although the other T4 tRNAs, including tRNA^{Leu} (which is normally processed from a dimeric precursor RNA containing tRNA^{Gln} and tRNA^{Leu}), are present (McClain 1979). Although an accumulated precursor RNA containing tRNA^{Gln} has not yet been isolated from these strains, the absence of mature tRNA^{Gln} indicates that RNase III is involved in its biosynthesis. If, as is suggested by in vitro transcription experiments (Goldfarb et al. 1978), all tRNA genes are cotranscribed, tRNA^{Gln} would be the first tRNA to be transcribed, since it maps to the 5' side of all other T4 tRNAs. Perhaps RNase III is involved in cleaving off a large leader region at the 5' end of this primary transcript; when this region is not endonucleolytically removed, the smaller precursor to tRNA^{Gln} is not formed. Evaluation of this hypothesis awaits isolation and characterization of a precursor RNA from RNase-III-mutant cells that contains the sequence of tRNA^{Gln}.

In *E. coli*, RNase III is apparently involved in the processing of rRNA transcription units, which contain tRNAs between the 16S and 23S rRNA sequences, and, in some cases, at the 3' sides of 5S rRNA sequences. In vitro processing of the 30S precursor to rRNA with RNase III or RNase P yields large tRNA precursors; when both RNase III and RNase P are used for the processing reaction, small precursor RNAs possessing mature 5' ends and extra sequences at the 3' ends are observed (Lund et al.; Young et al.; both this volume).

Yet another endoribonucleolytic activity, called RNase P4, has been partially purified by Bikoff and Gefter (1975). This enzyme, believed to be distinct from RNase P2 and RNase III, may be involved in cleavage of the primary transcript from the $tRNA_1^{Tyr}$ gene to yield a precursor RNA similar to that found in RNase P mutant cells.

With regard to endoribonucleases in eukaryotic cells, RNase-P-like activities have been identified in mammalian cells (Koseki et al. 1976) and silkworms (Garber et al. 1978). Although these activities were identified by virtue of their ability to accurately cleave, in vitro, the prokaryotic $tRNA_1^{Tyr}$ precursor RNA, it must be shown that these partially purified enzyme activities can also process their homologous eukaryotic precursor RNAs. This has been done with extracts of KB cells (Koseki et al. 1976) and silkworms (Garber et al. 1978). However, it has not been demonstrated that these enzymes generate the correct 5'-terminal sequence nor whether they participate in precursor RNA processing in vivo.

Exoribonucleases

Precursor RNAs often contain, at the 3' terminal, extra nucleotide residues either following the CCA sequence or replacing it. In the former case, the additional residues are removed, leaving the mature CCA sequence; in the latter case, the extra residues are removed to allow tRNA nucleotidyl transferase to synthesize the mature CCA sequence. In both instances, the responsible exonucleases must remove only precursor-RNA-specific residues, leaving the remainder of the tRNA molecule intact.

Crude extracts of *E. coli* are capable of removing nucleotides from the 3' terminals of precursor RNAs in vitro. However, several *E. coli* exoribonucleases are able to perform this action, and attempts to identify the enzymes responsible for processing of the 3' terminals of precursor RNAs in vivo have yielded conflicting results. Schedl et al. (1975) discovered an activity, RNase P3, that copurified with RNase II a $3' \rightarrow 5'$ exonuclease; although this enzyme can apparently generate the mature 3' terminal of $tRNA_1^{Tyr}$ in vitro, its role in the in vivo synthesis of tRNA has never been established and is seriously doubted. Sakano and Shimura (1975) described an exonuclease, designated RNase Q, capable of removing extra nucleotides from the 3' terminals of precursor RNAs; its relationship to RNase II is not known. An apparently distinct activity, identified by Bikoff et al. (1975), can perform this same exonucleolytic function.

Nevertheless, although these *E. coli* exonucleases are capable of 3'-end processing in vitro, not all of them may perform that function in vivo.

One enzyme known to participate in precursor RNA processing in vivo is the BN exonuclease. In *E. coli* mutant strains that lack this enzyme, some T4 precursor RNAs cannot be processed to mature tRNAs because extra nucleotides at the 3' terminals are not removed from them (Seidman and McClain 1975; Seidman et al. 1975b). BN exonuclease is required for the removal of extra nucleotides from the 3' ends of precursors to T4 tRNA^{Pro}, tRNA^{Ser}, and tRNA^{Ile} before the CCA terminal can be synthesized by tRNA nucleotidyl transferase. Schmidt and McClain (1978) have partially purified the BN exonuclease and have shown it to be capable of removing in vitro 3' residues from precursors to tRNA^{Pro}, tRNA^{Ser}, and tRNA^{Ile}. Since mutant cells that lack this enzyme are viable, with growth properties indistinguishable from wild-type *E. coli*, the BN enzyme may not function in the maturation of *E. coli* tRNAs. Furthermore, residues extraneous to the CCA sequence in certain bacteriophage precursor RNAs, namely, the precursor to T4 tRNA^{Thr} (W. H. McClain, unpubl.) and T2 tRNA^{Ser} (Seidman et al. 1975b), are removed in the strain that lacks BN exonuclease. These observations point to the existence of at least two 3' ribonucleases for the maturation of tRNA precursors, although there is a possibility that strain BN contains residual 3' ribonuclease activity sufficient to support the normal synthesis of some tRNAs. Nevertheless, the possible existence of two 3' ribonucleases makes logical sense; of the two putative 3' exonucleases, BN enzyme would normally function to remove from precursor RNAs nucleotides that occupy the position of the CCA sequence; a second 3' exonuclease would remove residues extraneous to the CCA sequence.

Recently, Ghosh and Deutscher (1978 and this volume) have reported the purification of an exonuclease, RNase D, that seems likely to be involved in the removal of nucleotides extraneous to the CCA sequence of precursor RNAs. Using synthetic precursor RNAs as substrates, they purified RNase D and showed that it is highly specific for precursor RNAs containing extra nucleotides at the 3' terminal and it has low activity on intact tRNA. The enzyme can also remove extra nucleotides from the 3' end of *E. coli* tRNA$_1^{Tyr}$ precursor RNA, but only after prior cleavage with RNase P. These findings suggest that the in vivo order of processing is RNase P cleavage followed by exonuclease action at the 3' terminal. Although preliminary results indicate that RNase D generates the mature 3' terminal of tRNA$_1^{Tyr}$, further experiments are required to determine whether the enzyme can quantitatively and specifically remove the extra 3' residues from a natural precursor RNA.

RNase D thus possesses those properties that are expected of a 3' processing exonuclease that removes extra residues, but stops at the CCA sequence. How might such specificity be achieved? Ghosh and Deutscher

(1978) showed that RNase D acts upon tRNA molecules with altered conformations. Perhaps a precursor RNA is recognized as an altered tRNA. Alternatively, the CCA sequence itself may confer resistance to the nuclease, since removal of these nucleotides from a precursor RNA will render it susceptible to degradation by RNase D.

Further insight into the features underlying processing enzyme specificity has been obtained through the study of mutant precursor RNAs whose maturation is blocked as a result of their mutations. McClain et al. (1975) isolated a series of T4 mutants defective in the production of tRNA^{Ser}. Nucleotide substitutions that block the production of tRNA^{Ser} are confined to one part of the proline-serine precursor RNA molecule containing the sequence of tRNA^{Ser}; this result indicates that the processing enzymes primarily interact with the residues of tRNA^{Ser}. Nucleotide substitutions in the amino acid stem of the tRNA^{Ser} portion of the proline-serine precursor RNA completely block RNase P action; however, in these mutants, the UAA$_{OH}$ sequence at the 3' terminal (see Fig. 1) is removed by the BN enzyme, although removal is not always complete. These results indicate that the integrity of the amino acid stem of tRNA^{Ser} is not necessary for the action of BN exonuclease. One might have speculated that the terminal of the amino acid stem would provide a reference point to guide the tailoring of the 3' terminal of the molecule, but the results mentioned above demonstrate that this part of the molecule neither limits nor directs the action of BN exonuclease. How BN exonuclease removes extra residues while preserving the remainder of the precursor RNA sequence is an interesting question for further study.

Additional studies have shown that nucleotide substitutions in virtually all segments of the tRNA^{Ser} portion of the proline-serine precursor RNA prevent BN exonuclease action. Presumably, these mutant precursor RNAs are not recognized by the BN enzyme, or, if they are recognized, the action of the enzyme is not limited, so that continued degradation of the molecule occurs. Certainly, not all of the residues identified in the mutant precursor RNAs participate directly in the exonuclease-precursor RNA complex. Rather, these results are an expression of the fact that tRNA molecules exhibit long-range order, so that alterations in one part of the molecule may be observed by a change in a property at another part of the molecule. With this interpretation comes an appreciation of the importance of normal precursor RNA conformation for its successful interaction with BN exonuclease.

tRNA Nucleotidyl Transferase

All tRNAs contain the 3'-terminal sequence CCA, which is required for the acceptor activity of mature tRNA. The biosynthetic origin of this sequence has been in question, since there exists an enzyme, tRNA nucleotidyl transferase, capable of adding the CCA sequence to those tRNA molecules from which it is absent (Deutscher 1973). The existence of this enzyme does not require the CCA sequence to be encoded in the DNA; in fact, nucleotide-sequence analysis of the genes encoding yeast tRNA^{Tyr} and tRNA^{Phe} demonstrated that the CCA sequence was not present in these tRNA genes, implying that tRNA nucleotidyl transferase or a similar activity is required for the biosynthesis of these tRNAs. However, the CCA sequence is present in many bacterial precursor RNAs (Altman and Smith 1971; Chang and Carbon 1975); thus, the question was raised of what role tRNA nucleotidyl transferase plays in precursor RNA processing.

The involvement of this enzyme in tRNA biosynthesis has been studied in *E. coli* mutants with reduced tRNA nucleotidyl transferase activity (*cca* mutants) (Deutscher et al. 1975; Mazzara and McClain 1977). The enzyme was shown to be required for the biosynthesis of some bacteriophage T4 tRNAs; in *cca* mutants, the production of some T4 tRNAs is blocked, and precursor RNAs lacking the 3'-terminal CCA residues accumulate (Seidman and McClain 1975; McClain et al. 1978). Only four of the eight T4 tRNAs require tRNA nucleotidyl transferase for biosynthesis; in the remainder, the 3'-terminal CCA sequence is transcriptionally derived. Further, the 3'-CCA residues of a given tRNA species in related T-even bacteriophages may have different biosynthetic origins. In T4, the CCA sequence of tRNA^{Ser} and tRNA^{Ile} is synthesized by tRNA nucleotidyl transferase, whereas this sequence is transcriptionally derived for T2 tRNA^{Ser} and T6 tRNA^{Ile}. These different modes of synthesis are a reflection of evolution of the tRNA genes; the selective pressures, if any, that produce these changes are unclear.

One speculation that accounts for different biosynthetic origins of 3'-CCA residues in the closely related T-even bacteriophages is that primitive tRNAs did not require 3'-CCA residues to function; the requirement for these residues evolved as the protein-synthetic machinery was refined. Initially, this requirement was met in the total tRNA population by tRNA nucleotidyl transferase, with the possible assistance of a nuclease to remove the residues that were not 3'-CCA residues. Subsequently, individual tRNA genes mutated to forms that encoded the 3'-CCA residues in their DNA sequences, thereby releasing the biosynthesis of those tRNAs from a requirement for one or more enzymes. By decreasing the number of biosynthetic steps involved, and thus permitting more rapid tRNA synthesis, this feature contributed to selection for tRNA genes encoding 3'-CCA residues. This latter conclusion is based on the fact that T2 tRNA^{Ser}, containing a transcriptionally derived 3'-CCA sequence, is synthesized at a faster rate than T4 tRNA^{Ser}, where 3'-CCA residues are added posttranscriptionally (Seidman et al. 1975a). In prokaryotes like *E. coli* and T-even bacteriophages, the majority of tRNA genes apparently encode 3'

CCA residues in their DNAs. In contrast, eukaryotic tRNA genes generally do not encode 3'-CCA residues, a situation that may simply reflect decreased evolutionary pressure for the selection of a 3'-CCA sequence in the tRNA genes of these organisms.

There is no evidence for a requirement for tRNA nucleotidyl transferase in *E. coli* tRNA biosynthesis. Aminoacylation studies of tRNAs isolated from *cca* mutant strains indicated that only a limited number of these tRNAs contain defective 3' terminals (Deutscher et al. 1975). Additional experiments showed that these defective terminals arose by degradation of the tRNA chains subsequent to biosynthesis (Deutscher et al. 1977; Mazzara and McClain 1977). In the presence of chloramphenicol, an antibiotic that inhibits exonucleolytic degradation at the 3' terminals of tRNAs while permitting continued tRNA synthesis, the proportion of defective tRNA molecules rapidly declined. Thus, in strains deficient in tRNA nucleotidyl transferase, tRNAs with intact CCA terminals are synthesized. These results strongly suggest that the primary function of tRNA nucleotidyl transferase in *E. coli* is not biosynthesis of the CCA sequence of tRNA, but possibly repair of defective tRNA species that arise as a result of nucleolytic degradation of 3' terminals. This conclusion must be stated with reservation, since it was based on studies using a mutant strain that contains approximately 2% residual tRNA nucleotidyl transferase activity. It is possible that the enzyme does participate in tRNA biosynthesis, but residual enzyme activity is sufficient to perform this function in the mutant cells. It must also be noted that tRNA nucleotidyl transferase may have an additional role, aside from its repair function, that is essential for normal growth of cells (Deutscher et al. 1974). This function has not yet been identified.

Modification Enzymes

After transcription of a tRNA sequence, specific nucleotides within the sequence are modified by rearrangements and substitutions of additional groups (McCloskey and Nishimura 1977). Modification of nucleotides is accomplished by a number of enzymes, several of which have been identified.

Exactly when during tRNA biosynthesis does nucleotide modification occur? Guthrie et al. (1973) reported that several T4 precursor RNAs lack Gm, although all the other nucleotide modifications found in mature tRNA were found in the precursor RNAs. *E. coli* precursor RNAs also lack this residue (Schaefer et al. 1973; Sakano et al. 1974). Thus, Gm modification occurs late during precursor RNA processing. Nevertheless, many precursor RNAs have been shown to contain modified nucleotides; thus, at least some nucleotide modifications occur at the precursor

RNA level. Furthermore, certain studies indicate that nucleotide modification is not a prerequisite to cleavage of precursor RNA. Cleavage of the precursor to tRNA$_1^{Tyr}$ by RNase P in vitro proceeds in the absence of complete nucleotide modification (Schaefer et al. 1973). In addition, tRNA molecules that are undermodified can be isolated from a variety of sources (see, e.g., Gefter and Russell 1969) indicating that cleavage of precursor RNAs to produce these tRNAs does not have an absolute requirement for the presence of modified nucleotides. However, it has not been experimentally demonstrated that undermodified precursor RNAs are cleaved at the same rate as normally modified molecules. It is possible that cleavage of undermodified precursor RNAs is less efficient.

Although several modification enzymes have been isolated, little is known about how they recognize the specific nucleotides they modify. Baguley et al. (1970) and Kuchino and Nishimura (1970) showed that tRNA methylases require specific regions of the cloverleaf structure of different tRNAs as well as specific nucleotide sequences. The importance of precursor RNA conformation for reaction with modification enzymes was further demonstrated by McClain and Seidman (1975), who reported that certain T4 precursor RNAs that contain base substitution mutations are undermodified, presumably because the conformation of the tRNA sequence within the precursor RNA is altered.

The function of modified nucleotides in tRNA has recently been reviewed (McCloskey and Nishimura 1977). Certain methylations or other aliphatic modifications affect tRNA function during protein synthesis (Gefter and Russell 1969; Björk and Neidhardt 1975; Laten et al., this volume). Thiolation and other modifications of uracil in the first position of the anticodon confer on tRNAs their specific codon recognition properties. Modified nucleotides may also be important in functions other than protein synthesis; a ψ modification in the anticodon stem is implicated in the regulation of certain operons (Singer et al. 1972).

REFERENCES

Aloni, Y., L. E. Hatlen, and G. Attardi. 1971. Studies of fractional HeLa cell metaphase chromosomes. II. Chromosomal distribution of sites for transfer RNA and 5S RNA. *J. Mol. Biol.* **56**:555.

Altman, S. and J. D. Smith. 1971. Tyrosine tRNA precursor molecule polynucleotide sequence. *Nat. New Biol.* **233**:35.

Altman, S., A. L. M. Bothwell, and B. C. Stark. 1975. Processing of *E. coli* tRNA^{Tyr} precursor RNA *in vitro*. *Brookhaven Symp. Biol.* **26**:12.

Bachmann, B. J., K. B. Low, and A. L. Taylor. 1976. Recalibrated linkage map of *Escherichia coli* K12. *Bacteriol. Rev.* **40**:116.

Baguley, B. C., W. Wehrli, and M. Staehelin. 1970. *In vitro* methylation of yeast serine transfer ribonucleic acid. *Biochemistry* **9**:1645.

Bernhardt, D. and J. E. Darnell. 1969. tRNA synthesis in HeLa cells: A precursor to tRNA and the effects of methionine starvation on tRNA synthesis. *J. Mol. Biol.* **42**:43.

Bikoff, E. K. and M. L. Gefter. 1975. In vitro synthesis of transfer RNA. I. Purification of required components. *J. Biol. Chem.* **250**:6240.

Bikoff, E. K., B. F. LaRue, and M. L. Gefter. 1975. In vitro synthesis of transfer RNA. II. Identification of required enzymatic activities. *J. Biol. Chem.* **250**:6248.

Björk, G. R. and F. C. Neidhardt. 1975. Physiological and biochemical studies on the function of 5-methyluridine in the transfer ribonucleic acid of *Escherichia coli*. *J. Bacteriol.* **124**:99.

Blatt, B. and H. Feldman. 1973. Characterization of precursors to tRNA in yeast. *FEBS Lett.* **37**:129.

Brenner, D. J., M. Y. Fournier, and B. P. Doctor. 1970. Isolation and partial characterization of the transfer ribonucleic acid cistrons from *Escherichia coli*. *Nature* **227**:448.

Burdon, R. H. 1975. Processing of tRNA precursors in higher organisms. *Brookhaven Symp. Biol.* **26**:138.

Burdon, R. H. and A. E. Clason. 1969. Intracellular location and molecular characteristics of tumour cell transfer RNA precursors. *J. Mol. Biol.* **39**:113.

Chang, S. and J. Carbon. 1975. The nucleotide sequence of a precursor to the glycine- and threonine-specific transfer ribonucleic acids of *Escherichia coli*. *J. Biol. Chem.* **250**:5542.

Chang, S. E. and J. D. Smith. 1973. Structure studies on a tyrosine tRNA precursor. *Nature New Biol.* **246**:165.

Clarkson, S. G., M. L. Birnsteil, and I. F. Purdom. 1973a. Clustering of transfer RNA genes of *Xenopus laevis*. *J. Mol. Biol.* **79**:411.

Clarkson, S. G., M. L. Birnsteil, and V. Serra. 1973b. Reiterated transfer RNA genes of *Xenopus laevis*. *J. Mol. Biol.* **79**:391.

Darnell, J. E. 1978. Implications of RNA·RNA splicing in evolution of eukaryotic cells. *Science* **202**:1257.

De Robertis, E. M. and M. V. Olson. 1979. Transcription and processing of cloned yeast tRNA genes microinjected into frog oocytes. *Nature* **278**:137.

Deutscher, M. P. 1973. Synthesis and functions of the –C–C–A terminus of transfer RNA. *Prog. Nucleic Acid Res. Mol. Biol.* **13**:51.

Deutscher, M. P., J. Foulds, and W. H. McClain. 1974. Transfer ribonucleic acid nucleotidyl transferase plays an essential role in the normal growth of *Escherichia coli* and in the biosynthesis of some bacteriophage T4 transfer ribonucleic acids. *J. Biol. Chem.* **249**:6696.

Deutscher, M. P., J. J.-C. Lin, and J. A. Evans. 1977. tRNA metabolism in *Escherichia coli* cells deficient in tRNA nucleotidyl transferase. *J. Mol. Biol.* **117**:1081.

Deutscher, M. P., J. Foulds, J. W. Morse, and R. H. Hilderman. 1975. Synthesis of the CCA terminus of transfer RNA. *Brookhaven Symp. Biol.* **26**:124.

Dunn, J. 1976. RNase III cleavage of single stranded RNA. The effect of ionic strength on the fidelity of cleavage. *J. Biol. Chem.* **251**:3807.

Garber, R. L., M. A. Q. Siddiqui, and S. Altman. 1978. Identification of precursor molecules to individual tRNA species from *Bombyx mori*. *Proc. Natl. Acad. Sci.* **75**:635.

Gefter, M. L. and R. L. Russell. 1969. A role of modifications in tyrosine transfer RNA: A modified base affecting ribosome binding. *J. Mol. Biol.* **39:**145.

Ghosh, R. K. and M. P. Deutscher. 1978. Identification of an *Escherichia coli* nuclease acting on structurally altered transfer RNA molecules. *J. Biol. Chem.* **253:**997.

Goldfarb, A., E. Seaman, and V. Daniel. 1978. Bacteriophage T4 tRNA operon: *In vitro* transcription and isolation of a polycistronic RNA product. *Nature* **273:**562.

Goodman, H. M., M. V. Olson, and B. D. Hall. 1977. Nucleotide sequence of a mutant eukaryotic gene: The yeast tyrosine-inserting ochre suppressor *SUP4-o*. *Proc. Natl. Acad. Sci.* **74:**5453.

Grigliatti, T. A., B. N. White, G. M. Tener, T. C. Kaufman, J. J. Holden, and D. T. Suzuki. 1973. Studies on the transfer RNA genes of *Drosophila*. *Cold Spring Harbor Symp. Quant. Biol.* **38:**461.

Guthrie, C. 1975. The nucleotide sequence of the dimeric precursor to glutamine and leucine transfer RNAs coded by bacteriophage T4. *J. Mol. Biol.* **95:**529.

Guthrie, C., J. G. Seidman, S. Altman, B. G. Barrell, J. D. Smith, and W. H. McClain. 1973. Identification of tRNA precursor molecules made by phage T4. *Nat. New Biol.* **246:**6.

Guthrie, C., J. G. Seidman, M. M. Comer, R. M. Bock, F. J. Schmidt, B. G. Barrell, and W. H. McClain. 1975. The biology of bacteriophage T4 transfer RNAs. *Brookhaven Symp. Biol.* **26:**106.

Hatlen, L. and G. Attardi. 1971. Proportion of the HeLa cell genome complementary to transfer RNA and 5S RNA. *J. Mol. Biol.* **56:**535.

Hawthorne, D. C. and U. Leupold. 1974. Suppressor mutations in yeast. *Curr. Top. Microbiol. Immunol.* **64:**1.

Hawthorne, D. C. and R. K. Mortimer. 1968. Genetic mapping of nonsense suppressors in yeast. *Genetics* **60:**735.

Hopper, A. K., F. Banks, and V. Evangelidis. 1978. A yeast mutant which accumulates precursor tRNAs. *Cell* **14:**211.

Ikemura, T. and H. Ozeki. 1977. Gross map location of *Escherichia coli* transfer RNA genes. *J. Mol. Biol.* **117:**419.

Knapp, G., J. S. Beckmann, P. F. Johnson, S. A. Fuhrman, and J. Abelson. 1978. Transcription and processing of intervening sequences in yeast tRNA genes. *Cell* **14:**221.

Koseki, A., A. L. M. Bothwell, and S. Altman. 1976. Identification of a ribonuclease P-like activity from human KB cells. *Cell* **9:**101.

Kuchino, Y. and S. Nishimura. 1970. Nucleotide sequence specificities of guanylate residue-specific tRNA methylases from rat liver. *Biochem. Biophys. Res. Commun.* **40:**306.

Mazzara, G. P. and W. H. McClain. 1977. Cysteine transfer RNA of *Escherichia coli*: Nucleotide sequence and unusual metabolic properties of the 3' C-C-A terminus. *J. Mol. Biol.* **117:**1061.

McClain, W. H. 1977. Seven terminal steps in a biosynthetic pathway leading from DNA to transfer RNA. *Accts. Chem. Res.* **10:**418.

———. 1979. A role for ribonuclease III in synthesis of bacteriophage T4 transfer RNAs. *Biochem. Biophys. Res. Commun.* **86:**718.

McClain, W. H. and J. G. Seidman. 1975. Genetic perturbations that reveal tertiary conformation of tRNA precursor molecules. *Nature* **257:**106.

McClain, W. H., B. G. Barrell, and J. G. Seidman. 1975. Nucleotide alterations in bacteriophage T4 serine transfer RNA that affect the conversion of precursor RNA into transfer RNA. *J. Mol. Biol.* **99:**717.
McClain, W. H., C. Guthrie, and B. G. Barrell. 1972. Eight transfer RNAs induced by infection of *Escherichia coli* with bacteriophage T4. *Proc. Natl. Acad. Sci.* **69:**3703.
McClain, W. H., J. G. Seidman, and F. J. Schmidt. 1978. Evolution of the biosynthesis of 3'-terminal C-C-A residues in T-even bacteriophage transfer RNAs. *J. Mol. Biol.* **119:**519.
McCloskey, J. A. and S. Nishimura. 1977. Modified nucleosides in transfer RNA. *Accts. Chem. Res.* **10:**403.
Nomura, M., E. A. Morgan, and S. R. Jaskunas. 1977. Genetics of bacterial ribosomes. *Annu. Rev. Genet.* **11:**297.
O'Farrell, P. Z., B. Cordell, P. Valenzuela, W. J. Rutter, and H. M. Goodman. 1978. Structure and processing of yeast precursor tRNAs containing intervening sequences. *Nature* **274:**438.
Ogden, R. C., J. S. Beckmann, J. Abelson, H. S. Kang, D. Söll, and O. Schmidt. 1979. *In vitro* transcription and processing of a yeast tRNA gene containing an intervening sequence. *Cell* **17:**399.
Olson, M. V., D. L. Montgomery, A. K. Hopper, G. S. Page, F. Horodyski, and B. D. Hall. 1977. Molecular characterization of the tyrosine tRNA genes of yeast. *Nature* **267:**639.
Piper, P. W., M. Wasserstein, F. Engbaek, K. Kaltoft, J. E. Celis, J. Zeuthen, S. Leibman, and F. Sherman. 1977. Nonsense suppressors of *Saccharomyces cerevisiae* can be generated by mutation of the tyrosine tRNA anticodon. *Nature* **262:**757.
Robertson, H. D., S. Altman, and J. D. Smith. 1972. Purification and properties of a specific *Escherichia coli* ribonuclease which cleaves tyrosine transfer ribonucleic acid precursor. *J. Biol. Chem.* **247:**5243.
Sakano, H. and Y. Shimura. 1975. Sequential processing of precursor tRNA molecules in *Escherichia coli*. *Proc. Natl. Acad. Sci.* **72:**3369.
Sakano, H., S. Yamada, T. Ikemura, Y. Shimura, and H. Ozeki. 1974. Temperature sensitive mutants of *Escherichia coli* for tRNA synthesis. *Nucleic Acids Res.* **1:**355.
Schaefer, K. P., S. Altman, and D. Söll. 1973. Nucleotide modification *in vitro* of the precursor of transfer RNA^{Tyr}. *Proc. Natl. Acad. Sci.* **70:**3626.
Schedl, P., P. Primakoff, and J. Roberts. 1975. Processing of *E. coli* tRNA precursors. *Brookhaven Symp. Biol.* **26:**53.
Schmidt, F. J. 1975. A novel function of *Escherichia coli* transfer RNA nucleotidyl transferase. *J. Biol. Chem.* **250:**8399.

Schmidt, F. J. and W. H. McClain. 1978. An *Escherichia coli* ribonuclease which removes an extra nucleotide from a biosynthetic intermediate of bacteriophage T4 proline transfer RNA. *Nucleic Acids Res.* **5**:4129.

Schmidt, F. J., J. G. Seidman, and R. M. Bock. 1976. Transfer ribonucleic acid biosynthesis. Substrate specificity of ribonuclease P. *J. Biol. Chem.* **251**: 2440.

Schweizer, E., C. MacKechnie, and H. O. Halvorson. 1969. The redundancy of ribosomal and transfer RNA genes in *Saccharomyces cerevisiae. J. Mol. Biol.* **40**:261.

Seidman, J. G. and W. H. McClain. 1975. Three steps in conversion of large precursor RNA into serine and proline transfer RNAs. *Proc. Natl. Acad. Sci.* **72**:1491.

Seidman, J. G., B. G. Barrell, and W. H. McClain. 1975a. Five steps in the conversion of a large precursor RNA into bacteriophage proline and serine transfer RNAs. *J. Mol. Biol.* **99**:733.

Seidman, J. G., F. J. Schmidt, K. Foss, and W. H. McClain. 1975b. A mutant of *Escherichia coli* defective in removing 3' terminal nucleotides from some transfer RNA precursor molecules. *Cell* **5**:389.

Singer, C. E., G. R. Smith, R. Cortese, and B. N. Ames. 1972. Mutant tRNA[His] ineffective in repression and lacking two pseudouridine modifications. *Nat. New Biol.* **238**:72.

Stark, B. C., R. Kole, E. J. Bowman, and S. Altman. 1978. Ribonuclease P: An enzyme with an essential RNA component. *Proc. Natl. Acad. Sci.* **75**:3717.

Valenzuela, P., A. Venegas, F. Weinberg, R. Bishop, and W. J. Rutter. 1978. Structure of yeast phenylalanine-tRNA genes: An intervening DNA segment within the region coding for the tRNA. *Proc. Natl. Acad. Sci.* **75**:190.

Vögeli, G., H. Grosjean, and D. Söll. 1975. A method for the isolation of specific tRNA precursors. *Proc. Natl. Acad. Sci.* **72**:4790.

Vögeli, G., T. S. Stewart, T. McCutchan, and D. Söll. 1977. Isolation of *Escherichia coli* precursor tRNAs containing modified nucleoside Q. *J. Biol. Chem.* **252**:2311.

Weber, L. and E. Berger. 1976. Base sequence complexity of the stable RNA species of *Drosophila melanogaster. Biochemistry* **15**:5511.

Wilson, J. H., J. S. Kim, and J. N. Abelson. 1972. Bacteriophage T4 transfer RNA. III. Clustering of the genes for T4 transfer RNAs. *J. Mol. Biol.* **71**: 547.

Young, R. A. and J. A. Steitz. 1978. Complementary sequences 1700 nucleotides apart from a ribonuclease III cleavage site in *Escherichia coli* ribosomal precursor RNA. *Proc. Natl. Acad. Sci.* **75**:3593.

In Vitro Synthesis of tRNA: Identification and Processing of Primary Transcripts

Violet Daniel, Menachem Zeevi, and Alexander Goldfarb
Biochemistry Department
Weizmann Institute of Science
Rehovot, Israel

It is now generally established that mature tRNA molecules from both prokaryotic and eukaryotic sources are not direct transcription products but arise from the cleavage of longer precursor molecules (Schafer and Söll 1974; Altman 1975; Smith 1976). tRNA biosynthesis appears to involve posttranscriptional events, such as nucleolytic cleavage of precursors, nucleotide modifications at specific residues, and in some cases enzymatic terminal addition of CCA_{OH}. Precursors to several RNAs specified by *Escherichia coli* and bacteriophage T4, which contain one or more tRNA sequences, have been isolated and characterized (Altman and Smith 1971; Barrell et al. 1974; Chang and Carbon 1975; Guthrie 1975; Sakano and Shimura 1975; Schedl et al. 1976; Ilgen et al. 1976). However, these tRNA precursors, which were shown to contain complete tRNA sequences plus extra nucleotides on the 5' and 3' ends and in the inter-tRNA spacer regions, seem to represent already partially cleaved products of an initially larger transcript, the nature of which remains unknown (Altman and Smith 1971; Barrell et al. 1974; Chang and Carbon 1975).

Transcription of tRNA genes in vitro by purified RNA polymerase has the advantage of producing completely unmodified tRNA precursors and makes it possible to study in detail the processing and modification of primary transcription products. The tRNA genes carried by the DNA of transducing bacteriophages $\phi 80 p s u^+ 3$ ($tRNA_1^{Tyr}$ su^+3 and su^-3) and $\lambda h 80 d g l y T s u^+ 36$ ($tRNA_2^{Tyr}$-$tRNA_2^{Gly}$ su^+36-$tRNA_3^{Thr}$) have been efficiently transcribed in vitro by purified *E. coli* RNA polymerase to produce high-molecular-weight tRNA precursors (Daniel et al. 1970; Grimberg and Daniel 1974). The tRNA precursors were processed by *E. coli* extract S100 to mature-sized tRNA (Daniel et al. 1975) and the ability of the synthesized tRNA to undergo base modification and aminoacylation was demonstrated (Zeevi and Daniel 1976). The transcription in vitro of the $tRNA_1^{Tyr}$ gene has also been studied in several other laboratories, using as template whole $\phi 80 p s u^+ 3$ DNA (Ikeda 1971; Zubay et al. 1971; Bikoff et al. 1975; Fournier et al. 1977; Kitamura et al. 1977) or DNA fragments (Kupper et al. 1975, 1978). Different values (up to 350 nucleotides) have been reported for the size of the in vitro transcript, all larger than that of the tRNA$_1^{Tyr}$ precursor (129 nucleotides) isolated in vivo by Altman and Smith (1971). The tRNA region of T4 DNA has also been extensively studied. It was shown to contain a cluster of tRNA genes coding for eight tRNA species and two other stable RNAs of unknown function (Abelson et al. 1974; Guthrie et al. 1974). The T4 tRNA genes have been transcribed in vitro and the unfractionated products processed with E. coli extracts to produce T4 tRNAs (Nierlich et al. 1973). Although it has been suggested that the tRNA genes may be organized in T4 as a single transcription unit (Kaplan and Nierlich 1975), the formation of a long polycistronic precursor could not be demonstrated.

In this paper we describe the transcription in vitro of large tRNA precursors and their processing by crude cell extracts (S100) and purified enzymes to mature-sized tRNA. An RNA molecule about 600 nucleotides long containing the tRNA$_1^{Tyr}$ sequence was isolated from $\phi 80psu^+3$ DNA transcripts, and the tRNA gene cluster tRNA$_2^{Tyr}$-tRNA$_2^{Gly}$-tRNA$_3^{Thr}$ carried by λ h80d$glyTsu^+36$ DNA was found to be transcribed as a polycistronic tRNA precursor. The tRNA gene cluster of bacteriophage T4 DNA was transcribed in vitro and a large RNA molecule approximately 5000 nucleotides long was isolated and found to contain the sequences of the eight T4 tRNAs. The findings provide direct evidence that the clustered tRNA genes form a single transcription unit. (Parts of this work have appeared elsewhere, Goldfarb et al. [1978].)

TRANSCRIPTION OF E. COLI tRNA GENES CARRIED BY TRANSDUCING BACTERIOPHAGES

Synthesis of tRNA$_1^{Tyr}$, tRNA$_2^{Tyr}$, tRNA$_2^{Gly}$, and tRNA$_3^{Thr}$ Precursors

$\phi 80psu^+3$ and λh80d$glyTsu^+36$ DNAs were transcribed in vitro by RNA polymerase in reaction mixtures containing α-^{32}P-labeled nucleoside triphosphates. The products were analyzed by polyacrylamide gel electrophoresis under denaturing conditions (Fig. 1). An RNA band approximately 600 nucleotides long was observed as a main product of $\phi 80psu^+3$ DNA transcription (Fig. 1, slot 1, band A). Transcription of λ h80d$glyTsu^+36$ DNA yielded two prominent RNA products (Fig. 1, slot 2, bands A, B) of about 500 and 215 nucleotides long, respectively. To determine whether these primary transcription products represent tRNA precursors, the RNA bands were extracted from the gel and the RNA purified, exhaustively digested by E. coli S100 extracts and fractionated by acrylamide gel electrophoresis. Figure 2 (slot 1) shows that RNA band A of the $\phi 80psu^+3$ DNA transcript yielded tRNA^{Tyr} upon processing. The tRNA^{Tyr} was further identified by the fingerprint of its RNase T1 digest. Similar treatment of the RNA bands isolated from the λh80d$glyTsu^+36$ DNA transcript (Fig. 1, slot 2) shows that band A (500 nucleotides long) is

Figure 1 Autoradiograph of acrylamide gel separation of RNA transcribed in vitro on $\phi 80psu^+3$ (*1*) and $\lambda h80dglyTsu^+36$ (*2*) DNAs. Each phage DNA (5 μg) was transcribed in 50-μl reaction mixtures containing: 0.05 M Tris-HCl (pH 7.9); 0.01 M $MgCl_2$; 0.050 M KCl; 1 mM dithiothreitol; 0.4 mM each of ATP, CTP, and GTP; 40 μM [α-^{32}P]UTP (2.5 Ci/mmole); 3 μg rho termination factor; and 15 units RNA polymerase. RNA polymerase (1600–8000 units/mg) was used in limiting amounts (5–10 units/μg). $\phi 80psu^+3$ DNA and rho factor were in excess. The synthesis was initiated after 5-min preincubation at 38°C by addition of the nucleoside triphosphates. After a 30-min incubation at 37°C the reaction was stopped by the addition of 1 μg DNase I followed by a 10-min incubation at 37°C. RNA was extracted by phenol, precipitated by ethanol, and fractionated by acrylamide gel electrophoresis on 40 × 20 cm 5% gel (20:1 acrylamide:bis acrylamide) containing 7 M urea in Tris-borate buffer (Peacock and Dingman 1968) at 175 V for 16–19 hr. 9S rabbit globin mRNA, 5S RNA, and tRNA^{Tyr} were run separately as reference markers and were located by staining the gel with Stainsall (Eastman).

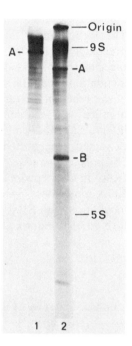

a polycistronic precursor of tRNA$_2^{Tyr}$, tRNA$_2^{Gly}$, and tRNA$_3^{Thr}$. This band was cleaved by S100 extract to produce two tRNA bands (Fig. 2, slot 2), which were identified by the fingerprints of their RNase T1 digests as tRNA$_2^{Tyr}$ and a mixture of tRNA$_2^{Gly}$ and tRNA$_3^{Thr}$. Figure 2 (slot 3) shows that RNA band B (about 215 nucleotides long) isolated from the $\lambda h80dgly$-Tsu^+36 DNA transcript yielded upon digestion two 4S RNA products, the larger was identified by fingerprint as tRNA$_2^{Tyr}$, whereas the smaller was not identified.

Figure 2 Analysis of the cleavage products obtained from isolated precursor RNA bands. RNA bands from the gel separation of in-vitro-synthesized $\phi 80psu^+3$ and $\lambda h80dglyTsu^+36$ RNAs were extracted with 0.5 M NaCl, 10 mM EDTA in 0.1 M Tris-HCl (pH 8.0). The RNA was then filtered through Sephadex G-25, precipitated by ethanol, and incubated at 37°C for 90 min with 2 mg/ml S100 extract. After phenol extraction and ethanol precipitation, the RNA was fractionated by acrylamide gel electrophoresis on 10 × 10-cm 5% gel containing 7 M urea as described in legend to Fig. 1. *E. coli* tRNA was run separately as reference marker. (*1*) Digest of RNA band A from $\phi 80psu^+3$ RNA (Fig. 1, slot 1); (*2,3*) digests of RNA bands A and B, respectively, from $\lambda h80dglyTsu^+36$ RNA (Fig. 1, slot 2).

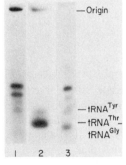

Processing of tRNA₁^Tyr Precursor

The formation of mature-sized tRNA₁^Tyr from the large 600-nucleotide precursor (Fig. 1, slot 1, band A) was studied by following the processing of the $\phi 80 psu^+ 3$ RNA transcripts by *E. coli* S100 extracts as a function of incubation time. The synthesized [α-^{32}P]RNA was extracted from a portion of the reaction mixture (zero time of processing). The rest of the reaction mixture was incubated with an *E. coli* S100 extract. At different time intervals (5, 10, 20, and 40 min) equal aliquots were removed; the RNA was extracted, precipitated by ethanol, and fractionated by electrophoresis on a 5% polyacrylamide 7 M urea gel. Figure 3A shows that the maturation of tRNA₁^Tyr is a stepwise process involving the formation of intermediate-sized RNA molecules. Concomitant with the disappearance of band A during incubation with S100 extract, additional RNA bands B, X, C, and D are formed. The intermediate character of these RNA bands was previously

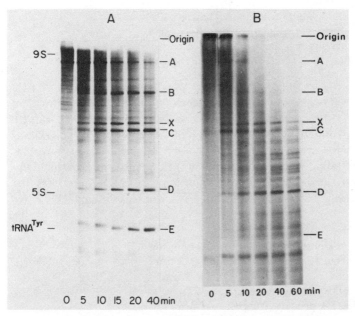

Figure 3 Cleavage products of RNA transcribed in vitro on $\phi 80 psu^+ 3$ DNA by crude S100 extract (*A*) and purified RNase P2 (*B*). $\phi 80 psu^+ 3$ DNA (20 μg) was transcribed in 200-μl reaction mixture as described in Fig. 1. The reaction was stopped by the addition of 20 μg actinomycin D and 4 μg DNase I. The RNA from a 30-μl aliquot was phenol-extracted to give the electrophoretic pattern for zero time of processing. The rest of the mixture was incubated at 37°C with 2 mg/ml S100 extract (*A*) or 30 μg/ml RNase P2 (*B*). Aliquots were removed at the times indicated, phenol was extracted, and the ethanol was precipitated and fractionated by acrylamide gel electrophoresis, as described in Fig. 1.

demonstrated by their extraction from the gel and subsequent cleavage by S100 extract to produce tRNA^{Tyr} (Daniel et al. 1975). RNA band D, about 130 nucleotides long, was shown by Daniel et al. (1975) to be similar to the monocistronic tRNA$_1^{Tyr}$ precursor isolated in vivo and processed at the 5' end by RNase P (Altman and Smith 1971). Figure 3A shows that tRNA$_1^{Tyr}$ (band E) appears only after the addition of S100 extract, increases in intensity with time, and reaches the size of mature tRNA^{Tyr} only after 15–20 minutes incubation. The fingerprint of a T1 digest of RNA from band E after 10 minutes incubation was identical with that of mature tRNA$_1^{Tyr}$, except that the 3'-end oligonucleotide contained two extra nucleotides, U and C (Zeevi 1978). The larger size of tRNA^{Tyr} (band E) at shorter periods of incubation is apparently due to an incomplete exonuclease 3'-end processing.

A specific RNase activity involved in the cleavage of polycistronic tRNA precursors at the spacer sequences separating the different tRNAs (RNase P2 [Schedl et al. 1976] or RNase O [Sakano and Shimura 1975; Shimura et al. 1978]) was recently described. The role of RNase P2 in the maturation of tRNA$_1^{Tyr}$ precursor (Fig. 1, slot 1, band A) was studied by following the processing of the in vitro ϕ80psu^+3 DNA transcripts by a purified RNase P2 preparation as a function of incubation time (Fig. 3B). It appears that RNase P2 cleaves the 600-nucleotide tRNA$_1^{Tyr}$ precursor and produces the same intermediate RNA bands as the crude *E. coli* S100 extracts (Fig. 3A). The tRNA^{Tyr} precursor (band D) accumulates during incubation with RNase P2 and, due to the absence of RNase P, very little or no mature tRNA$_1^{Tyr}$ is observed (Fig. 3B). A similar processing of the in vitro ϕ80psu^+3 DNA transcript by partially purified RNase P results in a different pattern of degradation, not involving the formation of the intermediate RNA molecules observed with RNase P2 or S100 extract, nor the formation of mature tRNA^{Tyr} (Zeevi 1978). It seems therefore reasonable to conclude that RNase P2 is the nuclease activity responsible for the initial processing of the primary transcription product of the tRNA$_1^{Tyr}$ to form the 130-nucleotide, intermediate-size precursor (Fig. 3B, band D). The 5' extra sequences of the precursor are cleaved by RNase P followed by a 3'-exonucleolytic trimming to produce mature tRNA$_1^{Tyr}$.

TRANSCRIPTION OF BACTERIOPHAGE T4 tRNA GENES

Figure 4 shows the organization of the tRNA gene cluster in bacteriophage T4 as established through the isolation and characterization of T4 deletion mutants lacking different sequences in this region (Wilson et al. 1972). To identify the primary transcription product of the T4 tRNA genes, we have transcribed in vitro the DNA from wild-type T4 and from T4 mutants containing deletions in the tRNA gene region and then compared the RNA products by electrophoresis on composite acrylamide-agarose gels. Figure 5 shows that a high-molecular-weight RNA

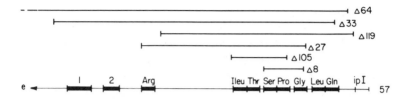

Mutant	Size of deletion base pairs	tRNA species									
		1	2	Arg	Ileu	Thr	Ser	Pro	Gly	Leu	Gln
psu$_b^-$ Δ64	5820±350										
psu$_b^-$ Δ33	4150±450	−	−	−	−	−	−	−	−	−	−
psu$_b^-$ Δ119	2070±190	+	+	+	−	−	−	−	−	−	−
psu$_b^-$ Δ27	1330±100	+	+	−	−	−	−	−	−	+	+
psu$_b^-$ Δ105	~200*	+	+	+	−	−	−	−	+	+	+
psu$_b^-$ Δ8	200±30	+	+	+	+	+	−	−	−	+	+

Figure 4 T4 tRNA gene region, modified from Wood and Revel (1976), Abelson et al. (1974), and Velten et al. (1976). The arrow indicates the direction of transcription. The table shows the production of T4 tRNAs by various T4 tRNA deletion mutants. Size of the deletions is from Wilson et al. (1972). (*) The size Δ105 is unknown, but on the basis of the tRNA pattern of the mutant, it should be about the size of Δ8. (Reprinted, with permission, from Goldfarb et al. 1978.)

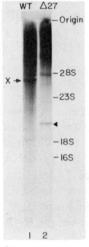

Figure 5 Autoradiograph of acrylamide gel separation of primary transcripts of T4 DNA with intact or deleted tRNA genes. 5.5 μg DNA from wild-type (WT) T4 and psu^-b deletion mutant Δ27 were transcribed by 20 μg RNA polymerase in 100-μl reaction mixtures containing 40 mM Tris-HCl (pH 7.9); 10 mM MgCl$_2$; 7 mM 2-mercaptoethanol; 250 mM KCl; 0.25 mM each of ATP, GTP, and UTP; and 0.125 mM of CTP (20 Ci/mmole). After 15-min incubation at 37°C, 5 μg DNase I and 25 μg carrier RNA were added. The RNA was extracted with phenol, precipitated with alcohol, dissolved, and fractionated by electrophoresis on 40 × 20-cm composite gel (2.2% acrylamide and 0.5% agarose) in Tris-borate buffer containing 0.2% SDS at 150 V for 18 hr. *E. coli* 23S and 16S [^{32}P]rRNA and mammalian 28S and 18S [^{32}P]rRNA were run separately as reference markers.

species (slot 1, band X) present among the transcripts of wild-type DNA is missing from the transcript of the deletion mutant psu^-bΔ27 DNA. Among the transcripts of the deletion mutant Δ27, we observe a new RNA species (slot 2, indicated by arrow) smaller in size than band X RNA of the wild type, which is probably the product of the partially deleted tRNA region.

To characterize the sites of initiation on the T4 DNA template, transcription was allowed to start in the presence of either ATP (UTP and CTP) or GTP (UTP and CTP). The complexes between RNA polymerase and T4 DNA that had not initiated transcription under these conditions were inactivated by cooling to 0°C, rifampicin was added to prevent reinitiation, and RNA synthesis was allowed to continue by the addition of the missing nucleotide and incubation at 37°C. Figure 6 shows the transcripts of T4 wild-type DNA that start with GTP (slot 2) and ATP (slot 3). It can be seen that RNA from band X is initiated by ATP.

PROCESSING OF A POLYCISTRONIC PRECURSOR OF T4 tRNAS

RNA band X from the gel of Figure 5 was extracted, processed by incubation with S100 extract for different periods of time, and the digests analyzed by electrophoresis on a 5% polyacrylamide 7 M urea gel. Figure 7 shows that band X is processed into several stable RNA species that migrate to the 4S region of the gel. In addition, a definite pattern of several RNA molecules of intermediate size can be observed.

To determine which T4 tRNA sequences are present in RNA band X,

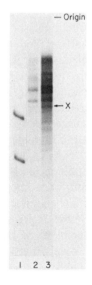

Figure 6 Primary transcripts of T4 DNA initiated with ATP or GTP. The conditions of transcription and electrophoresis were as in Fig. 5 except that after a 10-min preincubation at 37°C GTP, CTP, and UTP (slot 2) or ATP, CTP, and UTP (slot 3) were added to 0.125 mM each. After 5 min at 37°C the samples were cooled to 0°C and the missing nucleoside triphosphate together with rifampicin (2 μg/ml) and 40 μCi of [α-^{32}P]GTP were added and the mixtures incubated at 37°C for 30 min. Slot 1 shows the position of 23S (top band) and 16S (bottom band) [^{32}P]rRNA reference markers.

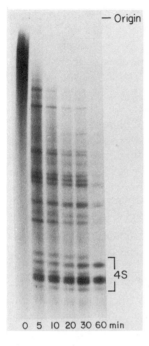

Figure 7 Autoradiograph showing the formation of stable RNA from the isolated precursor band X. RNA band X (Fig. 4) was eluted from the gel and incubated at 37°C with 2 mg/ml S100 extract. Aliquots were removed at times indicated and fractionated by electrophoresis on a 5% acrylamide gel containing 7 M urea in Tris-borate buffer. An aliquot taken before the digestion with S100 extract was run as zero time of processing. The partial degradation of RNA from band X observed for zero time of processing is due to the high specific activity of the material that results in fast radioactive damage of a molecule of this size.

we have compared the stable RNAs obtained from band X with the products of S100 processing of unfractionated RNA made on DNA containing intact or deleted tRNA region. The deletions used in this experiment remove different T4 tRNA genes (Fig. 4) and should result in disappearance of different tRNA species. The cleavage products were first fractionated on 10% acrylamide gel, and the 4S regions of the gel excised and applied to 20% acrylamide gel slabs for the second dimension of electrophoresis (Fig. 8). A comparison of the stable RNA patterns shows that at least five S100 cleavage products (1 to 5, Fig. 8) are common to band X and the unfractionated transcript of wild-type T4 DNA, and that deletions result in the disappearance of spots corresponding to different stable RNAs. Except Δ119, the stable RNA species 1 and 3 are present in all of the deletions. From the data in Figure 4 and the known length of T4 tRNAs (Abelson et al. 1974), we conclude that RNA species 1 is tRNA^{Leu} and species 3 is tRNA^{Gln}. Stable RNA species 2, which is affected by all four deletions used, seems to be tRNA^{Pro}, and stable RNA species 4, which is affected by all the deletions except deletion Δ105, is tRNA^{Gly}. Stable RNA species 5, which appears in all cases except deletion Δ27, seems to be tRNA^{Arg}. From these results we conclude that high-molecular-weight RNA band X is transcribed from tRNA region of T4 DNA and is a polycistronic tRNA precursor.

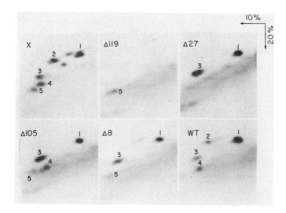

Figure 8 Two-dimensional electrophoresis of stable RNA products of isolated precursor band X and unfractionated transcripts of DNA with intact (WT) or deleted tRNA genes. DNA from the wild-type T4 psu^-b deletion mutants Δ119 and Δ105. Δ27 and Δ8 were transcribed in reaction mixtures as described in Fig. 5. A portion of the T4 transcript was extracted and fractionated by electrophoresis on gel to produce RNA band X as described in Fig. 5. The RNA band X was eluted from the gel and digested for 90 min at 37°C with E. coli S100 extract (5 mg/ml). The rest of the transcription mixtures were similarly digested by S100 extract. The digested RNAs were extracted with phenol, precipitated with alcohol, and fractionated by electrophoresis on 10% polyacrylamide gel (40 × 20 cm) in Tris-borate buffer with 0.2% SDS (300 V, 12 hr). 4S regions of the gel were cut out and applied onto 20% polyacrylamide slabs (15 cm long) for second dimension of electrophoresis at 200 V for 14 hr in Tris-borate buffer. (Reprinted, with permission, from Goldfarb et al. 1978.)

DISCUSSION

Studies with transducing phages carrying either the tRNA$_1^{Tyr}$ su^+3 and su^-3 or the tRNA$_2^{Tyr}$-tRNA$_2^{Gly}$-tRNA$_3^{Thr}$ gene cluster and with T4 tRNA genes have suggested an organization of tRNA gene clusters as single transcription units (Squires et al. 1973; Abelson et al. 1974; Ghysen and Celis 1974; Guthrie et al. 1974; Chang and Carbon 1975). However, large polycistronic tRNA precursors could not be isolated in vivo. Using the E. coli A49 mutant, which contains a temperature-sensitive RNase P, it was possible to demonstrate for the tRNA$_2^{Tyr}$-tRNA$_2^{Gly}$-tRNA$_3^{Thr}$ or T4 tRNA gene clusters the accumulation of precursors containing at most one or two tRNA sequences. Similarly, the largest precursor of tRNA$_1^{Tyr}$ obtained in the RNase P mutant was the partially processed 129-nucleotide precursor (Altman and Smith 1971).

The present results show that clusters of tRNA genes in prokaryotes are transcribed in vitro to produce polycistronic RNA precursors. The tRNA$_1^{Tyr}$ precursor transcribed in vitro from ϕ80p$su^{+,-}3$ DNA (Fig. 1, slot

1, band A) is a large molecule about 600 nucleotides long, which appears to be a substrate for the recently described RNase P2 (Schedl et al. 1976; Shimura et al. 1978). Digestion of $\phi 80psu^+3$ transcripts by S100 extracts produces intermediate RNA bands in addition to mature-sized tRNA (Fig. 3A). Some of these intermediates, previously shown to be derived from tRNA precursor band A (Daniel et al. 1975), are also produced by a purified RNase P2 preparation (Fig. 3B). Digestion with RNase P2, however, does not produce mature $tRNA_1^{Tyr}$; instead an RNA molecule about 130 nucleotides long (Fig. 3B, band D), similar to that isolated in vivo by Altman and Smith (1971), is accumulated. RNase P2 is therefore responsible for the first cleavage of the $tRNA_1^{Tyr}$ precursor to the 130-nucleotide intermediate that is then processed stepwise by RNase P and 3' exonuclease.

Transcription of the $tRNA_2^{Tyr}$-$tRNA_2^{Gly}$-$tRNA_3^{Thr}$ gene cluster on $\lambda h80dglyTsu^+36$ phage DNA produces a molecule about 500 nucleotides long (Fig. 1, slot 2, band A). This precursor molecule, which is processed to mature-sized tRNAs, is found to contain the sequences for all three tRNA species (Fig. 2, slot 2). The three tRNA genes carried by $\lambda h80dglyTsu^+36$ were reported to be clustered in a region of approximately 350 bp with the arrangement and orientation relative to transcription: 5'-$tRNA_2^{Tyr}$-$tRNA_2^{Gly}$-$tRNA_3^{Thr}$ (Chang and Carbon 1975; Sekya et al. 1976). Therefore, it seems that the polycistronic transcript synthesized in vitro is initiated at the $tRNA_2^{Tyr}$ promoter. The sequence corresponding to mature $tRNA_2^{Tyr}$ was found to be separated from $tRNA_2^{Gly}$-$tRNA_3^{Thr}$ by approximately 100 bp (Rossi et al. 1978). This, together with some observations suggesting a difference in the transcriptional properties of $tRNA_2^{Tyr}$ as compared to $tRNA_2^{Gly}$ and $tRNA_3^{Thr}$ genes (Grimberg and Daniel 1977), has led Rossi et al. (1978) to raise the question of the existence of a separate promoter in this region. The presence of a separate promoter for $tRNA_2^{Gly}$-$tRNA_3^{Thr}$ seems improbable, since we do not observe any other in vitro transcript containing $tRNA_2^{Gly}$ and $tRNA_3^{Thr}$ sequences, except the polycistronic one (Fig. 1, slot 2, band A). The $tRNA_2^{Tyr}$ sequence, however, is also found in a transcript that is approximately 215 nucleotides long (Fig. 1, slot 2, band B). This RNA band is cleaved by S100 extract to $tRNA_2^{Tyr}$ and another 4S RNA molecule (Fig. 2, slot 3).

The fact that we observe in vitro two transcripts containing $tRNA_2^{Tyr}$ is difficult to explain unless we assume that occasional termination of transcription occurs after the synthesis of $tRNA_2^{Tyr}$. This may be supported by the observation that a mixture of transcription products of $\lambda h80dglyTsu^+36$ digested with S100 yields more $tRNA_2^{Tyr}$ than $tRNA_2^{Gly}$ and $tRNA_3^{Thr}$ (Zeevi 1978). In vivo, no polycistronic precursor for the $tRNA_2^{Tyr}$-$tRNA_2^{Gly}$-$tRNA_3^{Thr}$ cluster was observed; instead, two smaller precursors were found to accumulate in temperature-sensitive RNase P mutants: a dimeric 170-nucleotide molecule containing $tRNA_2^{Gly}$ and $tRNA_3^{Thr}$ (Chang and Carbon 1975) and a molecule 205 nucleotides long that contains the tRNA$_2^{Tyr}$ sequence (Ilgen et al. 1976; Shimura et al. 1978). It is not clear, however, whether these RNA molecules are the result of the nucleolytic cleavage of a common polycistronic precursor. The tRNA$_2^{Tyr}$ precursor (205 nucleotides) isolated in vivo, apparently similar to our in vitro transcript (Fig. 1, slot 2, band B), could as well be transcribed from another chromosomal locus.

The isolation of the polycistronic T4 tRNA precursor makes the T4 tRNA operon a suitable system for the study of organization and expression of T4 genome, as well as the maturation process of tRNA. The T4 tRNA precursor (band-X RNA) is one of several high-molecular-weight in vitro T4 DNA transcripts. The apparent length of this precursor, as it was determined relative to rRNA markers in the composite gel (Fig. 5), is 4350 nucleotides. Since the electrophoresis was carried out under nondenaturing conditions, it is possible that the actual size of the precursor is much larger.

The T4 tRNA genes are clustered within a distance of 2.5 kb in T4 DNA. The closest known genes on the two sides of this region are *ipI* (internal protein) and *e* (endolysin), which are about 6 kb apart and are transcribed from right to left (Wood and Revel 1976), as shown in Figure 4. Black (1974) has suggested that genes *ipI* and *e* can be transcribed as a polycistronic mRNA. The T4 tRNA precursor described in the present paper may well include both the *ipI* and the *e* genes, although it is not yet proved. In any case, the precursor molecule is long enough to allow us to suggest that the tRNA genes are cotranscribed either with *ipI* or with *e*. The fact that the tRNA genes are located close to the *ipI* gene (Wood and Revel 1976) and a recent observation (J. Abelson, pers. comm.), obtained with a cloned segment of T4 DNA, showing that there is only one RNA polymerase binding site in the region near the *ipI* gene, support the idea that tRNA genes are transcribed together with *ipI*.

The complete nucleotide sequence of a DNA fragment about 500 bp long that includes genes from tRNA^{Gln} to tRNA^{Ile} was recently determined (J. Abelson, pers. comm.). It was found that the DNA sequences coding for tRNAs are situated adjacent to each other without separating spacer stretches. This implies that, due to the tandem location of stable RNA sequences, the processing of the high-molecular-weight transcript must give intermediate precursors up to 500 nucleotides long. As was shown in partial digestion experiments (Fig. 7), there is a set of several RNA intermediates with lengths ranging from one to several hundred nucleotides. The intermediates, as well as final stable 4S RNA products, appear simultaneously, even after short digestion periods. It seems likely that the processing endonucleases attack several specific cleavage sites at the same time on the high-molecular-weight transcript, generating random fragments combining two, three, four, or more adjacent tRNA sequences. At longer incubation times these fragments are further cleaved to final stable RNA products.

ACKNOWLEDGMENT

This research was supported in part by a grant from the United States–Israel Binational Science Foundation, Jerusalem, Israel.

REFERENCES

Abelson, J., K. Fukada, P. Johnson, H. Lamfrom, D. P. Nierlich, A. Otsuka, G. V. Paddock, T. C. Pinkerton, A. Sarabhai, S. Stahl, J. H. Wilson, and H. Yesian. 1974. Bacteriophage T4 tRNAs: Structure, genetics and biosynthesis. *Brookhaven Symp. Biol.* **26**:77.

Altman, S. 1975. Biosynthesis of transfer RNA in *Escherichia coli*. *Cell* **4**:21.

Altman, S. and J. D. Smith. 1971. Tyrosine tRNA precursor molecule polynucleotide sequence. *Nat. New Biol.* **233**:35.

Barrell, B. G., J. G. Seidman, C. Guthrie, and W. H. McClain. 1974. Transfer RNA biosynthesis: The nucleotide sequence of a precursor to serine and proline transfer RNAs. *Proc. Natl. Acad. Sci.* **71**:413.

Bikoff, E., B. F. La Rue, and M. L. Gefter. 1975. *In vitro* synthesis of transfer RNA. II. Identification of required enzymatic activities. *J. Biol. Chem.* **250**:6248.

Black, L. W. 1974. Bacteriophage T4 internal protein mutants: Isolation and properties. *Virology* **60**:166.

Chang, S. and J. Carbon. 1975. The nucleotide sequence of a precursor to the glycine and threonine specific transfer ribonucleic acids of *E. coli*. *J. Biol. Chem.* **250**:5542.

Daniel, V., J. I. Grimberg, and M. Zeevi. 1975. *In vitro* synthesis of tRNA precursors and their conversion to mature size tRNA. *Nature* **257**:193.

Daniel, V., E. Seaman, and A. Goldfarb. 1978. In vitro transcription and isolation of polycistronic RNA product of the T4 tRNA operon. *Nature* **273**:562.

Daniel, V., S. Sarid, J. S. Beckmann, and U. Z. Littauer. 1970. *In vitro* transcription of a transfer RNA gene. *Proc. Natl. Acad. Sci.* **66**:1260.

Fournier, M. J., L. Webb, and S. Tang. 1977. *In vitro* biosynthesis of functional *E. coli* su$_3^+$ tyrosine transfer RNA. *Biochemistry* **16**:3608.

Ghysen, A. and J. E. Celis. 1974. Joint transcription of two tRNA genes from *Escherichia coli*. *Nature* **249**:418.

Goldfarb, A., E. Seaman, and V. Daniel. 1978. *In vitro* transcription and isolation of a polycistronic RNA product of the T4 tRNA operon. *Nature* **273**:562.

Grimberg, J. I. and V. Daniel. 1974. *In vitro* transcription of three adjacent *E. coli* transfer RNA genes. *Nature* **250**:320.

―――. 1977. In vitro transcription of *E. coli* tRNA genes. *Nucleic Acids Res.* **4**:3743.

Guthrie, C. 1975. The nucleotide sequence of the dimeric precursor to glutamine and leucine transfer RNAs coded by bacteriophage T4. *J. Mol. Biol.* **95**:529.

Guthrie, C., J. G. Seidman, M. M. Comer, R. M. Bock, F. J. Smith, B. G. Barrell, and W. H. McClain. 1974. The biology of bacteriophage T4 transfer RNAs. *Brookhaven Symp. Biol.* **26**:106.

Ikeda, H. 1971. *In vitro* synthesis of tRNA^{Tyr} precursors and their conversion to 4S RNA *Nat. New Biol.* **234**:198.

Ilgen, C., L. I. Kirk, and J. Carbon. 1976. Isolation and characterization of large transfer ribonucleic acid precursors from *E. coli*. *J. Biol. Chem.* **251**:922.

Kaplan, D. A. and D. P. Nierlich. 1975. Initiation and transcription of a set of transfer RNA genes *in vitro*. *J. Biol. Chem.* **250**:934.

Kitamura, N., H. Ykeda, Y. Yamada, and H. Ishikura. 1977. Processing by ribonuclease II of the tRNA^{Tyr} precursor of *E. coli* synthesized *in vitro*. *Eur. J. Biochem.* **73**:297.

Kupper, H., R. Contreras, A. Landy, and H. G. Khorana. 1975. Promoter-dependent transcription of tRNA$_1^{Tyr}$ genes using DNA fragments produced by restriction enzymes. *Proc. Natl. Acad. Sci.* **72**:4754.

Kupper, H., T. Sekiya, M. Rosenberg, J. Egan, and A. Landy. 1978. A ρ-dependent termination site in the gene coding for tyrosine tRNA su$_3$ of *E. coli*. *Nature* **272**:423.

Nierlich, D. P., H. Lamfrom, A. Sarabhai, and J. Abelson. 1973. Transfer RNA synthesis *in vitro*. *Proc. Natl. Acad. Sci.* **70**:179.

Peacock, A. C. and W. C. Dingman. 1968. Molecular weight estimation and separation of ribonucleic acid by electrophoresis in agarose-acrylamide composite gels. *Biochemistry* **7**:668.

Rossi, J. J., W. Ross, J. Egan, D. J. Lipman, and A. Landy. 1979. Structural organization of *E. coli* tRNA^{Tyr} gene cluster in four different transducing phages. *J. Mol. Biol.* **128**:21.

Sakano, H. and Y. Shimura. 1975. Sequential processing of precursor tRNA molecules in *Escherichia coli*. *Proc. Natl. Acad. Sci.* **72**:3369.

Schafer, K. and D. Soll. 1974. New aspects in tRNA biosynthesis. *Biochimie* **56**:795.

Schedl, P., J. Roberts, and P. Primakoff. 1976. *In vitro* processing of *E. coli* tRNA precursors. *Cell* **8**:581.

Sekiya, T., R. Contreras, H. Küpper, A. Landy, and H. G. Khorana. 1976. *E. coli* tyrosine transfer RNA ribonucleic acid genes. *J. Biol. Chem.* **251**:5124.

Shimura, Y., H. Sakano, and F. Nagawa. 1978. Specific ribonucleases involved in processing of tRNA precursors of *E. coli*. *Eur. J. Biochem.* **86**:267.

Smith, J. D. 1976. Transcription and processing of transfer RNA precursors. *Prog. Nucleic Acid Res. Mol. Biol.* **16**:25.

Squires, C., B. Konrad, J. Kirschbaum, and J. Carbon. 1973. Three adjacent transfer RNA genes in *Escherichia coli*. *Proc. Natl. Acad. Sci.* **70**:438.

Velten, J., K. Fukada, and J. Abelson. 1976. In vitro construction of bacteriophage λ and plasmid DNA molecules containing DNA fragments from bacteriophage T4. *Gene* **1**:93.

Wilson, J. H., J. S. Kim, and J. Abelson. 1972. Bacteriophage T4 transfer RNA. III. Clustering of the genes for the T4 transfer RNAs. *J. Mol. Biol.* **71**:547.

Wood, W. B. and H. R. Revel. 1976. The genome of bacteriophage T4. *Bacteriol. Rev.* **40**:847.

Zeevi, M. 1979. "In vitro synthesis of tRNA." Ph.D. thesis, Weizmann Institute of Science, Rehovot, Israel.

Zeevi, M. and V. Daniel. 1976. Aminoacylation and nucleoside modification of *in vitro* synthesized transfer RNA. *Nature* **260**:72.

Zubay, G., L. C. Cheong, and M. Gefter. 1971. DNA directed cell-free synthesis of biologically active transfer RNA: su$_3^+$ tyrosyl-tRNA. *Proc. Natl. Acad. Sci.* **68**:2195.

tRNA Precursors in RNase P Mutants

Yoshiro Shimura, Hitoshi Sakano,* Shigeko Kubokawa, Fumikiyo Nagawa, and Haruo Ozeki
Department of Biophysics, Faculty of Science
Kyoto University
Kyoto 606, Japan

It is generally accepted that the transcription products of tRNA genes are unmodified and larger than mature size, and that these precursors are subsequently modified and processed to form mature tRNA molecules (Schäfer and Söll 1974; Altman 1975; Smith 1976; Perry 1976; Shimura and Sakano 1977). Although the tRNA precursors were first discovered in mammalian cells, most of our knowledge about the processing of tRNA precursors has come from *Escherichia coli*, where genetic analysis and nucleotide sequencing are easily applicable.

Ever since Altman and Smith (1971; Altman 1971) found a precursor molecule of su^+3 tRNA$_1^{Tyr}$ carrying extra nucleotides at both the 5' and 3' terminals of the tRNA molecule, much effort has been directed toward the characterization and processing of precursor molecules of tRNAs encoded by *E. coli* and phage T4. The sequences of several other precursors for *E. coli* and T4-encoded tRNAs have also been shown to contain extra nucleotides at both their 5' and 3' terminals (Barrell et al. 1974; Chang and Carbon 1975; Guthrie 1975). In addition, the presence of multimeric precursors that contain two or more tRNA sequences within a single molecule has been indicated (Guthrie et al. 1973; Schedl et al. 1974, 1976; Sakano and Shimura 1975, 1978; Ilgen et al. 1976). As to the specific nucleases involved in the cleavage of the tRNA gene transcripts, RNase P was the first enzyme shown to participate in the processing of *E. coli* tRNA precursors. This enzyme has been purified from the ribosome wash fraction of *E. coli* as an endonuclease that cleaves the su^+3 tRNA precursor at the 5' end of the tRNA sequence (Robertson et al. 1972). In addition to RNase P, other nucleases have been proposed to participate in the in vitro processing of tRNA precursors (Schedl et al. 1974, 1976; Bikoff and Gefter 1975; Bikoff et al. 1975; Sakano and Shimura 1975, 1978; Shimura et al. 1978).

The essential role of RNase P in tRNA biosynthesis was conclusively demonstrated by the isolation of temperature-sensitive mutants of *E. coli*

*Present address: Basel Institute for Immunology, Grenzacherstrasse 487, CH-4058 Basel, Switzerland.

defective in this nuclease activity (Schedl and Primakoff 1973; Sakano et al. 1974c). Although the mutants were selected for their inability to synthesize the su^+3 tRNA at the restrictive temperature, the mutation was shown to affect the synthesis of virtually all cellular and phage-encoded tRNAs (Sakano et al. 1974a,c; Ikemura et al. 1975; Sakano and Shimura 1978). It has been shown that many characteristic tRNA precursors accumulate in the mutants at high temperature. Analyses of these precursor molecules have provided much information on the biosynthesis of *E. coli* tRNAs. In addition, these studies have shed some light on the organization of *E. coli* tRNA genes. In this paper, we review our recent work on the RNase P mutants.

GENETICS OF RNase P MUTANTS

In the selection of temperature-sensitive mutants defective in tRNA biosynthesis, we used the phage selection method in which cells capable of synthesizing the su^+3 suppressor tRNA were counterselected by two virulent phages, BF23 and T6 (Sakano et al. 1974c). The principle of the phage selection method was previously applied in the isolation of mischarging mutants, as well as temperature-sensitive mutants of su^+3 amber suppressor gene (for review, see Shimura and Ozeki 1973). Among the mutants isolated, $ts241$ and $ts709$ have been shown to be defective in RNase P activity (Ikemura et al. 1975; Sakano and Shimura 1978).

Using phage T4 tRNA precursors accumulated in $ts241$, we have demonstrated that an S30 extract from this mutant has a temperature-sensitive endonuclease activity that cleaves the precursor molecules at the 5' ends of tRNA sequences (Sakano et al. 1974a). More recently, we have shown that when an extract of RNase P temperature-sensitive mutant was heated at 47°C for 30 minutes prior to incubation with the su^+3 tRNA precursor, the precursor molecule was hardly cleaved (Sakano and Shimura 1978). In contrast, an S30 extract from wild-type cells removed the 5' extra segment of the precursor efficiently even after the same treatment. Furthermore, according to Stark (1977), the partially purified RNase P preparations from $ts241$ and $ts709$ are more thermolabile than that of wild-type cells.

These mutants are temperature sensitive in their growth; they can grow at 30°C but not at 42°C. It is likely, therefore, that the temperature-sensitive growth is directly correlated to their thermolabile RNase P function. In addition, RNAs synthesized in these mutants show a characteristic pattern upon electrophoresis in a 10% polyacrylamide gel (Sakano et al. 1974c). When $ts241$ and $ts709$ were labeled with [^{32}P]orthophosphate at the restrictive temperature and the RNA fractionated by polyacrylamide gel electrophoresis, the electrophoretic gel patterns of RNA from the two mutants were essentially the same but different from that of wild-type cells (Fig. 1). The abnormal gel pattern of the mutant RNAs was not seen when the mutants were grown at 30°C. Analyses by two-dimensional gel electrophoresis showed that virtually none of mature *E. coli* tRNAs were synthesized in these mutants at high temperature (Sakano et al. 1974c; Ikemura et al. 1975; Sakano and Shimura 1978). As will be described later, most of the mutant-specific RNAs seen in Figure 1 are tRNA precursors.

The phenotypic properties of the mutants have been used to map the genes. First, we introduced various F′ episomes into the mutant strains to see which episomes restored the wild-type phenotype (Ozeki et al. 1974; Kubokawa et al. 1976). We have found that the mutational site of *ts*709 is covered by F′JCH5, an episome carrying the 62- to 70-minute region of the 100-minute *E. coli* genetic map (Fig. 2B). This episome does not rescue *ts*241 (Fig. 2B). On the other hand, the mutational site of *ts*241 is covered by F′140, carrying the 66- to 82-minute region, but not by F′141, carrying the 66- to 74-minute region (Fig. 2A). Both the F′140 and F′141 episomes cover the mutational site of *ts*709, as was suggested previously (Ozeki et al. 1974). The mutational site of *ts*241 must be between 74 and 82 minutes on the chromosome, whereas that of *ts*709 is between 66 and 70 minutes. We have designated the mutated RNase P gene in *ts*241 as *rnpA*, and that in *ts*709 as *rnpB* (Kubokawa et al. 1976).

The two RNase P genes have been mapped more precisely. The *rnpB* gene was mapped by cotransduction studies using phage P1. We have found that *rnpB* is cotransduced with *aspB* (68.4 min) at a frequency of 19.2% and with *argG* (68 min) at a frequency of 7.1%. From these results, we conclude that the *rnpB* gene maps at 69 minutes on the *E. coli* genetic map. The *rnpA* gene was cotransduced with neither *malT* (74 min) nor *xyl* (79 min) by phage P1. Transduction experiments could not be used to map the *rnpA* gene, because suitable genetic markers were not available

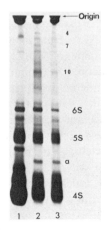

Figure 1 Fractionation of ^{32}P-labeled RNAs by electrophoresis in a 10% polyacrylamide gel. Cells, grown to a density of 2×10^8 cells/ml at 30°C, were transferred to 42°C and, after 10 min, carrier-free [^{32}P]orthophosphate was added to the cultures (0.3 mCi/ml). After 30 min, labeling was terminated by addition of an equal volume of warmed (45°C) water-saturated phenol. RNAs were extracted and electrophoresed as described previously (Sakano and Shimura 1978). After electrophoresis, the gel slab was autoradiographed. (*1*) Wild-type cells (strain 4273); (*2*) *ts*241; (*3*) *ts*709. The numbers of RNA bands correspond to the numbers of RNA spots in Fig. 3. Band a RNA is the same as spot a RNA of Fig. 3. (6S) 6S RNA; (5S) 5S rRNA.

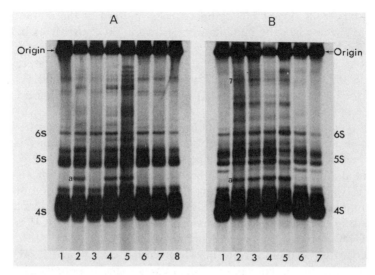

Figure 2 Fractionation of ^{32}P-labeled RNAs by electrophoresis in a 10% polyacrylamide gel. Cells were labeled with [^{32}P]orthophosphate, and RNAs were extracted and electrophoresed as described in Fig. 1. (*A*) Wild-type cells (strain 4273) (*1*); *ts*241 (*2*); *ts*241*ilv*$^-$*malT*$^-$/F'140 (*3*); *ts*241*ilv*$^-$*malT*$^-$/F'141 (*4*); *ts*709 (*5*); *ts*709/F'140 (*6*); *ts*709/F'141 (*7*); and wild-type cells (strain 4273) (*8*). (*B*) Wild-type cells (strain 4273) (*1*); a double mutant of *ts*241 and *ts*709 (*2*); *ts*241*argG*$^-$ (*3*); *ts*241*argG*$^-$/F'JCH5 (*4*): *ts*709*argG*$^-$ (*5*); *ts*709*argG*$^-$/F'JCH5 (*6*); and wild-type cells (strain 4273) (*7*). Accumulation of the multimeric tRNA precursors in the mutants was variable as described in the text. They are not clearly seen here except in *A* lane 5 (*ts*709). (7) Spot 7 RNA of Fig. 3; (a) spot a RNA of Fig. 3.

in that region. The location of *rnpA* was determined by interrupted mating experiments in which *ts*241*ilv*$^-$*malT*$^-$ was mated with AB312 (Hfr strain) at 32°C. The *rnpA*$^+$ recombinants appeared 3 minutes after *malT*$^+$ and 9 minutes ahead of *ilv*$^+$. From these results, we have concluded that the *rnpA* gene maps at about 77 minutes on the chromosome.

The mutational site of A49, an RNase P mutant isolated by Schedl and Primakoff (1973), was mapped and found to be in the *rnpA* gene. It should be pointed out that all of the RNase P mutants that have been independently isolated in our laboratory were mapped at either *rnpA* (e.g., *ts*1204) or *rnpB* (e.g., *ts*1040, *ts*1618, *ts*2418). We have not found an RNase P mutation that maps at a different locus. However, we have not tested these mutants by pairwise complementation. Therefore, we cannot exclude the possibility that there is more than one gene for each *rnp* locus.

We have constructed a double mutant of the *rnpA* and *rnpB* genes from *ts*241 and *ts*709 by transduction with phage P1. In this experiment, *ts*241*argG*$^-$ was used as recipient and *ts*709 as donor, and the double mutants of the two *rnp* genes were isolated among the $argG^+$ transductants. The RNA molecules synthesized in one of these double mutants at high temperature show the same electrophoretic gel pattern as those of the single mutants (Fig. 2B). It appears, therefore, that a mutation in either of the genes can block enzyme function resulting in accumulation of tRNA precursors.

At present we cannot give a precise biochemical explanation for the two RNase P genes. It is possible that the two genes specify different cellular components necessary for RNase P activity. Recently, it has been reported that RNase P contains an RNA component about 350 nucleotides in length (Stark et al. 1978). Consistent with this result is our observation that when cells were labeled with [^{32}P]orthophosphate and RNase P was purified according to the procedure described previously (Shimura et al. 1978), the enzyme preparation showed a discrete radioactive band upon electrophoresis in a 7% polyacrylamide gel (F. Nagawa and Y. Shimura, unpubl.). Furthermore, our preliminary experiments show that an RNA molecule about 350 nucleotides in length isolated from *ts*709 has a T1 fingerprint that is very similar but not identical to those of the corresponding RNA molecules isolated from the parental strain 4273 and *ts*241 (F. Nagawa et al., unpubl.). The details of these fingerprints are being analyzed and will be described (F. Nagawa et al., in prep.). Thus, there is a possibility that *ts*709 is a mutant of the RNA component of RNase P reported by Stark et al. (1978).

In addition to the two *rnp* genes, it is expected that there exist at least two or more genes that specify specific nucleases involved in the processing of tRNA precursors. However, these genes remain to be identified.

IDENTIFICATION AND STABILITY OF THE tRNA PRECURSORS ACCUMULATED IN *ts*241

As described above, RNA molecules synthesized in the RNase P mutants at 42°C show an electrophoretic gel pattern different from that of wild-type cells. The mutant-specific RNAs detected in *ts*241 have been most extensively characterized (Sakano and Shimura 1978).

When the cellular RNAs synthesized in this mutant at the restrictive temperature were fractionated by electrophoresis in a 10% polyacrylamide gel, slightly slower migration of 4S RNAs and the presence of new RNA bands of larger molecular weight were noted (Fig. 1). This abnormal gel pattern became indistinguishable from that of wild-type cells if the mutant RNA preparation had been treated with an S30 extract from *E. coli* Q13 prior to gel electrophoresis (Sakano and Shimura 1975). To characterize these mutant-specific RNAs, the RNA was purified by two-dimensional gel electrophoresis (Fig. 3) and analyzed by fingerprinting and in vitro cleavage reactions.

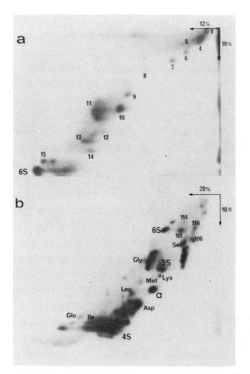

Figure 3 Two-dimensional separation of ^{32}P-labeled ts241 RNAs by polyacrylamide gel electrophoresis. Cells were labeled and RNA was extracted and electrophoresed as described previously (Sakano and Shimura 1978). For the first dimension (top to bottom), a 10% polyacrylamide gel was used. For the second dimension (right to left), a 12% gel was used for the separation of RNAs larger than 6S RNA (a), and a 20% gel for RNAs smaller than 6S RNA (b). Spots of multimeric tRNA precursors are numbered in the figure. Ser, Gly, Lys, Leu, Asp, Ile, Met, and Glu indicate the spots of monomeric precursors for tRNA$_3^{Ser}$, tRNA$_3^{Gly}$, tRNA^{Lys}, tRNA$_1^{Leu}$, tRNA$_1^{Asp}$, tRNA^{Ile}, tRNA^{mMet}, and tRNA$_2^{Glu}$, respectively. Spot a RNA is a monomeric precursor for tRNA$_1^{Asp}$. (6S) 6S RNA; (5S) 5S rRNA.

The slowly migrating 4S RNAs of Figure 3b are a mixture of many monomeric tRNA precursors. We have identified at least 30 species of tRNA precursors that belong to this class in ts709, an rnpB mutant (Ikemura et al. 1975). All of the monomeric precursors carry extra nucleotides at the 5' terminals of the corresponding tRNA molecules. The accumulation of these precursors at high temperature is a direct result of the RNase P block.

In addition to these monomeric precursors, the mutant-specific RNAs detected in the region between 4.5S and 5.5S of the 10% polyacrylamide gel are also monomeric precursors. Their molecular sizes were estimated to be 100–150 nucleotides on the basis of electrophoretic mobility. Fingerprint analyses have revealed that some of these RNAs contain nucleotide sequences identical to specific tRNAs of known nucleotide sequence. When these precursors were incubated with Q13 extract, they were cleaved into single 4S RNA bands whose fingerprint patterns resembled those of the mature tRNAs. Accordingly, we could assign some of these precursors to specific *E. coli* tRNAs. For instance, we have identified at least six precursor molecules including those for $tRNA_1^{Asp}$, $tRNA^{mMet}$, $tRNA^{Lys}$, $tRNA_1^{Leu}$, $tRNA_3^{Ser}$, and $tRNA_3^{Gly}$. The larger molecular sizes of these monomeric precursors are apparently due to the presence of longer stretches of extra nucleotides at the 5′ terminals of tRNA molecules. Most of the larger monomeric precursors have ribonucleoside triphosphates at their 5′ terminals, whereas most of the smaller monomeric precursors have ribonucleoside monophosphates at the 5′ ends. Therefore, the larger monomeric precursors must be derived from the 5′-proximal end of multimeric precursors, unless they are transcribed in a monomeric form. On the other hand, the majority of the smaller monomeric precursors appear to be derived from the internal or 3′-proximal regions of multimeric precursors, or else from the larger monomeric precursors as a result of processing reactions. It is worth noting that both large and small monomeric precursors were detected for some tRNA species. For example, the two types of monomeric precursors were identified for $tRNA_3^{Gly}$, $tRNA_1^{Leu}$, and $tRNA_1^{Asp}$. The larger monomeric precursors, like the smaller precursors, accumulate stably in the mutant as a direct result of the RNase P block.

As shown in Figure 3a, at least 15 RNA spots larger than 6S were found in the mutant RNA. The molecular sizes of these RNAs were estimated from their electrophoretic mobilities in a 5% polyacrylamide gel. All of these RNAs, except spots 5, 8, and 15 RNAs, contain modified nucleosides, such as T, ψ, and D. When the RNAs were individually eluted from the gel and treated with Q13 extract, most of them, except spots 5, 8, and 15 RNAs, were converted to 4S size. Some RNAs were converted to single 4S RNA bands (e.g., spot 9 RNA), whereas others yielded two or more 4S bands (e.g., spot 7 RNA), as judged by gel electrophoresis and fingerprinting. On the basis of these results, we concluded that these RNA molecules represent multimeric tRNA precursors in which more than two tRNA sequences are linked in tandem. Some multimeric precursors contain more than two different tRNA sequences, whereas in others two or more identical tRNA sequences are present within a single RNA chain.

In analyses of these multimeric tRNA precursors, the 4S products obtained from the individual precursors in the in vitro cleavage reactions were purified by gel electrophoresis and subjected to fingerprint analysis. The RNase T1 fingerprints of these products were compared with the T1 oligonucleotide maps of *E. coli* tRNAs whose tRNA sequences had been determined. At the same time we purified many mature tRNA molecules from wild-type cells by two-dimensional gel electrophoresis and compared their T1 oligonucleotide maps with those of the in vitro cleavage products. By this procedure, we have assigned many of the cleavage products of 4S size to specific *E. coli* tRNAs. For some precursors, the arrangement of tRNA sequences within a precursor molecule was determined through the analyses of the products of partial cleavage reactions (e.g., spot 7 RNA). In the case of spot 10 RNA, the arrangement was determined from sensitivity of the tRNA sequences (tRNA$_3^{Ser}$ and tRNA$_2^{Arg}$) to snake venom phosphodiesterase. The multimeric tRNA precursors thus far identified are listed in Table 1. For instance, spot 7 RNA of Figure 3a is a trimeric precursor having a long stretch of extra nucleotides at its 5' side. Its structure can be drawn schematically as follows.

Table 1 Multimeric tRNA precursors identified in *ts*241

Precursor[a]	Approximate chain length[b]	5' Terminal[c]	tRNA species included
1	450	pppG	mMet, X (unidentified), Gln 1
2	400	pppA	Val 1
3	370	pppA	Val 1
4	350	pppG	Leu 1
6	320	pppA	Val 1
7	300	pppG	mMet, X, Gln 1
9	240	pppG	Gly 3
10	230		Ser 3, Arg 2
11	230	pppG	Leu 1
12	220	pppG	mMet, X
13	220	pppA	Val 1
101	200		Tyr 2, Y (unidentified)
114	200		Ser 3, Arg 2
116	180		Val 2A, Val 2B
106	180		Lys, Val 1

[a]The precursor numbers correspond to the numbers of RNAs in Fig. 3.
[b]The chain lengths were estimated on the basis of electrophoretic mobilities in a 5% polyacrylamide gel. Markers of known chain lengths used to calibrate the gel included tRNA$_1^{Leu}$ (87 nucleotides), 5S RNA (120 nucleotides), 6S RNA (185 nucleotides), and G3 mRNA of phage fd (369 nucleotides, kindly provided by M. Takanami).
[c]Nucleoside tetraphosphates (pppAp and pppGp) were identified according to the procedures described by Cashel et al. (1969) and by Kramer et al. (1974), after complete digestion of tRNA precursors with RNase T2. pppAp and pppGp were identified by their mobilities relative to the markers ppppA, ppppG (Sigma), [γ-^{32}P]pppAp, and [γ-^{32}P]pppGp. The radioactive nucleoside tetraphosphates [γ-^{32}P]pppAp and [γ-^{32}P]pppGp were prepared by complete digestion with RNase T2 of the in-vitro-synthesized mRNAs of phage fd, which were labeled with [γ-^{32}P]ATP and [γ-^{32}P]GTP.

$$\text{pppG}\text{———}\text{tRNA}^{\text{mMet}}\text{———}\text{tRNA}^{\text{X}} \text{ (unidentified)}\text{———}\text{tRNA}_1^{\text{Gln}}\text{———} \quad (1)$$

This precursor molecule was remarkably enriched in the mutant at 42°C, when the cells irradiated by UV light were infected with $\lambda \text{p}su^-2_o$, a transducing phage carrying the *supE* gene isolated by Inokuchi et al. (1975) (see also Ozeki et al., this volume). Spot 10 RNA is a dimeric precursor containing the following.

$$(5')\text{———}\text{tRNA}_3^{\text{Ser}}\text{———}\text{tRNA}_2^{\text{Arg}}(3') \quad (2)$$

Spot 9 RNA is a precursor in which two or possibly three tRNA$_3^{\text{Gly}}$ sequences are linked in tandem. In addition to these relatively large multimeric precursors, we have identified some dimeric precursors in the 6S region (Fig. 3b), some of which are also listed in Table 1.

As noted in Table 1, some of the multimeric precursors identified have ribonucleoside triphosphates at their 5′ terminals, indicating that their 5′ leader sequences are intact. It is not known, however, whether they represent the initial transcripts of particular tRNA gene clusters or precursor forms that have already been processed at their 3′ terminals. The multimeric precursors containing three or more tRNA sequences within a molecule are not stable in the mutant. A few dimeric precursors that have long stretches of extra nucleotides at their 5′ sides are also not stable. In contrast, most of the dimeric precursors in the 6S region accumulate stably, as is the case with the dimeric precursor for phage T4 tRNAs (Sakano et al. 1974a; Guthrie 1975).

Quantitatively, the molar yields of the multimeric precursors containing three or more tRNA sequences are far less than those of the monomeric and most of the dimeric precursors that accumulate stably in the mutant. Furthermore, the number of tRNA species present in the multimeric precursors detected in the RNase P mutant is rather limited, as noted in Table 1. However, the majority of monomeric precursors seem to be derived from multimeric precursors as discussed previously. Thus, the multimeric precursors detected in *ts*241 probably represent a small portion of multimeric precursors actually produced in *E. coli* cells. In the RNase P mutant, the majority of multimeric precursors may be extremely unstable metabolically when this endonuclease activity is blocked, and processed very rapidly to monomeric or, less frequently, dimeric forms by the action of RNase O or some other endonuclease. It is possible that the multimeric precursors detected in the mutant are relatively more stable than the others. However, even in this case, accumulation of those precursors larger than dimers is only transient. Presumably, this is the reason for the fact that the majority of tRNA precursors exist in monomeric form when RNase P function is blocked.

A multimeric precursor is, of course, derived from a cluster of tRNA genes that are transcribed as a single unit. Our studies show that possibly many tRNA genes of *E. coli* form clusters on the chromosome and thus are transcribed as multimeric precursors. Although the multimeric precursors identified in ts241 are limited, they provide some information on the organization of E. coli tRNA genes.

PROCESSING OF tRNA PRECURSORS IN VITRO

As described earlier, all of the tRNA precursors detected in ts241 are cleaved at specific sites and converted to mature-sized molecules, if and when they are incubated with the S30 extract from E. coli Q13. Thus, the crude extract contains all nucleolytic activities for processing of the tRNA precursors. One processing enzyme is, of course, RNase P, which is required for the biosynthesis of essentially all E. coli tRNAs. In addition, several other nucleases have been reported to participate in the cleavage of tRNA precursors in vitro. Among these enzymes, there are endonucleases that cleave the spacer regions of multimeric precursors, and exonucleases that remove extra nucleotides at the 3' sides of tRNA precursors (Schedl et al. 1974, 1976; Bikoff and Gefter 1975; Bikoff et al. 1975; Seidman et al. 1975; Sakano and Shimura 1975, 1978; Kitamura et al. 1977; Shimura et al. 1978).

As mentioned previously, when an S30 extract from ts241 was incubated at 47°C for 30 minutes, its RNase P activity was completely inactivated (Sakano et al. 1974a; Sakano and Shimura 1978). In contrast, an S30 extract from wild-type cells still retained its RNase P activity after the same treatment. When the total RNAs synthesized in ts241 at 42°C were incubated with the heated mutant extract, most of the monomeric and dimeric precursors remained unaltered but the multimeric precursors larger than dimeric form disappeared. Detailed analyses showed that those large multimeric precursors were cleaved into smaller intermediates by the heated mutant extract, which could be further processed to mature-sized molecules if incubated with the crude extract from wild-type cells (Sakano and Shimura 1975, 1978). For instance, spot 7 RNA (Fig. 3), a trimeric precursor containing (5')tRNA^{mMet}-tRNA^X-tRNA$^{Gln}_1$(3'), was cleaved, upon incubation with the heated mutant extract, first at the spacer region between tRNA^{mMet} and tRNA^X to yield a monomeric precursor for tRNA^{mMet} and a dimeric precursor containing tRNA^X and tRNA$^{Gln}_1$. Subsequently, the dimeric precursor was cleaved at its spacer region to form two monomeric precursors for tRNA^X and tRNA$^{Gln}_1$. The monomeric precursors were not converted further to mature-sized molecules by the heated mutant extract, unless wild-type extract was applied. The same monomeric precursors have been detected in the mutant-specific RNAs. It is worth noting that these precursors were enriched appreciably in the mutant, as was the case with spot 7 RNA, when the cells were infected with λpsu^-2_o. The endonucleolytic activity that is present in the heated mutant extract and is responsible for the cleavage of multimeric precursors has been tentatively designated RNase O (Sakano and Shimura 1975) and has recently been purified from *E. coli* Q13 (Shimura et al. 1978). The mode of the cleavage reaction of spot 7 RNA observed with the crude extract has been supported by the cleavage experiments with the purified RNase O and RNase P preparations (Sakano and Shimura 1978). Spot 7 RNA could be cleaved by purified RNase P upon prolonged incubation to yield a dimer of tRNA^{mMet}-tRNA^{X} and a monomer of tRNA$_1^{Gln}$, which do not carry extra nucleotides at the 5' terminals of the tRNA sequences. But this reaction was extremely slow under the conditions employed, and it was hardly observed if purified RNase O was added with RNase P to the reaction mixture. It remains to be seen, however, whether RNase O actually participates in the processing of tRNA precursors in vivo.

On the other hand, the dimeric precursors that accumulate stably in the mutant are not cleaved by RNase O activity. Instead, they are cleaved directly by RNase P. For instance, spot 10 RNA, a dimeric precursor containing (5')tRNA$_3^{Ser}$-tRNA$_2^{Arg}$(3'), is cleaved by RNase P first at the RNase P site adjacent to the tRNA located at the 3'-terminal side, and subsequently at the site adjacent to the tRNA located at the 5'-terminal side of this precursor. This suggests that when a precursor contains two RNase P sites, the enzyme has a preference for the 3'-proximal site. Presumably, the 3'-terminal region plays a role in the precursor-enzyme interaction. These results are consistent with those of Altman et al. (1974) showing that the removal of several nucleotides from the 3' terminal of the su^+3 precursor inhibits the RNase P action. In the case of the dimeric precursor for tRNA^{Pro}-tRNA^{Ser} of phage T4, Seidman and McClain (1975) reported that the removal of the 3' extra nucleotides of this precursor and subsequent repairing of the CCA$_{OH}$ structure of tRNA^{Ser} located at the 3' side by tRNA-nucleotidyl transferase precedes the cleavage by RNase P at the site adjacent to tRNA^{Ser}. It remains to be examined whether the similar processing reaction takes place in the cleavage of *E. coli* dimeric precursors. It should be pointed out in this connection that spot 10 RNA contains the CCA sequence for both tRNAs.

In the case of the monomeric precursors we have studied, the cleavage of 5' leader sequences by RNase P precedes the removal of their 3' extra nucleotides (Sakano and Shimura 1975, 1978). The nuclease responsible for the latter processing reaction has been designated RNase Q (Sakano and Shimura 1975) and has recently been purified from *E. coli* Q13 (Shimura et al. 1978). In the case of a monomeric precursor for tRNA$_1^{Asp}$ (spot a RNA of Fig. 3b), the 3'-terminal CCA$_{OH}$ structure is generated by the action of RNase Q (F. Nagawa and Y. Shimura, unpubl.). Furthermore, our recent results show that RNase Q is different from RNase II and RNase P3, which have been reported by other investigators to participate in 3' processing of su^+3 tRNA precursors (Schedl et al. 1974; Bikoff and Gefter 1975). RNase Q is less active on single-stranded synthetic polyribonucleotides, such as polyadenylate, and is eluted from a DEAE-cellulose column at a higher salt concentration than RNase II. According to Bikoff and Gefter (1975), RNase P3 is eluted from the column at a lower salt concentration than RNase II. The RNase P3 preparation contained an appreciable amount of tRNA nucleotidyl transferase activity, which was essentially absent in our RNase Q preparation. We have also identified another nuclease that is similar in some respects to RNase Q (Shimura et al. 1978). The possibility that this enzyme could play a role in the 3' processing of tRNA precursors is currently being investigated. So far, genetic mutants defective in RNase Q activity have not been isolated. Therefore, as with RNase O, it also remains to be seen whether RNase Q is actually involved in tRNA biosynthesis in vivo.

It should be pointed out that the multimeric and monomeric precursors whose nucleotide modification is only partial are efficiently processed in the in vitro cleavage reactions. This shows that many of the modified nucleotides in tRNA do not play a role in the cleavage reactions per se. These observations are consistent with the results obtained by Schäfer et al. (1973). On the other hand, it has been proposed that modification of some nucleotides takes place at a distinct stage of the maturation process (Sakano et al. 1974b). For instance, the multimeric precursor for $tRNA_1^{Leu}$ (spot 4 RNA of Fig. 3a) contains T, ψ, and D, but not m^1G at the 38th residue from the 5' end of the tRNA. The latter modified nucleoside is present, however, in the monomeric precursors for $tRNA_1^{Leu}$, which are, in all likelihood, derived from the multimeric precursor molecule. Even in the monomeric precursors, Gm, which is present at the 18th residue of the tRNA, is totally missing. It appears that some nucleotide modifications (such as Tp, ψp, Dp, and Amp) occur at relatively early stages of the maturation process, whereas others (such as m^1Gp and Gmp) take place at later stages. Exact correlation between processing of tRNA precursors and nucleotide modification in cells has to be clarified in the future.

CONCLUDING REMARKS

The isolation of temperature-sensitive mutants defective in RNase P activity has provided proof that RNase P plays a key role in the biosynthesis of essentially all *E. coli* and phage-encoded tRNAs. Investigation of these mutants has also led to the identification of many tRNA precursors and a set of specific nucleases involved in the cleavage of the tRNA precursors in vitro. The stabilities of various types of tRNA precursor in the mutants at the restrictive temperature are consistent with the results obtained in the in vitro cleavage reactions of the precursors, and may provide some insight into the mode of the processing of tRNA precursors in vivo. Through the analyses of the in vitro cleavage reactions of the tRNA precursors, we have shown that at least three nucleases participate in the maturation of *E. coli* tRNAs. One of the remarkable features of the in vitro cleavage reactions is that these nucleases function on the tRNA precursors in a highly ordered fashion. The large multimeric precursors are initially cleaved by RNase O at the spacer regions before the precursors become susceptible to RNase P. Thus the function of RNase O is independent of that of RNase P. The monomeric precursors (and some dimeric precursors) are cleaved first by RNase P to form the correct 5′ ends of tRNA sequences and subsequently processed by RNase Q at their 3′ extra regions. This sequential model, proposed on the basis of the in vitro cleavage of the tRNA precursors, has to be tested by in vivo experiments.

All the tRNA precursors that we have characterized contain the CCA sequence. In many cases, the CCA sequence is followed by extra nucleotides. This is consistent with the results obtained by Deutscher et al. (1977), who showed that many *E. coli* tRNAs are formed in the mutants defective in tRNA nucleotidyl transferase. We have not detected any intervening sequences within tRNA sequences of the precursors so far examined. It has been shown that the yeast tRNA genes contain the intervening sequences at the 3′ sides of the anticodons of the tRNAs (Goodman et al. 1977; Valenzuela et al. 1978) and that the intervening sequences are transcribed to form the precursor molecules (Knapp et al. 1978). It appears that tRNA genes of prokaryotic cells are different from those of eukaryotic cells in this respect. However, more sequence analyses of the tRNA genes of prokaryotic cells are needed to clarify this problem.

At present we do not know whether the processing of tRNA precursors has some relevance to physiology of cells. However, in view of the fact that biologically functional tRNA molecules are produced from their precursors as the result of processing and modification, it is possible to assume that the processing may play a role, directly or indirectly, in determining the amounts of tRNA molecules in cells.

ACKNOWLEDGMENTS

We thank Y. Ryo for valuable discussions. This work was supported, in part, by a scientific research grant from the Ministry of Education of Japan.

REFERENCES

Altman, S. 1971. Isolation of tyrosine tRNA precursor molecules. *Nat. New Biol.* **229:** 19.

———. 1975. Biosynthesis of transfer RNA in *Escherichia coli. Cell* **4:** 21.

Altman, S. and J. D. Smith. 1971. Tyrosine tRNA precursor molecule polynucleotide sequence. *Nat. New Biol.* **233:** 35.

Altman, S., A. L. M. Bothwell, and B. C. Stark. 1974. Processing of *E. coli* tRNA^{Tyr} precursor RNA *in vitro*. *Brookhaven Symp. Biol.* **26**:12.

Barrell, B. G., J. G. Seidman, C. Guthrie, and W. H. McClain. 1974. Transfer RNA biosynthesis: The nucleotide sequence of a precursor to serine and proline transfer RNAs. *Proc. Natl. Acad. Sci.* **71**:413.

Bikoff, E. K. and M. L. Gefter. 1975. *In vitro* synthesis of transfer RNA. I. Purification of required components. *J. Biol. Chem.* **250**:6240.

Bikoff, E. K., B. F. La Rue, and M. L. Gefter. 1975. *In vitro* synthesis of transfer RNA. II. Identification of required enzymatic activities. *J. Biol. Chem.* **250**:6248.

Cashel, M., R. A. Lazzarini, and B. Kalgacher. 1969. An improved method for thin-layer chromatography of nucleotide mixtures containing ^{32}P-labeled orthophosphate. *J. Chromatogr.* **40**:103.

Chang, S. and J. Carbon. 1975. The nucleotide sequence of a precursor to the glycine- and threonine-specific transfer ribonucleic acids of *Escherichia coli*. *J. Biol. Chem.* **250**:5542.

Deutscher, M. P., J. Jung-Ching, and J. A. Evans. 1977. Transfer RNA metabolism in *Escherichia coli* cells deficient in tRNA nucleotidyl transferase. *J. Mol. Biol.* **117**:1081.

Goodman, H. M., M. V. Olson, and B. D. Hall. 1977. Nucleotide sequence of a mutant eukaryotic gene: The yeast tyrosine-inserting ochre suppressor SUP-o. *Proc. Natl. Acad. Sci.* **74**:5453.

Guthrie, C. 1975. The nucleotide sequence of the dimeric precursor to glutamine and leucine transfer RNAs coded by bacteriophage T4. *J. Mol. Biol.* **95**:529.

Guthrie, C., J. G. Seidman, S. Altman, B. G. Barrell, J. D. Smith, and W. H. McClain. 1973. Identification of tRNA precursor molecules made by phage T4. *Nat. New Biol.* **246**:6.

Ikemura, T., Y. Shimura, H. Sakano, and H. Ozeki. 1975. Precursor molecules of *Escherichia coli* transfer RNAs accumulated in a temperature-sensitive mutant. *J. Mol. Biol.* **96**:69.

Ilgen, C., L. L. Kirk, and J. Carbon. 1976. Isolation and characterization of large transfer ribonucleic acid precursors from *Escherichia coli*. *J. Biol. Chem.* **251**:922.

Inokuchi, H., F. Yamao, H. Sakano, and H. Ozeki. 1975. A tRNA cluster containing the suppressor *supE* of *Escherichia coli* (in Japanese). *Jpn. J. Genet.* **50**:466 (Abstr.).

Kitamura, N., H. Ikeda, Y. Yamada, and H. Ishikura. 1977. Processing by ribonuclease II of the tRNA^{Tyr} precursor of *Escherichia coli* synthesized *in vitro*. *Eur. J. Biochem.* **73**:297.

Knapp, G., J. S. Beckmann, P. F. Johnson, S. A. Fuhrman, and J. Abelson. 1978. Transcription and processing of intervening sequences in yeast tRNA genes. *Cell* **14**:221.

Kramer, R. A., M. Rosenberg, and J. A. Steitz. 1974. Nucleotide sequences of the 5' and 3' termini of bacteriophage T7 early messenger RNAs synthesized *in vivo*: Evidence for sequence specificity in RNA processing. *J. Mol. Biol.* **89**:767.

Kubokawa, S., H. Sakano, and H. Ozeki. 1976. The two genes controlling RNase P activity (in Japanese). *Jpn. J. Genet.* **51**:450 (Abstr.).

Ozeki, H., H. Sakano, S. Yamada, T. Ikemura, and Y. Shimura. 1974. Temperature sensitive mutants of *Escherichia coli* defective in tRNA biosynthesis. *Brookhaven Symp. Biol.* **26**:89.

Perry, R. P. 1976. Processing of RNA. *Annu. Rev. Biochem.* **45**:605.

Robertson, H. D., S. Altman, and J. D. Smith. 1972. Purification and properties of a specific *Escherichia coli* ribonuclease which cleaves a tyrosine transfer ribonucleic acid precursor. *J. Biol. Chem.* **247**:5243.

Sakano, H. and Y. Shimura. 1975. Sequential processing of precursor tRNA molecules in *Escherichia coli*. *Proc. Natl. Acad. Sci.* **72**:3369.

———. 1978. Characterization and *in vitro* processing of transfer RNA precursors accumulated in a temperature-sensitive mutant of *Escherichia coli*. *J. Mol. Biol.* **123**:287.

Sakano, H., Y. Shimura, and H. Ozeki. 1974a. Studies on T4 tRNA biosynthesis: Accumulation of precursor tRNA molecules in a temperature-sensitive mutant of *Escherichia coli*. *FEBS Lett.* **40**:312.

———. 1974b. Selective modification of nucleosides of tRNA precursors accumulated in a temperature-sensitive mutant of *Escherichia coli*. *FEBS Lett.* **48**:117.

Sakano, H., S. Yamada, T. Ikemura, Y. Shimura, and H. Ozeki. 1974c. Temperature sensitive mutants of *Escherichia coli* for tRNA synthesis. *Nucleic Acids Res.* **1**:355.

Schäfer, K. P. and D. Söll. 1974. New aspects in tRNA biosynthesis. *Biochimie* **56**:795.

Schäfer, K. P., S. Altman, and D. Söll. 1973. Nucleotide modification *in vitro* of the precursor of transfer RNA^{Tyr} of *Escherichia coli*. *Proc. Natl. Acad. Sci.* **70**:3626.

Schedl, P. and P. Primakoff. 1973. Mutants of *Escherichia coli* thermosensitive for the synthesis of transfer RNA. *Proc. Natl. Acad. Sci.* **70**:2191.

Schedl, P., P. Primakoff, and J. Roberts. 1974. Processing of *E. coli* tRNA precursors. *Brookhaven Symp. Biol.* **26**:53.

———. 1976. *In vitro* processing of *E. coli* tRNA precursors. *Cell* **8**:581.

Seidman, J. G. and W. H. McClain. 1975. Three steps in conversion of large precursor RNA into serine and proline transfer RNAs. *Proc. Natl. Acad. Sci.* **72**:1491.

Seidman, J. G., F. J. Schmidt, K. Foss, and W. H. McClain. 1975. A mutant of *Escherichia coli* defective in removing 3' terminal nucleotides from some transfer RNA precursor molecules. *Cell* **5**:389.

Shimura, Y. and H. Ozeki. 1973. Genetic study on transfer RNA. *Adv. Biophys.* **4**:191.

Shimura, Y. and H. Sakano. 1977. Processing of tRNA precursors in *Escherichia coli*. In *Nucleic acid-protein recognition* (ed. H. J. Vogel), p. 293. Academic Press, New York.

Shimura, Y., H. Sakano, and F. Nagawa. 1978. Specific ribonucleases involved in processing of tRNA precursors of *Escherichia coli*: Partial purification and some properties. *Eur. J. Biochem.* **86**:267.

Smith, J. D. 1976. Transcription and processing of transfer RNA precursors. *Prog. Nucleic Acid Res. Mol. Biol.* **16**:25.

Stark, B. C. 1977. "Further purification and properties of ribonuclease P from *Escherichia coli*." Ph.D. thesis, Yale University, New Haven, Connecticut.

Stark, B. C., R. Kole, E. J. Bowman, and S. Altman. 1978. Ribonuclease P: An enzyme with an essential RNA component. *Proc. Natl. Acad. Sci.* **75:** 3717.

Valenzuela, P., A. Venegas, F. Weinberg, R. Bishop, and W. J. Rutter. 1978. Structure of yeast phenylalanine-tRNA gene: An intervening DNA segment within the region coding for the tRNA. *Proc. Natl. Acad. Sci.* **75:** 190.

The Purification of 3' Processing Nucleases Using Synthetic tRNA Precursors

Ranajit K. Ghosh and Murray P. Deutscher
Department of Biochemistry
University of Connecticut Health Center
Farmington, Connecticut 06032

tRNA genes in all cells examined are transcribed initially as large precursor molecules that are converted to the mature species by the successive action of a variety of nucleases (Smith 1976). Two types of tRNA precursors, which differ in the location of the extra residues at the 3' terminal, have been identified. In one type, which appears to be prevalent in *Escherichia coli* (Altman and Smith 1971; Chang and Carbon 1975; Schedl et al. 1976), the CCA sequence is already present, and varying numbers of nucleotide residues follow this sequence. Since mutants defective in tRNA nucleotidyl transferase (Deutscher and Hilderman 1974) apparently have little, or no, defect in *E. coli* tRNA biosynthesis (Morse and Deutscher 1975; Deutscher et al. 1977), the nuclease that removes the extra residues must stop at the CCA sequence, possibly because the tRNA is immediately aminoacylated in vivo. The second type of tRNA precursor, which has been observed in phage-infected *E. coli* (Barrell et al. 1974) and presumably exists in eukaryotes (Goodman et al. 1977; Valenzuela et al. 1978), contains other nucleotides in place of all or part of the CCA sequence. The nuclease that removes these extra residues must stop at a point that would allow tRNA nucleotidyl transferase to fill in the CCA sequence. From the presumed properties of these putative nuclease activities, it might be expected that at least two different enzymes would be required.

Attempts to identify the enzymes responsible for 3' trimming of tRNA precursors have led to some confusion. Some mutants unable to carry out 3' processing appear to be defective in the 3' to 5' exonuclease ribonuclease II (RNase II) (Schedl and Primakoff 1973); this enzyme is present in partially purified preparations of 3' processing activity (Schedl et al. 1976). In addition, partially purified preparations of RNase II can remove extra 3' nucleotides from tRNA precursors (Bikoff et al. 1975; Kitamura et al. 1977). Another mutant strain, *E. coli* BN (Maisurian and Buyanovskaya 1973), is also defective in processing some tRNA precursors, although it processes others normally (Seidman et al. 1975). Studies of this mutant support the possibility that more than one enzyme may be involved in 3' processing. Furthermore, an additional exonuclease, RNase P3 (Bikoff et al. 1975), appears to remove extra 3' nucleotides from the tRNA^{Tyr} precursor more efficiently than RNase II. Shimura et al. (1978) have recently reported the identification of two exonucleases, termed RNase Q and RNase Y, which can remove nucleotides from the 3' end of tRNA precursors. However, since these putative processing nucleases have not been purified extensively or studied in any detail, it is not yet clear how these various enzyme activities are related and whether they actually function in processing.

The major difficulties in purifying and studying tRNA processing nucleases have been the very small amounts of substrate available and the use of cumbersome assays. These problems generally have precluded extensive purification and study of activities identified in cell extracts. To overcome these difficulties, we have developed procedures for preparing large amounts of synthetic tRNA precursors that can be used to assay 3' processing nucleases during purification by simple assay techniques. In this paper, we will describe the preparation and characterization of these synthetic tRNA precursors and their use in purification of several nucleases that act at the 3' end of tRNA precursors. The properties of these enzymes on a variety of substrates will be compared, and the relationship of these activities to previously described nucleases will be discussed.

PREPARATION AND CHARACTERIZATION OF SYNTHETIC tRNA PRECURSORS

Since two types of tRNA precursors exist, with or without the CCA sequence, it was necessary to prepare a synthetic precursor of each type to examine extracts for all 3' processing nucleases. The precursors made were tRNA-CCA[^{14}C]CC and tRNA-C[^{14}C]U. The first synthetic precursor contained two extra C residues following the CCA sequence and, therefore, possessed features of both the tRNA^{Tyr} (Altman and Smith 1971) and tRNA^{Thr} (Chang and Carbon 1975) natural precursors. The second synthetic precursor contained a U instead of a C within the CCA sequence and was identical to an intermediate in the biosynthesis of phage T4 tRNA^{Pro} (Seidman et al. 1975). Preparation of the synthetic tRNA precursors made use of the observation that purified preparations of rabbit liver tRNA nucleotidyl transferase could catalyze the anomalous incorporation of nucleotides within the CCA region or following this sequence (Deutscher 1972) (Fig. 1).

Details of the synthesis and purification of tRNA-CCA[^{14}C]CC and tRNA-C[^{14}C]U have recently been published (Deutscher and Ghosh 1978). Characterization of the synthetic tRNA precursors is shown in Table 1.

tRNA-C + [^{14}C] UTP $\xrightarrow{\text{tRNA nucleotidyl transferase}}$ tRNA-C-[^{14}C]U + PPi

tRNA-C-C-A + [^{14}C] CTP $\xrightarrow{\text{tRNA nucleotidyl transferase}}$ tRNA-C-C-A-[^{14}C]C-C + PPi

Figure 1 Reactions for preparation of synthetic tRNA precursors.

PURIFICATION OF POTENTIAL PROCESSING NUCLEASES

Purification of the various nucleases was followed by examining the ability of fractions to release acid-soluble radioactivity from several different labeled substrates. Trial purification experiments revealed that all fractions active with the synthetic tRNA precursors also could act on [^{32}P]tRNA treated with phosphodiesterase; thus, in the early stages of purification, the latter substrate was used to conserve synthetic precursors. In addition, ^{3}H-labeled poly(A) was used to assess RNase-II-like activity throughout purification. In the later purification steps, tRNA-C[^{14}C]U and tRNA-CCA[^{14}C]CC were also used to follow nuclease activities. Details of the assay procedure and of the purification of the putative processing nucleases have been described (Ghosh and Deutscher 1978b). The purification scheme resulted in the separation of two activities (fractions B and C) with the properties of RNase II (Singer and Tolbert 1965; Gupta et al. 1977) and a third activity identical to RNase D, which we described previously as a nuclease acting on tRNA molecules with altered structures (Ghosh and Deutscher 1978a). Other potential activities have not been purified sufficiently for study.

PROPERTIES OF THE PURIFIED ENZYMES

One measure of the specificity of each of the purified enzymes for the different substrates could be obtained from their relative enrichment with respect to the crude extract using each of the substrates. The data in Table 2 show that fractions B and C have been most highly enriched with

Table 1 Location of incorporated nucleotides in synthetic tRNA precursors

tRNA substrate	NTP[a]	Incorporation (nmoles/mg)	% Radioactivity in nucleoside	% Radioactivity in nucleotide	Residues added/chain
tRNA-C	[^{14}C]UTP	20.3	97	3	1.03
tRNA-CCA	[^{14}C]CTP	43.4	48	52	2.08

Samples were prepared as described. Material equivalent to about 100 μg of tRNA was hydrolyzed in 1 ml of 0.1 N NaOH at 70°C for 2 hr. After adjustment to pH 6 with HCl, the nucleosides and nucleotides were separated on small columns of Dowex 1-Cl (Deutscher 1973) and counted.

[a]NTP is nucleoside triphosphate.

Table 2 Enrichment of purified enzymes against various substrates

Enzyme	Substrate[a]			
	poly(A)	dt tRNA	tRNA-CU	tRNA-CCACC
Fraction B	250	29	33	150
Fraction C	780	250	66	360
RNase D	—	270	53	870

Each of the purified enzymes and the S30 fraction were assayed using ^{3}H-labeled poly(A) (98 nmoles nucleotide), [^{32}P]tRNA treated with diesterase (180 μg), tRNA-C[^{14}C]U (37 μg), and tRNA-CCA[^{14}C]CC (44 μg); and the specific activities calculated. The initial specific activities in the S30 fraction in this experiment were: poly(A), 2.7 units/mg; tRNA treated with diesterase (dt tRNA), 0.15 units/mg; tRNA-CU, 0.0015 units/mg; and tRNA-CCACC, 0.003 units/mg.
[a]The data presented are the increases in specific activity with respect to each substrate from the S30 fraction to the purified enzyme.

respect to hydrolysis of poly(A), whereas the RNase D fraction has lost all activity against poly(A) but has been highly enriched for activity against the synthetic precursor tRNA-CCACC. In contrast, none of the enzymes have been as well purified for activity against tRNA-CU as for the other substrates. Although none of the enzymes show absolute specificity for any substrate, these data suggest that RNase D may be the enzyme acting on tRNA precursors with extra residues after the CCA terminal. Furthermore, since the relative activity against tRNA-CU compared to other substrates decreased for each of the enzymes compared to the crude extract, it appears that a nuclease more specific for tRNA-CU has been removed during purification (Seidman et al. 1975).

Using poly(A) as substrate, fractions B and C had a broad pH optimum between 7 and 10. Both enzymes required a divalent cation for activity. Optimal activity was obtained at 1 mM Mg^{++}; Mn^{++} was 50% as effective, but Ca^{++} was inactive. The two enzymes also required a monovalent cation for activity. Optimal activity was obtained at 0.1 M K^+ or NH_4^+ for fraction B, and at 0.15 M cation for fraction C. Na^+ could not satisfy the monovalent cation requirement in either case. The m.w. of each fraction, determined by chromatography on Sephadex G-100, was about 90,000. These properties of fractions B and C are identical to those previously ascribed to RNase II (Gupta et al. 1977).

In contrast, using diesterase-treated tRNA as substrate, RNase D had a pH optimum between 9 and 10. The enzyme required 5 mM Mg^{++} for optimal activity; Mn^{++} was about 80% as effective, and Ca^{++} about 10%. RNase D was stimulated slightly by K^+ or NH_4^+, but was inhibited by Na^+. Essentially, identical requirements were found when tRNA-CCA[^{14}C]CC was used as substrate, except that K^+ inhibited slightly. The molecular weight of RNase D, determined on Sephadex G-100, was about 60,000. The properties of RNase D clearly distinguish this enzyme from fractions B and C and from other *E. coli* nucleases.

The actions of fractions B and C and RNase D on the substrates tRNA-CC[^{14}C]A, tRNA-C[^{14}C]U, and tRNA-CCA[^{14}C]CC are shown in Figure 2. Each enzyme's activity has been normalized to the same amount of hydrolysis using diesterase-treated [^{32}P]tRNA as substrate. Each of the enzymes is relatively inactive against intact tRNA and against the synthetic precursor tRNA-CU, whereas they display substantial activity against the other synthetic precursor, tRNA-CCACC. Thus, each of these enzymes is capable of removing extra nucleotides following the CCA sequence, with the greatest specificity being shown by RNase D. In the case of the latter enzyme, intact tRNA is digested at <3% of the rate of tRNA-CCACC, as might be expected for a 3' processing nuclease.

PRODUCTS OF RNase D ACTION ON tRNA-CCA[^{14}C]CC

We have previously shown that RNase D is an exonuclease acting at the 3' end of RNA and releasing 5' mononucleotides (Ghosh and Deutscher 1978a). As expected, the only radioactive product of RNase D action on

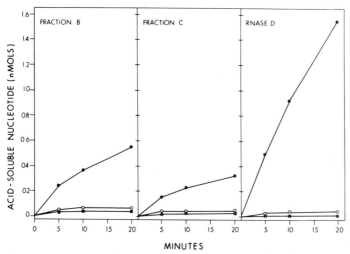

Figure 2 Rate of hydrolysis of intact tRNA and synthetic tRNA precursors with fractions B and C and RNase D. Assays were performed as described except that no KCl was present in the RNase D assays. Approximately 70 μg of each tRNA substrate was present and varying amounts of enzyme, depending on the substrate. Samples were incubated at 37°C for the times indicated, and radioactivity made acid soluble was determined. All rates have been normalized to 7 milliunits of enzyme using 70 μg of diesterase-treated [^{32}P]tRNA as substrate. (●) tRNA-CCA[^{14}C]CC; (○) tRNA-CC[^{14}C]A; (■) tRNA-C[^{14}C]U.

tRNA-CCA[^{14}C]CC was [^{14}C]CMP, as determined by paper chromatography or by conversion to [^{14}C]C after alkaline phosphatase treatment. This was true early in the reaction when only 10% of the radioactivity was made acid soluble and also when 90% of the tRNA had been degraded.

Of more interest was the structure of the tRNA product. If RNase D were a 3' processing nuclease, it should remove the extra 3' residues from tRNA-CCA[^{14}C]CC, but have no further action on the tRNA, as already suggested by the data in Figure 2. This was shown conclusively by the increase in amino acid acceptor activity after treatment of the precursor with RNase D (Table 3). The presence of tRNA nucleotidyl transferase led to a slightly greater increase, suggesting that on some chains part of the CCA sequence was removed. However, in this experiment, aminoacylation was carried out after RNase D action. Presumably, in vivo tRNAs would be aminoacylated as soon as the extra residues were removed, preventing any further digestion by RNase D. Since RNase D removes the terminal A of tRNA very slowly (Fig. 2), aminoacylation should prevent any digestion of the CCA sequence. In fact, in vivo studies have shown that tRNA nucleotidyl transferase probably is not required for *E. coli* tRNA biosynthesis (Morse and Deutscher 1975; Deutscher et al. 1977; Mazzara and McClain 1977).

ACTION OF NUCLEASES ON A NATURAL tRNA PRECURSOR

Once purified nucleases were obtained, it was of interest to test their specificity against a natural tRNA precursor. For this purpose we used the *E. coli* tRNA^{Tyr} precursor, which contains 41 extra nucleotides at the 5' end and 3 extra residues at the 3' end (Altman and Smith 1971). Our initial

Table 3 Amino acid acceptor activity of tRNA-CCACC after treatment with RNase D

Additions	Acceptor activity (pmoles)
None	75
RNase D	183
RNase D + tRNA nucleotidyl transferase	217

Reaction mixtures contained in 0.1 ml: 10 mM Tris-Cl (pH 7.5), 5 mM MgCl$_2$, 35 μg of tRNA-CCA[^{14}C]CC, and approximately 70 milliunits of RNase D. After incubation for 1 hr at 37°C, the following components were added (final concentrations): 1 mM ATP, 0.2 mM ^{14}C-labeled amino acid mixture (15 amino acids, each at 78 cpm/pmole), and an *E. coli* aminoacyl-tRNA-synthetase mixture. Aminoacylation was carried out for an additional 10 min at 37°C, and acid-precipitable radioactivity determined. tRNA nucleotidyl transferase (0.12 unit) was added where indicated.

studies measured the release of acid-soluble radioactivity from ^{32}P-labeled precursor. The data presented in Table 4 demonstrate that RNase D removes nucleotides from the tRNA precursor, but only after prior cleavage with RNase P. This order of processing is the same as observed in vivo (Schedl et al. 1976). In contrast, fractions B and C act on the precursor without RNase P action, resulting in extensive release of mononucleotides.

Preliminary results, obtained in collaboration with S. Altman (unpubl.), indicate that RNase D acts on both the tRNA fragment and the 5' fragment after cleavage of the precursor with RNase P. Fingerprint analysis suggests that RNase D action on the tRNA fragment leads to some production of the mature 3' terminal. However, further experiments are required to determine whether RNase D can quantitatively and specifically remove the extra 3' residues from a natural precursor.

DISCUSSION

The results presented here demonstrate the feasibility of using tRNA nucleotidyl transferase to prepare large amounts of well-defined, specifically labeled, synthetic tRNA precursors. Since the only nucleotides labeled are those that would be removed by processing nucleases, these tRNA precursors are ideal substrates for identification and purification of such enzymes using simple assays for production of acid-soluble radioactivity. In addition, since large amounts of the synthetic precursors can be made, a variety of enzymological studies can be carried out with these substrates that are not possible with the natural precursors, which have to be used at concentrations far below their K_m values.

Table 4 Action of various nucleases on tRNA$_1^{Tyr}$ precursor

Additions	Acid-soluble radioactivity (cpm/30 min)
None	<10
RNase D	<10
RNase P	16
RNase P + RNase D	250
Fraction B	840
Fraction C	807

Assays were performed using 5200 cpm of tRNA precursor, approximately 10 milliunits of RNase D, and 2 milliunits of fractions B and C. RNase P (3 μg) was used where indicated.

This initial analysis has led to the separation of four activities that are capable of removing nucleotides from the 3' end of synthetic precursor tRNAs. One of these activities, termed fraction A, was not well purified. Its requirements were similar to fractions B and C, but its substrate specificity differed. Further work is necessary to determine whether this activity is a distinct enzyme.

Fractions B and C, on the other hand, had enzymatic properties and molecular weights identical to those previously ascribed to RNase II (Singer and Tolbert 1965; Gupta et al. 1977), and one of the activities, fraction C, was highly purified. However, it is not yet clear why we repeatedly observed two easily separable peaks of this activity on hydroxyapatite columns. The resolution of two peaks of activity was probably not due to proteolytic digestion of RNase II, since all buffers contained the protease inhibitor, phenylmethylsulfonyl fluoride. However, since the two fractions could be separated, they must differ in structure. Further studies are necessary to resolve this question.

The properties of RNase D indicate that it is a distinct enzyme and a likely candidate for processing the 3' terminal of *E. coli* tRNA precursors in vivo. Our previous results with *E. coli* mutants lacking tRNA nucleotidyl transferase (Deutscher and Hilderman 1974) suggested that whatever the active 3' processer in vivo was, it would stop without removing residues from the CCA sequence (Morse and Deutscher 1975; Deutscher et al. 1977). The very low activity of RNase D on intact tRNA is consistent with this observation. Furthermore, the observation that RNase D action on tRNA-CCACC regenerates amino acid acceptor activity also suggests that this enzyme is involved in processing. Most importantly, the action of RNase D on a natural tRNA precursor followed the same order of processing as found in vivo. Of all the potential processing nucleases detected in this study, only RNase D required prior RNase P action to remove nucleotides from the precursor.

We have previously shown that RNase D acts on tRNA molecules with altered structures (Ghosh and Deutscher 1978a). The fact that this enzyme also displays high specificity for a tRNA precursor containing extra residues following the CCA sequence suggests that such a precursor is recognized as an altered tRNA. Once the extra residues are removed, the tRNA becomes relatively resistant to the action of the enzyme, suggesting that the native structure is significantly different from the precursor. Alternatively, the CCA sequence itself may confer resistance to the nuclease, since removal of these residues with snake venom phosphodiesterase regenerates an efficient substrate. Surprisingly, the other type of tRNA precursor, tRNA-CU, which is missing part of the CCA sequence, is relatively resistant to RNase D. Although the explanation for this observation in terms of tRNA structure is not yet clear, the results support the suggestion (Seidman et al. 1975) that two different nucleases would be required to process the two different types of tRNA precursors.

The relationship of RNase D to other suggested 3' processing nucleases remains to be determined. However, from the limited description of these other enzymes, we suspect that RNase D may be the same as RNase P3 (Bikoff et al. 1975) and RNase Q (Shimura et al. 1978). Although the RNase-II-like enzymes, fractions B and C, can remove nucleotides from the 3' end of tRNA precursors, their lack of specificity on the natural precursor suggests that they are not involved in the final 3' processing of tRNA. However, a surprising feature of RNase II action that complicates this interpretation is that once the 5' fragment of the tRNA precursor has been cleaved off, the enzyme's action becomes much more limited. This is apparent from studies with the synthetic precursor, with the natural precursor (R. K. Ghosh and M. P. Deutscher, unpubl.; S. Altman, unpubl.), and from the work of others (Bikoff et al. 1975; Schedl et al. 1976). Shimura et al. (1978) have described a nuclease, termed RNase Y, which behaves similarly and may be RNase II. Thus, a definite conclusion regarding RNase II involvement must await further work. Since RNase D initially purifies with RNase II and has similar enzymatic properties, any future studies must be done carefully to distinguish between these activities.

The results presented here show that the use of synthetic tRNA precursors greatly simplifies the identification and purification of potential processing nucleases. Coupled with the use of other substrates, such as poly(A), these studies have been able to distinguish among the various activities and relate them to known enzymes. In addition, we have been able to attain a high degree of purification for several of these enzymes. Studies are now in progress to determine the purity of each of these enzymes and their mode of action on natural tRNA precursors.

ACKNOWLEDGMENTS

This work was supported by grant GM-16317 from the National Institutes of Health. The technical assistance of Rong-Chang Ni is greatly appreciated. We thank Sidney Altman (Yale University) for providing us with RNase P and the [^{32}P]tRNA^{Tyr} precursor.

REFERENCES

Altman, S. and J. D. Smith. 1971. Tyrosine tRNA precursor molecule polynucleotide sequence. *Nat. New Biol.* **233**:32.

Barrell, B. G., J. G. Seidman, C. Guthrie, and W. H. McClain. 1974. Transfer RNA biosynthesis: The nucleotide sequence of a precursor to serine and proline transfer RNAs. *Proc. Natl. Acad. Sci.* **71**:413.

Bikoff, E. K., B. F. LaRue, and M. L. Gefter. 1975. In vitro synthesis of transfer RNA. II. Identification of required enzyme activities. *J. Biol. Chem.* **250:**6248.

Chang, S. and J. Carbon. 1975. The nucleotide sequence of a precursor to the glycine- and threonine-specific transfer ribonucleic acids of *Escherichia coli*. *J. Biol. Chem.* **250:**5542.

Deutscher, M. P. 1972. Extents of normal and anomalous nucleotide incorporation catalyzed by transfer ribonucleic acid nucleotidyl transferase. *J. Biol. Chem.* **247:**469.

———. 1973. Anomalous adenosine monophosphate incorporation catalyzed by rabbit liver transfer ribonucleic acid nucleotidyl transferase. *J. Biol. Chem.* **248:**3116.

Deutscher, M. P. and R. K. Ghosh. 1978. Preparation of synthetic tRNA precursors with tRNA nucleotidyl transferase. *Nucleic Acids Res.* **5:**3821.

Deutscher, M. P. and R. H. Hilderman. 1974. Isolation and partial characterization of *Escherichia coli* mutants with low levels of transfer ribonucleic acid nucleotidyl transferase. *J. Bacteriol.* **118:**621.

Deutscher, M. P., J. J. C. Lin, and J. A. Evans. 1977. Transfer RNA metabolism in *Escherichia coli* cells deficient in tRNa nucleotidyl transferase. *J. Mol. Biol.* **117:**1081.

Ghosh, R. K. and M. P. Deutscher. 1978a. Identification of an *Escherichia coli* nuclease acting on structurally altered transfer RNA molecules. *J. Biol. Chem.* **253:**997.

———. 1978b. Purification of potential 3' processing nucleases using synthetic tRNA precursors. *Nucleic Acids Res.* **5:**3831.

Goodman, H. M., M. V. Olson, and B. D. Hall. 1977. Nucleotide sequence of a mutant eukaryotic gene: The yeast tyrosine-inserting ochre suppressor SUP4-0. *Proc. Natl. Acad. Sci.* **74:**5453.

Gupta, R. S., T. Kasai, and D. Schlessinger. 1977. Purification and some novel properties of *Escherichia coli* RNase II. *J. Biol. Chem.* **252:**8945.

Kitamura, N., H. Ikeda, Y. Yamada, and H. Ishikura. 1977. Processing by ribonuclease II of the tRNA^{tyr} precursor of *Escherichia coli* synthesized *in vitro*. *Eur. J. Biochem.* **73:**297.

Maisurian, A. N. and E. A. Buyanovskaya. 1973. Isolation of an *Escherichia coli* strain restricting bacteriophage suppressor. *Mol. Gen. Genet.* **120:**227.

Mazzara, G. P. and W. H. McClain. 1977. Cysteine transfer RNA of *Escherichia coli*: Nucleotide sequence and unusual metabolic properties of the 3' C-C-A terminus. *J. Mol. Biol.* **117:**1061.

Morse, J. W. and M. P. Deutscher. 1975. Apparent non-involvement of transfer RNA nucleotidyl transferase in the biosynthesis of *Escherichia coli* suppressor transfer RNAs. *J. Mol. Biol.* **95:**141.

Schedl, P. and P. Primakoff. 1973. Mutants of *Escherichia coli* thermosensitive for the synthesis of transfer RNA. *Proc. Natl. Acad. Sci.* **70:**2091.

Schedl, P., J. Roberts, and P. Primakoff. 1976. In vitro processing of *E. coli* tRNA precursors. *Cell* **8:**581.

Seidman, J. G., F. J. Schmidt, K. Foss, and W. H. McClain. 1975. A mutant of *Escherichia coli* defective in removing 3' terminal nucleotides from some transfer RNA precursor molecules. *Cell* **5:**389.

Shimura, Y., H. Sakano, and F. Nagawa. 1978. Specific ribonucleases involved in processing of tRNA precursors of *Escherichia coli*. *Eur. J. Biochem.* **86:**267.

Singer, M. F. and G. Tolbert. 1965. Purification and properties of a potassium-activated phosphodiesterase (RNase II) from *Escherichia coli*. *Biochemistry* **4**:1319.

Smith, J. D. 1976. Transcription and processing of transfer RNA precursors. *Prog. Nucleic Acid Res.* **16**:25.

Valenzuela, P., A. Venegas, F. Weinberg, R. Bishop, and W. J. Rutter. 1978. Structure of yeast phenylalanine-tRNA genes: An intervening DNA segment within the region coding for the tRNA. *Proc. Natl. Acad. Sci.* **75**:190.

Aspects of Rnase P Structure and Function

Sidney Altman, Emma J. Bowman,* Richard L. Garber,† Ryszard
Kole,‡ Raymond A. Koski,§ and Benjamin C. Stark**
Department of Biology
Yale University
New Haven, Connecticut 06520

Both genetic and biochemical studies have shown that Rnase P is an endoribonuclease that is necessary for the processing of all precursor tRNAs in *Escherichia coli* (for reviews, see Smith 1976; McClain 1977; Altman 1978a,b). Some important features of the pathways of RNA processing are illustrated by the manner in which the substrates for this enzyme were first isolated, and by the way the enzyme itself was first identified. In this paper we shall detail some of the chronology of studies of Rnase P and the nature of the interaction of the enzyme with its substrates. In addition we will summarize some biochemical studies of Rnase-P-like activities in tRNA biosynthesis in eukaryotes.

PATHWAY ANALYSIS

The analysis of the metabolism or catabolism of small molecules in microorganisms has traditionally relied on the availability of mutant enzymes in the pathway. The absence of wild-type enzyme made possible the accumulation of its immediate substrate in the pathway. In the analysis of tRNA biosynthesis, in addition to mutant enzymes, one can also isolate mutants in the initial substrates of the pathway, tRNA precursor molecules, since these are macromolecules with nucleotide sequences encoded in DNA. Certain mutations in tRNA genes yield transcripts that interact less efficiently than the wild-type transcripts with the tRNA processing enzymes (Altman et al. 1974). As a result, these gene transcripts temporarily accumulate in vivo. These mutants can be isolated most easily in cells carrying suppressor tRNA genes by looking for loss of suppression. The loss of the suppressing phenotype is frequently correlated with the inability of the host cell to make adequate

---

Present addresses: *Department of Human Genetics, Yale University School of Medicine, New Haven, Connecticut 06510; †Department of Cell Biology, Roche Institute of Molecular Biology, Nutley, New Jersey 07110; §Institut für Molekular Biologie der Universitat Zurich, 8057 Zurich, Switzerland; **Department of Botany, Washington State University, Pullman, Washington 99163. ‡Permanent address: Institute of Biochemistry and Biophysics, Polish Academy of Sciences, Warsaw, Poland.

quantities of suppressor tRNA through defects in the tRNA biosynthetic pathway (Abelson et al. 1970; Smith et al. 1970). Such weakly suppressing or nonsuppressing mutants in the tRNA^{Tyr} su^+3 gene of *E. coli* were examined several years ago for gene transcripts that might be so altered that they were not processed into mature, functional tRNA. The mutant gene transcripts could only be isolated by extracting cells quickly with phenol (Altman 1971; Altman and Smith 1971). The need for a fast extraction technique shows that the majority of mutant gene transcripts are rapidly degraded in vivo by scavenging ribonucleases. The precursor to wild-type or su^+3 tRNA^{Tyr} accumulates in very small amounts even with the fast extraction technique, presumably because it is efficiently processed into mature tRNA. Some mutants make only about 10% of the normal amounts of mature tRNA. In these cases, the mutant gene transcripts can perhaps adopt a conformation, albeit with a lower rate constant, that is recognized efficiently by processing enzymes.

Isolated mutant gene transcripts can be used as substrates for an in vitro search for processing enzymes. In this way, RNase P activity was identified in crude extracts of *E. coli* (Altman and Smith 1971; Robertson et al. 1972). The levels of RNase P used in vitro are probably in great excess, thereby providing complete cleavage of mutant substrates under standard incubation conditions. An important observation made with the *E. coli* system, as well as with crude extracts of mammalian tissue culture cells, chick tissue, and silkworm tissue (Koski et al. 1976; Garber and Altman 1979; E. J. Bowman and S. Altman, unpubl.), is that even in these crude extracts the biologically correct endonucleolytic cleavage is detectable. (The ease with which RNase P cleavage is observed may be partly due to the relative resistance of tRNA to nonspecific degradation.) This indicates that the substrates and intracellular enzymes are designed such that the canonical natural substrates in the tRNA biosynthetic pathway are acted upon with tremendous specificity by the enzymes of the pathway in crude extracts. This is the case not only with the action of RNase P on tRNA precursor genes but also with respect to $3' \rightarrow 5'$ exonucleolytic trimming activities (Robertson et al. 1972).

The availability of mutant precursor tRNA substrates for studies in vitro soon led to proof that these molecules were indeed cleaved less efficiently than wild-type substrates (Altman et al. 1974). In addition, it was possible to remove extra nucleotides from their 3' terminals by either chemical or enzymatic means. The resultant altered substrates were cleaved less efficiently than precursor tRNA containing the CCA sequence, suggesting that this sequence might be important for RNase P cleavage. The involvement of the CCA sequence in RNase P cleavage of bacteriophage T4 tRNA precursors both in vitro and in vivo has been demonstrated (Schmidt et al. 1976; McClain 1977).

Some natural precursors are undermodified and can serve as substrates for nucleotide modification enzymes. Although the nucleotide modifications appear to play no role in the sequence of nucleolytic cleavages, there does seem to be a difference in the efficiency of modification of certain nucleotides depending upon whether the substrate is the precursor before or after cleavage by RNase P (Schaefer et al. 1973). This is one of the indications that the tRNA moiety of the precursor is in a slightly different conformation from the tRNA moiety of cleaved precursor or mature tRNA (Chang and Smith 1973).

A great advance in the study of tRNA biosynthetic pathways occurred with the isolation of E. coli mutants with thermosensitive RNase P function. These mutants were selected by Schedl and Primakoff (1973) and Shimura, Ozeki, and their co-workers (Sakano et al. 1974) for their inability to make tRNA at restrictive temperatures. Some of the selected mutants were shown to have temperature-sensitive RNase P function in vitro. Furthermore, it has been shown that precursors to most if not all tRNAs accumulate in these mutants at restrictive temperatures, suggesting that only one enzyme is needed for maturation of the 5' terminals of tRNA molecules (Schedl et al. 1974; Ikemura et al. 1975). Of course, the availability of these mutants enabled the isolation, as implied, of many more substrates for RNase P. Nucleotide sequence analysis of these substrates showed, surprisingly, that there is no sequence homology around the RNase P cleavage sites (aside from the ubiquitous CCA sequence at the 3' terminals; Altman 1978a,b). Furthermore, since the mutations in the substrates that affect sensitivity to RNase P cleavage occur far from the cleavage site (e.g., A15, A25, and A31, as shown in Fig. 1), there is strong circumstantial evidence that this endonuclease recognizes tertiary structure of its substrates. For example, the A15 mutation of tRNA^{Tyr} su^+3 alters the folding kinetics of the tRNA moiety such that it remains in an incorrect conformation part of the time (Smith 1974; Leon et al. 1977). Thus, the rate of RNase P action on this precursor substrate may be slowed because the correct conformation for binding to the substrate is not available as frequently as it is with the wild type.

In addition to the tRNA precursor molecules, RNase P can cleave other substrates. One of these, the precursor to 4.5S RNA, a molecule of unknown function in E. coli (Griffin 1975; Bothwell et al. 1976a), also accumulates at restrictive temperatures in RNase P temperature-sensitive mutants. RNase P, in vitro, can generate the mature molecule from the precursor. The other substrate is a small RNA, M3, coded for by ϕ80 (Pieczenik et al. 1972; Bothwell et al. 1976b). Both these reactions proceed in vitro at a rate about 5% or lower than the reaction with precursor tRNA. The two molecules have similar secondary structures. We regard these reactions as secondary reactions of RNase P, presumably ones in which the enzyme is recognizing the junction between long double-strand regions and single-strand regions.

Figure 1 Nucleotide sequence of a precursor to *E. coli* tRNA$_1^{Tyr}$ showing modifications found in the mature molecule, the position and nature of various mutations decreasing the tRNA yield, and the anticodon su^+3 mutation (C35) that does not alter tRNA yield. The arrow pointing inward is the site of RNase P cleavage, and extra 3'-terminal nucleotides are boxed.

CHARACTERIZATION OF RNase P

To gain more insight into the mechanism of substrate recognition by RNase P, we embarked several years ago upon an attempt to purify this enzyme to homogeneity. This has been a long process, made so in part by the cumbersome assay for RNase P activity that involves gel electrophoresis and subsequent autoradiography of reaction products. Nevertheless, we have found this kind of assay essential, since examination of reaction products in a gel reveals most clearly the presence of contaminating exo- or endoribonuclease activities. Of course, once homogeneously pure enzyme is in hand, it will be convenient to have a faster, less explicit assay for RNase P activity (e.g., based on phosphatase sensitivity of a ^{32}P-labeled phosphate group at the 5' terminal of the mature tRNA sequence) to undertake kinetic studies of enzyme function. The current RNase-P-purification procedure relies, except for one step, upon methods that minimize strong ionic interactions (Stark et al. 1978; Kole and Altman 1979). The sequence of steps we currently employ are, starting from S30 preparation: DEAE-Sephadex, Sepharose 4B, Sephadex G-200, and two n-octyl Sepharose column chromatographies in succession. This yields RNase P purified about 2000-fold from crude cell extracts. When analyzed in SDS-polyacrylamide gels, we see that the purest material we have contains, in addition to a prominent protein band called C5, a prominent band of nucleic acid called M2 RNA, which is determined to be RNA by several criteria, and several smaller, minor RNA species, which are probably breakdown products of M2 RNA (Fig. 2). RNase P preparations are enriched for M2 RNA as the purification procedure is followed through. We have not found a method of eliminating this RNA species from any RNase P preparation.

An additional fractionation step, which we have not yet incorporated into large-scale purification procedures, involves buoyant density gradient centrifugation.

The buoyant density of RNase P in CsCl is 1.71 g/ml indicating that the enzyme may be part of a ribonucleoprotein complex. Figure 3 shows buoyant density determinations of RNase P both at the first n-octyl Sepharose column (top panel) and DEAE-Sephadex stages of purification (bottom panel, separate experiment). Note that the less pure enzyme is separated from at least 50% of the protein. It is surprising, however, that so much protein and associated RNA does, in fact, appear at a high density position in CsCl. Some of the RNA is of relatively low molecular weight (S. Altman, unpubl.), indicating that not just ribosome core particles make up the ribonucleoprotein of high buoyant density in *E. coli*.

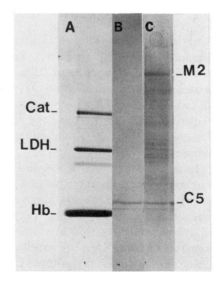

Figure 2 SDS-polyacrylamide gel analysis of RNase P purified through the scheme mentioned in the text. Peak column fractions from the second n-octyl Sepharose column were dialyzed against water, lyophilized, and loaded onto a 7–15% linear gradient gel as described in Kole and Altman (1979). (*Lane A*) Standards: Cat is catalase ($M_r = 58,000$); LDH is lactate dehydrogenase ($M_r = 36,000$); Hb is hemoglobin ($M_r = 16,000$). (*Lane B*) RNase P purified as described above. Lanes *A* and *B* show the gels overstained with Coomassie brilliant blue. (*Lane C*) Same sample as in lane *B* but the gel has been overstained with methylene blue after staining with Coomassie brilliant blue. The dots near the bottom of the gels indicate the position of pyronin Y marker dye.

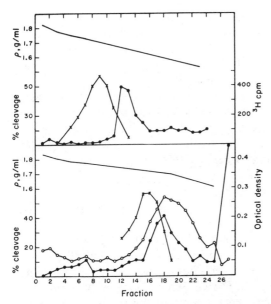

Figure 3 Buoyant density gradient centrifugation of RNase P. (*Top*) RNase P purified through the *n*-octyl Sepharose step was layered on a preformed CsCl gradient as described (Stark et al. 1978) and centrifuged for 22 hr at 45,000 rpm and 3°C. *Drosophila virilis* satellite [^{3}H]DNA was added as marker. (x) RNase P activity; (●) ^{3}H cpm. (*Bottom*) RNase P purified through the DEAE-Sephadex step was centrifuged as described above except that the preformed gradient was made of equal steps of 60%, 56%, and 52% CsCl solutions. (x) RNase P activity; (○) OD_{260}; (●) OD_{595}. Protein concentration was determined by the method of Bradford (1976) and monitored at OD_{595}. OD_{260} values are twice those shown on the right-hand ordinate.

This is a matter worthy of further investigation. Based on an empirical formula used by Spirin (1969), we estimate that if RNase P is a ribonucleoprotein, it is at most 23% protein and at least 77% RNA. This means that the chain length of M1 (M2) must be 300 nucleotides or more. In fact, ^{32}P-labeled RNase P yields an RNA moiety that has an RNase T1 fingerprint similar to band IX RNA characterized by Ikemura and Dahlberg (1973), which they estimated to be in the size range of 300–400 nucleotides.

In functional studies of RNase P we have shown that both micrococcal nuclease and pancreatic RNase A will inactivate the enzyme (Stark et al. 1978). The enzyme preparations we have used to inactivate RNase P have been tested for the absence of protease activity by their inability to inactivate enzymes that are known to consist solely of protein. The experiments with RNase A have also been carried out in the presence and absence of phenylmethylsulfonyl fluoride, an inhibitor of serine proteases, with no difference in results. We have no evidence of protease contamination in either the micrococcal nuclease or RNase A preparations we have used. Furthermore, different batches of enzyme with which we pretreat RNase P, obtained from different sources, have all been effective.

Although both the biochemical and functional experiments indicate that RNase P may be made up of RNA and protein, further proof requires reconstitution of active enzyme from its protein and RNA moieties.

The protein and RNA components of RNase P preparations can be separated by column chromatography in buffers containing 7 M urea. This can be achieved on urea-Sephadex G-200 columns, where the separation is based on the relatively large size of the RNA moiety (m.w. about 115,000) and the small size of the protein moiety (about 20,000) or on ion exchange columns (Fig. 4). In urea-DEAE-Sephadex columns, the basic protein component flows through the column while the RNA is tightly bound and the reverse is true in urea-CM-Sephadex columns. We have taken various RNA-free protein fractions from these columns and added to them M2 RNA that was purified to homogeneity (Kole and Altman 1979). These components were mixed together in solutions containing 7 M urea at 4°C, and then dialyzed to remove the urea. Aliquots from the

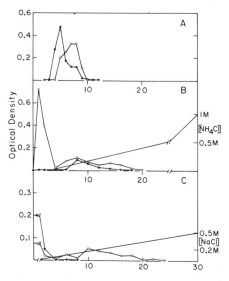

Figure 4 Column chromatography of RNase P in buffers containing 7 M urea. Chromatography was carried out as described in Kole and Altman (1979). (*A*) Urea-Sephadex G-200 column; (*B*) urea-DEAE-Sephadex column; (*C*) urea-CM-Sephadex column. (○) OD_{595}, protein concentration as measured by the method of Bradford (1976); (●) OD_{260}, RNA concentration; (——) salt concentration. (Reprinted, with permission, from Kole and Altman 1979.)

mixtures were then assayed for RNase P activity. The results of one such experiment are shown in Figure 5, in which it is apparent that the M2 RNA alone, or the protein components alone, have no RNase P or any other ribonuclease activity on the precursor tRNA substrate. However, when M2 RNA is mixed with protein fractions containing the C5 protein, RNase P activity can be recovered in 100% yield. Bulk tRNA, 5S, or a mixture of 16S and 23S rRNAs, does not give positive results in this reconstitution mixture (Fig. 5). We have further shown that partially degraded M2 RNA will also function in the reconstitution mixtures but we do not know, for example, if nicked M2 RNA, held together by hydrogen bonds is still functional or if, for example, the ends of the molecule are necessary for function (Kole and Altman 1979).

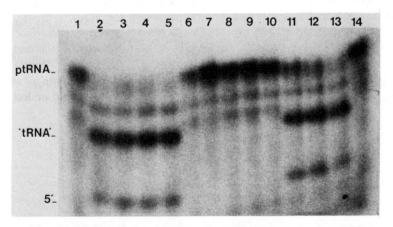

Figure 5 Reconstitution of RNase P activity from protein purified by urea-DEAE-Sephadex chromatography and various RNAs. Flow-through protein from a urea-DEAE-Sephadex column (Fig. 4B), 7.2 µg, and 3 µg of the appropriate RNA were mixed for reconstitution, as described in Kole and Altman (1979), and assayed for RNase P activity. (ptRNA) The position of intact precursor to tRNA^{Tyr}; ('tRNA') the RNase P cleavage products containing the mature tRNA sequence; (5') the 5' proximal extra 41 nucleotides. (*Lane 1*) No enzyme in assay mixture with tRNA^{Tyr} precursor substrate; (*lane 2*) excess of RNase P in assay mixture; (*lanes 3–5*) M2 RNA added in the reconstitution mixture—2 µl, 5 µl, and 10 µl, respectively, of the reconstitution mixture assayed for activity; (*lane 6*) buffer containing 7 M urea and no RNA added to reconstitution mixture—10 µl of mixture assayed for activity; (*lane 7*) M2 RNA alone—3 µg, assayed for activity; (*lanes 8–10*) 16S and 23S rRNAs, 5S RNA, and bulk tRNA, respectively, added to reconstitution mixtures—10 µl of each mixture tested; (*lanes 11–13*) RNase P, purified through the G-200 step, was treated with 7 M urea as in reconstitution mixtures and then dialyzed directly against buffer without urea—0.5 µl, 2 µl, and 5 µl were tested in these assays, respectively; (*lane 14*) same as lane *1*. (Reprinted, with permission, from Kole and Altman 1979.)

POSSIBLE FUNCTION OF RNA IN RNase P ACTION

If RNA has a direct role in RNase P action, i.e., assuming that the purified enzyme complex is indeed made up of a protein and an RNA component, one can postulate two possible roles for the RNA in the action of the enzyme. First, M2 RNA may serve to stabilize the C5 protein in an active, catalytic conformation. If this is the case, then the nature of this protein-RNA interaction might be of a kind comparable to that of rRNA with ribosomal proteins and will be of great interest. On the other hand, the M2 RNA itself may be involved in substrate recognition, i.e., in placing the enzyme complex in a suitable orientation on its various substrates such that accurate cleavage at the 5' terminals of mature tRNA sequences always occurs. As we stated above, mutations that affect the ability of RNase P to cleave certain precursor tRNAs affect positions that determine the secondary and tertiary structures of the molecule (mutations in the anticodon do not alter RNase P action) (Altman and Smith 1971; Smith 1976; Altman et al. 1974). Thus, RNase P, through its RNA moiety, might interact with some of the invariant nucleotides (shared by all tRNAs in *E. coli*; Rich and RajBhandary 1976) in precursor tRNA substrates to position the enzyme on the substrate. This would enable the nucleolytic protein moiety to cleave accurately the 5' extra nucleotides from the remainder of the molecule.

These ideas must be considered as purely conjectural at the moment and will serve as working hypotheses for future experiments. Additionally, since we have used in the reconstitution experiments protein subfractions from RNase P preparations that are only 30–50% pure with respect to C5 protein, it is possible to invoke indirect effects of M2 RNA to account for our results to date. We regard such mechanisms, which postulate basic proteins masking the catalytic activity of C5 protein (a basic protein itself) and being unmasked in the presence of M2 RNA, as unlikely explanations of our data.

RNase-P-LIKE ACTIVITIES IN EUKARYOTES

RNase-P-like activities exist in eukaryotic cells, as well as in *E. coli* (Altman and Robertson 1973; Koski et al. 1976; Garber and Altman 1979; E. J. Bowman and S. Altman, in prep.). These activities can be detected in crude extracts of eukaryotic tissue by mixing any bacterial tRNA precursor with the extract and analyzing the reaction products by gel electrophoresis. The cleavage specificity of these eukaryotic RNase-P-like activities on the precursor tRNA substrates is identical to that of *E. coli* RNase P. Since no mutants in RNase P function exist in eukaryotic organisms, it is difficult to rigorously demonstrate the need for RNase P in the biosynthesis of tRNA. We can show, however, that subcellular fractions from the tissue of the silkworm *Bombyx mori* that contain RNase-P-like activity and $3' \rightarrow 5'$ exonuclease can process individual silkworm precursor tRNA molecules so that they comigrate in two-dimensional gel electrophoresis experiments with mature tRNAs from the same tissue (Garber and Altman 1979). In addition, purified RNase P from *E. coli* can cleave individual silkworm and nematode precursor tRNAs to generate the 5' terminals of the mature tRNA sequences (Garber and Altman 1979; J. D. Smith, pers. comm.). Of course, the size maturation of certain precursor tRNAs may also require the action of an enzyme that eliminates intervening sequences and ligates together the appropriate tRNA fragments (Knapp et al. 1978; O'Farrell et al. 1978). Our studies of tRNA biosynthesis in eukaryotes strongly suggest the need for both an RNase-P-like activity and a $3' \rightarrow 5'$ exonuclease activity, as in *E. coli*, in tRNA maturation.

ACKNOWLEDGMENTS

We thank L. Atkins, R. M. Gershon, and L. Kornreich for technical assistance. R. L. G., R. A. K., and B. C. S. were recipients of predoctoral training grants. E. J. B. was a recipient of a postdoctoral training grant. This research was supported by U. S. Public Health Service grant GM-19422 to S. A.

REFERENCES

Abelson, J. N., L. M. Barnett, A. Landy, R. L. Russel, and J. D. Smith. 1970. Mutant tyrosine transfer ribonucleic acids. *J. Mol. Biol.* **47:** 15.

Altman, S. 1971. Isolation of tyrosine tRNA precursor molecules. *Nat. New Biol.* **229:** 19.

———. 1978a. Transfer RNA biosynthesis. In *Biochemistry of nucleic acids II* (ed. B. F. C. Clark), vol. 17, p. 19. University Park Press, Baltimore, Maryland.

———. 1978b. Biosynthesis of tRNA. In *Transfer RNA* (ed. S. Altman), p. 48. MIT Press, Cambridge, Massachusetts.

Altman, S. and H. D. Robertson. 1973. RNA precursor molecules and ribonucleases in *E. coli*. *Mol. Cell Biochem.* **1:** 83.

Altman, S. and J. S. Smith. 1971. Tyrosine tRNA precursor molecule polynucleotide sequence. *Nat. New Biol.* **233:** 35.

Altman, S., A. L. M. Bothwell, and B. C. Stark. 1974. Processing of *E. coli* tRNA^{Tyr} precursor RNA *in vitro*. *Brookhaven Symp. Biol.* **26:** 12.

Bothwell, A. L. M., R. L. Garber, and S. Altman. 1976a. Nucleotide sequence and *in vitro* processing of a precursor molecule to *Escherichia coli* 4.5S RNA. *J. Biol. Chem.* **251:** 7709.

Bothwell, A. L. M., B. C. Stark, and S. Altman. 1976b. Ribonuclease P substrate specificity cleavage of a bacteriophage ϕ80-induced RNA. *Proc. Natl. Acad. Sci.* **73:** 1912.

Bradford, M. M. 1976. A rapid and sensitive method for the quantitation of microgram quantities of protein utilizing the principle of protein-dye binding. *Anal. Biochem.* **72:**248.

Chang, S. E. and J. D. Smith. 1973. Structural studies on a tyrosine tRNA precursor. *Nat. New Biol.* **246:**165.

Garber, R. L. and S. Altman. 1979. In vitro processing of *B. mori* precursor tRNA molecules. *Cell* **17:**389.

Griffin, B. E. 1975. Studies and sequences of *Escherichia coli* 4.5S RNA. *J. Biol. Chem.* **250:**5426.

Ikemura, T. and J. E. Dahlberg. 1973. Small ribonucleic acids of *Escherichia coli*. *J. Biol. Chem.* **248:**5024.

Ikemura, T., Y. Shimura, H. Sakano, and H. Ozeki. 1975. Precursor molecules of *Escherichia coli* transfer RNAs accumulated in a temperature-sensitive mutant. *J. Mol. Biol.* **96:**69.

Knapp, G., J. S. Beckmann, P. I. Johnson, S. A. Fuhrman, and J. Abelson. 1978. Transcription and processing of intervening sequences in yeast tRNA genes. *Cell* **14:**221.

Kole, R. and S. Altman. 1979. Reconstitution of RNase P activity from inactive RNA and protein. *Proc. Natl. Acad. Sci.* **76:**3795.

Koski, R. A., A. L. M. Bothwell, and S. Altman. 1976. Partial purification and characterization of a ribonuclease P-like activity from human KB cells. *Cell* **9:**101.

Leon, V., S. Altman, and D. M. Crothers. 1977. Influence of A15 mutation on the conformational energy balance in *E. coli* tRNA^{Tyr}. *J. Mol. Biol.* **113:**253.

McClain, W. H. 1977. Seven terminal steps in a biosynthetic pathway leading from DNA to transfer RNA. *Accts. Chem. Res.* **10:**418.

O'Farrell, P. Z., B. Cordell, P. Valenzuela, W. J. Rutter, and H. M. Goodman. 1978. Structure and processing of yeast precursor tRNAs containing intervening sequences. *Nature* **274:**438.

Pieczenik, G., B. G. Barrell, and M. L. Gefter. 1972. Bacteriophage ϕ80-induced low molecular weight RNA. *Arch. Biochem. Biophys.* **152:**152.

Rich, A. and U. L. RajBhandary. 1976. Transfer RNA: Molecular structure, sequence and properties. *Annu. Rev. Biochem.* **45:**805.

Robertson, H. D., S. Altman, and J. D. Smith. 1972. Purification and properties of a specific *Escherichia coli* ribonuclease which cleaves a tyrosine transfer ribonucleic acid precursor. *J. Biol. Chem.* **247:**5243.

Sakano, H., S. Yamada, T. Ikemura, Y. Shimura, and H. Ozeki. 1974. Temperature sensitive mutants of *Escherichia coli* for tRNA synthesis. *Nucleic Acids Res.* **1:**355.

Schaefer, K., S. Altman, and D. Söll. 1973. In vitro nucleotide modification of the *E. coli* tRNA^{Tyr} precursor. *Proc. Natl. Acad. Sci.* **70:**3626.

Schedl, P. and P. Primakoff. 1973. Mutants of *Escherichia coli* temperature sensitive for the biosynthesis of transfer RNA. *Proc. Natl. Acad. Sci.* **70:**2091.

Schedl, P., P. Primakoff, and J. Roberts. 1974. Processing of *E. coli* tRNA precursors. *Brookhaven Symp. Biol.* **26:**53.

Schmidt, F. J., J. G. Seidman, and R. M. Bock. 1976. Transfer RNA biosynthesis: Substrate specificity of RNase P. *J. Biol. Chem.* **251:**2440.

Smith, J. D. 1974. Mutants which allow accumulation of transfer tRNA^{Tyr} precursor molecules. *Brookhaven Symp. Biol.* **26**:1.

———. 1976. Transcription and processing of transfer RNA precursors. *Prog. Nucleic Acid Res. Mol. Biol.* **16**:25.

Smith, J. D., L. M. Barnett, S. Brenner, and R. L. Russell. 1970. More mutant tyrosine transfer ribonucleic acids. *J. Mol. Biol.* **54**:1.

Spirin, A. S. 1969. Informosomes. *Eur. J. Biochem.* **10**:20.

Stark, B. C., R. Kole, E. J. Bowman, and S. Altman. 1978. Ribonuclease P: An enzyme with an essential RNA component. *Proc. Natl. Acad. Sci.* **75**:3717.

Biochemical Characterization of RNase P: A tRNA Processing Activity with Protein and RNA Components

Christine Guthrie and Robert Atchison
Department of Biochemistry and Biophysics
University of California, San Francisco
San Francisco, California 94143

RNase P is an *Escherichia coli* endonuclease that was originally identified by its ability to generate a 5' mature product from the precursor to the psu^+3 amber suppressor derived from tRNA$_1^{Tyr}$ (Robertson et al. 1972). Based on this observation, selection techniques were devised that allowed the isolation of temperature-sensitive mutants defective in RNase P function (Schedl and Primakoff 1973; Ozeki et al. 1974). Genetic analyses have established that mutations conferring thermolabile RNase P activity can arise at two distinct loci, *rnpA* at 77 minutes and *rnpB* at 67 minutes (Kubokawa et al. 1976). A large number of tRNA precursors accumulate in these mutants at the nonpermissive temperature (Schedl et al. 1974; Ikemura et al. 1975; Ilgen et al. 1976; Sakano and Shimura 1978), suggesting that RNase P is required for the synthesis of all *E. coli* tRNAs. When bacteriophage T4 infections of RNase P mutants are performed at high temperature, none of the eight phage-specific tRNAs are synthesized; two dimeric precursors and one monomeric precursor accumulate (Abelson et al. 1974; Guthrie 1975). The RNA profiles of the two genetic classes appear to be identical (Shimura et al., this volume).

A number of mono- and dimeric precursors have been shown to be cleaved accurately by RNase P in vitro (Altman and Smith 1971; Barrell et al. 1974; Guthrie 1975; Chang and Carbon 1975). Comparison of nucleotides at and around these cleavage sites reveals that enzymatic recognition is not sequence-specific. A variety of data are consistent with the prediction that RNase P recognizes some universal features of tRNA conformation. These include results from biochemical analysis of mutants selected for functional inactivation of individual *E. coli* and T4 tRNA species (Seidman et al. 1974; McClain et al. 1975; Guthrie and McClain 1979), as well as chemical modification (Chang and Smith 1973), thermal denaturation (Leon et al. 1977), and modified nucleotide analysis (McClain and Seidman 1975) of mutationally altered substrates. Using nuclease S1, which is single-strand-specific, as a probe for the conformation of a T4-coded dimeric precursor that can be isolated in nonmutant form, we have shown that the dimer comprises two domains within which the specific residue interactions must closely resemble those found in the tRNA molecules (Manale et al. 1979). Furthermore, the nucleotides comprising both RNase P cleavage sites apparently participate in secondary and/or tertiary interactions. Thus, catalysis clearly entails more than simple accessibility of the cleavage sites to the enzyme.

It now appears that this enzyme may act on more structurally diverse substrates than monomeric or dimeric precursors that are either 3' mature (bacteriophage T4) or immature (*E. coli*). We have recently analyzed the processing of *E. coli* 30S pre-rRNA transcripts, which contain the sequences for several tRNA genes between the sequences of 16S and 23S rRNAs. The results show that RNase P can cleave 30S pre-rRNA endonucleolytically at the 5' tRNA terminals to generate larger precursors to 16S and 23S rRNAs (Lund et al., this volume).

To account for the impressive specificity of RNase P with respect to apparent substrate diversity and conformational discrimination, we considered the possibility that these properties reflected the cooperation of a number of protein components, some of which might participate in other stages of protein synthesis and thus also "recognize" tRNA structure. Our ultimate objective is the identification of the minimal set of components that can accurately perform the endonucleolytic cleavages required for 5' maturation of a T4 dimeric precursor.

We report here the properties of a partially purified RNase P preparation. In addition to four major protein components, the active fractions generated by a number of different methods also contain substantial amounts of RNA. This RNA is also heterogeneous; although predominantly 4S, the population includes larger species (6S–10S). There appears to be a high affinity between the RNA and protein components. Dissociation of this apparent ribonucleoprotein (RNP) complex can be achieved by precipitation of the RNA released from EDTA-unfolded enzyme with polyethyleneimine (PEI). The RNA-free proteins have no detectable maturation activity. We have been able to regenerate approximately 30% activity by the addition of RNA phenol-extracted from crude enzyme preparations. Interpretations of this requirement for RNA in in vitro reconstitution of RNase P function are discussed.

RESULTS

DEAE-Sephadex chromatography of ammonium-sulfate-precipitated material washed off ribosomes at 0.2 M NH_4Cl generates an activity profile as shown in Figure 1. When the bacteriophage-T4-coded dimeric precursor to $tRNA^{Gln}$ and $tRNA^{Leu}$ is used as a substrate, three types of activity are revealed.

1. Elution with 0.35 M NH$_4$Cl generates fractions that prevent the substrate from entering a 10% polyacrylamide gel (Bind 1 and Bind 2 in Fig. 1). Material that appears to "bind" the substrate is also eluted at 0.8 M salt (Bind 3).
2. Material eluting at 0.35 M salt behind the major OD$_{260}$ peak exhibits nonspecific nuclease activity.
3. RNase P activity, identified by the production of species with the electrophoretic mobilities characteristic of tRNA^{Gln} and tRNA^{Leu}, elutes at a position coincident with the major OD$_{260}$ peak at 0.8 M salt.

Fractions containing binding and RNase P activities were pooled as indicated in Figure 1 and assayed after concentration as shown in Figure 2. If SDS is added to samples containing Bind 2 prior to gel electrophoresis (lane 5), the substrate is no longer found at the origin but comigrates with the untreated precursor; no cleavage activity is detected. Similar results are obtained with Bind 1 and Bind 3 (data not shown).

When RNase P is rechromatographed on a second DEAE-Sephadex column using a linear gradient (0.3–0.8 M NH$_4$Cl), the active fractions elute at 0.45 M salt. It appears that perhaps only 50% of the RNase P activity applied to the column is recovered; moreover, the binding and nuclease activities observed in the step fractionation are regenerated (data not shown).

A number of alternative and/or additional purification procedures, including column chromatography on Sephadex G-200 and gel electrophoresis under nondenaturing conditions, have all failed to result in fractionation of the activity or appreciable increase in purity. The latter techniques suggest that the active enzyme is large ($\geq$200,000 daltons), but broad activity profiles were often observed. Analysis of active fractions by SDS gel electrophoresis reveals a heterogeneous collection of proteins visualized by Coomassie staining. Although the number and relative stoichiometry of these components varies with different procedures as well as with individual preparations of enzyme, the most highly purified samples contain from four to six prominent bands. It was subsequently found that these fractions also contain substantial amounts of RNA. Staining with ethidium bromide (EtdBr) reveals a population of molecules; although predominantly 4S, larger species (6S–10S) are also present. A typical profile of RNase P after step elution from DEAE-Sephadex is shown in Figure 3, A and B (lane 3).

The continued association of RNA and proteins through many fractionation procedures suggests that the RNA has a high (if not specific) affinity for these proteins. Moreover, many properties of this enzyme (such as high affinity for DEAE and heterogeneous distribution on sizing) are consistent with its existence as an RNP complex. Although the observed stoichiometry of RNA to protein suggested that most if not all of the RNA is fortuitously associated, identification of the protein components of RNase P clearly necessitated removal of the nucleic acid.

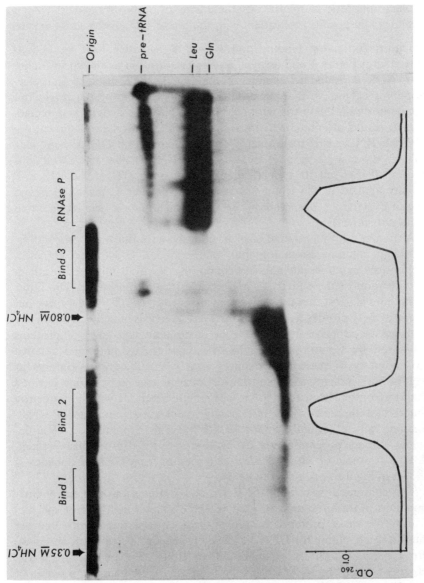

Figure 1 (See facing page for legend.)

The RNA could not be dissociated by column chromatography on DEAE-Sephadex in the presence of 6 M urea, or after extensive dialysis against 2 M LiCl and 4 M urea, a standard method for extracting ribosomal proteins (Spitnik-Elson 1965). Extraction with 66% acetic acid (Hardy et al. 1969), a more stringent procedure for stripping proteins from rRNA, yielded an inactive RNA-free protein fraction. This result at least suggested that the RNA-protein association was not covalent; the loss of enzymatic function was not particularly surprising. It was clearly crucial to find a procedure that would allow quantitative removal of RNA under conditions that could be reasonably expected to preserve protein activity.

We have now developed a method in which the RNA and protein are first dissociated by dialyzing the enzyme against EDTA, in the absence of salt or Mg^{++}. The unfolded enzyme is then dialyzed against high salt (2 M NH_4Cl), followed by the addition of PEI. After twofold dilution of the salt concentration, the RNA-PEI precipitate is removed by centrifugation. Supernatant proteins are concentrated by ammonium sulfate precipitation and resuspended in a buffer containing 1 M NH_4Cl. It can be seen in Figure 3 that the proteins produced by this method (lane 4) are free of RNA as detected by EtdBr staining. The Coomassie profile is qualitatively indistinguishable from that of the control sample (lane 5), which is treated identically except for the addition of PEI. In this particular experiment, approximately 100% of the proteins subjected to the PEI treatment were recovered after ammonium sulfate precipitation; we generally recover 50–80% of the proteins. We have found that recovery is markedly strain dependent; RNase P prepared from the RNase I$^-$ strain MRE600, subjected to the identical treatment, routinely gives only $\simeq$10% protein recovery.

As seen in Figure 4 (lane 2), the control sample retains essentially full activity when assayed for the ability to process the pre-tRNA$^{Gln+Leu}$

Figure 1 Activity and OD_{260} profile of DEAE-Sephadex column chromatography of RNase P. Preparation of subcellular fractions was by the method of Robertson et al. (1972). Material washed off ribosomes at 0.2 M NH_4Cl was precipitated with ammonium sulfate (30–50% w/v) and applied to DEAE-Sephadex A-50 in 0.02 M NH_4Cl. The column was washed with 2 volumes of the same buffer and then eluted with 5 volumes each of buffer containing the indicated salt concentrations. Fractions were assayed for activity for 30 min at 37°C essentially as described previously (Robertson et al. 1972) using as substrate the dimeric precursor to bacteriophage T4 tRNA^{Gln} and tRNA^{Leu} isolated as reported in Guthrie (1975). The positions of the unreacted substrate (pre-tRNA) and the 5' mature cleavage products (Leu, Gln) are indicated. In the presence of some fractions, the substrate remains at the origin of the 10% acrylamide gel. The last two lanes on the right contain control RNase P and a sham reaction, respectively. Fractions were pooled as indicated by the overhead brackets and concentrated by dialysis against 50% glycerol.

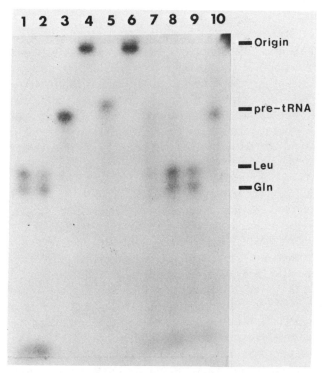

Figure 2 Activity of concentrated pooled fractions from the DEAE-Sephadex column shown in Fig. 1. Lane *6* contains Bind 2; lane *5* contains the identical reaction mixture, but was made 1% in SDS and heated for 2 min at 100°C prior to electrophoresis. Lanes *1, 2,* and *3* contain, in addition to Bind 2, 0.06 OD_{260} unit of RNA obtained by phenol extraction of the material washed off ribosomes at 0.2 M NH_4Cl and used for subsequent purification of RNase P. The RNA was added to the sample in lane *3* after termination of the reaction; the mixture was then treated with SDS. Lane *4* is identical to lane *3* except for the addition of SDS. Lanes *7, 8,* and *9* contain 0.01, 0.1, and 1.0 μl, respectively, of the concentrated RNase P pool (0.1 μl of this material contained 0.06 OD_{260} unit). Lane *10* is a sham reaction.

precursor. In contrast, the RNA-free proteins produced by PEI treatment exhibit no detectable cleavage activity (lane 4); when the sample is loaded in SDS (as shown), the precursor migrates in the sham position (lane 1). The addition of RNA, obtained by phenol extraction of the ribosomal salt wash fraction used as the starting material for DEAE-Sephadex purification of RNase P, results in the partial restoration of activity. As shown in Figure 4 (lane 3), in this particular experiment the activity was low compared to that exhibited by an identical amount of control protein (lane 2). Examination of the gel profile reveals a product with a mobility

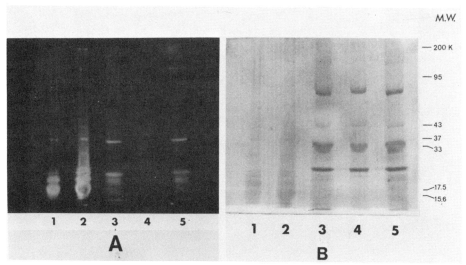

Figure 3 RNA and protein components of RNase P. Electrophoresis on a 10% polyacrylamide gel in the presence of SDS was carried out by the method of Laemmli (1970). The gels were stained with EtdBr and photographed under UV light to visualize RNA (*A*) and then stained with Coomassie brillant blue to visualize proteins (*B*). (M.W.) The mobilities of reference proteins of given molecular weights (in daltons). (*1*) RNA prepared by phenol extraction of the ribosomal salt wash (0.2 M NH$_4$Cl) used for the purification of RNase P; (*2*) RNA phenol-extracted from the low salt (0.02 M NH$_4$Cl) postribosomal supernatant (S100); (*3*) RNase P purified by DEAE-Sephadex chromatography as shown in Fig. 1; (*4*) RNA-free proteins generated by treatment of RNase P with PEI. Enzyme was first dialyzed against 0.005 M EDTA, 0.02 M Tris-HCl (pH 7.5) for 14 hr and then against the same buffer containing 0.001 M EDTA and 2.0 M NH$_4$Cl for 8 hr. PEI stock solution prepared as described by Burgess (Jendrisak and Burgess 1975) was added to the dialysed enzyme sample in 2.0 M NH$_4$Cl, and the mixture was incubated for 10 min with gentle agitation. The concentration of NH$_4$Cl was then reduced to 1.0 M and the mixture was incubated for 20 min. The precipitate was removed by centrifugation, and the material in the supernatant was precipitated by addition of 0.6 g/ml of ammonium sulfate, harvested by centrifugation, and resuspended in buffer containing 1.0 M NH$_4$Cl. (*5*) A control enzyme sample subjected to every step of this procedure except the addition of PEI.

slightly faster than that of tRNA^{Leu}, but little material migrating in the position of tRNA^{Gln}. We have previously shown that when the dimeric precursor is treated with limiting amounts or partially heat-inactivated preparations of RNase P, only a single cleavage event occurs (Guthrie 1975); the 5' mature tRNA^{Leu} moiety is generated, whereas failure to mature the 5' terminal of tRNA^{Gln} results in the production of a fragment containing tRNA^{Gln} sequences and the 5' terminal of the precursor.

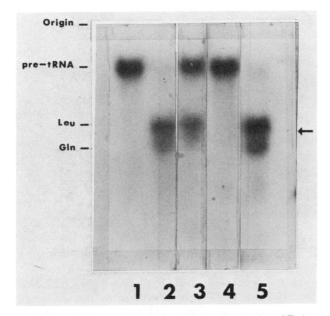

Figure 4 Activity of RNA-free RNase P proteins. (*5*) A standard preparation of RNase P purified by chromatography on DEAE-Sephadex; (*4*) proteins generated by treatment of RNase P with PEI; (*3*) same sample plus 0.07 OD$_{260}$ unit of RNA phenol-extracted from a crude enzyme fraction; (*2*) control enzyme treated identically except for the addition of PEI. These samples contain equivalent amounts of protein, as determined by Coomassie staining of the SDS gels shown in Fig. 3. (*1*) A sham reaction. All samples were heated in the presence of 1% SDS prior to electrophoresis. (←) Position of the intermediate cleavage product still containing the 5′ terminal of the precursor.

$$\begin{array}{c} \text{RNase P} \\ \downarrow \\ \text{p} \!-\!\!|\!-\!\text{tRNA}^{\text{Gln}}\!-\!\!|\!-\!\text{tRNA}^{\text{Leu}}\!-\!\!-\!\text{OH} \\ \downarrow \quad \text{p}\!-\!\!-\!\text{tRNA}^{\text{Leu}}\!-\!\!-\!\text{OH} \\ \text{p}\!-\!\!|\!-\!\text{tRNA}^{\text{Gln}}\!-\!\!-\!\text{OH} \end{array} \quad (1)$$

Despite the low efficiency of reconstitution observed in this experiment, as shown in Figure 5 (lane 11), the substrate was completely cleaved when incubation was extended from 30 to 60 minutes. The products from this reaction were eluted from the gel and analyzed by two-dimensional electrophoresis following digestion with RNase T1 or RNase A (data not shown). This has confirmed the presence of the predicted 5′ mature tRNA terminals.

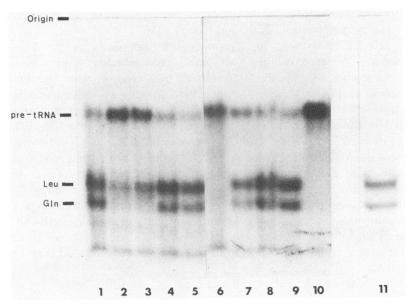

Figure 5 Efficiency of reconstitution of RNase P activity from RNA and inactive RNA-free proteins. Activity was determined by titrating either a standard amount of enzyme-derived RNA with increasing amounts of proteins generated by PEI treatment (*lanes 2–5*) or a standard amount of protein with increasing amounts of RNA (*lanes 7–9*). (*1*) Control, mock PEI-treated enzyme; (*2, 3, 4,* and *5*) one-, two-, five-, and tenfold equivalent amounts, respectively, of RNA-free protein, plus 0.06 OD_{260} unit of RNA; (*6*) fivefold equivalent of RNA-free protein alone; (*7, 8,* and *9*) fivefold excess of protein plus 0.015, 0.03, and 0.06 OD_{260} units of RNA, respectively; (*10*) 0.06 OD_{260} unit of RNA alone; (*11*) scale-up of the reaction shown in lane 3 of Fig. 4. Enzyme components were increased threefold, substrate was increased tenfold, and the reaction was allowed to proceed for 60 min at 37°C. All of the samples, except that in lane *11*, were treated with SDS prior to electrophoresis.

Using other preparations of PEI proteins we have been able to obtain efficiencies of reconstitution approaching 30%. As shown in Figure 5, this efficiency is dependent on the concentration of both protein and RNA. Maximum activity was obtained with 0.25 µg of PEI proteins and 0.06 OD_{260} unit of RNA (Fig. 5, lanes 4 and 9). It should be noted that these experiments are performed simply by incubating the proteins in the presence of the RNA. We have never observed an enhancement of activity by preincubation of the components at high ionic strength or temperature. Similarly, whereas initial experiments were performed by mixing RNA and protein in the presence of urea, followed by dialysis into high salt (no urea), this procedure is not necessary for the production of enzymatic function. Thus, the regeneration of RNase P activity does not appear to resemble the reconstitution of ribosomes from inactive RNA and protein components (Traub and Nomura 1969).

As a preliminary step in determining the specificity of the RNA requirement in reconstitution, we have asked whether several other species can substitute for the RNA derived from crude enzyme fractions. As shown in Figure 6 (lane 1), when PEI proteins are incubated with substrate in the absence of any added RNA and the reaction mixture is loaded in the absence of SDS, the precursor does not enter the 10% polyacrylamide gel. With the addition of 0.06 OD$_{260}$ unit of enzyme-derived RNA (lane 2), 5' mature tRNA^{Gln} and tRNA^{Leu} are produced, together with the intermediate cleavage product described above. An equivalent amount of poly(U) (lane 3) is ineffective in generating the mature cleavage products; the substrate remains at the origin of the gel. The same result is found with the addition of purified E. coli tRNA^{Leu} (lane 5). If unfractionated E. coli tRNA is used (lane 4), the substrate now migrates in the sham position (lane 7); no cleavage products can be detected.

These results suggest that the requirement for RNA in the reconstitution exhibits at least some specificity. They also draw our attention to the binding phenomenon, and prompted us to ask the possible relationship of this activity of the RNA-free PEI proteins to that seen in the

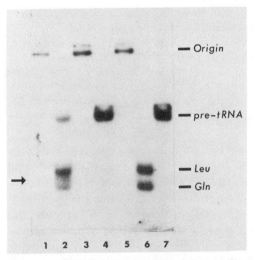

Figure 6 Specificity of the RNA required for reconstitution of RNase P activity. (Lanes 1–5) RNA-free proteins derived by PEI treatment of RNase P; (lanes 2–5) contain 0.06 OD$_{260}$ unit of enzyme-derived RNA (2), poly(U) (3), commercial E. coli tRNA (4), and purified E. coli tRNA^{Leu} (5); (lane 6) control RNase P; (lane 7) a sham reaction. None of these samples were treated with SDS. (→) Position of the intermediate cleavage product described in the text.

DEAE step fractionation of undissociated RNase P. As shown in Figure 2 (lanes 1 and 2), when enzyme-derived RNA is added to Bind 2, the substrate is cleaved to generate products with the mobilities of tRNA^{Gln} and tRNA^{Leu}. Titration experiments suggest that RNase P activity unmasked in this way can account for 30–50% of the total activity applied to the column. The addition of this RNA after termination of the reaction and prior to gel electrophoresis (Fig. 2, lane 4) has no effect. In other words, the RNA must be present during the reaction for cleavage to occur. In other experiments we have shown by electrophoretic analysis of phenol-extracted reaction mixtures that the material bound at the origin consists solely of intact substrate and does not contain mature cleavage products. We also have asked whether other RNA species are capable of unmasking RNase P activity from Bind 2. *E. coli* tRNA and a variety of ribopolymers were added at concentrations equivalent to the amount of RNA used in the experiment shown in Figure 2; in no case did we detect material with the mobility of the correct cleavage products.

Thus, the generation of RNase P activity from both Bind 2 and the inactive, RNA-free PEI proteins requires the participation of some specific RNA molecules. We note that the molecular-weight profile of the RNA phenol-extracted from the ribosomal salt wash used to prepare RNase P (Fig. 3, lane 1) is very similar to that of the DEAE-purified enzyme (Fig. 3, lane 3); the major difference is that the former contains greater amounts of 4S material and is, in fact, essentially indistinguishable from RNA phenol-extracted from the low salt-washed ribosomal supernatant (S100; Fig. 3, lane 2). Nevertheless, there must be some significant difference in the population of RNA molecules derived from the ribosomal salt wash, since S100 RNA is not effective in restoring activity to PEI proteins.

DISCUSSION

We have shown that partially purified preparations of RNase P contain both protein and RNA species. In attempting to identify the minimal number of components required for accurate cleavage of a phage-T4-coded dimeric precursor, we have found that the activity behaves as an RNP complex in which a heterogeneous population of RNA molecules exhibit a high, if not specific, affinity for a small collection of cell proteins. Dissociation of this complex cannot be achieved by many conventional techniques known to disrupt RNA-protein interactions and requires prior unfolding by dialysis against EDTA in the absence of salt or Mg^{++}. The RNA can be quantitatively removed by precipitation with PEI in the presence of 1 M NH_4Cl; at this salt concentration the proteins do not bind to the highly basic PEI. At the same time, this ionic strength is both necessary and sufficient to maintain the solubility of these proteins in the absence of RNA. We believe this to be a crucial feature of the dissociation technique; the solubility of RNase P is strikingly salt-dependent, and is reduced markedly after removal of the associated RNA.

The control sample retains essentially full activity, but the RNA-free protein fraction generated by this procedure cannot effect 5' maturation of the precursor. Substrate incubated with these proteins remains at the origin of a 10% acrylamide gel. If SDS is added to the reaction mixture prior to electrophoresis, the precursor migrates in the position of the sham; no cleavage products are detected. Processing activity can be partially restored to these inactive proteins by the addition of RNA phenol-extracted from the ribosomal salt wash used for purification of the enzyme by step elution from DEAE-Sephadex. Reconstitution efficiencies approaching 30% of the control activity have been obtained. Sequence analysis of the products has confirmed that the reconstituted enzyme cleaves with high fidelity.

The profile of enzyme-associated RNA is complex; 6S and 10S components are found in addition to predominant 4S species. In preliminary attempts to fractionate the RNA derived from crude enzyme preparations, we have observed that the high-molecular-weight fraction is at least five to ten times more active (per OD_{260} unit) than the total RNA population in reconstituting activity (R. Atchison and C. Guthrie, in prep.). This finding is consistent with experiments presented here in which we have shown that neither poly(U) nor *E. coli* tRNA are competent to inspire processing activity, and indicates that there is a specificity requirement for the RNA.

Our results demonstrate that one or more RNA species are required for restoration of enzymatic function to an inactive, RNA-free protein fraction. We wish to emphasize that the significance of these findings is currently unclear. This fraction contains at least four prominent protein components, as visualized by Coomassie staining of 10% SDS gels. The functional basis of this heterogeneity must be clarified before we can determine whether the apparent RNA requirement has physiological relevance.

Although the RNA may indeed play a direct role in substrate recognition and/or catalysis, our present findings cannot eliminate the possibility that the apparent requirement is an in vitro artifact of our purification or assay procedures. In the absence of added RNA, substrate is prevented from entering a 10% acrylamide gel. Since it would appear that this protein fraction contains one or more components with a high affinity for the substrate, it is possible that in the absence of an RNA competitor, the substrate is inaccessible to the catalytic activity. The putative inhibitor might become fortuitously associated with RNase P during extraction. We have shown that ion exchange chromatography generates several fractions that bind the substrate; one of these (Bind

2) can give rise to RNase P activity when assayed in the presence of exogenous enzyme-derived RNA. Alternatively, this type of model could accommodate interpretations in which the RNA plays a legitimate but indirect role in cleavage in vivo. As mentioned at the beginning of this paper, the recognition specificity of RNase P could be accounted for by the participation of several components with the capacity to recognize a tRNA structure. The presumptive binding proteins could function as either general or substrate-specific recognition factors, which prevent the maturation of tRNA transcripts in the absence of a "regulatory" RNA that has a higher affinity for these proteins than does the substrate. It will be of obvious interest to determine whether mutants at one of the two *rnp* loci in fact synthesize an RNA species responsible for thermolabile enzyme function.

Altman (Stark et al. 1978) has recently claimed that an RNA molecule is an essential subunit of RNase P. In light of the arguments presented here, we feel it is premature to draw such a conclusion in the absence of data demonstrating efficient reconstitution of enzymatic activity from individual, homogeneous RNA and protein components. We are currently pursuing such experiments in an attempt to rigorously test the possibility that 5' maturation of tRNA precursors involves the participation of one or more RNA species in a regulatory or catalytic capacity.

ACKNOWLEDGMENTS

We gratefully acknowledge the participation of Elizabeth Bikoff and Gary Dean in early phases of this work. We thank Bruce Alberts, Richard Burgess, Robert Lehman, and Patrick O'Farrell for valuable advice. We are particularly indebted to Roger Garrett for suggesting the use of basic competitors. Finally, we would like to express our gratitude to John Abelson, Diane Colby, William McClain, and Harry Noller for encouraging us by their continual intellectual interest in this project. This work was supported by grants PCM-76-21474 from the National Science Foundation and GM-21119 from the National Institutes of Health.

REFERENCES

Abelson, J., K. Fukada, P. Johnson, H. Lamfrom, D. Nierlich, A. Otsuka, G. Paddock, T. Pinkerton, A. Sarabhai, S. Stahl, J. Wilson, and H. Yesian. 1974. Bacteriophage T4 tRNAs: Structure, genetics and biosynthesis. *Brookhaven Symp. Biol.* **26**:77.

Altman, S. and J. D. Smith. 1971. Tyrosine tRNA precursor molecule polynucleotide sequence. *Nat. New Biol.* **233**:35.

Barrell, B. G., J. G. Seidman, C. Guthrie, and W. H. McClain. 1974. Transfer

RNA biosynthesis: The nucleotide sequence of a precursor to serine and proline transfer RNAs. *Proc. Natl. Acad. Sci.* **71:**413.
Chang, S. and J. Carbon. 1975. The nucleotide sequence of a precursor to the glycine- and threonine-specific transfer ribonucleic acids of *Escherichia coli. J. Biol. Chem.* **250:**5542.
Chang, S. E. and J. D. Smith. 1973. Structural studies on a tyrosine tRNA precursor. *Nat. New Biol.* **246:**165.
Guthrie, C. 1975. The nucleotide sequence of the dimeric precursor to glutamine and leucine tRNAs coded by bacteriophage T4. *J. Mol. Biol.* **95:**529.
Guthrie, C. and W. H. McClain. 1979. Rare transfer RNA essential for phage growth: Nucleotide sequence comparison of normal and mutant T4 tRNA^{Ile}. *Biochemistry* **18:**3786.
Hardy, S., C. G. Kurland, P. Voynow, and G. Mora. 1969. The ribosomal proteins of *E. coli.* I. Purification of the 30S ribosomal proteins. *Biochemistry* **8:**2897.
Ikemura, T., Y. Shimura, H. Sakano, and H. Ozeki. 1975. Precursor molecules of *Escherichia coli* transfer RNAs accumulated in a temperature-sensitive mutant. *J. Mol. Biol.* **96:**69.
Ilgen, C., L. Kirk, and J. Carbon. 1976. Isolation and characterization of large transfer ribonucleic acid precursors from *Escherichia coli. J. Biol. Chem.* **251:**922.
Jendrisak, J. and R. Burgess. 1975. A new method for the large-scale purification of wheat germ DNA-dependent RNA polymerase II. *Biochemistry* **14:**4645.
Kubokawa, S., H. Sakano, and H. Ozeki. 1976. Two genes for RNase P in *E. coli. Jpn. J. Genet.* **51:**420.
Laemmli, U. K. 1970. Cleavage of structural proteins during the assembly of the head of bacteriophage T4. *Nature* **227:**680.
Leon, V., S. Altman, and D. M. Crothers. 1977. Influence of the A15 mutation on the conformational energy balance in *Escherichia coli* tRNA^{Tyr}. *J. Mol. Biol.* **113:**253.
Manale, A., C. Guthrie, and D. Colby. 1979. S1 nuclease as a probe for the conformation of a dimeric tRNA precursor. *Biochemistry* **18:**77.
McClain, W. and J. Seidman. 1975. Genetic perturbations that reveal conformation of tRNA precursor molecules. *Nature* **257:**106.
McClain, W. H., B. G. Barrell, and J. G. Seidman. 1975. Nucleotide alterations in bacteriophage T4 serine transfer RNA that affect the conversion of precursor RNA into transfer RNA. *J. Mol. Biol.* **99:**717.
Ozeki, H., H. Sakano, S. Yamada, T. Ikemura, and Y. Shimura. 1974. Temperature-sensitive mutants of *Escherichia coli* defective in tRNA biosynthesis. *Brookhaven Symp. Biol.* **26:**89.
Robertson, H., S. Altman, and J. D. Smith. 1972. Purification and properties of a specific *Escherichia coli* ribonuclease which cleaves a tyrosine transfer ribonucleic acid precursor. *J. Biol. Chem.* **247:**5243.
Sakano, H. and Y. Shimura. 1978. Characterization and *in vitro* processing of transfer RNA precursors accumulated in a temperature-sensitive mutant of *Escherichia coli. J. Mol. Biol.* **123:**287.
Schedl, P. and P. Primakoff. 1973. Mutants of *Escherichia coli* thermosensitive for the synthesis of transfer RNA. *Proc. Natl. Acad. Sci.* **70:**2091.
Schedl, P., P. Primakoff, and J. Roberts. 1974. Processing of *E. coli* tRNA precursors. *Brookhaven Symp. Biol.* **26:**53.

Seidman, J. G., M. M. Comer, and W. H. McClain. 1974. Nucleotide alterations in the bacteriophage T4 glutamine transfer RNA that affect ochre suppressor activity. *J. Mol. Biol.* **90:**677.

Spitnik-Elson, P. 1965. The preparation of ribosomal protein from *Escherichia coli* with lithium chloride and urea. *Biochem. Biophys. Res. Commun.* **18:**557.

Stark, B. C., R. Kole, E. J. Bowman, and S. Altman. 1978. Ribonuclease P: An enzyme with an essential RNA component. *Proc. Natl. Acad. Sci.* **75:**3717.

Traub, P. and M. Nomura. 1969. Structure and function of *E. coli* ribosomes. VI. Mechanism of assembly of 30S ribosomes studied *in vitro*. *J. Mol. Biol.* **40:**391.

rRNA and tRNA Processing Signals in the rRNA Operons of *Escherichia coli*

Richard A. Young, Richard J. Bram, and Joan A. Steitz
Department of Molecular Biophysics and Biochemistry
Yale University
New Haven, Connecticut 06510

It has been known for some time that the three rRNAs of *Escherichia coli* are transcribed in the order 16S, 23S, and 5S into a single precursor molecule, which is then cleaved and trimmed by a host of specific ribonucleases. Precursors of bacterial tRNAs have long been the subject of studies aimed at pinpointing those aspects of their structures that dictate the precise processing events required to fashion a functional tRNA molecule. However, only relatively recently were tRNA genes identified within the *E. coli* rRNA operons (Lund et al. 1976; see also Morgan et al., this volume). The presence of such tRNA genes requires the processing pathways for both types of stable RNA molecules to be productively coordinated within a growing cell.

To examine rRNA and tRNA processing sites in the primary rRNA transcript of *E. coli*, we have determined selected sequences from two of the seven *E. coli* rRNA operons. These are *rrnD* and *rrnX* (a hybrid operon) carried by the transducing phages λ*daroE* and λ*dilv5*, respectively. So far we have analyzed the region preceding the 16S gene, the entire 16S–23S spacer, and the region between the 23S and 5S gene. The location of these DNA stretches in the two rRNA operons is shown in Figure 1. Features involved in the recognition of the rRNA transcript by known rRNA and tRNA processing enzymes are summarized below. A description of the DNA analysis, as well as a more thorough discussion of the sequences obtained, has been presented elsewhere (Young and Steitz 1978; Young et al. 1979).

RNase III RECOGNITION OF SITES BEYOND THE 16S AND 23S TERMINALS

RNase III was originally identified as an *E. coli* endonuclease capable of completely cleaving double-stranded RNA to fragments about 15 nucleotides long (Crouch 1974; Robertson and Dunn 1975). Later the enzyme

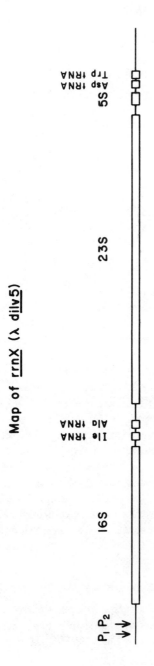

Figure 1 Map to scale of the rRNA operons *rrnX* and *rrnD*. (□) Genes; (——) spacer regions; (P_1 and P_2) transcription start sites in each operon. The lengths in nucleotide pairs of the transcribed spacer regions are: 284 (*rrnX*) and 293 (*rrnD*) from P_1 to 16S, 176 from P_2 to 16S, 68 from 16S to tRNA$_1^{Ile}$, 42 from tRNA$_1^{Ile}$ to tRNA$_{1B}^{Ala}$, 174 from tRNA$_{1B}^{Ala}$ to 23S, 92 from 23S to 5S, and 57 from 5S to tRNA$_1^{Asp}$ (*rrnX*). (See Young and Steitz 1978, 1979; Young et al. 1979.)

was discovered to be responsible for the site-specific processing of bacteriophage T7 mRNAs (Dunn and Studier 1973a) and for the extremely rapid conversion of the 30S rRNA primary transcript to more immediate precursors of the 16S, 23S (Dunn and Studier 1973b; Nikolaev et al. 1973), and 5S (Ginsburg and Steitz 1975) rRNAs. The first RNase III cleavage sites examined, those in the phage T7 early mRNA precursor (Robertson et al. 1977; Rosenberg and Kramer 1977; Oakley and Coleman 1977), showed a highly conserved secondary, rather than primary, structure. That is, each sequence folds into a hairpin loop consisting of two helical segments containing 9–11 bp and separated by an internal loop (bubble), within which chain scission occurs on one or both sides.

When analyzing the *rrn* region preceding the 16S gene, we looked for sequences capable of assuming a comparable truncated, hairpin loop structure. Previously, the 17S product generated in vitro by RNase III action on the 30S pre-rRNA had been characterized with respect to its terminal and non-16S oligonucleotides (Ginsburg and Steitz 1975). However, no local hairpin loops could be formed from sequences surrounding the 17S terminals. Rather, we discovered that DNA sequences preceding the 16S gene are complementary to those after 16S, which is some 1600 nucleotides distant. Apparently, these two regions come together in the primary rRNA transcript to form a perfect 26-bp stem that is then recognized and cleaved by RNase III (as indicated in Fig. 2A). It is unclear whether pairing between distant portions of nuclear transcripts is essential for mRNA or rRNA processing in eukaryotic cells.

We also find that sequences at the two ends of the 23S gene are complementary (Fig. 2B). Here, the exact sites of RNase III cleavage are not yet known. However, the idea that the two structures illustrated in Figure 2 indeed form in the pre-rRNA and provide recognition elements for RNase III is strongly supported by a number of previous observations.

1. Giant secondary structure loops surround 16S and 23S sequences when single-stranded rDNA is examined by electron microscopy under partially denaturing conditions (Wu and Davidson 1975).
2. RNase III is unable to cleave a transcript synthesized from a DNA fragment that includes *rrn* promoter sequences but ends within the 16S gene (M. Cashel, pers. comm.).
3. Double-stranded RNAs that are approximately 20 bp long and contain the oligonucleotides predicted in Figure 2 can be isolated from total *E. coli* RNA and cleaved by RNase III in vitro (Robertson and Barany 1978; H. D. Robertson, pers. comm.).
4. The appearance of the mature 5' and 3' terminals of 23S rRNA within structure B (Fig. 2B) is consistent with the finding that the RNase III-generated, 23S-containing product of 30S pre-rRNA contains no large oligonucleotides not present in mature 23S (D. Charny et al., unpubl.).

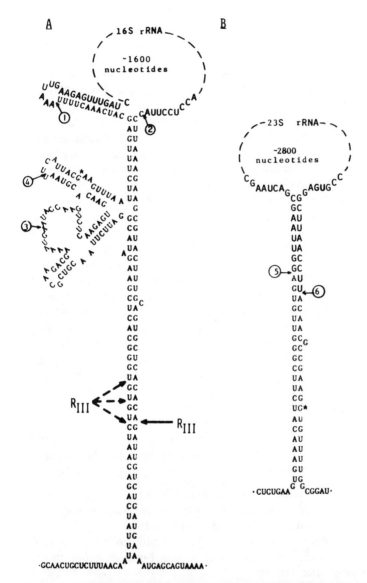

Figure 2 Proposed secondary structures surrounding RNase III cleavage sites in the 30S pre-rRNA from *rrnD*. The nucleotide pairing shown was calculated to be the most stable configuration according to Gralla and Crothers (1973). Endonuclease sites (circled numerals 1–5) are described in the text. (Bold type) Mature rRNA sequences; (*) sequence differences between *rrnD* and *rrnX*. At both positions, the nucleotide in *rrnX* is A. (*A*) (⎯⎯→), sites of RNase III cleavage to generate 17S rRNA; (--→) uncertainty concerning which of the three possible bonds is cut to generate the 5' end of 17S. (*B*) Proposed secondary structure surrounding 23S rRNA in the 30S pre-rRNA of *rrnD*.

The realization that apparently the same active site of RNase III can cleave completely double-stranded RNA, the phage T7 precursor mRNA regions described above, and the two structures pictured in Figure 2 reopens the question of what this endonuclease recognizes in a substrate molecule. No strict sequence homology either at the terminals generated or in adjacent sequences is evident. In common are the appearance of helical regions of at least 9 bp both 5′ and 3′ to each point of scission and the cleavage of bonds nearly opposite each other in the proposed RNA secondary structure. Perhaps merely the length of the double-stranded RNA stem aligns the enzyme for exact cleavage. Much more work on RNase III recognition sites is clearly needed.

OTHER PROCESSING SITES IN 16S AND 23S PRECURSORS

Figure 2 also indicates bonds in 30S pre-rRNA that are cut by several additional ribonucleases. Two of these enzymes are known to participate in the in vivo maturation of 16S rRNA. The circled 1 in Figure 2 is RNase M16 (Dahlberg et al. 1978), which generates the mature 5′ end of 16S rRNA and the circled 2 is an enzyme that cleaves 3′ to 16S sequences (Hayes and Vasseur 1976). Two other endoribonucleases (Fig. 2, circled 3 and 4) that act in vitro or when the normal pathway of 16S maturation is obstructed are described in detail elsewhere (Lund and Dahlberg 1977; Dahlberg et al. 1978). Presumably one or two additional enzymes, which so far are totally uncharacterized, act to produce the mature 5′ and 3′ ends of 23S rRNA (Fig. 2, circled 5 and 6). In no case do we know what features of the rRNA primary or secondary structure are involved in these nuclease recognition events.

FEATURES OF THE 16S–23S SPACER REGION

Both *rrnD* and *rrnX* contain the genes for $tRNA_1^{Ile}$ and $tRNA_{1B}^{Ala}$ in their 16S–23S spacer. The two spacer sequences differ in only one out of 437 bp, yet the fact that *rrnD* and *rrnX* diverge in their promoter regions argues that these two spacer regions are distinct but highly conserved. (See Young et al. 1979 for further discussion.)

From the distances given in Figure 1, it is apparent that both tRNA genes are clustered within the first 60% of the spacer DNA. The mature CCA terminal is encoded, as is true of other *E. coli* tRNA genes. However, the 42-nucleotide distance between the two tRNAs is significantly longer than observed in other dimeric tRNA precursors (Altman 1978).

We do not understand the function of the distal 40% of the spacer DNA, although extensive homologies with comparable portions of the 16S–23S spacer from *rrnE*, which contains tRNA$_2^{Glu}$, suggest an important role for this region in rRNA or tRNA biogenesis (Morgan et al., this volume). Conceivably, the conserved nucleotides are involved in cleavage

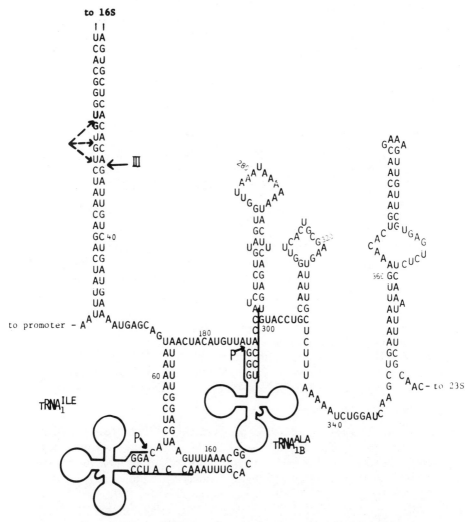

Figure 3 Possible RNA secondary structure of the *rrnD* 16S–23S rRNA spacer transcript. Nucleotide pairing shown was calculated as described in Fig. 2. Spacer residues 22–409 are shown. (P) Site of cleavage by RNase P. (Reprinted, with permission, from Young et al. 1979.)

of the spacer transcript by a new endonuclease recently identified by Gegenheimer and Apirion (1978).

Most striking is the high AT content (63%) of the noncoding portions of the spacer DNA, compared to 60% GC for the two tRNA genes. Perhaps this contrasting base composition facilitates the folding of the nascent tRNA molecules into the proper secondary and tertiary structures required for their processing from the 30S rRNA primary transcript.

Figure 3 illustrates a possible secondary structure for the *rrnD* spacer RNA. Note that the presumed site of RNase P action occurs in a single-stranded region 5' to tRNA$_1^{Ile}$ but within a double-stranded stem 5' to tRNA$_{1B}^{Ala}$. Both situations have been reported in previously examined *E. coli* and phage T4 tRNA precursors (Bothwell et al. 1976). Unlike most in vivo conditions, where approximately equal amounts of tRNA$_1^{Ile}$ and tRNA$_{1B}^{Ala}$ are produced (Ikemura and Nomura 1977), in vitro action of purified RNase P on spacer RNA yields significantly fewer cleavages 5' to tRNA$_1^{Ile}$ than to tRNA$_{1B}^{Ala}$ (E. Lund et al., pers. comm.). Perhaps this differential susceptibility results from the formation of the unusual structure with three converging helices at the end of tRNA$_1^{Ile}$ pictured in Figure 3. Whatever the cause, the contrast between in vivo and in vitro tRNA$_{1B}^{Ala}$ processing underscores the fine tuning of the intricate pathways of sequential cleavage by which both tRNA and rRNA molecules are produced in normal *E. coli*.

ACKNOWLEDGMENT

This work was supported by grant AI-10243 from the National Institutes of Health.

REFERENCES

Altman, S. 1978. Transfer RNA biosynthesis. In *International review of biochemistry. Biochemistry of nucleic acids II* (ed. B. F. C. Clark), Vol. 17, p. 19. University Park Press, Baltimore.

Bothwell, A. L. M., B. C. Stark, and S. Altman. 1976. Ribonuclease P substrate specificity: Cleavage of a bacteriophage φ80-induced RNA. *Proc. Natl. Acad. Sci.* **73:** 1912.

Crouch, R. J. 1974. Ribonuclease III does not degrade deoxyribonucleic acid-ribonucleic acid hybrids. *J. Biol. Chem.* **250:** 3050.

Dahlberg, A. E., H. Tokimatsu, M. Zahalak, F. Reynolds, P. C. Calvert, A. B. Rabson, E. Lund, and J. E. Dahlberg. 1978. Processing of the 5' end of *E. coli* 16S ribosomal RNA. *Proc. Natl. Acad. Sci.* **75:** 3598.

Dunn, J. J. and F. W. Studier. 1973a. T7 early RNAs are generated by site specific cleavage. *Proc. Natl. Acad. Sci.* **70:** 1559.

———. 1973b. T7 early RNAs and *E. coli* ribosomal RNAs are cut from large precursor RNAs *in vivo* by ribonuclease III. *Proc. Natl. Acad. Sci.* **70:** 3296.

Gegenheimer, P. and D. Apirion. 1978. Processing of rRNA by RNase P: Spacer tRNAs are linked to 16S rRNA in an RNase P RNase III mutant strain of *E. coli. Cell* **15:** 527.

Ginsburg, D. and J. A. Steitz. 1975. The 30S ribosomal precursor RNA from *E. coli. J. Biol. Chem.* **250:** 5647.

Gralla, J. and D. M. Crothers. 1973. Free energy of imperfect nucleic acid helixes. *J. Mol. Biol.* **73:** 497.

Hayes, F. and M. Vasseur. 1976. Processing of the 17S *E. coli* precursor RNA in the 27S pre-ribosomal particle. *Eur. J. Biochem.* **61:** 433.

Ikemura, T. and M. Nomura. 1977. Expression of spacer tRNA genes in ribosomal RNA transcription units carried by hybrid ColE1 plasmids in *E. coli. Cell* **11:** 779.

Lund, E. and J. E. Dahlberg. 1977. Spacer transfer RNAs in ribosomal RNA transcripts of *E. coli*: Processing of 30S ribosomal RNA *in vitro. Cell* **11:** 247.

Lund, E., J. E. Dahlberg, L. Lindahl, S. R. Jaskunas, P. P. Dennis, and M. Nomura. 1976. Transfer RNA genes between 16S and 23S rRNA genes in rRNA transcription units of *E. coli. Cell* **7:** 165.

Nikolaev, N., L. Silengo, and D. Schlessinger. 1973. Synthesis of a large precursor to ribosomal RNA in a mutant of *E. coli. Proc. Natl. Acad. Sci.* **70:** 3361.

Oakley, J. J. and J. E. Coleman. 1977. Structure of a promoter for T7 RNA polymerase. *Proc. Natl. Acad. Sci.* **74:** 4366.

Robertson, H. D. and F. Barany. 1978. Enzymes and mechanisms in RNA processing. In *Proceedings of the 12th FEBS Congress*, p. 285. Pergamon Press, Oxford.

Robertson, H. D. and J. J. Dunn. 1975. Ribonucleic acid processing activity of *E. coli* ribonuclease III. *J. Biol. Chem.* **250:** 3050.

Robertson, H. D., E. Dickson, and J. J. Dunn. 1977. A nucleotide sequence from a ribonuclease III processing site in bacteriophage T7 RNA. *Proc. Natl. Acad. Sci.* **74:** 822.

Rosenberg, M. and R. A. Kramer. 1977. Nucleotide sequence surrounding a ribonuclease III processing site in bacteriophage T7 RNA. *Proc. Natl. Acad. Sci.* **74:** 984.

Wu, M. and N. Davidson. 1975. Use of gene 32 protein staining of single-strand polynucleotides for gene mapping by electron microscopy: Application to the $\phi 80_3 ilvsu^+7$ system. *Proc. Natl. Acad. Sci.* **72:** 4506.

Young, R. A. and J. A. Steitz. 1978. Complementary sequences 1700 nucleotides apart form a ribonuclease III cleavage site in *E. coli* ribosomal precursor RNA. *Proc. Natl. Acad. Sci.* **75:** 3593.

Young, R. A. and J. A. Steitz. 1979. Tandem promoters direct *E. coli* ribosomal RNA synthesis. *Cell* **17:** 225.

Young, R. A., R. Macklis, and J. A. Steitz. 1979. Sequence of the 16S–23S spacer region in two ribosomal operons of *E. coli. J. Biol. Chem.* **254:** 3264.

RNA Processing in an *Escherichia coli* Strain Deficient in Both Rnase P and Rnase III

Hugh D. Robertson and Edward G. Pelle*
The Rockefeller University
New York, New York 10021

William H. McClain
Department of Bacteriology
University of Wisconsin
Madison, Wisconsin 53706

It is now evident that primary transcripts of most prokaryotic and eukaryotic RNAs contain extra regions of unknown function interspersed among regions destined to occur in mature molecules. The role of RNA processing enzymes is the correct removal of the extra regions leading to the formation of functional mature RNAs. The most intensively studied RNA processing reactions are those carried out by *Escherichia coli* Rnase III (Robertson et al. 1968; Robertson and Dunn 1975; Dunn 1976), *E. coli* Rnase P (Robertson et al. 1972; Stark et al. 1978), and *Bacillus subtilis* Rnase M5 (Sogin et al. 1977; Meyhack et al. 1978). All three of these enzymes retain the ability to cleave their natural substrates properly in vitro, even following extensive purification of both the enzyme and the RNA precursor molecule.

It is difficult to identify and study the mechanism of various *E. coli* RNA processing enzymes without adopting a comprehensive approach that includes both the genetic and the biochemical definitions of a proposed RNA processing activity. This complication is evident in two of the approaches that have been recently employed to study this problem: several additional enzymatic activities that have been characterized biochemically from *E. coli* and tested on various RNA precursor substrates (Sakano and Shimura 1975; Schedl et al. 1976; Ghosh and Deutscher 1978) require further identification; in addition, mutations in well-characterized *E. coli* mutant strains lacking Rnase III or Rnase P and the resulting changes in RNA metabolism have been correlated with potential new RNA processing activities (Seidman et al. 1975; Gegenheimer et al.

---

*Present address: Department of Hematology, New York University School of Medicine, New York, New York.

1977; Studier 1975). In the present report, we will illustrate a genetic/biochemical approach to the identification of RNA processing activities. First, we will describe the construction of an *E. coli* strain that carries previously characterized mutations in both RNase P and RNase III, and then show: (1) how this strain can be used to detect further specific RNA processing activities in *E. coli*; (2) how subcellular extracts of this strain and its parent *E. coli* strains can be used to characterize these new activities and compare them to those previously reported by others; and (3) how the RNA precursor molecules accumulating in the strain that carries both RNase III and RNase P mutations can be used to isolate oligonucleotides from regions that undergo specific processing for studying the detailed cleavage mechanism of the enzyme RNase III.

CONSTRUCTION OF THE RNase III⁻, RNase P⁻ STRAIN

The strategy was as follows. The strain carrying mutations in two RNA processing genes was constructed by Hfr-mediated recombination between the two single-mutant strains. The F⁻ strain is mutant for RNase P. Progeny F⁻ strains that received the RNase III mutant allele of the male were selected as recombinants for a nutritional marker (*glyA*) that is linked to the RNase III locus. The recombinants were then tested to verify that they carried both RNase mutant alleles.

The following parental strains were utilized: The male was *E. coli* strain BL214 and the female was *E. coli* strain A49. The genotype of BL214 is: HfrP045, Sm^s, *rnc-105* (the RNase III allele), *glyA*⁺, *rel*-1?, *uraP119* (Studier 1975). The genotype of A49 is: F⁻, Sm^r, *rnp-49* (the RNase P allele), *glyA*, *arg*, *lacZ*, *trp* (Schedl and Primakoff 1973). Each strain was grown by gentle shaking in rich broth to about 2×10^8 cells/ml. Strain BL214 was grown at 37°C and strain A49 at 30°C. For the genetic cross, equal volumes of the cultures were mixed and gentle shaking at 39°C was continued for 1 hour. The cells were then washed with minimal medium twice by centrifugation and spread on glucose-minimal-medium plates supplemented with streptomycin, arginine, thymine, and tryptophan. This allowed selection of F⁻*glyA*⁺ recombinants. Plates were incubated at 30°C for 2 days. A 10^4-fold increase in the number of colonies was observed on plates prepared from the cross mixture compared to plates prepared from self crosses.

To carry out preliminary characterization of the recombinants, small broth cultures were prepared from colonies arising on the plate containing the cross mixture. Initial testing for the *rnc-105* allele involved examining cell-free extracts for inability to solubilize ³H-labeled poly(rA·U); only extracts containing RNase III solubilize this substrate (Robertson 1968). Extracts from 8 of 11 isolates did not solubilize poly(rA·U). Initial testing for the *rnp-49* allele involved examining colony-forming ability at 30°C and 42°C; *rnp-49* cells grew at 30°C but not at 42°C. Seven of 11 cultures did not grow at 42°C.

Four of 11 isolates were defective in their ability to solubilize poly(rA·U) and to form colonies at 42°C. One of these isolates, designated ABL1, was picked for further study as described below.

THE EXISTENCE OF ADDITIONAL SPECIFIC RNA PROCESSING ACTIVITIES IN *E. coli*

Growth of strain ABL1 was carried out under various conditions and the RNA present after pulse labeling with ^{32}P was characterized. Analysis of RNA (pulse-labeled with ^{32}P) for periods of from 5 to 20 minutes shows that the strain carrying two mutations accumulates both 30S rRNA precursor and a group of tRNA precursors at 42°C, as demonstrated by polyacrylamide gel analysis on 2% or 3–20% gradient gels (see Fig. 1). An interesting feature of the behavior of this strain is that, at high temperature, where both RNase III and RNase P are lacking, the 30S rRNA precursor continues to be cleaved in vivo. Thus, at least a third endonucleolytic RNA processing activity must be capable of involvement in the rRNA processing pathway, despite the ability of both RNase III and RNase P to cleave isolated 30S rRNA precursor molecules (see Lund et al.; Apirion et al.; both this volume). We conclude that the *E. coli* strains that carry multiple mutations of RNA processing enzymes can be useful in guiding further research in this area. However, it is not sufficient to characterize these hypothetical additional RNA processing activities solely by listing the novel RNA species that accumulate on polyacrylamide gels under various conditions. It is also necessary to characterize the enzymes present in these cells by biochemical techniques.

CHARACTERIZATION OF ADDITIONAL RNA PROCESSING ACTIVITIES IN SUBCELLULAR EXTRACTS OF STRAIN ABL1

In carrying out biochemical tests for the presence of potentially novel *E. coli* RNA processing enzymes, it should be noted that the previous literature describes several cases of apparently novel RNase activities. Some of these activities have, in fact, been found to be caused by previously characterized enzymes. For example, in the case of the reported RNase O activity, which was thought to be involved in early steps of *E. coli* tRNA maturation (Sakano and Shimura 1975), further research has revealed that RNase III is, in fact, responsible for carrying out this activity (Shimura and Sakano 1977). Another activity reportedly involved in early steps of tRNA maturation is RNase P2, which was detected in RNase P mutants by Schedl et al. (1973; 1976) and which has been partially purified. Any endonucleolytic activity found subsequently would,

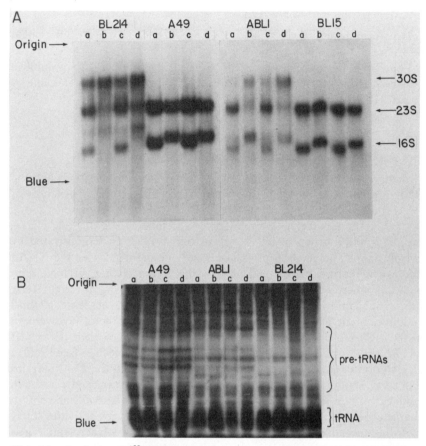

Figure 1 Analysis of [^{32}P]RNA grown in strains of BL15, A49, BL214, and ABL1. Cells were grown in modified peptone broth (Robertson and Jeppesen 1972) to an A_{260} of 0.4 at a temperature of 34°C. Four 10-ml aliquots of each strain divided off and two of these made 400 μg/ml in chloramphenicol. 2.5 mCi of ^{32}P-labeled phosphoric acid were added and incubation continued for an additional 10 min. 0.1-ml samples were added to 1 ml of chilled 0.03 M sodium phosphate (pH 7.2) containing 0.02 M sodium cyanide and the cells collected by centrifugation at 10,000g. Cells were resuspended and prepared for electrophoresis according to the method of Dunn and Studier (1973) and the samples run on 2% acrylamide/0.5% agarose or 3–20% acrylamide gradient gels. (*A*) 2% acrylamide/0.5% agarose gel electrophoresis of four bacterial strains. Each strain was grown with or without 400 μg/ml chloramphenicol and at 34°C or 42°C as follows: 34°C without chloramphenicol (*a*); 34°C with chloramphenicol (*b*); 42°C without chloramphenicol (*c*); and 42°C with chloramphenicol (*d*). Origin and bromphenol blue locations are indicated along with positions of 30S, 23S, and 16S RNAs. 25,000 cpm/slot were added. (*B*) 3–20% gradient polyacrylamide gel electrophoresis of strain A49, ABL1, and BL214. Aliquots of these strains were analyzed as above on 3–20% gradient gels. Lanes *a–d* under each strain refer to the same conditions as described for *A*. Positions of the origin of electrophoresis, bromphenol blue marker, mixed tRNAs, and pre-tRNAs are indicated.

therefore, have to be compared with RNase P2. Finally, Dunn (1976), Westphal and Crouch (1975), and Paddock et al. (1976) have reported that RNase III itself is capable of anomalous behavior, particularly under low salt or high enzyme:substrate conditions. Therefore, any candidate for a new endonucleolytic RNase in *E. coli* must also be compared to RNase III under a variety of conditions.

Although Figure 1 reveals that enzymatic activities remain in strain ABL1, which can cleave the 30S rRNA precursor molecule, this substrate is a rather large and complex molecule with which to carry out initial characterization of a new activity. In addition, this molecule has a high turnover rate in vivo. We have, therefore, chosen the *E. coli* tRNA^{Tyr} precursor molecule, normally an RNase P substrate (Robertson et al. 1972), for these initial tests. Strain ABL1 was first shown to lack RNase P and RNase III activity by standard assays (Robertson et al. 1968, 1972). Figure 2 shows cleavage of the 129-base tRNA^{Tyr} precursor by ribosome wash fractions of strain ABL1 and its parental strains A49 and BL214. It is evident that A49 and ABL1 lack RNase P but possess another endonuclease that cleaves the substrate to give several bands not related to mature tRNA^{Tyr}.

Figure 2 also shows evidence that, at sufficient concentration, RNase III is capable of introducing cleavages into the tRNA^{Tyr} precursor molecule. To gain evidence about these activities, the experiments shown in Figure 3 were carried out. Three different preparations of RNase III at various states of purity were used to treat the tRNA^{Tyr} precursor with and without addition of competitor double- or single-stranded RNA. All three preparations of RNase III, including that purified nearly to homogeneity on affinity columns of agarose bound to poly(I)·poly(C) (Dunn 1976), show evidence of cleaving the tRNA precursor to give three bands larger than the tRNA band produced by RNase P (Fig. 3, lane 12) and several bands smaller than the RNase-P-specific 41-base 5' fragment of the tRNA precursor. Careful comparison between these bands (best observed in lanes 3, 6, and 9) with those produced by extracts of strain ABL1 (lane 2) suggests that the most prominent of the three products produced by RNase III has a mobility intermediate between that of the two prominent bands produced by ABL1 extracts. This same mobility difference is evident in Figure 2, where both A49 and ABL1 extracts produce bands (lanes 6 and 8) that differ from those produced by RNase III either at 42°C (lane 11) or 37°C (lane 13).

The action of RNase III upon tRNA^{Tyr} precursor molecules at high enzyme:substrate ratio occurs most efficiently at salt concentrations above 0.1 M NH$_4$Cl (Figs. 2 and 3), suggesting that it represents the secondary cleavage reported by Westphal and Crouch (1975). Results of competitive inhibitor experiments support this notion, since single-stranded RNA is at least as efficient as double-stranded RNA in inhibiting the reaction (Dunn 1976). The significance of this reaction to tRNA matura-

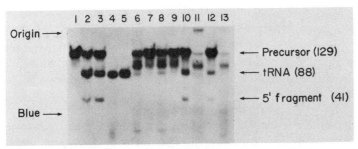

Figure 2 Digestion of *E. coli* tRNA^{Tyr} precursor molecule by subcellular fractions of strains BL15, A49, BL214, and ABL1, as well as by purified RNase III and RNase P. Subcellular fractions of the four strains were prepared according to Robertson et al. (1972) through the stage of 0.2 M NH$_4$Cl ribosome washing. These 0.2 M NH$_4$Cl ribosome washes were then used to test for remaining enzymatic activities against 10^4 cpm of ^{32}P-labeled *E. coli* tRNA^{Tyr} precursor under two sets of conditions (described below). One-µl aliquots of each 0.2 M NH$_4$Cl wash fraction were added to 20-µl reactions and incubated for 30 min. All (but fractions *12* and *13*) were preincubated for 15 min at 42°C prior to the addition of substrates to inactivate residual temperature-sensitive RNase P when present. Reactions were subjected to polyacrylamide gel electrophoresis on 10% gels as before (Robertson et al. 1972) and autoradiography was carried out utilizing Dupont Cronex, 2 X-ray film. (*1*) Control: no added enzyme, conditions used were 0.01 M Tris-HCl (pH 7.6), 0.13 M NH$_4$Cl, 5% sucrose; (*2*) BL15 extract, conditions as in *1*; (*3*) BL15 extract, conditions used were 0.01 M Tris-HCl (pH 7.6), 0.02 M NH$_4$Cl (or less), 0.005 M MgCl$_2$, 5% sucrose; (*4*) BL214 extract, conditions as in *1*; (*5*) BL214 extract, conditions as in *3*. Both strains BL15 and BL214 contain RNase P and are carrying out the RNase P reaction. The reaction is much more complete in BL214 extracts because of a threefold excess of protein in the BL214 extract, in comparison with BL15. (*6*) A49 extract, conditions as in *1*; (*7*) A49 extract, conditions as in *3*; (*8*) ABL1 extract, conditions as in *1*; (*9*) ABL1 extract, conditions as in *3*. A49 and ABL1 lack RNase P, as expected, but show evidence of low levels of other cleavage activity. (*10*) RNase P reaction with purified enzyme, conditions as in *1*; (*11*) RNase III purified through DEAE-cellulose (5 units), conditions as in *1*; (*12* and *13*) repeat reactions of *10* and *11*, except that the 42°C preincubation was omitted and the digestions were carried out at 37°C. Positions of the origin of electrophoresis and the bromphenol blue marker dye are indicated. In addition, the position of intact tRNA^{Tyr} precursor, the tRNA-containing RNase P product, and the 5' extra fragment are also indicated, along with their chain lengths in nucleotides.

tion awaits further investigation, although there have been several earlier suggestions that RNase III could be involved early in the tRNA maturation pathway (Shimura and Sakano 1977).

The action of ABL1 extracts on the tRNA^{Tyr} precursor probably represents another enzyme, such as RNase P2 (Schedl et al. 1976). The products that accumulate in the absence of RNase P and RNase III in

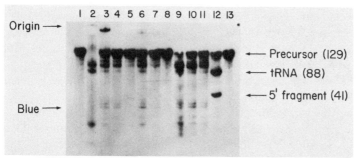

Figure 3 RNase III digestion of the *E. coli* tRNA^{Tyr} precursor molecule. Digestions were set up and carried out as described in Fig. 2. Reactions in *1* and *2* were preincubated for 15 min at 42°C, and then carried out at 42°C; all other digestions were for 30 min at 37°C. All incubations were under conditions described for lane 3 in Fig. 2. (*1*) Control, no enzyme; (*2*) one µl of ABL1 0.2 M NH$_4$Cl ribosomal wash; (*3*) RNase III purified through DEAE-cellulose (5 units); (*4*) same as *3*, but with 2 µg of *Penicillium chrysogenum* double-stranded RNA; (*5*) same as *3*, but with 5 µg of bacteriophage f2 single-stranded RNA; (*6*) RNase III purified through DEAE-cellulose and phosphocellulose (5 units); (*7*) same as *6*, but with *P. chrysogenum* RNA as in *4*; (*8*) same as *6*, but with f2 RNA as in *5*; (*9*) RNase III purified through DEAE-cellulose and phosphocelluloses, and then through poly(I-C) agarose (5 units); (*10* and *11*) contain poly(I-C) agarose-purified enzyme and RNAs as in *4* and *5*; (*12*) RNase P; (*13*) control, no enzyme.

ABL1 extracts, and the salt requirements of the reaction (Fig. 2), are similar to those reported by Schedl et al. (1976). Fingerprinting analysis of the two most prominent ABL1-specific bands that are larger than tRNA^{Tyr}, along with the most prominent bands produced by RNase III cleavage (see Figs. 2 and 3), is being carried out in collaboration with S. Altman (Yale University), to characterize the difference between these two reactions. We conclude that at least one additional specific endonucleolytic RNase exists in *E. coli* and that this activity could be responsible for the continued processing of rRNA precursors that goes on in strain ABL1 even at high temperatures.

ISOLATION AND ACCURATE IN VITRO PROCESSING OF SHORT RNA REGIONS THAT SPECIFY RNase III CLEAVAGE IN VIVO

Based on previous investigations of RNase III interactions with its three identified substrates (Dunn and Studier 1973; Ginsburg and Steitz 1975; Studier 1975; Lund and Dahlberg 1977; Rosenberg and Kramer 1977; Robertson and Barany 1978) (rRNA precursors, phage T7 mRNA precursors, and double-stranded RNA), we have proposed a mechanism of action for this enzyme (Robertson and Barany 1978). We also suggested that double-stranded regions in 30S rRNA precursors could contain the specific sites cleaved during RNase III processing (Robertson and Barany 1978). To test our hypothesis, we have characterized highly structured RNase-resistant regions that can be isolated from rRNA precursors that accumulate in *E. coli* cells carrying RNase III⁻ mutations (Robertson and Barany 1978). We will describe some properties of one of these isolated regions, a 23-bp RNA segment that contains the terminals of 17S rRNA.[1] Our experiments show that the RNA duplex contains two complementary strands that originate over 1600 bases apart in the rRNA precursor. Our experiments also show that the isolated small RNA retains all of the information necessary for RNase III to recognize and cleave at the correct locations.

Figure 4 depicts the characterization of a set of small RNAs relatively resistant to the single-strand-specific T2 and pancreatic RNases (in 0.1–0.25 M NaCl) that can be isolated from [^{32}P]RNA from ABL1 cells using a technique described previously (Robertson et al. 1977; Robertson and Barany 1978). Upon electrophoresis on 16% polyacrylamide gels (Fig. 4A), RNA from the mutant strain is found to contain several well-defined bands (a, b, and c) migrating slower than the blue marker, whereas the RNA from wild-type *E. coli* contains only less-well-defined material running with or ahead of the blue marker. Thus, rRNA precursor molecules of *E. coli* contain a small number of regions that contain extensive Watson-Crick base pairing and that can be isolated in high yield from mutants deficient in processing enzymes as 20-bp to 30-bp segments following pancreatic and RNase T2 digestion in high salt.

Figure 4B depicts an RNase T1 fingerprint of total RNA from *E. coli* ABL1, revealing a complexity expected for the 6000 bases of rRNA and rRNA precursor (Robertson and Jeppesen 1972; Robertson et al. 1977). Figure 4C depicts an RNase T1 fingerprint of the structured regions prepared from ABL1 RNA as described above. Under the low-salt conditions of the exhaustive RNase T1 digestion used for fingerprinting analysis, the structured regions are fully digested, but at a slower rate than is the case for single-stranded RNA. Evidence for this can sometimes be seen, as in Figure 4D, where two of the larger oligonucleotides occur as double spots, part of each still retaining the 2′,3′-cyclic phosphate form that is intermediate in RNase T1 action. The major conclusion of Figure 4C, however, is its striking simplicity in comparison to Figure 4B. Thus, the structured regions represent a highly specific subset of the total 30S RNA. If the RNA from band b from a 16% polyacrylamide gel, such as that illustrated in Figure 4A, is first eluted and then subjected to

---

[1] As proposed by Ginsburg and Steitz (1975), the term 17S rRNA will be used to denote the intermediate in 16S rRNA synthesis that arises in vivo and accumulates in chloramphenicol-treated cells as well as the RNA species containing the 16S rRNA sequence produced in vitro by RNase III cleavage of the 30S rRNA precursor. These two RNA species share common 5′- and 3′-terminal sequences (Ginsburg and Steitz 1975).

RNase T1 fingerprinting, the pattern shown in Figure 4D is obtained. It is evident that only two of the four prominent oligonucleotides seen in Figure 4C (1 and 2 in figure) are present in substantial yield. Similar analyses of bands a and c lead to the conclusion that these bands also represent very small and specific subsets of the total *E. coli* rRNA precursor sequence (data not shown).

The large T1 oligonucleotides 1–4 from bands a and b were subjected to sequence analysis. Oligonucleotide 2 from band b contains 8 bases identical with the 3′ terminal of 17S rRNA, whereas oligonucleotide 3 from band a contains 9 bases identical to the 3′ terminal of mature 23S rRNA (H. D. Robertson, in prep.). All of the RNase T1 and pancreatic-RNase-resistant oligonucleotides from band b were subjected to sequence analysis (data not shown). In addition, two individual strands were found to be present in band b and their separation by fingerprinting using an unhydrolyzed mixture of yeast RNA in the second dimension (Brownlee and Sanger 1969) is shown in Figure 6C. When these separated strands were eluted and subjected to pancreatic RNase and RNase T1 fingerprinting analysis, specific subsets of the band b oligonucleotides were obtained (data not shown). Figure 5 shows our proposed sequence for the two strands of band b, which we have depicted in a double-stranded configuration. Also indicated in Figure 5 are regions on both strands that are homologous to 5′ and 3′ terminals of 17S rRNA.

The experiments summarized in Figure 6 were carried out to explore the function of isolated bands a–c. Figure 6A shows electrophoresis analysis on a 16% polyacrylamide gel of bands a–c before and after treatment with RNase III. It is evident that all three of the isolated bands can be cleaved by the enzyme. In particular, about 80% of the material in band b appears to have been reduced in size upon RNase III treatment (Fig. 6A, 3 and 4). Portions of the reactions illustrated in Figure 6A were also subjected to two-dimensional analysis, as shown in Figure 6, C and D. It is evident that the two prominent spots shown in Figure 6C (representing the two separated strands of band b) are largely replaced by a set of four prominent smaller oligonucleotides (Fig. 6D), with about 20% of the material remaining in the position of the starting material. Repeated tests have shown that the same four products are obtained when band b is treated by RNase III, and that bands a and c also yield characteristic patterns when their RNase III cleavage products are analyzed in this way (data not shown). When the two strands of band b were separated as in Figure 6C and then eluted and treated separately with RNase III, no cleavage was detected.

We wanted to identify the sites within the sequence of band b where RNase III cleavage occurs in vitro. The fingerprint shown in Figure 6B demonstrates that most of the material comprising the two prominent RNase T1 products 1 and 2 disappears following RNase III treatment of

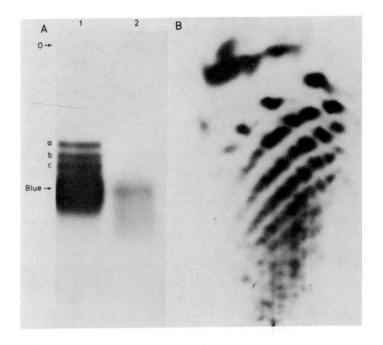
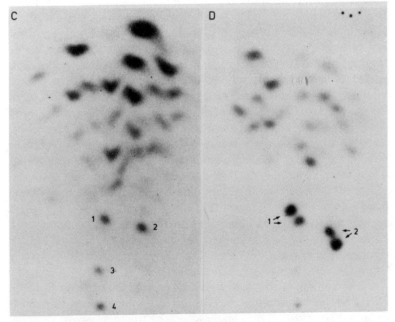

Figure 4 (See facing page for legend.)

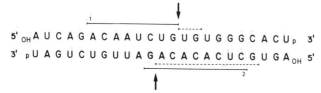

```
        ↓
5'ₒₕA U C A G A C A A U C U G U G U G G G C A C Uₚ  3'
3' ₚU A G U C U G U U A G A C A C A C U C G U G Aₒₕ  5'
                                                    2
        ↑
```

Figure 5 Sequence and structure of band b isolated from *E. coli* rRNA precursors. RNA sequence analysis of RNase T1 and pancreatic RNase-resistant oligonucleotides of band b and its two separated strands was carried out. These data, which are to be reported in full elsewhere (H. D. Robertson et al., in prep.), together with published results (Ginsburg and Steitz 1975; Young and Steitz 1978) allowed us to draw the conclusions illustrated here. The two strands are depicted in an antiparallel orientation to illustrate the potential for 23 bp. (———) Positions in the sequence of RNase T1-resistant oligonucleotides 1 and 2; (– – –) sequences identical to previously identified 5' and 3' terminals of 17S rRNA; (↑) positions of RNase III cleavage of isolated band b in vitro, which are also positions of in vivo processing of rRNA precursors. In this orientation, the upper strand of band b would be connected to the lower strand by an intervening sequence of over 1600 nucleotides that would lie to the right of the band as depicted here.

Figure 4 Preparation of structured regions from *E. coli* rRNA precursors. [^{32}P]RNA was isolated from *E. coli* strains BL15 and ABL1. RNA (3×10^8 cpm) from each strain was treated with RNase T2 (2 units/ml) and pancreatic RNase (0.5 µg/ml) in 0.2 ml of 0.1 M Tris-HCl, 0.2 M NaCl for 48 min at 37°C. (*A*) Autoradiograph of analysis of the resulting RNase-resistant fractions (3×10^6 cpm in each case) on 16% polyacrylamide gels: (*1*) RNA from ABL1 strain; (*2*) RNA from BL15; (O) origin of electrophoresis; (Blue) bromphenol blue marker; (a, b, and c) positions of bands characteristic of strain ABL1 that were eluted for further analysis. *B, C,* and *D* illustrate autoradiographs of RNase T1 fingerprints carried out as described by Brownlee and Sanger (1969) on RNA from strain ABL1 at different stages of purification. (*B*) 10^6 cpm of ^{32}P-labeled whole cellular RNA; (*C*) 10^6-cpm aliquot for the RNA shown in *A*, lane *1*; (*D*) 5×10^5 cpm of purified band b from the gel illustrated in *A*. (1, 2, 3, and 4) Positions of prominent RNase-T1-resistant oligonucleotides characteristic of the mutant strain. In *D*, the doublets observed for 1 and 2 are the result of slight underdigestion that leaves some material with 2',3'-cyclic-phosphate end groups (Barrell 1971). In each case, the origin of the two-dimensional analysis is located at the lower right of the illustration.

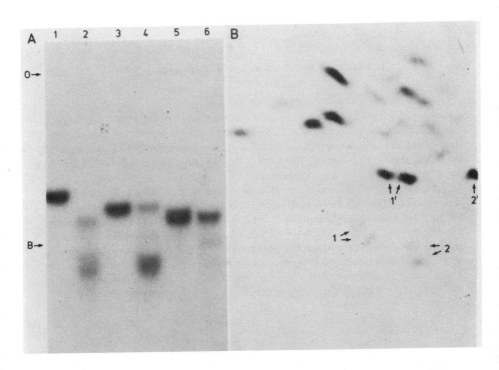

Figure 6 RNase III cleavage bands of a, b, and c isolated from *E. coli* precursors. Bands a–c were eluted from the 16% polyacrylamide gel in Fig. 4A and portions (10^5 cpm) were incubated with 5 units of RNase III under buffer conditions optimal for RNase III cleavage of double-stranded RNA (in 0.01 M Tris-HCl [pH 7.6], 0.13 M NH$_4$Cl, 0.01 M MgCl$_2$, 5% sucrose). Control aliquots were also incubated in RNase III assay buffer alone. (*A*) 16% polyacrylamide gel analysis of bands a–c with and without RNase III treatment: band a alone (*1*); band a + RNase III (*2*); band b alone (*3*); band b + RNase III (*4*);

band b and is replaced by two new spots labeled 1' and 2' (see Figs. 4D and 6B). As shown in Figure 5, oligonucleotide 1 is in the lower strand of band b, whereas 2 is in the upper. Thus, RNase III cleaves once in each strand. Two-dimensional analysis (Robertson and Dunn 1975) of alkaline digests of the RNase III digestion products of band b yielded two 5'-terminal end groups: pGp and pUp. Furthermore, analysis of the RNase-T1-resistant oligonucleotides produced from RNase-III-treated band b (Fig. 6B) revealed the presence of pGp and pUG. The cleavage points indicated in Figure 5 are the only ones fully commensurate with all of this data. The cleavage point in the lower strand is unambiguous. There remains some chance, however, that the upper strand, as shown in Figure 5, could be cleaved between the C and the U, two positions to the left of the cleavage position shown. Further experiments are in progress concerning

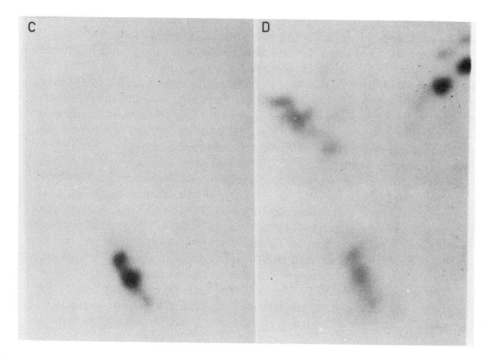

Figure 6 (continued) band c alone (5); band c + RNase III (6). (*B*) RNase T1 fingerprint of 10^4 cpm of band b following RNase III treatment; (1 and 2) oligonucleotides also indicated in Figs. 1 and 2; (1 and 2) new locations of 1 and 2 following RNase III cleavage. (*C*) Separation of the two strands of band b (not treated by RNase III; see *A*, lane 3) by fingerprinting; an unhydrolyzed mixture of yeast RNA was employed in the second dimension; the two prominent spots correspond to the two strands of band b (see Fig. 5). (*D*) Following digestion by RNase III, band b was fingerprinted under conditions identical to *C*.

this point. Similar analysis of band a reveals a substantial shift in mobility of oligonucleotide 4 following RNase III treatment (data not shown).

These experiments demonstrate that rRNA precursor molecules of *E. coli* contain several small predominantly double-stranded RNA regions that can be isolated by virtue of their relative resistance to single-strand-specific nucleases (see Fig. 4). The previously reported nucleotide sequences of the 5′ and 3′ terminals of 17S rRNA (Ginsburg and Steitz 1975) occur, one each, within the two strands of the band b sequence reported here (see Fig. 5). Furthermore, these sequences are contained within the DNA sequence proposed by Young and Steitz for regions on both sides of the gene encoding 17S rRNA of *E. coli* (Young and Steitz 1978). Twenty-three bases corresponding to the upper strand of band b are in the DNA sequence located to the left of the 16S region, whereas a region of DNA to the right of the 16S coding region is identical to the lower strand of band b (Young and Steitz 1978; see also Fig. 5).

Our results represent the first direct evidence for long-range functional association of RNA regions in a precursor molecule. Since the RNase III reaction with band b continues to occur in the absence of the intervening 16 rRNA sequence with undiminished specificity, we conclude that association between widely separated regions of rRNA precursors is critical in forming signals for proper RNase III cleavage. Thus, our hypothesis (Robertson and Barany 1978) that the structured regions in 30S rRNA precursors contain RNase III cleavage sites is confirmed. Previous experiments on the structural association of rRNA terminals at the RNA level have been carried out on 23S rRNA by Branlant et al. (1976), who proposed a base-paired structure for the 23S rRNA 5' and 3' terminals from the results of partial T1 digestion. In addition, Gegenheimer et al. (1977) have proposed that helical structures involving the terminals of 16S and 23S rRNA may form in rRNA precursors. This proposal was based, in part, on electron microscopic studies of DNA encoding *E. coli* rRNA (Wu and Davidson 1975). The sequence analysis by Young and Steitz (1978) of DNA regions at the terminals of rRNA cistrons also led them to speculate that the 5' and 3' terminals of 17S rRNA interact to form a helical structure.

DISCUSSION

As has been the case with other aspects of RNA processing in *E. coli*, the search for additional endonucleases in strains lacking RNase III and RNase P is complicated by the need to reconcile earlier genetic and biochemical data. Using strain ABL1, we have illustrated an approach to this problem. In the absence of RNase III and RNase P, at least one additional activity is present that may be the previously reported RNase P2 (Schedl et al. 1976). In particular, by comparison with the data reported by Schedl et al. (1976), we plan to test whether an activity is present in ABL1 and responsible for the in vivo cleavages that we observe, which are indeed identical to those of RNase P2. We have also detected, in RNase III, an endonucleolytic activity that, under secondary digestion conditions, can cleave the tRNA^{Tyr} precursor. This activity could also operate in vivo. Finally, we have illustrated the use of strain ABL1 in isolating active RNA processing sites from RNA precursors that accumulate under nonpermissive conditions. The RNase III processing sites so isolated should allow us to investigate the activity of this enzyme and provide a prototype for the isolation of oligonucleotides containing processing signals specific for other enzymes.

ACKNOWLEDGMENTS

This work was supported in part by grants from the U.S. National Science Foundation and the American Cancer Society. We thank S. Altman (Yale University) for his kind gift of tRNA^{Tyr} precursor and for helpful discussions. We also thank J. J. Dunn (Brookhaven National Laboratory) for providing a sample of RNase III for comparative purposes and for helpful discussions.

REFERENCES

Barrell, B. G. 1971. Fractionation and sequence analysis of radioactive nucleotides. In *Procedures in nucleic acid research* (ed. G. L. Cantoni and D. R. Davies), vol. 2, p. 751. Harper and Row, New York.

Branlant, C., J. S. Widada, A. Krol, and J.-P. Ebel. 1976. Extensions of the known sequences at the 3′ and 5′ ends of 23S ribosomal RNA from *E. coli*, possible base pairing between these 23S RNA regions and 16S ribosomal RNA. *Nucleic Acids Res.* **3:**1671.

Brownlee, G. G. and F. Sanger. 1969. Chromatography of ^{32}P-labeled oligonucleotides on thin layers of DEAE-cellulose. *Eur. J. Biochem.* **11:**395.

Dunn, J. J. 1976. RNase III cleavage of single-stranded RNA. *J. Biol. Chem.* **251:**3807.

Dunn, J. J. and F. W. Studier. 1973. T7 early RNAs and *E. coli* ribosomal RNAs are cut from large precursor RNAs in vivo by ribonuclease III. *Proc. Natl. Acad. Sci.* **70:**3296.

Gegenheimer, P., N. Watson, and D. Apirion. 1977. Multiple pathways for primary processing of ribosomal RNA in *E. coli. J. Biol. Chem.* **252:**3064.

Ghosh, R. K. and M. P. Deutscher. 1978. Purification of potential 3′ processing nucleases using synthetic tRNA precursors. *Nucleic Acids Res.* **5:**3831.

Ginsburg, D. and J. A. Steitz. 1975. The 30S ribosomal precursor RNA from *E. coli. J. Biol. Chem.* **250:**5647.

Lund, E. and J. E. Dahlberg. 1977. Spacer transfer RNAs in ribosomal RNA transcripts of *E. coli*: Processing of 30S ribosomal RNA in vitro. *Cell* **11:**247.

Meyhack, B., B. Pace, O. C. Uhlenbeck, and N. R. Pace. 1978. Use of T4 RNA ligase to construct model substrates for a ribosomal RNA maturation endonuclease. *Proc. Natl. Acad. Sci.* **75:**3045.

Paddock, G. V., K. Fukada, J. Abelson, and H. D. Robertson. 1976. Cleavage of T4 species I ribonucleic acid by *E. coli* ribonuclease III. *Nucleic Acids Res.* **3:**1351.

Robertson, H. D. and F. Barany. 1978. Enzymes and mechanisms in RNA processing. In *Proceedings of the 12th FEBS Congress*, p. 285. Pergamon Press, Oxford.

Robertson, H. D. and J. J. Dunn. 1975. Ribonucleic acid processing activity of *E. coli* ribonuclease III. *J. Biol. Chem.* **250:**3050.

Robertson, H. D. and P. G. N. Jeppesen. 1972. Extent of variation in three related bacteriophage RNA molecules. *J. Mol. Biol.* **68:**417.

Robertson, H. D., S. Altman, and J. D. Smith. 1972. Purification and properties of a specific *E. coli* ribonuclease which cleaves a tyrosine transfer ribonucleic acid precursor. *J. Biol. Chem.* **247**:5243.

Robertson, H. D., E. Dickson, and W. Jelinek. 1977. Determination of nucleotide sequences from double-stranded regions of HeLa cell nuclear RNA. *J. Mol. Biol.* **115**:571.

Robertson, H. D., R. E. Webster, and N. D. Zinder. 1968. Purification and properties of ribonuclease III from *E. coli*. *J. Biol. Chem.* **243**:82.

Rosenberg, M. and R. A. Kramer. 1977. Nucleotide sequence surrounding a ribonuclease III processing site in bacteriophage T7 RNA. *Proc. Natl. Acad. Sci.* **74**:984.

Sakano, H. and Y. Shimura. 1975. Sequential processing of precursor tRNA molecules in *E. coli*. *Proc. Natl. Acad. Sci.* **72**:3369.

Schedl, P. and P. Primakoff. 1973. Mutants of *E. coli* thermosensitive for the synthesis of transfer RNA. *Proc. Natl. Acad. Sci.* **70**:2091.

Schedl, P., J. Roberts, and P. Primakoff. 1976. In vitro processing of *E. coli* tRNA precursors. *Cell* **8**:581.

Seidman, J. G., B. G. Barrell, and W. H. McClain. 1975. Five steps in the conversion of a large precursor RNA into bacteriophage proline and serine transfer RNAs. *J. Mol. Biol.* **99**:733.

Shimura, Y. and H. Sakano. 1977. Processing of tRNA precursors in *Escherichia coli*. In *Nucleic acid-protein recognition* (ed. H. J. Vogel), p. 293. Academic Press, New York.

Sogin, M. L., B. Pace, and N. R. Pace. 1977. Partial purification and properties of a ribosomal RNA maturation endonuclease from *B. subtilis*. *J. Biol. Chem.* **252**:1350.

Stark, B., R. Kole, E. Bowman, and S. Altman. 1978. Ribonuclease P: An enzyme with an essential RNA component. *Proc. Natl. Acad. Sci.* **75**:3713.

Studier, F. W. 1975. Genetic mapping of a mutation that causes ribonuclease III deficiency in *E. coli*. *J. Bacteriol.* **124**:307.

Young, R. A. and J. A. Steitz. 1978. Complementary sequences 1700 nucleotides apart form a ribonclease III cleavage site in *E. coli* ribosomal precursor RNA. *Proc. Natl. Acad. Sci.* **75**:3593.

Westphal, H. and R. J. Crouch. 1975. Cleavage of adenovirus messenger RNA and of 28S and 18S ribosomal RNA by RNase III. *Proc. Natl. Acad. Sci.* **72**:3077.

Wu, M. and N. Davidson. 1975. Use of gene 32 protein staining of single-strand polynucleotides for gene mapping by electron microscopy. *Proc. Natl. Acad. Sci.* **72**:4506.

Processing of Spacer tRNAs from rRNA Transcripts of *Escherichia coli*

Elsebet Lund and James E. Dahlberg
Department of Physiological Chemistry
University of Wisconsin
Madison, Wisconsin 53706

Christine Guthrie
Department of Biochemistry and Biophysics
University of California, San Francisco
San Francisco, California 94143

In *Escherichia coli*, the genes for 16S, 23S, and 5S RNA are cotranscribed from seven operons (Kiss et al. 1977; for review, see Nomura et al. 1977). A primary transcript, 30S pre-rRNA, accumulates in mutants that lack functional RNase III (Dunn and Studier 1973; Nikolaev et al. 1973). In addition to the sequences for the mature rRNAs, this 30S pre-rRNA contains the sequences for a number of tRNAs (Lund and Dahlberg 1977). In four rRNA operons, the gene for $tRNA_2^{Glu}$ is located between the 16S and 23S rRNA genes; in the other three operons this spacer region contains the genes for both $tRNA_1^{Ile}$ and $tRNA_{1B}^{Ala}$ (Lund et al. 1976; Morgan et al. 1977; Ikemura and Nomura 1977). The biological significance of this genetic organization is unclear.

In cells containing functional RNase III, 30S pre-rRNA is not observed; presumably precursors to the RNAs and tRNAs are normally generated by processing of the nascent 30S pre-rRNA concurrently with its transcription (Hamkalo and Miller 1973; for reviews, see Pace 1973; Perry 1976). The total number and order of reactions involved in the eventual maturation of these precursors is currently unknown, but a likely presumption is that enzymes that process other cell tRNAs are involved in the maturation of spacer tRNAs. Furthermore, since RNase III strains are viable and do produce mature, functional ribosomes (Kindler et al. 1973), it seems probable that enzymes involved in the processing of tRNA could also provide an alternative route to rRNA maturation (Gegenheimer et al. 1977; Gegenheimer and Apirion 1978). As a first step toward evaluating these hypotheses, we have analyzed two enzymes known to have important roles in rRNA and tRNA maturation for their ability to process 30S pre-rRNA in vitro.

RNase III (Robertson et al. 1968) is responsible for generation of $pre16S_{III}$ rRNA, $pre23S_{III}$ rRNA, and $pre5S_{III}$ rRNA (Dunn and Studier

123

1973; Nikolaev et al. 1973; Ginsburg and Steitz 1975; Hayes et al. 1975; Young and Steitz 1978), as well as mature-length bacteriophage T7 mRNA (Dunn and Studier 1973; Robertson and Dunn 1975). The role of this enzyme in the processing of some tRNA precursors is presently unclear (Sakano and Shimura 1978). RNase P (Robertson et al. 1972) has been shown to cleave a number of *E. coli* and bacteriophage T4 tRNA precursors endonucleolytically to produce the mature 5' tRNA terminals (Altman and Smith 1971; Guthrie et al. 1973; Schedl and Primakoff 1973; Sakano et al. 1974; Guthrie 1975; Seidman et al. 1975a; Chang and Carbon 1975; Ikemura et al. 1975; Ilgen et al. 1976; Schedl et al. 1976; Sakano and Shimura 1978).

We report here that both RNase III and RNase P are capable of cleaving intact 30S pre-rRNA at several distinct sites in vitro. Since digestion with RNase III plus RNase P produces spacer tRNA precursors that are immature at their 3' ends, it is clear that these enzymes are not sufficient by themselves for complete tRNA maturation. Finally, comparison of the sequences of these RNase III and RNase P cleavage products with the sequences of rDNA spacer regions (i.e., the several hundred nucleotides between the structural genes for 16S and 23S rRNAs) (Young et al. 1979; Morgan et al., this volume) has allowed us to determine the sites of cleavage by these two enzymes.

METHODS

RNase P was partially purified from *E. coli* 514 through the DEAE-Sephadex A-50 column chromatography step of Robertson et al. (1972). The active fractions from the column were further purified by application to a second DEAE-Sephadex column and elution with a linear gradient from 0.3 M to 0.8 M NH_4Cl. The active fractions, which eluted at 0.45 M salt, were pooled and stored in 10% glycerol at −20°C in the presence of phenylmethylsulfonyl fluoride. Further details will be described elsewhere (Guthrie and Atchison, this volume).

The 30S pre-rRNA used as substrate was isolated from cells of the RNase III⁻ strain AB301-105 as reported earlier (Lund and Dahlberg 1977). This RNA was digested and analyzed as described in the figure legends.

RESULTS

As shown in Figure 1a, digestion of 30S pre-rRNA with either RNase III or RNase P generates two discrete, large fragments. The relative mobilities of these large digestion products are not identical; fragments resulting from RNase P treatment migrated significantly more slowly than those produced by RNase III. This difference in product size indicates that the

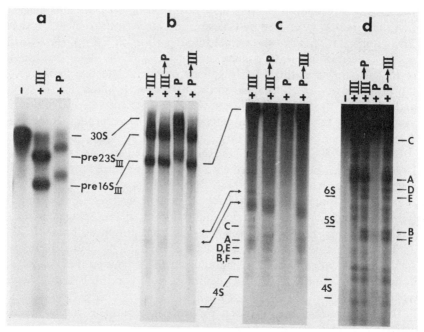

Figure 1 Cleavage of ^{32}P-labeled 30S pre-rRNA in vitro by RNase III and RNase P. After incubation of the RNA with the indicated enzymes, the fragments produced were analyzed by electrophoresis in 2.25% (*a*), 2.5% (*b*, *c*), or 10% (*d*) polyacrylamide-containing gels and visualized by autoradiography. (*a*) Large RNAs produced by treatment with RNase III or RNase P alone. (−) 30S pre-rRNA incubated in the absence of enzymes. Conditions of digestion were 0.3–1 μg rRNA in 10–20 μl of 0.02 M Tris-HCl (pH 7.6), 0.01 M MgCl$_2$, 0.13 M NH$_4$Cl, 0.001 M β-mercaptoethanol, 5% glycerol at 37°C for 30 min. The amounts of the enzymes used were sufficient to give complete digestion, as determined by preliminary titrations of each preparation. Reactions were terminated by addition of 1/10 volume 1% SDS, 0.2 M EDTA. Note that RNase P products migrate significantly more slowly than RNase III products. (*b*) Sequential cleavage of 30S pre-rRNA with RNases III and P. After treatment with either enzyme alone, the RNA samples were divided into two aliquots, one of which was further digested with the other enzyme. Conditions of each digestion were as described for *a*. Note that addition of RNase P before or after treatment with RNase III did not appear to affect the mobilities of the large RNAs, but it did alter the mobilities of several smaller molecules denoted by the arrows to the right. (*c*) Longer exposure of the gel shown in *b*, expanded to illustrate the production and cleavage of intermediate-sized fragments. Note that RNase P treatment alone does not appear to generate any intermediate-sized fragments, but the enzyme alters the mobilities of several RNase III-generated fragments. The letters indicate mobilities of six fragments studied in *d* and Figs. 2 and 3. Arrows indicate fragments that were generated by RNase III and altered by RNase P but not studied further in this work. (*d*) Separation of 4S-8S fragments generated by RNase III and RNase P treatment alone or sequentially. Fragments are designated as in *c*.

two enzymes recognize different sites in the 30S pre-rRNA and suggests that the products generated by RNase P contain RNase III cleavage sites. This interpretation was substantiated by digestion of 30S pre-rRNA with both RNases. As shown in Figure 1b, regardless of the order of addition of the two enzymes, the large fragments produced had the mobilities of the products generated by digestion with RNase III alone, i.e., pre16S$_{III}$ and pre23S$_{III}$. These results further show that RNase III recognition in vitro does not require the entire 30S pre-rRNA structure (cf. Robertson et al., this volume).

Digestion of 30S pre-rRNA with RNase P alone generated only the two large products, whereas cleavage with RNase III produced several 4S–10S fragments (Fig. 1c,d). Many of these RNase-III-generated fragments were characterized previously by RNase T1 oligonucleotide fingerprinting (Ginsburg and Steitz 1975; Lund and Dahlberg 1977). Several of these RNA fragments disappeared or were significantly reduced in yield after additional treatment with RNase P; concomitantly, several new and smaller fragments appeared. Generation of these new products required both RNase III and RNase P, as seen most clearly in the 10% gel shown in Figure 1d.

To characterize the small RNA products generated by the combined action of RNase III and RNase P, these fragments were purified by two-dimensional polyacrylamide gel electrophoresis (Ikemura and Dahlberg 1973) as shown in Figure 2, a and b. Each of the 4S–8S RNA fragments generated by RNase III and purified as in Figure 2a was tested for its ability to serve as a substrate for RNase P in vitro. As expected from a comparison of a and b of Figure 2, only two fragments (A and C) were cleaved by RNase P (Fig. 2c and data not shown). (As indicated in Fig. 1, b and c, two or more larger RNase-III-generated fragments also appeared to be substrates for RNase P, but they did not enter the 10% polyacrylamide gel and were not studied there. These larger fragments may have been derived from the 3′ ends of some 30S pre-rRNA transcripts that contain the sequences for three additional tRNAs [Lund and Dahlberg 1977; Ikemura and Nomura 1977; Morgan et al. 1978; E. Lund and J. E. Dahlberg, in prep.].)

The substrates (A and C) and products (B and D–F) of the RNase P reactions shown in Figure 2c were characterized by analysis of RNase T1 digestion products. The RNase T1 fingerprints of these six fragments are shown in Figure 3. Oligonucleotides characteristic of tRNA$_2^{Glu}$ were observed in fragments A and B; those of tRNA$_1^{Ile}$ were found in fragments C, D, and E; and those of tRNA$_{1B}^{Ala}$ were found in fragments C and F. All of these fragments contained oligonucleotides in addition to those derived from the mature tRNAs, indicating that the substrates and their products together comprised two classes of in vitro intermediates in tRNA maturation. Therefore, we denoted the fragments generated by

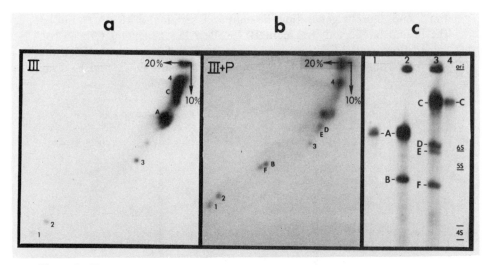

Figure 2 Two-dimensional polyacrylamide gel analysis (10–20%) of 4S–8S fragments produced by treatment of 30S rRNA with RNase III alone (*a*) or with a mixture of RNase III plus RNase P (*b*). The numbers refer to fragments present in both digests; the letters refer to fragments that were altered by treatment with RNase P. (Fragments 1, 2, 3, 4, A, and C correspond to fragments h, k, m, x, p, and y, respectively, of Figs. 9 and 10 of Lund and Dahlberg [1977].) (*c*) One-dimensional (10% polyacrylamide) gel analysis of products formed by treatment of fragments A or C (purified from the gel shown in *a*) with RNase P. The identities of substrates and products in the various panels were confirmed by RNase T1 digestion and fingerprinting as in Fig. 3.

RNase III as preGlu$_{III}$ (A) and preIle-Ala$_{III}$ (C), and the RNAs produced by further treatment with RNase P as pGlu (B), pIle (l and s, the longer and shorter versions, with or without the expected AUUGp sequence at the 5′ end) (D and E), and pAla (F).

The sequences of the RNase T1 oligonucleotides shown in Figure 3 are in excellent agreement with those predicted from the DNA nucleotide sequences determined recently by Young et al. (1979) for the tRNA$_1^{Ile}$-tRNA$_{1B}^{Ala}$-containing spacer region of *rrnD* and *rrnX* and by Morgan et al. (this volume) for the tRNA$_2^{Glu}$-containing spacer region of *rrnE*. Examination of those DNA sequences allowed us to assign the various precursor-specific oligonucleotides to the 5′ or 3′ sides of the mature tRNA sequences. These oligonucleotides are indicated in Figure 3 by the numbers 5 or 3, respectively; oligonucleotides derived from the mature tRNAs are not labeled.

A comparison of the results from Figure 3, a and b, showed that preGlu$_{III}$ (fragment A) contained precursor-specific sequences at both its 5′ and 3′ ends. Treatment with RNase P removed the extra sequences at the 5' end, thereby generating the mature 5' terminal of tRNA$_2^{Glu}$, but left the 3' end unaffected (fragment B). preIle-Ala$_{III}$ (fragment C) also contained precursor-specific nucleotides at its 5' and 3' terminals. Digestion of preIle-Ala$_{III}$ generated three products (fragments D, E, and F). Frag-

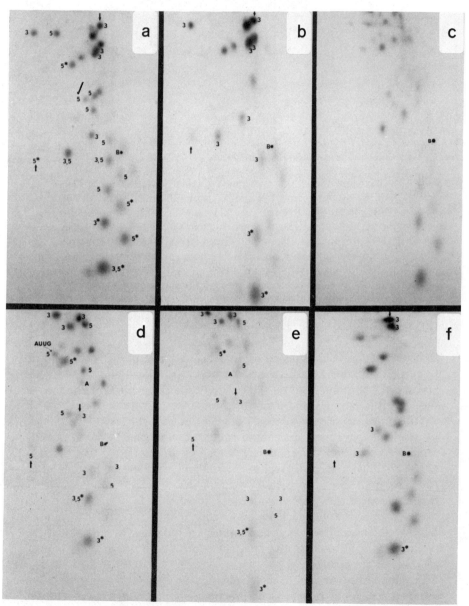

Figure 3 (See facing page for legend.)

ment F contained the mature 5' terminal of tRNA$_{1B}^{Ala}$ but, again, still retained precursor-specific sequences at the 3' terminal. The other two products (fragments D and E) contained the tRNA^{Ile} sequence, but these products still had precursor-specific sequences at both the 5' and 3' terminals. These data thus allowed the determination of the sites of cleavage by RNases III and P, which generated the 5' and 3' ends of fragments A–F. These sequences are shown in Figure 4.

At the 5' ends of the pre-tRNAs$_{III}$, we expected to find sequences corresponding to those contiguous with the 3' terminal of pre16S$_{III}$ (Ginsburg and Steitz 1975). In fact, the results only partially fulfilled this expectation. Two fragments (D and E) correspond to the 5' terminal of preIle-Ala$_{III}$; comparison of the RNase T1 digests of these products (Fig. 3d,e) revealed that the sequence AUUGp was present in fragment D but absent from fragment E. As can be seen in Figure 4a, this oligonucleotide should have been present if, in fact, a single RNase III cleavage event generated the 3' terminal of pre16S$_{III}$ and the 5' terminal of preIle-Ala$_{III}$. For the same reason, this oligonucleotide should have been present in preGlu$_{III}$; as in the case of the pIle(s) (fragment E), this sequence was absent. The facts that both forms of pIle (l and s) were generated and that the sequences (and secondary structures) spanning the region between the 3' terminal of pre16S$_{III}$ and the 5' terminals of both preGlu$_{III}$ and pIle were identical made it difficult to find a simple resolution to this apparent paradox.

The RNase T1 oligonucleotide UCUCACACA$_{OH}$ has been found in

Figure 3 RNase T1 fingerprints of intermediate products shown in Fig. 2c. The fragments were eluted from the gel and digested with RNase T1; then the resulting oligonucleotides were separated by two-dimensional paper electrophoresis (Sanger et al. 1965; Peters et al. 1977). Each oligonucleotide was further characterized by redigestion with pancreatic RNase (Adams et al. 1969; Barrell 1971). *a–f* refer to fragments A–F of Fig. 2. In *a* and *b*, oligonucleotides with no number next to them are those present in tRNA$_2^{Glu}$; oligonucleotides of tRNA^{Ile} are unmarked in *d* and *e*; and oligonucleotides of tRNA$_{1B}^{Ala}$ are unmarked in *f*. (5' and 3') Oligonucleotides predicted to be adjacent to the 5' and 3' ends of the respective RNAs, based on the DNA sequences of the rRNA spacer regions (Young et al. 1979; Morgan et al., this volume). (*) Oligonucleotides present both in the tRNA and in the regions immediately adjacent to it. (A) AUAAGp, an oligonucleotide that is not found in mature tRNA^{Ile} but is present in pIle (l and s) due to lack of nucleotide modification. This finding is consistent with the very low yield of AUt^{6}AAGp, the fully modified form of this oligonucleotide, in fragments D and E. The 5'-terminal (↑) and 3'-terminal (↓) oligonucleotides of each fragment are indicated. The position of AUUGp, an oligonucleotide that is expected to be in fragments A, B, and E but is not, is indicated in *d*. Electrophoresis was from right to left and from top to bottom in the first and second dimensions, respectively. (B) The position of the blue dye, xylene cyanol FF.

pre16S$_{III}$ rRNA (Ginsburg and Steitz 1975), but oligonucleotides from the extended form of this sequence (UCUCACACAGp and AUU$_{OH}$) have not been isolated. Since the 3' terminal of the pre16S$_{III}$ generated in our experiments was not analyzed, we cannot rule out the possibility that, under our conditions, the 5' terminals of preGlu$_{III}$ and pIle(s) (fragment

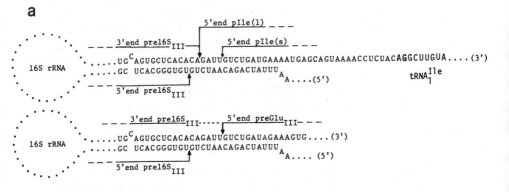

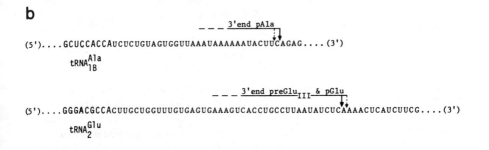

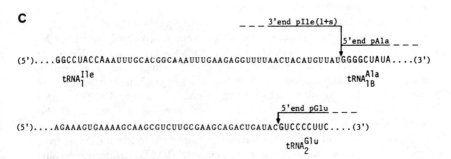

Figure 4 (See facing page for legend.)

E) are generated four nucleotides to the 3' side of the normal terminal of pre16S$_{III}$ by RNase III cleavage. Nor can we say whether these terminals are the result of two sequential cleavages, the first by RNase III at the correct in vivo site and the second by RNase III or by a contaminating nuclease that removes the four extra nucleotides during sample isolation. We also cannot rule out the unlikely possibility that there is a sequence heterogeneity around this site and that only specific subsets of pre16S$_{III}$ and pre-tRNA$_{III}$ molecules survive the processing.

DISCUSSION

We have shown that either RNase P or RNase III can cleave intact 30S pre-rRNA at several distinct sites. RNase P appears to generate only two large fragments, but RNase III generates a number of small (4–10S) products in addition to pre16S$_{III}$ and pre23S$_{III}$. Some of these products are, in turn, substrates for RNase P. From the fingerprint analysis shown in Figure 3 and the comparison of nucleotide sequences as indicated in Figure 4, we have been able to determine the RNase III and RNase P cleavage sites. These are summarized schematically in Figure 5.

In both kinds of spacers (those containing tRNA$_2^{Glu}$ and those containing tRNA$_1^{Ile}$ and tRNA$_{1B}^{Ala}$), cleavage by RNase III was found to occur predominantly at three sites: (1) at or near the 3' end of pre16S$_{III}$ rRNA, (2) at or near the 5' end of pre23S$_{III}$ rRNA, and (3) between the 3' end of the tRNAs and the 5' end of pre23S$_{III}$ rRNA. As indicated in Figure 5, there appear to be several less favored sites of RNase III cleavage distal

Figure 4 Cleavage sites for RNase III and RNase P as determined from terminal sequences of the fragments generated. Sequences shown are those predicted from the DNA sequences of *rrnD* and *rrnE* (Young et al. 1979; Morgan et al., this volume). The ends of the fragments were determined from analysis of terminal oligonucleotides shown in Fig. 3. The 5' and 3' ends of pre16$_{III}$ are as reported elsewhere (Ginsburg and Steitz 1975; Young and Steitz 1978; Dahlberg et al. 1978). (*a*) 5' ends of pIle and preGlu$_{III}$ (fragments D, E, and A). The 5' end of preIle-Ala$_{III}$ was not determined but it probably corresponds to the end of pIle. In the case of pIle(s) and preGlu$_{III}$, we cannot distinguish between the possibilities that two RNase III cleavages occurred 4 nucleotides apart or that only one cleavage occurred, four nucleotides to the 3' side of the 3' terminal of pre16S$_{III}$. (*b*) 3' Ends of pAla, preGlu$_{III}$, and pGlu. The two arrows indicate that the side on which the nucleotide cleavage occurred cannot be determined. Although the 3' end of preIle-Ala$_{III}$ was not analyzed, the size of this fragment indicates that its 3' end is the same as that of pAla. (*c*) RNase P cleavage sites in the spacer region of 30S pre-rRNA. Cleavage occurred at the sites corresponding to the 5' ends of mature tRNA$_{1B}^{Ala}$ and tRNA$_2^{Glu}$, but not at the 5' end of tRNA$_1^{Ile}$. These cleavage sites were inferred from digestion of RNase III cleavage products with RNase P as shown in Fig. 2c.

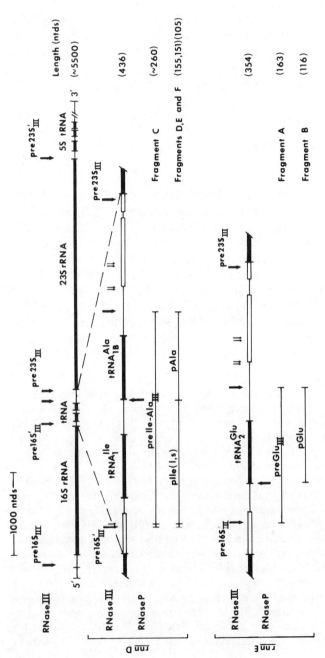

Figure 5 Schematic representation of cleavage sites for RNase III (↓) and RNase P (↑) in the rRNA transcripts of *E. coli*. The spacer regions of *rrnD* and *rrnE* genes are expanded tenfold. (■) Sequences found in mature rRNAs and tRNAs; (——) precursor-specific sequences; (□) precursor-specific sequences that are in common between *rrnD* and *rrnE* genes; (↓) sites at which cleavage by RNase III occurs in some but not all cases (E. Lund and J. E. Dahlberg, in prep.). Additional cleavage by RNase P appears to occur near the 3' end of some (30S) rRNA transcripts (see Fig. 2b,c), but they are not indicated because their exact positions are unknown.

to the major site that generates the pre-tRNA$_{III}$ precursors and proximal to the site generating pre23S$_{III}$. Fragments arising from this region were characterized previously (Lund and Dahlberg 1977) and are discussed in detail elsewhere (E. Lund and J. E. Dahlberg, in prep.).

The sites between the 3′ end of the tRNAs and the 5′ end of pre23S$_{III}$-rRNA that generate the 3′ terminals of the pre-tRNAs$_{III}$ are of interest when compared with the sites at or near the 3′ end of pre16S$_{III}$ rRNA generating the respective 5′ terminals. The latter fall within regions of identical sequence and secondary structure in both types of spacers and have the primary and secondary structural features observed for RNase III cleavage sites in other substrates. In contrast (as seen more clearly in Fig. 4), the RNase III cleavages generating the pre-tRNA$_{III}$ 3′ terminals occur at sites that share no sequence homology with each other; nor are they the same number of nucleotides away from the 3′ terminals of the respective mature tRNA sequences. Thus, it remains to be seen what cues RNase III to cleave at these places, and what, if any, structural features are shared by these sites and other characterized RNase III sites.

Each spacer region appears to be cleaved just once by RNase P. In the case of the monomeric preGlu$_{III}$ precursor, this cleavage generates a precursor (pGlu) that has the correct 5′ terminal of the mature tRNA but is still immature at the 3′ terminal. The dimeric preIle-Ala$_{III}$ precursor is also cleaved only once by RNase P to generate the mature 5′ terminal of tRNA$_{1B}^{Ala}$. The significance of this observation is unclear for several reasons. It is known that a number of dimeric precursors from both T4 and *E. coli* are cleaved by RNase P at two sites to generate the 5′ terminals of both tRNA moieties. In some cases, there is evidence for a preferential order of cleavage in vitro, such that cleavage at the 3′ distal tRNA occurs before that at the 5′ proximal tRNA (Guthrie 1975; Schmidt and McClain 1978; Sakano and Shimura 1978). However, this preference is ordinarily significant only under conditions of very limiting enzyme (Guthrie 1975; C. Guthrie and R. Atchison, unpubl.).

Although we cannot rule out the possibility that 5′-terminal maturation of tRNA$_1^{Ile}$ is accomplished by an enzyme other than RNase P, we consider this extremely unlikely. A plausible explanation is that the structure of the preIle-Ala$_{III}$ RNA makes the 5′ terminal of tRNA$_1^{Ile}$ inaccessible to RNase P. In fact, Young et al. (1979) have recently proposed a secondary structure for this region, which is certainly compatible with this hypothesis. Since approximately equimolar yields of tRNA$_{1B}^{Ala}$ and tRNA$_1^{Ile}$ are synthesized in vivo (Ikemura and Nomura 1977; T. Ikemura, pers. comm.), we conclude that if preIle-Ala$_{III}$ is in fact a biosynthetic intermediate, 5′-terminal maturation of the tRNA$_1^{Ile}$ moiety must be preceded or accompanied by the action of one or more additional nucleases. Since none of the spacer tRNAs have mature 3′ terminals, it is clear that at least one more enzyme is required for maturation of these tRNAs (cf. Bikoff et al. 1975; Seidman et al. 1975b; Schedl et al. 1976; Sakano and Shimura 1978). Of course, it is also possible that preIle-Ala$_{III}$ is not the immediate precursor to tRNA$_1^{Ile}$ in wild-type cells. As mentioned earlier, transcription of rRNA and spacer tRNA genes is probably tightly coupled with the processing of these RNAs; perhaps tRNA$_1^{Ile}$ is ordinarily processed by RNase P before transcription of tRNA$_{1B}^{Ala}$ is completed.

In conclusion, we have shown that RNase P can achieve accurate 5'-end maturation of at least two spacer tRNAs in vitro. Furthermore, our demonstration that RNase P can cleave intact 30S pre-rRNA in vitro is at least consistent with the idea that the presence of tRNA genes in rRNA spacers might provide cells with alternative routes for rRNA biosynthesis. At the same time, since (1) the products of RNase P cleavage of 30S pre-rRNA require further 5' and 3' trimming to produce mature rRNA and (2) in the absence of RNase III no spacer tRNA intermediates accumulate, it is clear that the participation of other enzymatic activities would be required.

ACKNOWLEDGMENTS

We thank R. Crouch (National Institute of Child Health and Human Development, Bethesda, Maryland) for generously providing us with purified RNase III, and express our appreciation to R. E. Atchison (University of California, San Francisco) for his preparation of RNase P and for valuable advice. We also thank R. Young, J. A. Steitz, E. Morgan, L. Post, T. Ikemura, and M. Normura for allowing us to see their DNA sequences prior to publication. This work was supported by grant PCM-77-07357 from the National Science Foundation to J. E. D. and E. L. and grants PCM-76-21474 from the National Science Foundation and GM-21119 from the National Institutes of Health to C. G.

REFERENCES

Adams, J. M., P. G. N. Jeppesen, F. Sanger, and B. G. Barrell. 1969. Nucleotide sequence from the coat protein cistron of R17 bacteriophage RNA. *Nat. New Biol.* **223:** 1009.

Altman, S. and J. D. Smith. 1971. Tyrosine tRNA precursor molecule polynucleotide sequence. *Nature* **233:** 35.

Barrell, B. G. 1971. Fractionation and sequence analysis of radioactive nucleotides. In *Procedures in nucleic acid research* (ed. G. L. Cantoni and D. R. Davies), vol. 2, p. 751. Harper and Row, New York.

Bikoff, E., B. F. LaRue, and M. L. Gefter. 1975. *In vitro* synthesis of transfer RNA. II. Identification of required enzymatic activities. *J. Biol. Chem.* **250:** 6248.

Chang, S. and J. Carbon. 1975. The nucleotide sequence of a precursor to the glycine- and threonine-specific transfer ribonucleic acids of *Escherichia coli*. *J. Biol. Chem.* **250**:5542.

Dahlberg, A. E., J. E. Dahlberg, E. Lund, H. Tokimatsu, A. B. Rabson, P. C. Calvert, F. Reynolds, and M. Zahalak. 1978. Processing of the 5' end of *Escherichia coli* 16S ribosomal RNA. *Proc. Natl. Acad. Sci.* **75**:3598.

Dunn, J. J. and F. W. Studier. 1973. T7 early RNAs and *Escherichia coli* ribosomal RNAs are cut from large precursor RNAs *in vivo* by ribonuclease III. *Proc. Natl. Acad. Sci.* **70**:3296.

Gegenheimer, P. and D. Apirion. 1978. Processing of rRNA by RNase P: Spacer tRNAs are linked to 16S rRNA in an RNase P RNase III mutant of *E. coli*. *Cell* **15**:527.

Gegenheimer, P., N. Watson, and D. Apirion. 1977. Multiple pathways for primary processing of ribosomal RNA in *Escherichia coli*. *J. Biol. Chem.* **252**:2064.

Ginsburg, D. and J. A. Steitz. 1975. The 30S ribosomal precursor RNA from *Escherichia coli*. *J. Biol. Chem.* **250**:5647.

Guthrie, C. 1975. The nucleotide sequence of the dimeric precursor to glutamine and leucine transfer RNAs coded by bacteriophage T4. *J. Mol. Biol.*, **95**:529.

Guthrie, C., J. G. Seidman, B. G. Barrell, S. Altman, J. D. Smith, and W. H. McClain. 1973. Identification of tRNA precursor molecules made by phage T4. *Nat. New Biol.* **246**:6.

Hamkalo, B. A. and O. L. Miller. 1973. Visualization of genetic transcription. In *Gene expression and its regulation* (ed. F. T. Kenney et al.), p. 63. Plenum Press, New York.

Hayes, F., M. Vasseur, N. Nikolaev, D. Schlessinger, J. S. SriWidada, A. Krol, and C. Branlant. 1975. Structure of 30S pre-ribosomal RNA of *E. coli*. *FEBS Lett.* **56**:85.

Ikemura, T. and J. E. Dahlberg. 1973. Small ribonucleic acids of *Escherichia coli*. *J. Biol. Chem.* **248**:5024.

Ikemura, T. and M. Nomura. 1977. Expression of spacer tRNA genes in ribosomal RNA transcription units carried by hybrid Col E1 plasmids in *Escherichia coli*. *Cell* **11**:729.

Ikemura, T., Y. Shimura, H. Sakano, and H. Ozeki. 1975. Precursor molecules of *Escherichia coli* transfer RNAs accumulated in a temperature-sensitive mutant. *J. Mol. Biol.* **96**:69.

Ilgen, C., L. L. Kirk, and J. Carbon. 1976. Isolation and characterization of large transfer ribonucleic acid precursors from *Escherichia coli*. *J. Biol. Chem.* **251**:922.

Kindler, P., T. U. Keil, and P. H. Hofschneider. 1973. Isolation and characterization of a ribonuclease III deficient mutant of *Escherichia coli*. *Mol. Gen. Genet.* **126**:53.

Kiss, A., B. Sain, and P. Venetianer. 1977. The number of rRNA genes in *Escherichia coli*. *FEBS Lett.* **79**:77.

Lund, E. and J. E. Dahlberg. 1977. Spacer transfer RNAs in ribosomal RNA transcripts of *E. coli*: Processing of 30S ribosomal RNA *in vitro*. *Cell* **11**:247.

Lund, E., J. E. Dahlberg, L. Lindahl, S. R. Jaskunas, P. P. Dennis, and M. Nomura. 1976. Transfer RNA genes between 16S and 23S rRNA genes in rRNA transcription units of *E. coli*. *Cell* **7**:165.

Morgan, E. A., T. Ikemura, and M. Nomura. 1977. Identification of spacer tRNA genes in individual ribosomal RNA transcription units of *Escherichia coli. Proc. Natl. Acad. Sci.* **74:**2710.
Morgan, E. A., T. Ikemura, L. Lindahl, A. M. Fallon, and M. Nomura. 1978. Some rRNA operons in *E. coli* have tRNA genes at their distal ends. *Cell* **13:**335.
Nikolaev, N., L. Silengo, and D. Schlessinger. 1973. A role for ribonuclease III in processing of ribosomal ribonucleic acid and messenger ribonucleic acid precursors in *Escherichia coli. J. Biol. Chem.* **248:**7967.
Nomura, M., E. A. Morgan, and S. R. Jaskunas. 1977. Genetics of bacterial ribosomes. *Annu. Rev. Genet.* **11:**297.
Pace, N. R. 1973. Structure and synthesis of the ribosomal ribonucleic acid of prokaryotes. *Bacteriol. Rev.* **37:**562.
Perry, R. P. 1976. Processing of RNA. *Annu. Rev. Biochem.* **45:**605.
Peters, G. G., F. Harada, J. E. Dahlberg, W. Haseltine, A. Panet, and D. Baltimore. 1977. The low molecular weight RNAs of Moloney murine leukemia virus: Identification of the primer for RNA-directed DNA synthesis. *J. Virol.* **21:**1031.
Robertson, H. D. and J. J. Dunn. 1975. Ribonucleic acid processing activity of *Escherichia coli* ribonuclease III. *J. Biol. Chem.* **250:**3050.
Robertson, H. D., S. Altman, and J. D. Smith. 1972. Purification and properties of a specific *Escherichia coli* ribonuclease which cleaves a tyrosine transfer ribonucleic acid precursor. *J. Biol. Chem.* **247:**5243.
Robertson, H. D., R. E. Webster, and N. D. Zinder. 1968. Purification and properties of ribonuclease III from *Escherichia coli. J. Biol. Chem.* **243:**82.
Sakano, H. and Y. Shimura. 1978. Characterization and *in vitro* processing of transfer RNA precursors accumulated in a temperature-sensitive mutant of *Escherichia coli. J. Mol. Biol.* **123:**287.
Sakano, H., S. Yamada, T. Ikemura, Y. Shimura, and H. Ozeki. 1974. Temperature sensitive mutants of *Escherichia coli* for tRNA synthesis. *Nucleic Acids Res.* **1:**355.
Sanger, F., G. G. Brownlee, and B. G. Barrell. 1965. A two-dimensional fractionation procedure for radioactive nucleotides. *J. Mol. Biol.* **13:**373.
Schedl, P. and P. Primakoff. 1973. Mutants of *Escherichia coli* thermosensitive for the synthesis of transfer RNA. *Proc. Natl. Acad. Sci.* **70:**2091.
Schedl, P., J. Roberts, and P. Primakoff. 1976. In vitro processing of *E. coli* tRNA precursors. *Cell* **8:**581.
Schmidt, F. J. and W. H. McClain. 1978. Transfer RNA biosynthesis. Alternate orders of ribonuclease P cleavage occur *in vitro* but not *in vivo. J. Biol. Chem.* **253:**4730.
Seidman, J. G., B. G. Barrell, and W. H. McClain. 1975a. Five steps in the conversion of a large precursor RNA into bacteriophage proline and serine transfer RNAs. *J. Mol. Biol.* **99:**733.
Seidman, J. G., F. J. Schmidt, K. Foss, and W. H. McClain. 1975b. A mutant of *Escherichia coli* defective in removing 3' terminal nucleotides from some transfer RNA precursor molecules. *Cell* **5:**389.
Young, R. A. and J. A. Steitz. 1978. Complementary sequences 1700 nucleotides

apart from a ribonuclease III cleavage site in *Escherichia coli* ribosomal precursor RNA. *Proc. Natl. Acad. Sci.* **75:** 3593.

Young, R. A., R. Macklis, and J. A. Steitz. 1979. Sequence of the 16S–23S spacer region in two ribosomal RNA operons of *Escherichia coli. J. Biol. Chem.* **254:** 3264.

Processing of rRNA and tRNA in *Escherichia coli*: Cooperation between Processing Enzymes

David Apirion, Basanta K. Ghora, Greg Plautz,
Tapan K. Misra, and Peter Gegenheimer
Department of Microbiology and Immunology
Washington University School of Medicine
St. Louis, Missouri 63110

rRNAs of *Escherichia coli* are transcribed from a number of polycistronic transcription units, each of which codes for 16S, 23S, and 5S rRNA (for a review, see Pace 1973). Genes for some tRNAs are located within rRNA transcription units, either in the central spacer or in the 3' trailer region (Lund et al. 1976; Lund and Dahlberg 1977; Morgan et al. 1977, 1978; Gegenheimer and Apirion 1978). Processing of rRNA transcripts, then, must include tRNA processing steps. Inasmuch as intact polycistronic transcripts of rRNA are not detectable in wild-type bacterial cells, the first processing cleavage events must take place while the polycistronic precursor is still being transcribed (see Pace 1973; Gegenheimer and Apirion 1975; Gegenheimer et al. 1977; Hoffman and Miller 1977).

Mutant strains of *E. coli* that are defective in rRNA or in tRNA processing enzymes have been isolated. These include the mutations *rnc-105*, *rne-3071*, and *rnp-49*, which affect the enzymes RNase III, RNase E, and RNase P, respectively (Kindler et al. 1973; Apirion and Watson 1975; Apirion 1978; Apirion and Lassar 1978; Schedl and Primakoff 1973).

This discussion describes the structural analysis and in vitro processing of rRNA processing intermediates accumulated in mutant strains defective in the enzymes RNase III, RNase E, and RNase P, singly or in combinations, and demonstrates that all of these enzymes participate in production of mature cellular rRNA and tRNAs, including tRNA species not cotranscribed with rRNA. These studies also demonstrate the existence of another RNA processing enzyme, RNase F. These four enzymes accomplish most and perhaps all the major endonucleolytic rRNA and tRNA processing steps in *E. coli*.

RNA ACCUMULATION IN RNA PROCESSING MUTANTS

We examined the large and small RNAs synthesized by strains bearing various RNA processing mutations. Cultures of strains lacking RNase III (*rnc*) or containing thermolabile RNase P (*rnp*) or RNase E (*rne*), either singly or in all the possible combinations, were labeled at 30°C and 43°C, and their RNA species fractionated on 5%/10% tandem polyacrylamide gels and on 3% polyacrylamide slab gels, as shown in Figures 1 and 2, respectively. (The construction of some of those strains was already described, and further details will be deferred to future publications. For the construction of *rncrne* strains, see Apirion [1978] and Apirion and Lassar [1978], and for the construction of *rncrnp* strains, see Gegenheimer and Apirion [1978].) The wild-type strain produces the typical stable molecules tRNA, 5S rRNA, and 4.5S, 6S, and 10S RNAs, as well as other, unstable species (Fig. 1). This strain also synthesizes mature 16S (m16S) and mature 23S (m23S) rRNA, and pre16Sa and pre23Sa rRNA precursors (Fig. 2). The *rnc* strain synthesizes all these RNAs, except that pre16Sb and pre23Sb replace the wild-type pre16Sa and pre23Sa molecules; additionally, one detects new species of 8S (Fig. 1), 18S, 25S, and 30S (Fig. 2; see Gegenheimer et al. 1977; Gegenheimer and Apirion 1978). The pre16Sb and pre23Sb precursors each contain at their terminals extra sequences not found in wild-type pre16Sa or pre23Sa (D. Apirion et al., unpubl.). 18S RNA consists of pre16S and 5' leader sequences; 25S RNA contains pre23S and 5S sequences (Ghora and Apirion 1979a); and 30S represents unprocessed primary transcripts of the rRNA gene cluster (Ginsburg and Steitz 1975; Lund and Dahlberg 1977).

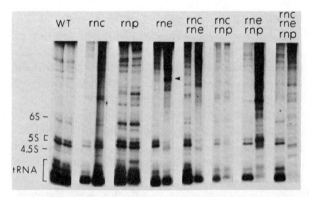

Figure 1 Small RNAs from processing mutant strains. Strains with the indicated genotypes were grown in 1-ml cultures; 0.2-ml aliquots were labeled with 10–100 µCi $^{32}P_i$, harvested, lysed, and fractionated on 5%/10% tandem polyacrylamide gels according to the method described by Gegenheimer et al. (1977). The first of each pair of lanes represents cells labeled for 30 min at 30°C; the second lane shows RNAs labeled for 30 min after the culture was incubated at 43°C for 40 min. An autoradiograph of the gel is presented. The 5% portion of the gel was removed prior to autoradiography. (→) 9S molecule.

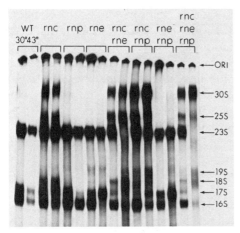

Figure 2 Large rRNAs from processing mutant strains. Lysates of the strains described in Fig. 1 were fractionated on 3% polyacrylamide gels (Gegenheimer et al. 1977). Details are identical to these in Fig. 1, except that labeling was for 25 min.

The *rnp* mutant exhibits normal, large rRNAs and displays a complex pattern of small RNAs at the nonpermissive temperature (Fig. 1). Levels of mature-sized tRNA are reduced and 4.5S RNA is not present, but several new species appear. Some of the new molecules, however, are also present to some extent even at the permissive temperature, suggesting that the mutant RNase P is partially inactive even at 30°C.

The *rne* strain produces large rRNAs like those of the wild type. Its small RNAs are also similar, except that at the nonpermissive temperature no 5S rRNA is detectable; instead, a prominent new 410-base 9S species appears (Fig. 1). Fingerprint analysis (Ghora and Apirion 1979b) did not reveal appreciable amounts of 5S sequences in the RNA at the 5S region at 43°C or any 9S sequences at 30°C.

Next we examined RNA metabolism in strains bearing pairs of mutations. Each double mutant strain shows some features predicted for a combination of the two single-mutant phenotypes. The *rncrne* strain failed to accumulate 5S RNA at 43°C (Fig. 1) and did not produce either 9S or 23S rRNA; rather it accumulated a 25S rRNA (Fig. 2; Apirion and Lassar 1978; Apirion 1978), which contains the sequences of pre23S and of pre5S rRNA (Ghora and Apirion 1979a). The *rncrnp* strain synthesizes most of the new RNAs seen in *rnp* strains. Production of mature-sized tRNA, however, is obviously less than in the *rnp* single mutant. Examination of the large rRNAs (Fig. 2) reveals the presence of all the molecules found in *rnc* strains and a new 19S RNA species that contains the sequences of 18S linked to spacer tRNAs (see below). The *rnernp* strain shows the novel *rnp*-specific bands as well as the cessation of 5S rRNA accumulation seen in the *rne* strain. Again, tRNA accumulation at 43°C is less than that of either parental strain.

The triple mutant strain *rncrnernp* shows, at the nonpermissive temperature (43°C), a striking reduction in synthesis of mature-sized tRNA molecules, as compared with the *rnp* parental or with the *rnp* in combination with *rnc* or *rne* mutations, and fails to accumulate 5S rRNA. Metabolism of the large rRNAs reflects all three mutations: the 19S rRNA is found, as in *rncrnp* strains, and 25S is formed at the expense of 23S, as observed in *rncrne* strains.

rRNA PRECURSORS ARE PROCESSED BY RNase III AND RNase E

In vivo, RNase III is the enzyme that releases first pre16S and then pre23Sa rRNA precursors from the growing rRNA transcript because in *rnc* strains these species are missing (Gegenheimer et al. 1977; D. Apirion et al., unpubl.). Since metabolism of pre5S rRNA is normal in the absence of RNase III, another activity termed RNase E (Gegenheimer et al. 1977), which can process pre5S rRNA from the rRNA transcript, must exist in *E. coli*. To identify this enzyme, temperature-sensitive mutants were isolated from RNase III⁻ strains and their rRNA metabolism was examined (Apirion 1978). One such strain accumulated 25S rRNA at the nonpermissive temperature instead of 23S rRNA and 5S rRNA (Apirion and Lassar 1978). Introduction of the *rnc*⁺ allele into this latter strain restored normal metabolism of large rRNAs, but at the nonpermissive temperature 5S rRNA still was not produced and 9S rRNA accumulated instead (Figs. 1 and 4; see also Fig. 4 in Apirion and Lassar 1978). It was shown that 9S contains 5S rRNA sequences (Ghora and Apirion 1978, 1979b). These observations suggested that the mutant strains that fail to produce normal levels of 5S RNA are defective in RNase E.

To verify that RNase E is the enzyme that excises pre5S sequences, we examined the processing in vitro of transcripts of two pre5S-containing rRNAs, the 9S and 25S molecules accumulated in *rne* and in *rnernc* strains, respectively (Figs. 1 and 2). We found RNase E activity in the salt wash of wild-type ribosomes and, after further purification through several columns, it cleaves 9S rRNA to pre5S rRNA (Ghora and Apirion 1978, and unpubl.). Digestion of 25S with an RNase E preparation also yields two 5S-size products representing the major sequence variants of pre5S rRNA (Ghora and Apirion 1979a). The products generated in vitro from 9S or 25S RNAs are authentic pre5S rRNAs, as evidenced by comparing T1 fingerprints and RNase A redigestion products of pre5S processed in vitro with those of authentic pre5S obtained in vivo from chloramphenicol-treated cells (Ghora and Apirion 1978, 1979a). Since the pre5S RNA generated from 9S contains minor oligonucleotides representative of all the sequence variants of *E. coli* 5S rRNA, we conclude that 9S RNA represents transcripts of all the actively transcribed rRNA operons of *E. coli* and that pre5S molecules from all of these operons are released from the nascent rRNA transcripts by RNase E.

RNase P PARTICIPATES IN THE PROCESSING OF SPACER tRNAs

As Figure 2 shows, an *rncrnp* strain that is RNase III$^-$ and temperature sensitive for RNase P accumulates at the nonpermissive temperature a new 19S RNA species, which contains pre16S rRNA sequences covalently linked to spacer tRNA sequences (Gegenheimer and Apirion 1978). The presence of spacer tRNA was demonstrated by in vitro processing of 19S RNA with cell extracts. Among the products obtained after fractionation by two-dimensional gel electrophoresis was a species (spot 1 RNA) whose RNase T1 fingerprint and RNase A redigestion products were identical with those of tRNA$_2^{Glu}$; another product (spot 2 RNA) was tentatively identified as a mixture of tRNA$_1^{Ile}$ and tRNA$_{1B}^{Ala}$ (Gegenheimer and Apirion 1978). These are the three tRNAs that are known to reside in spacer regions between 16S and 23S rRNA genes in the various rRNA operons (Lund et al. 1976). Recently, we found that RNase P can process in vitro three 4S size molecules from 19S RNA. Fingerprint and redigestion analysis of these spots confirm that they correspond to tRNA$_2^{Glu}$, tRNA$_1^{Ile}$, and tRNA$_{1B}^{Ala}$. These experiments suggest that the tRNAs in the 19S RNA are located at or near the 3' end of the molecule.

The structure of 19S RNA suggests that it is removed from the growing rRNA transcript by an endonuclease that recognizes and processes the 3' ends of tRNA sequences in the larger transcripts (see Bikoff et al. 1975). Since in the absence of RNase III, RNase E, and RNase P the nascent rRNA transcript is still cleaved by this endoribonuclease to 19S and 25S rRNAs (see Fig. 2), it must be distinct from these three previously described activities. We refer to this new activity as RNase F. It is possible, of course, that it could be identical with previously described nucleases. At present, we do not know whether exonucleolytic trimming of the 3' end takes place after RNase F introduces its cut in the nascent rRNA chain (see Fig. 4, cut 3).

COOPERATION BETWEEN tRNA AND rRNA PROCESSING ENZYMES

Since RNase P helps cleave spacer tRNAs from the rRNA transcripts, we wondered whether there was a more general phenomenon of interaction between RNase P, originally identified as processing tRNA, and RNase III and RNase E, which are known to process rRNA. We asked whether mutations in RNase III and RNase E might affect the processing of tRNA, just as a mutation in RNase P affects rRNA maturation. The data in Figure 1 provided an indication that tRNA maturation might be more severely restricted in strains lacking RNase III and/or RNase E as well as RNase P, compared with the single RNase P mutant. Therefore, we examined tRNA accumulation in all of these strains by fractionation of their small RNAs in the 4S to 5S range using two-dimensional gel electrophoresis. We analyzed RNA synthesized at 30°C or 43°C. The RNA synthesized at 30°C appeared to be normal and therefore only RNA synthesized at 43°C is shown in Figure 3. Figure 3a shows the tRNAs of a ^{32}P-labeled wild-type strain. The tRNAs of the *rnc* strain, which are displayed in Figure 3b, are all apparently normal. The *rne* strain in c shows normal tRNAs but lacks mature 5S rRNA. The *rnp* strain in d produces an appreciable amount of mature- or almost mature-sized tRNA molecules as compared to wild-type cells. These molecules might arise by other enzymatic cleavages close to the RNase P site or by residual RNase P activity. The 4S–5S RNAs of the *rncrne* strains, displayed in Figure 3e, differ little from the RNAs of the single *rne* strain. Combination of the *rnc* allele with *rnp* does not significantly change the pattern of tRNA accumulation from that of the single *rnp* strain, as seen in Figure 3f. Most dramatically, however, an *rnprne* double mutant shows a drastic reduction in mature tRNA species and a number of new 5S-size species, as seen in Figure 3g. A further decrease in tRNA maturation is found when all of these mutations are combined; this is demonstrated in Figure 3h. The *rnprncrne* mutant accumulates few, if any, mature tRNA molecules. These findings suggest that RNases III, E, and P all participate in the processing of tRNA precursors. The lack of processing in the multiple mutant strains is genuine and does not result from an effect of processing on transcription, since in the triple mutant strain a large variety of RNA molecules longer than tRNA accumulate at the nonpermissive temperature (see Figs. 1 and 2).

From the triple mutant strain, further mutants were isolated in which only 30S rRNA is accumulated at 43°C. The frequency of occurrence of such mutants suggests that they are in single genes. Hence, these mutants affect the activity we have defined as RNase F, suggesting that this is the only enzyme, besides RNases III, E, and P, that cuts the nascent rRNA transcript. These quadruple mutants synthesize even fewer tRNA-like molecules at the nonpermissive temperature than does the triple mutant, but they still accumulate larger RNA molecules in the range of 100–600 nucleotides, as well as the 30S rRNA. This observation suggests that the transcription apparatus of *E. coli* does not require positive feedback signals from transitory, processed RNA molecules.

The results presented here and elsewhere (Gegenheimer et al. 1977; Gegenheimer and Apirion 1978; Apirion and Lassar 1978; Ghora and Apirion 1978, 1979a,b) describe all the processing reactions required for production of the monomeric rRNA precursors pre16S, pre23S, pre5S

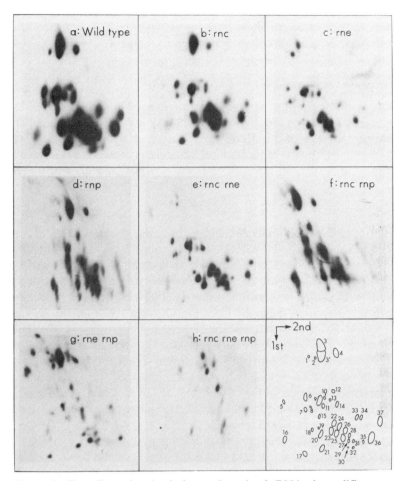

Figure 3 Two-dimensional gel electrophoresis of tRNAs from different processing mutants. Cultures (1 ml each) of strains with the indicated genotype were grown, labeled, lysed, and phenol extracted according to the method described by Gegenheimer and Apirion (1978) for cultures to be analyzed on two-dimensional gels. Cultures were shifted from 30°C to 43°C for 40 min prior to labeling with 200 μCi $^{32}P_i$ for a further 60 min. Approximately 1×10^6 cpm of each sample were fractionated: the first dimension was 10% acrylamide and the second was 20% acrylamide, 7 M urea. Electrophoretic procedures used were those described by Ikemura and Dahlberg (1973) and Gegenheimer and Apirion (1978); gels were exposed for about 3 days. The last panel contains the interpretation of *a*. Spots 3, 3', and 4 are the three forms of 5S rRNA; spot 17 is tRNA$_2^{Glu}$.

rRNA, and spacer tRNAs from the nascent rRNA transcript (Fig. 4). The enzymes involved are RNase III, RNase E, RNase P, and RNase F. As yet, we have not identified the activities responsible for processing of distal trailer tRNAs from the transcript. It is likely, however, that these activities will be similar to those that process other tRNAs, specifically spacer tRNAs, and they are thus designated in Figure 4.

PROCESSING IN WILD-TYPE CELLS

The detailed sequence of processing events discussed here is illustrated by the model presented in Figure 4. An RNA polymerase molecule that initiates transcription of an rRNA gene cluster or operon continues to synthesize the various components of the polycistronic rRNA primary transcript, that is, pre16S, pre23S, and pre5S rRNA, spacer, and trailer tRNAs, and leader and spacer sequences. Before transcription is terminated, however, processing endonucleases are already removing monomeric rRNA precursors from the nascent transcript. As polymerase molecules complete synthesis of pre16S rRNA, the inverted complementary sequences flanking the m16S transcript anneal to form a double-stranded stem from which m16 sequences loop out. As Figure 4 shows, the stem so formed is susceptible to endonucleolytic cleavage by RNase III, which cuts within it (cuts 1A and 1B) to release pre16S precursor rRNA plus a 5' leader fragment from the growing RNA chain. As the spacer region is synthesized, tRNAs are removed by endonucleolytic cleavage at the 3' side (cut 3) by an activity that has not yet been isolated (RNase F) and at the 5' end by RNase P (cut 2). This 3' cut probably precedes RNase P cleavage; such an order is observed in tRNA processing in general. Trimming of the 3' end, perhaps by RNase D (Ghosh and Deutscher 1978) or RNase P3 (Bikoff et al. 1975), may also be required to produce mature spacer tRNAs.

As the RNA polymerase now completes transcription of 23S genes, RNase III excises pre23S sequences (cuts 4A and B), again by cleaving in the double-stranded stem formed by complementary sequences surrounding m23S. A portion of the internal spacer (between cuts 3 and 4A), from which tRNAs have already been removed, is also generated by this cleavage. The distal portion of the gene cluster is still being transcribed. As they are formed, pre5S rRNA sequences are excised by RNase E (cuts 5A and B), and distal or trailer tRNAs are removed by other cleavages (cuts 9 and 10), which could be caused by the same enzymes as those that process spacer tRNAs.

These primary processing events give rise to precursor forms of 16S, 23S, and 5S rRNAs, whose maturation most likely occurs in ribonucleoprotein particles. The processing of pre16S to m16S by RNase M16 (Meyhack et al. 1974; Hayes and Vasseur 1976) is probably endonucleolytic (cuts 6A and B). Similarly, cuts 7A and B are proposed to mature pre23S to m23S rRNA. On the other hand, pre5S rRNA might be trimmed to m5 exonucleolytically (Jordan et al. 1971) by RNase M5 (Fig. 4, cut 8). (For an enzyme that can mature pre5S to m5S in *Bacillus subtilis*, see Sogin et al. 1977.)

PROCESSING IN MUTANT STRAINS

The origin of rRNA species seen in the various mutants described in this article can readily be described by reference to the model shown in Figure 4. In strains lacking the processing endonuclease RNase III, scission of nascent rRNA transcripts is initiated by enzymes that cut in the spacer region to remove tRNA sequences (cuts 2 and 3). 18S RNA is not the immediate product of these cleavages, for it does not contain the spacer sequences proximal to cut 2 (D. Apirion et al., unpubl.). It is likely, therefore, that after RNase P cleavage of the transcript, these extra spacer sequences are very rapidly trimmed, giving rise to 18S RNA (see Fig. 4C). Production of pre16Sb rRNA requires removal of the 5' leader sequence from 18S RNA, probably by nonspecific nucleases that leave the duplex stem intact (Fig. 4A). The product, a slightly larger-than-normal pre16Sb, is matured to m16S rRNA by the maturation enzyme RNase M16 (cuts 6A and B).

RNase E cleavage of the nascent rRNA transcript would generate pre5S rRNA and pre23S-like molecules containing extra spacer sequences at both ends that would be rapidly and nonspecifically digested to a resistant double-stranded stem, giving rise to the pre23Sb molecule of the *rnc* strain. This pre23Sb is further processed to mature m23S rRNA by the maturation enzyme RNase M23 (cuts 7A and B).

The 25S species is seen even though RNase E is active in this strain; we can thus conclude that although RNase E excision of pre5S from nascent 9S RNA (cuts 5A and B) is so rapid that no uncleaved 9S RNA is detected in wild-type cells (Ghora and Apirion 1979b), the addition of pre23S sequences to nascent 9S transcripts severely impairs the efficiency of pre5S cleavage by RNase E. In an analogous fashion, 30S RNA transcripts are also detected in the RNase III⁻ cell, even though RNases P, E, and F are all present. Notice also in Figure 2 that the ratio of the unprocessed rRNA species, 25S and 30S, to the processed species, 23S and 16S, is increased as the number of processing enzymes is decreased (see also Gegenheimer et al. 1977).

When RNase E is inactivated in the *rne* mutant, pre5S rRNA is not removed from the distal portion of the transcript, which instead accumulates as 9S RNA (Fig. 4D). If RNase III cleavage is also not performed, as in the *rncrne* strain at the nonpermissive temperature, pre5S sequences are found linked to pre23S rRNA via the intermediate spacer RNA, and 25S RNA accumulates at the expense of pre23S and pre5S (Fig. 4E). In these strains, no pre23S rRNA is detected.

Spacer tRNAs are linked to 18S RNA, giving 19S RNA, if RNase P is inactivated in an *rnc* strain (Fig. 4F). Inactivation of RNase III, RNase P, and RNase E allows the production only of 25S and of 19S RNA (Fig. 4G). Uncleaved 30S RNA is also synthesized. It appears, then, that in *rncrnernp* strains, an endonuclease that cuts at the 3' terminal of pre-

PROCESSING MAP OF RIBOSOMAL RNA

A. SECONDARY STRUCTURE AND CLEAVAGE SITES

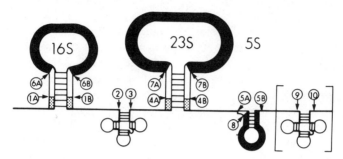

Cuts 1,4. RNase III   2,9. RNase P   3,10. RNase F.   5. RNase E
6. RNase M16   7. RNase "M23"   8. RNase "M5"

B. PROCESSING IN WILD-TYPE STRAINS

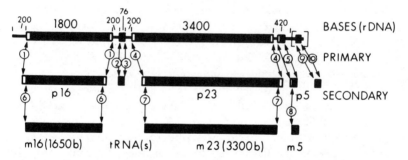

C. PROCESSING IN RNase III⁻ STRAINS

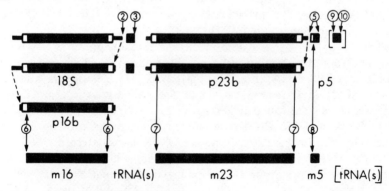

Figure 4 (See facing page for legend.)

D. PRIMARY CUTS IN RNase E⁻ STRAINS AT 43°

E. PRIMARY CUTS IN RNase III⁻RNase E⁻ STRAINS AT 43°

F. PRIMARY CUTS IN RNase III⁻RNase P⁻STRAINS AT 43°

G. PRIMARY CUTS IN RNase III⁻ RNase P⁻ RNase E⁻ STRAINS AT 43°

Figure 4 Processing of rRNA in *E. coli*. (*A*) Proposed structure and cleavage sites of the rRNA transcript. Secondary structures of pre16S and pre23S were based on the electron micrographic data of Wu and Davidson (1975) and that of tRNAs on the tRNA$_{su^+3}^{Tyr}$ precursor (Altman 1975) and of 5S on the proposal of Fox and Woese (1975). Distal (trailer) tRNAs are bracketed, since not all rDNAs contain them. (→) Cleavage sites positioned according to both known sequences of mature RNAs and the observed lengths or oligonucleotide contents of processing intermediates described here; (①–⑩), cutting events, numbers refer to the enzyme involved; (A and B) two (or more) separate cuts are required; (■) mature rRNA sequences; (□) precursor-specific sequences removed during secondary processing steps; (▨) sequences found only in pre16Sb and pre23Sb of RNase III⁻ cells; (———) (except for tRNAs) nonconserved sequences discarded during primary processing. Enzymes are discussed in the text. (*B*) The first line shows the transcriptional map of a representative rDNA unit. Base distances are noted above the map. The primary and secondary cuts, numbered as in *A*, are shown above the products they generate. Open and solid segments are as in *A*. (*C*) Details are as in *A* and *B*, except that the transcriptional map is not shown. (--→) Trimming by unidentified nucleases. (*D–G*) Processing in other mutant strains. Products of primary cleavage in other mutant strains are shown under the cuts that produce them. (Details as in *A* and *B*.) This figure is based on data derived from the work of a number of laboratories (see text).

tRNA sequences (RNase F) cleaves nascent rRNA chains into proximal 19S and distal 25S portions, which might also contain the trailer tRNAs. The sizes of 19S (about 2250 bases) and of 25S (about 4000 bases) are similar to those of products expected from a single cleavage of 30S (about 6250 bases).

OTHER PROCESSING ENZYMES

The present study indicates that the four major primary processing endoribonucleases of *E. coli* are RNase III, E, F, and P. The observation that few monomeric tRNAs are produced in an RNase P^- RNase E^- strain (Fig. 3) suggests that RNase E might be identical to the activity called RNase P2 (Schedl et al. 1976), since both of these enzymes should be able to process multimeric tRNA precursors.

Secondary processing of pre16S, pre23S, and pre5S to m16S, m23S, and m5S is not affected by mutations in the above four enzymes. These steps must be carried out by other activities: RNases M16, M23, and M5.

PROCESSING AND DEGRADATIVE RIBONUCLEASES

When various parameters pertaining to RNA stability were measured in the triple-mutant *rnc rne rnp* strain at the permissive and nonpermissive temperatures, it was found that they were similar at both temperatures and not very different from those measured in normal strains (D. Apirion et al., unpubl.). Therefore, it seems that processing endoribonucleases do not normally participate in degradative reactions in the cell.

PRIMARY AND SECONDARY PROCESSING

All the processing steps affected by the *rnc, rne,* and *rnp* mutations described here can take place in the absence of protein synthesis; this is supported by the fact that it is possible to carry out in vitro cleavages with these enzymes using naked RNA molecules as substrates.

In the cell, however, the growing rRNA chain does not present itself to the processing enzymes as a naked RNA molecule but rather as a ribonucleoprotein particle. The secondary processing enzymes, those designated M16, M23, and M5 in Figure 4, are inhibited when protein synthesis is blocked and in vitro studies indicate that their substrates are not naked RNA molecules but ribonucleoprotein particles (Meyhack et al. 1974; Hayes and Vassuer 1976).

With regard to specificity and order of processing events, these studies show that primary processing endonucleases will recognize their sites with only a certain influence from neighboring sequences. That is, RNase E will excise pre5S sequences in vivo either from nascent 9S RNA in RNase III⁺ cells or from these transcripts linked to pre23S RNA in RNase III⁻ cells, but the efficiency of pre5S cleavage is reduced in the latter situation. RNase P and RNase F cleavage of spacer tRNAs also seems less efficient when pre16S sequences are not previously removed by RNase III because uncleaved 30S transcripts are detected in RNase III⁻ RNase P⁺ or RNase III⁻ RNase P⁻ strains (Fig. 2).

PROCESSING SIGNALS

The observed involvement of RNase P in processing of rRNA transcripts in vivo and in vitro, albeit by recognition of tRNA sequences, and the effects demonstrated here of RNase III and RNase E mutations on in vivo accumulation of tRNAs suggest that these three enzymes cannot be classified simply as tRNA processing enzymes or as rRNA processing enzymes; rather the suggestion is that all three are involved in the processing of various RNA precursors. It is possible, then, that processing of widely differing transcripts involves recognition of only a few key features.

Many RNAs are characterized by a high degree of secondary structure, which is apparently crucial to their function. Enzymes that process precursor RNAs seem to recognize secondary structure. Three sequenced RNase P substrates (Altman and Smith 1971; Bothwell et al. 1976a,b) share similar arrangements of stem-and-loop structures; RNase P cleavage occurs near the stem of the mature molecule. RNase III also recognizes double-stranded RNA regions (see Paddock et al. 1976; Robertson et al. 1977; Rosenberg and Kramer 1977).

As shown in Figure 4A, we propose that RNases E and F also cleave near potential duplex stems. We suggest that endonucleolytic processing of RNA involves recognition of stem-and-loop structures, and cleavage near single-strand/double-strand junctions. However, the recognition site for each processing endoribonuclease is unique. This is evident from the fact that each of the RNA processing mutants shows a unique pattern of RNA molecules and the RNA terminals generated by one of the enzymes (RNase III, for instance) are not found in an RNase III⁻ strain.

Bearing in mind that features recognized in many different substrates by a given RNA processing enzyme are likely to have similar but not identical secondary and tertiary structures, we can offer a general principle governing enzymatic recognition and processing of RNA transcripts. A relatively limited number of processing enzymes, principally endonucleases, with stringent but not sequence-specific substrate specificities, acting on a modest repertoire of variations in a few fundamentally distinctive processing signals, may be capable of carrying out a broad range of different, highly specific, and potentially well-ordered processing steps, including rather intricate pathways for processing of RNA transcripts.

Considering the material discussed here, the following conclusions can be made:

1. Transcription of RNA is independent of its processing.
2. Primary processing of rRNA and tRNA occurs during transcription.
3. Most RNA transcripts are processed by more than a single endoribonuclease.
4. The four enzymes responsible for most endonucleolytic processing of stable RNA transcripts in *E. coli* are RNases III, E, F, and P.
5. Ribonucleases, by and large, fulfill either processing or degradative functions.
6. Processing endonucleases are highly specific and each performs a unique function. Their recognition sites may be composed of unique combinations of relatively simple yet distinctive features of secondary and tertiary structure.
7. The efficiency, but not the specificity, of processing cuts is affected by the size of the substrate.
8. A certain amount of flexibility exists in the order of initial processing events, but the final steps must be preserved.

SUMMARY

Primary endonucleolytic cleavage of rRNA and tRNA transcripts in *E. coli* is accomplished by four major processing enzymes: RNase III, RNase E, RNase P, and a new activity designated RNase F. These enzymes act on nascent rRNA transcripts to give pre16S, pre23S, pre5S rRNA, and tRNA species. Inactivation of one or more of these enzymes results in the accumulation in vivo of RNA intermediates, which can be processed in vitro by these enzymes. Most tRNAs are also produced by the action of these enzymes, since, in their absence, few if any monomeric tRNA species accumulate in vivo.

ACKNOWLEDGMENTS

We are most grateful to John Dunn (Brookhaven National Laboratory) for supplying us with RNase III and Sid Altman (Yale University) and Bill McClain (University of Wisconsin, Madison) for providing us with RNase P. Ned Watson and Joel Anderson (Washington University School of Medicine, St. Louis, Missouri) assisted in the construction of strains. This work was supported by National Science Foundation grant 76-81665 and National Institutes of Health grant GM-25579.

REFERENCES

Altman, S. 1975. Biosynthesis of transfer RNA in *Escherichia coli*. *Cell* **4:** 21.
Altman, S. and J. D. Smith. 1971. Tyrosine tRNA precursor molecule polynucleotide sequence. *Nat. New Biol.* **233:** 35.
Apirion, D. 1978. Isolation, genetic mapping, and some characterization of a mutation in *Escherichia coli* that affects the processing of ribonucleic acid. *Genetics* **90:** 659.
Apirion, D. and A. B. Lassar. 1978. A conditional lethal mutant of *Escherichia coli* which affects the processing of ribosomal RNA. *J. Biol. Chem.* **253:** 1738.
Apirion, D. and N. Watson. 1975. Mapping and characterization of a mutation in *Escherichia coli* that reduces the level of ribonuclease III specific for double-stranded ribonucleic acid. *J. Bacteriol.* **124:** 317.
Bikoff, E. K., B. F. LaRue, and M. L. Gefter. 1975. In vitro synthesis of transfer RNA. II. Identification of required enzymatic activities. *J. Biol. Chem.* **250:** 6248.
Bothwell, A. L. M., R. L. Garber, and S. Altman. 1976a. Nucleotide sequence and in vitro processing of a precursor molecule to *Escherichia coli* 4.5S RNA. *J. Biol. Chem.* **251:** 7709.
Bothwell, A. L. M., B. C. Stark, and S. Altman. 1976b. Ribonuclease P substrate specificity: Cleavage of a bacteriophage 80-induced RNA. *Proc. Natl. Acad. Sci.* **73:** 1912.
Fox, G. E. and C. R. Woese. 1975. 5S RNA secondary structure. *Nature* **256:** 505.
Gegenheimer, P. and D. Apirion. 1975. *Escherichia coli* ribosomal ribonucleic acids are not cut from an intact precursor molecule. *J. Biol. Chem.* **250:** 2407.
―――. 1978. Processing of ribosomal RNA by RNase P: Spacer tRNAs are linked to 16S rRNA in an RNase P RNase III mutant strain of *E. coli*. *Cell* **15:** 527.
Gegenheimer, P., N. Watson, and D. Apirion. 1977. Multiple pathways for primary processing of ribosomal RNA in *Escherichia coli*. *J. Biol. Chem.* **252:** 3064.
Ghora, B. K. and D. Apirion. 1978. Structural analysis and in vitro processing to p5 rRNA of a 9S RNA molecule isolated from an *rne* mutant of *E. coli*. *Cell* **15:** 1055.
―――. 1979a. 5S rRNA is contained within a 25S rRNA which accumulates in mutants of *Escherichia coli* defective in processing of ribosomal RNA. *J. Mol. Biol.* **127:** 507.
―――. 1979b. Identification of a novel RNA molecule in a new RNA processing mutant of *Escherichia coli* which contains 5S rRNA sequences. *J. Biol. Chem.* **254:** 1951.
Ghosh, R. K. and M. P. Deutscher. 1978. Purification of potential 3' processing nucleases using synthetic tRNA precursors. *Nucl. Acids Res.* **5:** 3831.
Ginsburg, D. and J. A. Steitz. 1975. The 30S ribosomal precursor RNA from *Escherichia coli*: A primary transcript containing 23S, 16S, and 5S sequences. *J. Biol. Chem.* **250:** 5647.
Hayes, F. and M. Vasseur. 1976. Processing of 17S *Escherichia coli* precursor RNA in the 27S pre-ribosomal particle. *Eur. J. Biochem.* **61:** 433.
Hoffman, S. and O. L. Miller, Jr. 1977. Visualization of ribosomal ribonucleic

acid synthesis in a ribonuclease III-deficient strain of *Escherichia coli. J. Bacteriol.* **132:**718.

Ikemura, T. and J. E. Dahlberg. 1973. Small ribonucleic acids of *Escherichia coli*. I. Characterization by polyacrylamide gel electrophoresis and fingerprint analysis. *J. Biol. Chem.* **248:**5024.

Jordan, B. R., B. G. Forget, and R. Monier. 1971. A low molecular weight ribonucleic acid synthesized by *Escherichia coli* in the presence of chloramphenicol: Characterization and relation to normally synthesized 5S ribonucleic acid. *J. Mol. Biol.* **55:**407.

Kindler, P., T. U. Keil, and P. H. Hofschneider. 1973. Isolation and characterization of a ribonuclease III deficient mutant of *Escherichia coli. Mol. Gen. Genet.* **126:**53.

Lund, E. and J. E. Dahlberg. 1977. Spacer transfer RNAs in ribosomal RNA transcripts of *E. coli*: Processing of 30S ribosomal RNA in vitro. *Cell* **11:**247.

Lund, E., J. E. Dahlberg, L. Lindahl, S. R. Jaskunas, P. P. Dennis, and M. Nomura. 1976. Transfer RNA genes between 16S and 23S rRNA genes in rRNA transcription units of *E. coli. Cell* **7:**165.

Meyhack, B., I. Meyhack, and D. Apirion. 1974. Processing of precursor particles containing 17S rRNA in a cell free system. *FEBS Lett.* **49:**215.

Morgan, E. A., T. Ikemura, and M. Nomura. 1977. Identification of spacer tRNA genes in individual ribosomal RNA transcription units of *Escherichia coli. Proc. Natl. Acad. Sci.* **74:**2710.

Morgan, E. A., T. Ikemura, L. Lindahl, A. M. Fallon, and M. Nomura. 1978. Some rRNA operons in *E. coli* have tRNA genes at their distal ends. *Cell* **13:**335.

Pace, N. R. 1973. Structure and synthesis of the ribosomal RNA of prokaryotes. *Bacteriol. Rev.* **37:**562.

Paddock, G. C., K. Fukada, J. Abelson, and H. D. Robertson. 1976. Cleavage of T4 species I ribonucleic acid by *Escherichia coli* ribonuclease III. *Nucleic Acids Res.* **3:**1351.

Robertson, H. D., E. Dickson, and J. J. Dunn. 1977. A nucleotide sequence from a ribonuclease III processing site in bacteriophage T7 RNA. *Proc. Natl. Acad. Sci.* **74:**822.

Rosenberg, M. and R. A. Kramer. 1977. Nucleotide sequence surrounding a ribonuclease III processing site in bacteriophage T7 RNA. *Proc. Natl. Acad. Sci.* **74:**984.

Schedl, P. and P. Primakoff. 1973. Mutants of *Escherichia coli* thermosensitive for the synthesis of transfer RNA. *Proc. Natl. Acad. Sci.* **70:**2091.

Schedl, P., J. Roberts, and P. Primakoff. 1976. *In vitro* processing of tRNA precursors. *Cell* **8:**581.

Sogin, M. L., B. Pace, and N. R. Pace. 1977. Partial purification and properties of a ribosomal RNA maturation endonuclease from *Bacillus subtilis. J. Biol. Chem.* **252:**1350.

Wu, M. and N. Davidson. 1975. Use of gene 32 protein staining of single-strand polynucleotides of gene mapping by electron microscopy: Application to the 80d$_3$*ilvsu*$^+$7 system. *Proc. Natl. Acad. Sci.* **72:**4506.

The Interaction of RNase M5 with a 5S rRNA Precursor

Norman R. Pace, Bernd Meyhack, Bernadette Pace, and Mitchell L. Sogin
National Jewish Hospital and Research Center
and Department of Biochemistry, Biophysics and Genetics
University of Colorado Medical Center
Denver, Colorado 80206

As is evident from several other papers in this volume, considerable information regarding the posttranscriptional processing of tRNA is accumulating. Although most RNA molecules, in both prokaryotes and eukaryotes, are the products of more or less extensive posttranscriptional metabolism, detailed studies of these processes have been possible only with tRNA and 5S rRNA of *Bacillus subtilis*. Although not directly involved with tRNA, studies of the processing of 5S rRNA are relevant in this area of study in that much of this methodology is immediately applicable to the study of the maturation of the precursors of tRNA. Moreover, there is the common goal of understanding the nature of protein-polynucleotide interactions. In brief, we have isolated the endonuclease responsible for the terminal maturation of 5S rRNA in *B. subtilis* and devised procedures for altering the precursor RNA substrate to explore the polynucleotide recognition elements utilized by this highly specific enzyme.

THE MATURATION OF *B. SUBTILIS* 5S rRNA

Because of its simple structure (about 120 nucleotides long), 5S rRNA is useful for exploring the details of posttranscriptional RNA metabolism. At the outset of these studies, we felt that the 5S rRNA of *Escherichia coli* was not likely to be a suitable model, since the immediate precursor of 5S rRNA in this organism is maximally only 3 nucleotides larger than the mature form (Monier et al. 1970) and such limited precursor-specific length might not be of functional significance. We therefore undertook a search for an organism utilizing more complex 5S rRNA metabolism than that of *E. coli*.

Examination of the immediate precursors of the mature rRNA species in both prokaryotes and eukaryotes generally is straightforward. Since the substrates for the maturation nucleases generally are ribonucleoprotein particles, inhibition of protein synthesis affects the accumulation of the precursor RNA. Using this fact, we scanned several prokaryotes for appropriately sized precursors of 5S rRNA and chose *B. subtilis* for detailed study (Pace et al. 1973). *B. subtilis* produces several structurally distinct precursors of 5S rRNA. These fall into three size classes and presumably derive from different genes; the organism probably possesses about seven compound 16S-23S-5S rRNA transcriptional units (Pace and Pace 1971). The nucleotide sequence of one of the precursor classes (Sogin et al. 1976), which we term pre5A, is shown in Figure 1, folded into its probable secondary structure according to Fox and Woese (1975). The pre5A rRNA class consists of two structural populations, which differ by a single nucleotide residue (position 156). The other precursor classes, pre5B (about four molecules of about 150 nucleotides) and pre5C (one molecule of about 240 nucleotides), differ from pre5A in the lengths and structural details of their precursor-specific appendages (N. R. Pace et al. 1973, and unpubl.).

During the initial characterization of 5S rRNA precursors from various bacterial species, we noted that in *B. subtilis*, in contrast to other organisms examined, small amounts of mature 5S rRNA (m5S rRNA) were produced even though protein synthesis was completely halted (Pace et al. 1973). This suggested that the relevant maturation nuclease in *B. subtilis* did not have a rigorous requirement for a ribonucleoprotein substrate, and that we could use the naked pre-RNA as a substrate in the isolation of the enzyme. This proved successful and considerable purification of the enzyme, which we term RNase M5, has been reported (Sogin et al. 1977). The reaction catalyzed by RNase M5 is shown in the gel autoradiograph of Figure 2. The ^{32}P-labeled pre5A rRNA

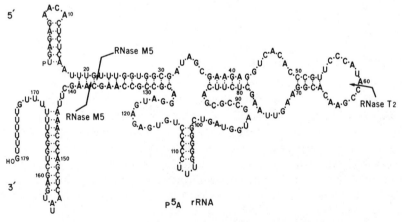

Figure 1 The structure of pre5A rRNA. The probable secondary structure of pre5A (Sogin et al. 1976) is formed by folding complementary bases in the mature segment of the nucleotide sequence into the secondary structure suggested by Fox and Woese (1975). RNase M5 removes the precursor-specific segments F1 and F2 from the 3' and 5' terminals, respectively. (RNase M5→) Cleavage sites.

precursor is cleaved endonucleolytically to yield concomitantly three products, m5S rRNA, a 42-nucleotide fragment (F1) derived from the 3' terminal of pre5A rRNA, and a 21-nucleotide fragment (F2) derived from the 5' end. The phosphodiester bonds cleaved in pre5A rRNA are indicated in Figure 1: the enzymatic scission generates 3'-OH and 5'-phosphoryl groups. It is not yet clear whether RNase M5 makes sequential or simultaneous scissions in the pre5A rRNA molecule, but the enzyme does not seem to release the precursor from its surface until the RNA is fully mature. The enzyme consists of two readily separable components, α and β. The β component is capable of binding pre5A rRNA to a membrane filter, but no scission of the RNA occurs until α, the presumed catalytic component, is added (B. Pace, unpubl.).

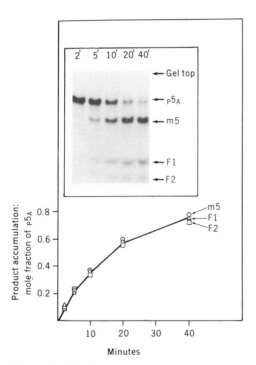

Figure 2 The cleavage of pre5A rRNA by RNase M5. ^{32}P-labeled pre5A rRNA (750,000 cpm) in 1 ml of reaction mixture (0.01 M Tris-HCl [pH 7.3], 0.01 M MgCl$_2$, 0.001 M dithiothreitol, 30% w/v glycerol) containing 50 μg β component and 20 μg α component (Sogin et al. 1977) was incubated at 37°C. Aliquots withdrawn at intervals were adjusted to contain 0.01 M EDTA and 0.5% SDS and reaction products were resolved by electrophoresis through a 10% polyacrylamide slab gel (Sogin et al. 1977). Radioactive bands were located by autoradiography and excised for monitoring ^{32}P content by liquid scintillation counting. The inset is the autoradiograph from which the quantities of reactants were derived. (Reprinted, with permission, from Meyhack et al. 1978.)

We have examined the susceptibilities of several types of RNA to RNase M5, including tRNA precursors, the immediate precursor of mature 16S rRNA, poly[r(AU)], and duplex RNA from *Penicillium chrysogenum*. These latter two RNA species are excellent substrates for the RNase III of *E. coli* (Robertson and Dunn 1975). However, so far we have not identified any RNase M5 substrates other than precursors of m5S rRNA. Since this enzyme is so fastidious and the natural substrates and products are sufficiently simple in structure to be manipulable, we have undertaken to determine the recognition features employed in this protein-polynucleotide interaction.

INVOLVEMENT OF PRECURSOR SEGMENTS IN THE RNase M5 REACTION

The two bonds in pre5A rRNA that are cleaved by RNase M5 are distant from one another in nucleotide sequence, but juxtaposed if the molecule is folded into its probable secondary structure (Fig. 1). We have pointed out (Sogin et al. 1977) that it seems unlikely that a single catalytic site could effect both cleavages simultaneously because the phosphodiester bonds acted upon would present opposite chemical polarities to the catalytic amino acids. Independent placement of the two susceptible phosphodiester bonds at a single catalytic site with the same chemical polarity would require intervening rotation of the pre5A rRNA substrate relative to the enzyme surface. On the assumption that equivalent polynucleotide recognition elements are utilized for both scissions, we had hoped to define these elements by identifying features common to the two substrate sites. We have discussed several features of the pre5A rRNA, which, because of the low probability of their common occurrence in relation to both of the cleaved bonds, might serve as sites of recognition for RNase M5 (Sogin et al. 1977). These features include the duplex region within which the susceptible phosphodiester bonds lie, two hexamer sequences (UGAGAG at positions 1–6 and 116–121 in Fig. 2) disposed with twofold rotational symmetry about the susceptible bonds, and two regions of translational symmetry (sequence repeats) that surround the susceptible bonds.

Since each of the three structural features in the pre5A molecule is associated with one or both of the precursor-specific segments, we initially focused on the involvement of these segments in the reaction (Meyhack et al. 1977). The notion was to test reaction intermediates, lacking either the 5' or the 3' precursor element, as substrates for the enzyme. The absence of information essential to recognition by RNase M5 should be reflected in a reduced rate of cleavage of the deficient substrate by the enzyme.

We were unable to recover the cleavage intermediates required for these studies from the RNase M5 reactions. However, it was possible to construct the necessary molecules from fragments derived from the precursor and mature rRNA molecules. The secondary structures of pre5A and m5S rRNAs are sufficiently compact so that under the appropriate mild digestion conditions, RNase T2 cleaves a single phosphodiester bond in pre5A rRNA at A60 (see Fig. 1) and in m5S rRNA at the corresponding position. Upon electrophoretic resolution in a denaturing polyacrylamide gel, the half, or cleaved, molecules are recovered in good yield; we have documented the isolation and analysis of these fragments (Meyhack et al. 1977).

Although the isolated pre5A rRNA halves proved unsusceptible to RNase M5, annealing of the two halves restored full substrate capacity. Aspects or structural consequences of secondary structure are therefore required by the enzyme. Since substrate capacity is restored by annealing the halves and therefore is not influenced by the broken phosphodiester bond at the apex of the molecule, we were able to generate artificial RNase M5 reaction intermediates and test the role of the precursor segments in the recognition process (Meyhack et al. 1977). The substrate lacking the 5' precursor segment (F2) was constructed by annealing the pre5A 3' half with the 5' half of m5S rRNA; this construct is designated pre5A-3'/m5S-5'. The alternate substrate, lacking the 3' precursor-specific sequence (F1), was generated from the alternate pre5A and m5S rRNA halves. The various precursors then were presented to RNase M5 and the reaction products scored by gel electrophoresis. Figure 3 compiles an autoradiograph of such a gel and the derived reaction kinetics.

It is evident from Figure 3 that the restored pre5A or the artificial intermediate containing the 5' precursor segment are as competent as the native pre5A precursor in the interaction with RNase M5. However, the constructed substrate lacking the 5' precursor segment but containing the 3' precursor segment is considerably diminished in susceptibility to RNase M5. Clearly the 5' precursor-specific segment contributes information to the RNase-M5–pre5A-rRNA interaction. These observations also suggest that, if the enzyme in fact makes sequential cleavages in pre5A rRNA, F1 is released prior to F2.

THE INFORMATIONAL ROLE OF THE 5' PRECURSOR SEGMENT

F2, the 5' precursor segment of pre5A rRNA, contributes to all of the features postulated to be involved in pre5A rRNA recognition by RNase M5. It is included in the duplex stalk containing the susceptible bonds, and it possesses elements of rotational and translational symmetry. However, the two symmetry features (the UGAGAG sequence at residues 1–6 and UUUG at residues 18–21; see Fig. 1) are located several nucleotides apart. So, we examined the relative contributions of these features by constructing test substrates with 5'-precursor-specific appendages of different nucleotide lengths and sequences (Meyhack et al.

1978). These we produced by joining *B. subtilis* m5 rRNA to synthetic oligonucleotides of defined structure, using bacteriophage T4 RNA ligase. The RNA ligase elicited in *E. coli* infected by phage T4 is capable of catalyzing the ATP-dependent formation of 3'-5' phosphodiester bonds between 5'-phosphorylated donor oligonucleotides and oligoribonucleotide acceptors with 3'-OH groups (Walker et al. 1975). In initial studies, we

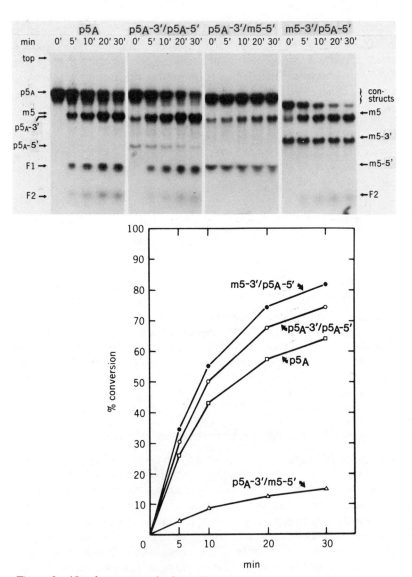

Figure 3 (*See facing page for legend.*)

employed as RNA ligase donor the 5' half of m5S rRNA (produced by limited RNase T2 digestion as described above), which contains a 5'-phosphoryl group derived from native m5 rRNA and a 3'-phosphoryl group as a result of RNase T2 action. The presence of the 3'-phosphate prevents cyclization or self-addition of the donor. As RNA ligase acceptors, we used a variety of short, synthetic oligomers, chosen on the basis of availability and of nucleotide sequence to provide relatively unambiguous tests for the involvement of the symmetry features in the RNase M5 reaction. We detail elsewhere the synthesis and structural analysis of these molecules (Meyhack et al. 1978).

Following their isolation and characterization, the ligase products, which consisted of the 5' half of m5S rRNA with synthetic oligomers appended to the 5' end, were annealed with the complementary 3' half of the pre5A pre-rRNA to yield constructs differing from the native substrate with regard to the structure of the 5' terminal, precursor-specific segment. These constructs were then furnished to RNase M5 and the release of F1 scored by gel electrophoresis; the derived kinetics are summarized in Figure 4. It is evident that the character of the 5' precursor segment markedly influences the rate of F1 release. Thus, as shown above, the restored pre5A molecule (curve pre5A-5') exhibits a maximum rate of F1 release, whereas the absence of any 5' precursor segment (m5S-5') results in poor substrate capacity. However, the entirety of the 5' precursor segment is not required for maximum rate of cleavage by RNase M5; mere addition of UUUG to the m5S-5' fragment restores the full susceptibility of the substrate to cleavage (curve [UUUG] m5S-5').

Figure 3 The influence of the precursor-specific segments on the RNase M5 reaction. The 5' and 3' half molecules of pre5A and m5S rRNAs were isolated following limited RNase T2 digestion and then reannealed to generate test substrates, all as described in the text and detailed by Meyhack et al. (1977). The substrate constructs were incubated with RNase M5 as described in Fig. 2, and samples were withdrawn from the reactions at the indicated times. Reaction products than were separated on a polyacrylamide slab gel and, following drying of the gel, detected by autoradiography. (*Top*) Autoradiograph of the gel; appropriate regions of the gel were excised and monitored for radioactivity. Radioactivities present in the gel positions corresponding to the various products at zero time were subtracted from the value observed in samples drawn at subsequent times. Radioactivity occupying the m5S region of zero time slots represents the pre5A-3' fragment or, in the case of the m5S-3'/pre5A-5' construct, some material that annealed to yield a conformation hydrodynamically distinct from the native construct and therefore different in electrophoretic mobility. Maturation products are labeled m5, F1, and F2. Some fragments that did not reanneal are labeled as such. (*Bottom*) Percent conversion of several precursor constructs to m5S rRNA calculated from data obtained from the same gel as shown at top. (Reprinted, with permission, from Meyhack et al. 1977.)

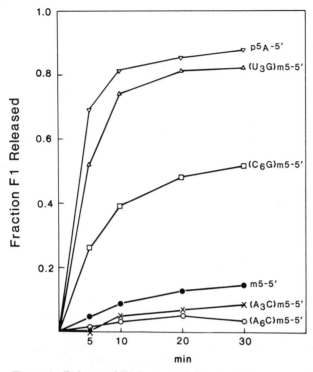

Figure 4 Release of F1 from partially artificial precursors. The m5S-5' fragments with various oligonucleotides appended to the 5' terminals were annealed with ^{32}P-labeled pre5A-3' fragment. Equivalent molar amounts of these substrates were individually incubated with RNase M5 in the standard maturation assay for different times and the products were resolved by gel electrophoresis. After autoradiography, radioactive bands were excised from the dried gel and monitored for ^{32}P content; release of F1 from pre5A-3' was scored by the appearance of the 77-nucleotide m5S-3' fragment. The curves in the figure are designated according to the 5' half molecules that were annealed to pre5A-3' to generate the test substrates. (Reprinted, with permission, from Meyhack et al. 1978.)

Therefore, at least one of the elements of twofold rotational symmetry, the UGAGAG sequence residing at the 5' end of the pre5A molecule, is not required by RNase M5 for the release of F1 at optimal rate. Furthermore, translational symmetry is not required for good substrate activity; substitution of UUUG by CCCCCCG as the 5' precursor segment also yields a quite effective substrate with respect to the release of F1 (curve [CCCCCCG] m5S-5'). Not all 5' precursor structures potentiate the release of F1 by RNase M5, however. The synthetic 5' precursor structures AAAC and AAAAAAC, which cannot restore duplex character to the substrate site by hydrogen bonding to nucleotide residue C137

(adjacent to F1 in pre5A rRNA), do not facilitate recognition and release of the F1 fragment by RNase M5 (curves [AAAC] m5S-5' and [AAAAAAC] m5S-5').
Clearly the entirety of the 5' precursor segment of pre5A rRNA is not required for optimal rate of RNase M5 release of F1. We next inspected the features within F2, the 5' precursor segment, that might be required for its own release (Meyhack et al. 1978). This was approached by examining the susceptibilities of the synthetic 5' precursor segments discussed above to release by RNase M5. The various synthetic oligomers were condensed, using RNA ligase, to uniformly ^{32}P-labeled 5' half of m5S rRNA, and the products were annealed with the nonradioactive 3' half of m5S rRNA. Then the test substrates were presented to RNase M5 at high enzyme-to-substrate ratios and reaction products analyzed on denaturing gels. An autoradiograph of the relevant gel lanes is shown in Figure 5. It is evident that the m5S-5' fragment is separable from that containing the added, synthetic precursor-specific oligonucleotides (minus RNase M5 lanes). The individual RNase M5 subunits, α and β, are

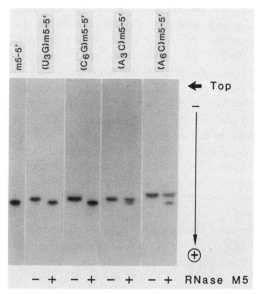

Figure 5 Release of synthetic precursor segments from artificial precursors. Uniformly ^{32}P-labeled m5S-5' fragments with nonradioactive 5'-terminal oligonucleotides as indicated were annealed to nonradioactive m5S-3' fragment and then incubated in the standard maturation assay with (+) or without (−) RNase M5. Reaction products were resolved by gel electrophoresis; autoradiographs of the gel lanes are shown. ^{32}P-labeled m5S-5' is a marker for the expected RNase M5 reaction product. (Reprinted, with permission, from Meyhack et al. 1978.)

incapable of effecting maturation (data not shown). However, upon presentation of the substrates to the RNase M5 holoenzyme (lanes plus RNase M5) each is wholly or in part reduced to the length of the 39-nucleotide, 5' half of m5S rRNA. The efficiencies of cleavage of the various substrates qualitatively follow the same pattern as the facilitation of F1 release by the corresponding synthetic 5' sequence. That is, when reannealed with the m5S-3' half, (UUUG) m5S-5' and (CCCCCCG) m5S-5' are completely cleaved by RNase M5, whereas (AAAC) m5S-5' and (AAAAAAC) m5S-5' are marginally acceptable.

Thus, the elements of rotational and translational symmetry are not required for either of the cleavage events. Only one of the 5' precursor-specific nucleotides, G21, is important in the RNase M5 interaction leading to catalytic action at both susceptible bonds. The most significant recognition features utilized by the enzyme in its search for the substrate, therefore, must lie within the mature domain of the precursor rRNA.

RNase M5 ACTION ON pre5A ABBREVIATED IN THE MATURE DOMAIN

In the course of defining conditions for the limited digestion of pre5A and m5S rRNA, to obtain fragments for the construction of the substrates discussed above, we discovered a variety of partial pre5A molecules that were stable to manipulation, even though not covalently intact. Many of these were missing portions of the m5S rRNA sequence and so offered a series of test substrates for determining regions of the m5S domain of the precursor that are required by RNase M5. The detailed analysis of such partial precursor molecules will be presented elsewhere (Meyhack and Pace 1978), but the structures and susceptibilities to RNase M5 of the most useful of the partial molecules are shown in Figure 6.

Species-2 RNA is cleaved by RNase M5, even though nucleotides 95–121 are deleted by limited RNase T1 digestion. This deleted portion of the precursor, therefore, cannot be important to the RNase-M5–pre5A recognition process. However, species-4 RNA, which is only slightly more abbreviated than species-2 RNA, is not susceptible to maturation cleavage. Species-4 RNA differs from species-2 RNA in lacking residues 92–94 (AUG) and 122–125 (UAGG); either or both sequences could be responsible for the loss in substrate capacity. To resolve the importance of these two sequences in the recognition process, we reannealed appropriate fragments isolated by denaturing species-2 and species-4 RNAs. Construct 4A/2B (Fig. 6b), which lacks the A92-U93-G94 element, proved to be an excellent substrate. However, the alternate construct (Fig. 6d, construct 2A/4B), which lacks the sequence U122-A123-G124-G125, is not susceptible to RNase M5. Clearly the region encompassed by the sequence U122-A123-G124-G125 is important to substrate recognition. The mere presence of UAGG plus the substrate region of the

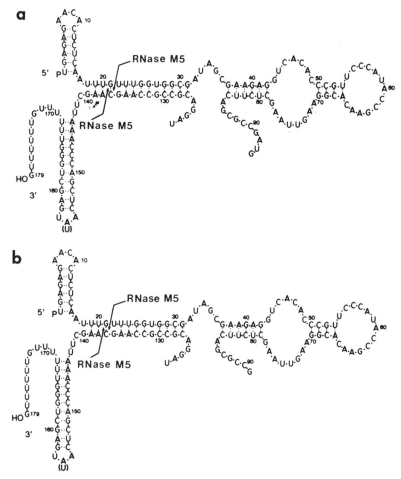

Figure 6 The structures of partially deleted pre5A rRNA test substrates. As outlined in the text and detailed elsewhere (Meyhack and Pace 1978), a collection of pre5A fragments were isolated from limited RNase T1 digests of ^{32}P-labeled pre5A rRNA. The construct molecules were generated by annealing appropriate fragments isolated following denaturation of certain of the partially deleted RNA species. (→) Bonds cleaved in susceptible substrates. (*a*) Species 2; (*b*) construct 4A/2B; (*c*) species 4; (*d*) construct 2A/4B; (*e*) species 7; (*f*) construct pre5A-5'/2B. (Reprinted, with permission, from Meyhack and Pace 1978.)

precursor is not sufficient for recognition, however. Annealing the 2B fragment to the above-discussed 5' half of pre5A rRNA, which contains the 5'-terminal 60 nucleotides of the molecule, does not yield a susceptible substrate (Fig. 6f).

The most reasonable and encompassing interpretation of these results is that RNase M5 recognizes an array of features coordinated by the

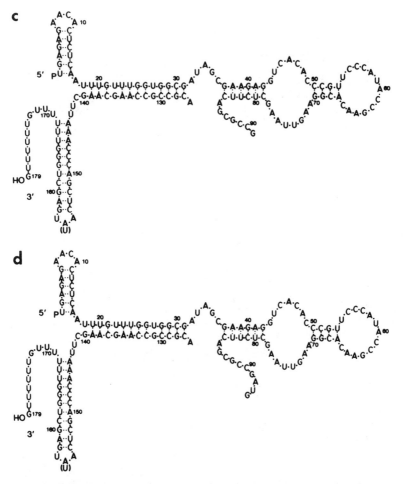

Figure 6 (Continued)

overall conformation of the mature domain of potential substrates, and that the enzyme neglects residues 95–125, which include the conspicuous prokaryotic loop. The relative contributions of the strongly ionic, phosphodiester backbone and base-specific, hydrogen-bonded donor-acceptor elements to the conformation-dependent array remain to be evaluated. However, it seems likely that both are involved. As a first approach toward exploring the base specificity of the RNase M5 reaction, we have begun to append synthetic oligonucleotides onto mature 5S rRNA from organisms other than *B. subtilis* and testing these as substrates for *B. subtilis* RNase M5. For example, we have added UUUG, UUUA, and AAAA to the 5' end of *E. coli* m5S rRNA, the nucleotide sequence of which is shown in Figure 7. The residues that differ from the m5S rRNA

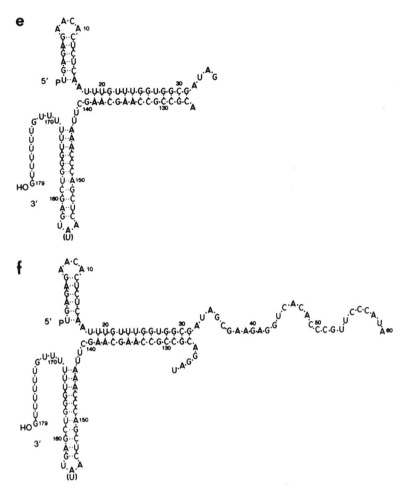

Figure 6 (*Continued*)

of *B. subtilis* are shown in bold letters. Although considerably different in nucleotide sequence, the *E. coli* and *B. subtilis* m5S rRNA molecules almost certainly are equivalent in conformation. They are functionally interchangeable when reconstituted into 50S ribosome subunits of *Bacillus stearothermophilus* (Wrede and Erdman 1973) or *E. coli* (E. Matthews, pers. comm.). In the Fox and Woese (1975) conformation, the *B. subtilis* and *E. coli* m5S rRNA molecules exhibit common sequences in unpaired regions, but almost complete sequence divergence in their duplex components (compare Fig. 7 with the mature domain of the pre5A molecule shown in Fig. 1). Yet, the substrates constructed from *E. coli* m5S rRNA are not cleaved by *B. subtilis* RNase M5. Therefore, nucleotide sequence, as well as conformation, must be important to the overall binding-

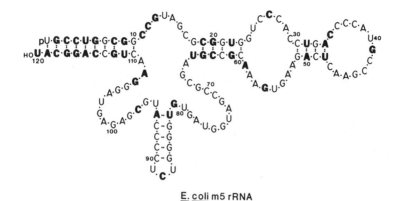

Figure 7   The structure of *E. coli* m5S rRNA. The nucleotide sequence of *E. coli* MRE600 m5S rRNA, as determined by Brownlee et al. (1968), is arranged in the Fox and Woese (1975) secondary structure. Residues at which the m5S rRNA of *E. coli* and *B. subtilis* differ are indicated by bold letters.

cleavage process. However, it is evident that we must accumulate information on the tertiary structure of the m5S rRNA molecule if we are to understand this protein-polyribonucleotide interaction.

A CONVENIENT SUBSTRATE FOR RNase M5 AND OTHER MATURATION NUCLEASES

The high degree of substrate specificity of RNase M5, and processing endonucleases in general, renders their purification and study difficult. Assays for specific cleavages are necessarily cumbersome, expensive, and poorly adaptable to detailed, quantitative analysis. These assays involve submitting labeled pre-RNA to endonuclease preparations and scoring digestion products following polyacrylamide gel electrophoresis. In addition to this technical problem, it is important to consider that substrate cleavage by RNase M5 involves at least two recognition events. The enzyme must first bind the pre-RNA substrate; it must then identify and act upon the susceptible phosphodiester bonds. Thus far, we have been able to examine only the overall cleavage reaction, that is, the capacity of a test substrate to undergo cleavage by RNase M5. To explore in detail the parameters of this RNA-protein interaction, we require careful kinetic analyses that are not readily obtainable by the rather imprecise gel technology for scoring maturation products. Consequently, we have been concerned with the development of a useful assay system for rigidly specific maturation endonucleases. The finding noted above that m5S rRNA containing a synthetic oligomer at its 5' terminal is a substrate for RNase M5 permitted the invention of a convenient assay (Meyhack et al.

1978). We first isolated nonradioactive half molecules of m5S rRNA by limited digestion with RNase T2, then used RNA to append the nonradioactive, synthetic oligomer UUUG to the 5' terminal of the isolated 5' half of m5S rRNA. Next, the 5' terminal of the construct was labeled by polynucleotide kinase and [γ-^{32}P]ATP, and this product was then annealed with the nonradioactive 3' half of m5S rRNA to yield a substrate that, upon exposure to RNase M5, releases acid-soluble, ^{32}P-labeled UUUG. As shown in Figure 8, this partially synthetic substrate is as susceptible to cleavage by RNase M5 as is the native substrate, and the

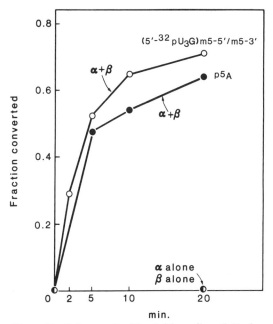

Figure 8 Release of acid-soluble radioactivity from a terminally labeled, partially synthetic RNase M5 substrate. As outlined in the text and detailed elsewhere (Meyhack et al. 1978), nonradioactive UUUG was added to the 5' half of *B. subtilis* m5S rRNA by the action of T4 RNA ligase. Following labeling at the 5' terminal with polynucleotide kinase and [γ-^{32}P]ATP, the 5'-^{32}P-labeled pUUUG-m5S-5' half was annealed with equimolar m5S-3' fragment and incubated in the standard maturation assay with RNase M5 holoenzyme or the individual RNase M5 subunits as indicated. Aliquots withdrawn at the indicated intervals were precipitated with 10% trichloroacetic acid and collected on membrane filters; acid-soluble radioactivities are expressed as a fraction of total input radioactivity. In a parallel reaction, uniformly ^{32}P-labeled pre5A rRNA was incubated with RNase M5 and the products were resolved by gel electrophoresis. Radioactive bands were excised from the dried gel and monitored for ^{32}P content. The mole fractions of total input substrate converted to m5S rRNA are plotted against incubation time. (Reprinted, with permission, from Meyhack et al. 1978.)

enzyme action requires both components α and β. We analyze elsewhere this synthetic substrate in detail and prove that the released product is 5'-^{32}P-labeled pUUUG$_{OH}$ (Meyhack et al. 1978).
In addition to providing a convenient assay for the relevant enzymes, the procedure permits the construction of substrates of known specific radioactivity for the characterization of the kinetic parameters of processing endonuclease interaction with nearly native substrates. The isolation of pre-RNA molecules from cells generally involves pulse-labeling after treatment of cultures with inhibitors of RNA maturation or, alternatively, the labeling of mutant populations under conditions not permitting growth. Only trace amounts of the RNA precursors can generally be isolated and these are usually of uncertain purity and, therefore, of unknown specific radioactivity. Consequently, it has not been possible to explore the detailed reaction kinetics and hence the reaction mechanism of any maturation endonuclease. Such studies are now feasible, employing substrates constructed from the mature RNA species as described here.

ACKNOWLEDGMENTS

This work was supported by National Institutes of Health grants GM-20147 and GM-22367. N.R.P. is the recipient of National Institutes of Health research career development award GM-00189.

REFERENCES

Brownlee, G. G., F. Sanger, and B. G. Barrell. 1968. The sequence of 5S ribosomal RNA. *J. Mol. Biol.* **34:**379.
Fox, G. E. and C. R. Woese. 1975. 5S RNA secondary structure. *Nature* **256:**505.
Meyhack, B. and N. R. Pace. 1978. Involvement of the mature domain in the *in vitro* maturation of *Bacillus subtilis* precursor 5S ribosomal RNA. *Biochemistry* **17:**5804.
Meyhack, B., B. Pace, and N. R. Pace. 1977. Involvement of precursor-specific segments in the *in vitro* maturation of *Bacillus subtilis* precursor 5S ribosomal RNA. *Biochemistry* **16:**5009.
Meyhack, B., B. Pace, O. Uhlenbeck, and N. R. Pace. 1978. Use of T$_4$ RNA ligase to construct model substrates for a ribosomal RNA maturation endonuclease. *Proc. Natl. Acad. Sci.* **75:**3045.
Monier, R., J. Feunteun, B. Forget, B. Jordan, M. Reynier, and F. Verricchio. 1970. 5S RNA and the assembly of bacterial ribosomes. *Cold Spring Harbor Symp. Quant. Biol.* **34:**139.
Pace, B. and N. R. Pace. 1971. Gene dosage for 5S ribosomal RNA in *E. coli* and *B. megaterium*. *J. Bacteriol.* **150:**142.
Pace, N. R., M. L. Pato, J. McKibbin, and C. W. Radcliffe. 1973. Precursors of 5S ribosomal RNA in *Bacillus subtilis*. *J. Mol. Biol.* **75:**619.

Robertson, H. D. and J. J. Dunn. 1975. Ribonucleic acid processing activity of *Escherichia coli* ribonuclease III. *J. Biol. Chem.* **250**:3050.

Sogin, M. L., B. Pace, and N. R. Pace. 1977. Partial purification and properties of a ribosomal RNA maturation endonuclease from *Bacillus subtilis*. *J. Biol. Chem.* **252**:1350.

Sogin, M. L., N. R. Pace, M. Rosenberg, and S. M. Weissman. 1976. The nucleotide sequence of a 5S ribosomal RNA precursor from *Bacillus subtilis*. *J. Biol. Chem.* **251**:3480.

Walker, G. C., O. C. Uhlenbeck, E. Bedows, and R. I. Gumport. 1975. T_4-induced RNA ligase joins single-stranded oligonucleotides. *Proc. Natl. Acad. Sci.* **72**:122.

Wrede, P. and V. A. Erdmann. 1973. Activities of *B. stearothermophilus* 50S ribosomes reconstituted with prokaryotic and eukaryotic 5S RNA. *FEBS Lett.* **33**:315.

Enzymatic Removal of Intervening Sequences in the Synthesis of Yeast tRNAs

Richard C. Ogden, Gayle Knapp, Craig L. Peebles,
Hyan S. Kang*, Jacques S. Beckmann†, Peter F. Johnson,
Shella A. Fuhrman,‡ and John N. Abelson
Departments of Chemistry and Biology‡
University of California, San Diego
La Jolla, California 91093

The phenomenon of noncolinearity between a gene and its mature product has been shown to be a general one in the eukaryotic world. This discovery raised the question of how the cell removes the intervening sequences in the biosynthesis of RNA. Some answers to this question are presented here. The discovery by Hopper et al. (1978) that yeast tRNA precursors accumulate in a mutant strain ($ts136$) has considerably facilitated the study of the RNA splicing reaction. This mutant, isolated by Hutchison et al. (1969), defines the $ma1$ gene of yeast. It is presumed to be defective in a step in RNA transport from nucleus to cytoplasm. At the nonpermissive temperature, the 35S rRNA precursor accumulates (Hopper et al. 1978), the appearance of mRNA in the cytoplasm is halted, poly(A)-containing RNA accumulates in the nucleus (Shiokawa and Pogo 1974), and a particular subset of tRNA precursors accumulates (Knapp et al. 1978).

The separation of those tRNA precursors that accumulate in $ts136$ has been accomplished by two-dimensional polyacrylamide gel electrophoresis. A typical two-dimensional separation is shown in Figure 1. Originally the precursor-specific spots were identified by hybridization of the RNA to a set of *Escherichia coli* recombinant plasmid clones, each of which carries one or more yeast tRNA genes (Beckmann et al. 1977). Five of the RNAs (spots indicated in Fig. 1) hybridized to clones that have been identified as containing genes for tRNA^{Tyr}, tRNA^{Phe}, tRNA$_3^{Leu}$, tRNA$_{UCG}^{Ser}$, and tRNA^{Trp}. These identifications have been subsequently confirmed by RNA sequence analysis. Four other RNAs hybridized to clones containing tRNA genes of unknown specificity. Identifications of these tRNA precursors is currently under investigation using various RNA sequencing techniques.

Present addresses: *Department of Microbiology, Seoul National University, College of Natural Sciences, Seoul, 151, Korea; †Agricultural Research Organization, The Volcani Center, Institute of Field and Garden Crops, Bet Dagan, Israel.

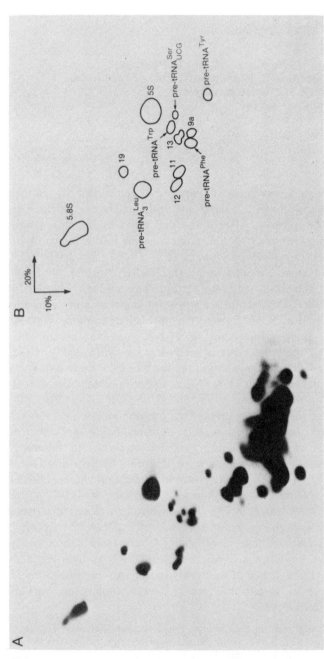

Figure 1 Two-dimensional polyacrylamide gel electrophoresis of ^{32}P-labeled RNA isolated from the yeast *ma1* mutant. ^{32}P-labeled RNA was prepared using a temperature-sensitive diploid, M304, that is homozygous for the *ma1* locus (as described in Ogden et al. [1979]). A typical preparation yields approximately 1–2 mCi of labeled RNA that is purified by electrophoresis in a 10% polyacrylamide-4 M urea gel, followed by electrophoresis in a second dimension using a 20% polyacrylamide gel. Autoradiography revealed the positions of the different pre-tRNAs. The yields of individual ^{32}P-labeled pre-tRNAs recovered by elution were about 2–25 × 10^6 cpm. (*A*) Autoradiograph of the polyacrylamide gel separations. (*B*) Diagrammatic representation and identification of the ^{32}P-labeled pre-tRNAs. (Adapted from Kang et al. 1979.)

Earlier investigations by Fradin et al. (1975) using in vivo pulse-labeling of yeast cell cultures showed an accumulation of about 27 putative tRNA precursors that could be processed to 4S-sized material using a yeast S30 extract (Blatt and Feldmann 1973). Comparison of the number of potential precursors with those accumulated in ts136 suggests that there is a complex set of precursors that can be detected during short pulse-labeling. Of this set only a specific, highly restricted subset accumulates in ts136.

An investigation of the characteristics of this subset of precursors leads to the conclusion that all nine precursors that accumulate in ts136 share common features. All nine contain precursor-specific sequences near the center of the molecules. These sequences range in size from 14 nucleotides (pre-tRNA^{Tyr}) to approximately 60 nucleotides (spot 19) and, where evidence could be obtained, they have been shown to be faithful transcripts of the intervening sequence in the gene (see below). These precursors have also already been partially matured. It is likely that all of them have the same 5' and 3' terminals as their mature tRNAs. The presence of the CCA_{OH} at the 3' terminal in particular indicates that these precursors are intermediates between the primary transcription product and mature molecules because in no case has a eukaryotic tRNA gene been shown to encode the CCA_{OH} sequence. Thus, the CCA_{OH} end must be added posttranscriptionally by tRNA nucleotidyl transferase. In addition, several of the modified bases observed in the mature tRNA are found in the precursors. The modified bases of the D and TψC loops are usually present in high partial molar yields. On the other hand, modifications of bases in the anticodon loop are generally not observed. For example, i^6A in tRNA^{Tyr}, the Y base in tRNA^{Phe}, and 2'-O-methyl modifications are not found. One exception to this is that low yields of ψ have been detected in the anticodon of pre-tRNA^{Tyr}.

tRNA PRECURSORS ARE FAITHFUL TRANSCRIPTS

The nucleotide sequences of genes containing intervening sequences have been determined for four tRNAs. Sequences of three of the eight tRNA^{Tyr} genes have been reported by Goodman et al. (1977). Each gene contained an intervening sequence of 14 bp. In addition to the three genes, the sequence of an ocher allele ($sup4$-o) of one them was determined and the expected anticodon change from GTA to TTA was observed. This provided strong proof that a functioning gene had been sequenced. In three tRNA^{Phe} genes, intervening sequences of 18 or 19 bp were found (Valenzuela et al. 1978). One tRNA^{Trp} gene and one tRNA$^{Leu}_3$ gene have been sequenced (H. S. Kang et al., in prep.). The tRNA^{Trp} gene contains an intervening sequence of 34 bp; the tRNA$^{Leu}_3$ gene has one of 32 bp.

The RNA sequences of the tRNA precursors were compared to the sequences of these genes. The original analyses for pre-tRNA^{Tyr} and pre-tRNA^{Phe} were reported previously (Knapp et al. 1978). The results of those analyses were as stated above. The intervening sequences as predicted by the sequences of the genes were present and the 5' and 3' ends were identical to those of mature tRNA^{Tyr} or tRNA^{Phe} (see also O'Farrell et al. [1978], who performed a similar analysis with pre-tRNA^{Tyr}). Subsequent to the analyses of pre-tRNA^{Tyr} and pre-tRNA^{Phe}, characterizations of pre-tRNA^{Trp} and pre-tRNA$_3^{Leu}$ have been completed (Ogden et al. 1979; H. S. Kang et al., in prep.). The conclusions are similar: The oligonucleotides predicted by the DNA sequences are found in the precursors and the only precursor-specific sequences are in the middle of the molecules; the 5' and 3' ends have been processed to mature terminals; and modifications of some bases were detected in positions as summarized above. Etcheverry et al. (1979) have determined the nucleotide sequence of the intervening sequence in pre-tRNA$_{UCG}^{Ser}$. In this precursor there are 19 nucleotides in the intervening sequence. It is important to note that this latter case is the only one in which a complete sequence of the intervening sequence in the precursor has been obtained. In all other cases the results strongly corroborate the DNA sequence but the actual sequences of some of the oligoribonucleotides and some of the overlaps have not been obtained.

Figure 2 shows the nucleotide sequences of the five identified pre-tRNAs. In addition, the similarities observed among these sequences are summarized in this figure. The sequences of pre-tRNA^{Tyr} and pre-tRNA^{Phe} are drawn in secondary structures that have maximized base pairing as determined by the calculation of the most favorable free energy using the rules derived by Tinoco et al. (1973) and Borer et al. (1974). The structures of pre-tRNA^{Tyr} and pre-tRNA$_{UCG}^{Ser}$ are consistent with studies of the nuclease sensitivity of the precursors (O'Farrell et al. 1978; Etcheverry et al. 1979; respectively). In each case, the intervening sequence is located at the 3' side of the base immediately adjacent to the anticodon. The acceptor stem and the D and TψC stems and loops appear as they are found in the tRNA cloverleaf structure. The anticodon stem is intact but it is augmented by a second helical region. This second helix always includes the anticodon that base pairs with a complementary region in the intervening sequence. This helix sometimes is separated from the anticodon stem by one or two unpaired bases that could form bulge or interior loops (Tinoco et al. 1973) in an otherwise continuously base-stacked stem region. In two of the precursors, pre-tRNA^{Trp} and pre-tRNA$_3^{Leu}$, which contain large intervening sequences, a majority of the "additional" nucleotides are located at the 3' end of the intervening sequence at the position of the interior loops. These additional nucleotides can be contained in a hairpin-loop structure that has a substantial base-paired stem. In addition to the variable interior loop another variable hairpin loop is formed at the end of the second helix. Unifying features of these structures that may be important in recognition of the precursor substrate by the RNA splicing system will be discussed below.

IN VITRO MATURATION OF PRECURSORS CONTAINING INTERVENING SEQUENCES

All nine of the precursors that accumulate at the nonpermissive temperature in ts136 are substrates for splicing, a processing reaction in which the intervening sequence is removed intact and the ends of the "half-tRNA-sized molecules" are joined to form a mature tRNA molecule. Although some progress towards the purification of the enzymes involved in RNA splicing has been made, we have been able to characterize many of the properties of the splicing system using a yeast ribosomal wash fraction (prepared as described by Peebles et al. 1979).

Figure 3 illustrates the time course of in vitro splicing for the precursor of yeast tRNA^{Phe}. Concomitantly with the disappearance of the precursor, formation of the mature-sized tRNA is observed. At early times the transient appearance of smaller RNA products is seen. These RNAs migrate in the polyacrylamide gel with the mobility of half-tRNA-sized molecules. The size and time course of appearance of these smaller RNAs suggested that they were likely to be the halves generated by excision of the intervening sequence. As evidence presented below will show, it is probable that these halves are the true intermediates in the splicing reaction. The extent of utilization of precursor with the concurrent appearance in excellent yield of mature tRNA (96% of the theoretical maximum in the case presented here) implies that nearly all of the precursor was competent for processing and that, even in the crude extract, neither random degradation nor abortive splicing pathways consume a significant fraction of the RNA precursor.

An investigation of the role of ATP in the splicing reaction is shown in Figure 4. The RNA substrate used in this experiment has been designated as "19" (see Fig. 1) and has been shown to be a precursor to an unidentified tRNA. It is the largest tRNA precursor yet identified and one of the more abundant. In the presence of 1 mM ATP (Fig. 4d) the results of incubation with the yeast ribosomal wash fraction are similar to those shown in Figure 3 for pre-tRNA^{Phe}. The precursor is rapidly converted to the tRNA product. In addition, a second band, migrating slightly faster than the mature-sized tRNA, is observed. This RNA is likely to be the excised intervening sequence and is approximately 60 nucleotides in length. In the absence of ATP (Fig. 4a) the precursor is utilized at a similar rate; however, the pattern of product formation is

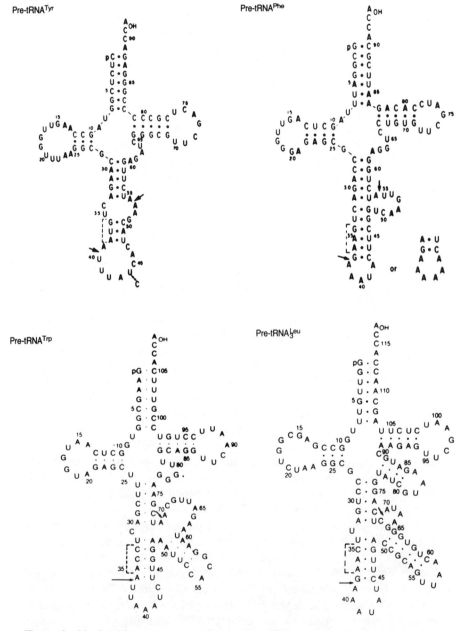

Figure 2 Nucleotide sequences of five yeast tRNA precursors. The nucleotide sequences of the precursors have all been arranged in secondary structures similar to those derived for pre-tRNA^{Tyr} and pre-tRNA^{Phe}. In the cases of pre-tRNA^{Trp} and pre-tRNA$_3^{Leu}$, additional hairpin helices in the intervening sequence contribute to the favorable free energy of the secondary structures.

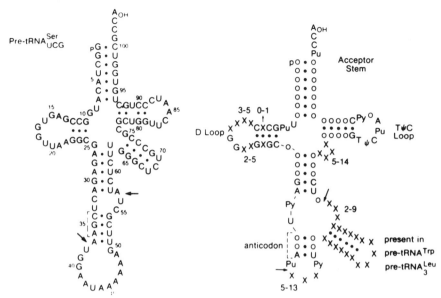

Figure 2 (continued) Splice points are indicated (→). A composite secondary structure that summarizes the constant and variable positions of nucleotides is also shown. (○) Variable nucleotides in the mature portion of the precursor; (x) variable positions in the intervening sequence; (Pu, Py) positions where a purine or pyrimidine, respectively, is conserved in the structure. Regions with variable numbers of nucleotides are indicated adjacent to the structure. (Adapted from Kang et al. 1979.)

very different. No tRNA product is formed. Instead, half-tRNA-sized products are formed along with the 60-nucleotide-long molecule. At intermediate levels of ATP a mixture of mature-sized tRNA and halves appears. The rate and extent of formation of the tRNA product increase with increasing ATP concentration, whereas the production of halves decreases proportionally. The rate and extent of utilization of the precursor is independent of the ATP concentration. This experiment suggests that the endonucleolytic stage of the in vitro splicing reaction is not dependent on ATP and can be uncoupled from the ATP-dependent ligation step.

These results also presented a method for the preparation of the half-tRNA-sized molecules that are presumed to be intermediates in the splicing reaction. The intermediates are prepared by incubating the tRNA precursors with the yeast ribosomal wash fraction in the absence of ATP. The products of these incubations are shown in Figure 5. The predominant products of the ATP-independent reactions are the half-tRNA-sized molecules and the intervening sequences. These products were used in two types of experiments. First, the ability of the half-tRNA-sized molecules to

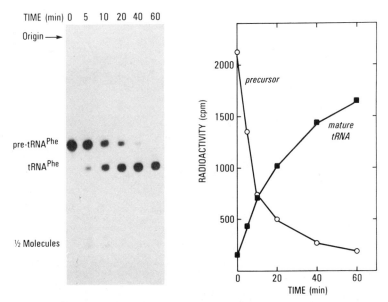

Figure 3 Time course of the in vitro splicing reaction. Aliquots (1 μl) from an in vitro splicing reaction (10 μl) containing ^{32}P-labeled pre-tRNA^{Phe} (105,000 cpm) were removed after various times of incubation at 30°C and analyzed by polyacrylamide gel electrophoresis (reaction conditions described by Peebles et al. 1979). The autoradiograph (*left*), served as a guide for slicing the gel. Fractions were counted without scintillant (40% efficiency) and cpm plotted directly. Identification of pre-tRNA^{Phe}, tRNA^{Phe}, and half-tRNA-sized molecules ($\frac{1}{2}$ molecules) has been confirmed by fingerprint analysis of these products. (Reprinted, with permission, from Peebles et al. 1979.)

be ligated by the ribosomal wash fraction was tested. Second, sequence analysis of the halves and intervening sequences was conducted to determine the structure of the products of the splicing endonuclease and to infer the structure of the substrates of the splicing ligase.

Figure 6 demonstrates that half-tRNA-sized molecules, prepared as described in Figure 5, can be ligated in the presence of ATP and ribosomal wash fraction. Both half-tRNA-sized molecules must be present for the reaction to yield tRNA products. Characterization of the products by fingerprint analysis has shown that they are indeed covalently rejoined as in the mature tRNA. Thus, it has been demonstrated that the splicing reaction can be divided operationally into two reactions: an endonucleolytic step in which the precursor is cleaved twice in an ATP-independent reaction to yield the intervening sequence and half-tRNA-sized molecules and a ligation step that requires ATP and produces the mature-sized tRNA product. However, it is important to remember that these two reactions may actually occur via a concerted mechanism when they are not experimentally uncoupled in vitro.

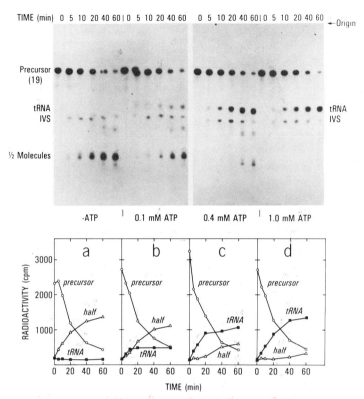

Figure 4 ATP is required for the in vitro splicing reaction. The reaction mixture (10 μl) contained pre-tRNA (19, 92,000 cpm) and the indicated concentrations of ATP (reaction conditions described by Peebles et al. [1979]). Aliquots (1 μl) were taken after various periods of incubation at 30°C, subjected to gel electrophoresis and autoradiography, and analyzed quantitatively as for Fig. 3. Identification of intervening sequence (IVS) and half-tRNA-sized molecules (½ molecules) depends on mobility and minor nucleotide analysis. (Reprinted, with permission, from Peebles et al. 1979.)

Characterization of the half-tRNA-sized molecules and intervening sequences has proved to be equally important to an understanding of the mechanism of the splicing reaction. There are several possible ways in which the excision of the intervening sequence could occur. The structure of the excised intervening sequence would reflect the way in which it had been removed. These possible structures are enumerated for pre-tRNA^{Tyr} in Figure 7. First, the reaction might proceed via a concerted mechanism so that the intervening sequence is a circle. This could happen if the splicing occurred via concerted reaction with a reciprocal exchange at the two points of cleavage. Alternatively, the product could be a unique linear molecule. There are two possible classes of linear products: those

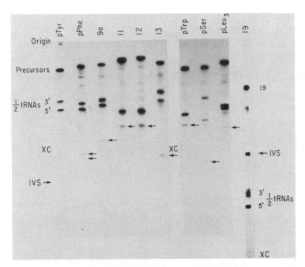

Figure 5 Preparative cleavage of tRNA precursors. Cleavage of the nine tRNA precursors was performed using conditions described by Knapp et al. (1979). The RNA species were identified by sequence analysis, mobility, and/or minor nucleotide analysis. (Adapted from Kang et al. 1979.)

with 5'-phosphate and 3'-OH terminals and those with 5'-OH and 3'-phosphate terminals. In both of these classes there are three possible sequence permutations in the case of pre-tRNA^{Tyr} because of the repeated sequence at the cleavage sites. Another possibility is that a mixture of all three linear sequences would be produced in the splicing reaction.

Digestion of the intervening sequence from pre-tRNA^{Tyr} with RNase A can distinguish among the various possibilities shown in Figure 7. In this case all of the RNase A oligonucleotides can be uniquely separated by one-dimensional electrophoresis on DEAE-cellulose paper at pH 3.5. The result of this analysis is shown in Figure 8. The product cannot be a circle because the predicted oligonucleotide, GpApApUp, is not detected. Instead, the unique linear product terminates with the sequence, GpApAp (1b in Fig. 7). No products with 5'-phosphate terminals were observed. Furthermore, the structure of the intervening sequence is the same whether it was excised in the presence or absence of ATP. Fingerprint analysis of the intervening sequences excised from pre-tRNA^{Phe}, pre-tRNA$^{Ser}_{UCG}$, pre-tRNA^{Trp}, and pre-tRNA$^{Leu}_3$ indicate that they, too, are unique linear molecules with 5'-OH and 3'-phosphate terminals.

If the cleavage reaction is a simple scission of two phosphodiester bonds, the previous data imply that the half-tRNA-sized molecules produced in the ATP-independent reaction should also have 3'-phosphate and 5'-OH terminals. Fingerprint analysis of the five precursors whose

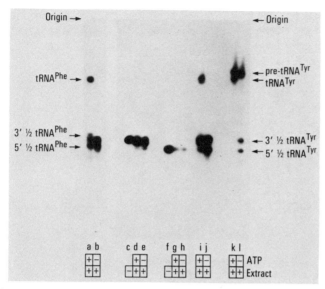

Figure 6 Half-tRNA molecules are substrates for ligation. Reaction mixtures (1.5 μl) contained either the 3' half of tRNA^{Phe} (4000 cpm) plus the 5' half of tRNA^{Phe} (4000 cpm; a and b); the 3' half of tRNA^{Tyr} (5000 cpm; c, d, and e); the 5' half of tRNA^{Tyr} (5000 cpm; f, g, and h); the 3' half of tRNA^{Tyr} (5000 cpm) plus the 5' half of tRNA^{Tyr} (5000 cpm; i and j); or pre-tRNA^{Tyr} (5700 cpm; k and l). Preincubation at 42°C and 37°C (5 min each) of the reaction mixture without added extract was to allow formation of correct secondary structures. Extract was replaced by buffer A (Peebles et al. 1979) plus 10 mM MgCl$_2$ and 0.3 M KCl in c and f; ATP was omitted from b, e, h, j, and l. Incubation was stopped after 40 min at 30°C and the samples were analyzed by polyacrylamide gel electrophoresis and autoradiography. Half-tRNA molecules were prepared from pre-tRNA molecules in a large-scale reaction similar to l and products identified by fingerprint analysis. (Reprinted, with permission, from Peebles et al. 1979.)

sequences are shown in Figure 2 has shown this to be the case. These results are summarized in Figure 9. It can be seen that in all five of the precursors the splice point is immediately adjacent to the nucleotide at the 3' side of the anticodon (the nucleotide that is often hypermodified in the mature tRNA) and cleavage leaves 5'-OH and 3'-phosphate terminals.

MECHANISM OF THE tRNA SPLICING REACTION

These results have strong implications concerning the mechanism of precursor cleavage and subsequent ligation. Several features of one mechanism that may be constructed are novel when compared with other processing enzymes. First, in all endonucleolytic RNA processing enzymes previously described, the scission of the phosphodiester chain

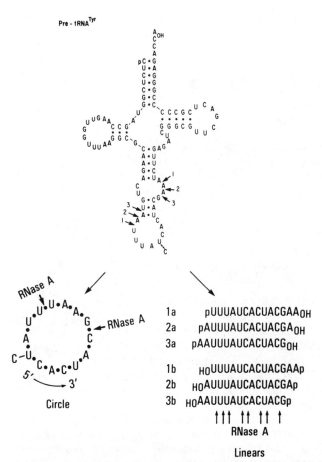

Figure 7 Possible structures of the excision product. Numbered arrows indicate the three possible excision points to which the linear sequence permutations correspond. Unnumbered arrows indicate RNase A cleavage sites. (Reprinted, with permission, from Knapp et al. 1979.)

produces 5'-phosphate and 3'-OH ends. This, of course, is not a chemical necessity; there are many endonucleases (e.g., RNase A and RNase T1) that produce 3'-phosphate terminals. However, these enzymes are generally considered to be degradative enzymes. The position of the terminal phosphate left by the splicing endonuclease is also surprising with respect to the ligation step that must then occur. To splice the ends the ligase must join a 3'-phosphate to a 5'-OH. This requirement is not a feature of the T4 RNA ligase or of the known DNA ligases that specifically join 5'-phosphate to 3'-OH terminals. We have shown that the 3'-phosphate is required for the ligase reaction (Knapp et al. 1979). 5'-Half-tRNA-sized molecules from which the 3'-phosphate has been

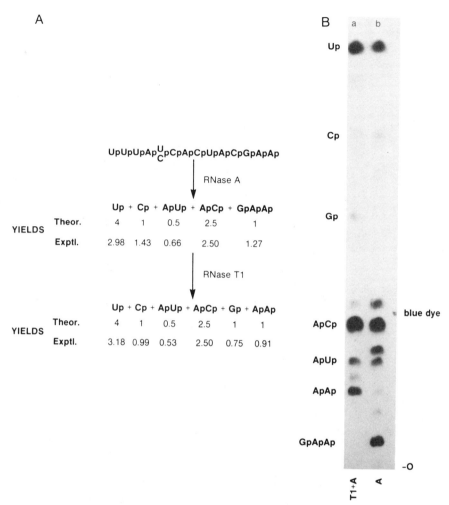

Figure 8 Analysis of the intervening sequence from pre-tRNA^{Tyr}. (A) The theoretical and experimental yields of the RNase A and RNase A plus RNase T1 digestion products. (B) Autoradiograph of the products of combined RNase A and RNase T1 digestion (a) or RNase A digestion (b) separated on DEAE-cellulose paper at pH 3.5. (Reprinted, with permission, from Knapp et al. 1979.)

selectively removed are no longer substrates in the ATP-dependent joining reaction.

It is interesting to speculate whether or not yeast pre-tRNA splicing might share common features with the splicing of mRNA precursors in other eukaryotes. It is probable that with regard to the initial recognition and excision of the intervening sequence, the pre-tRNA splicing system will be unique (see discussion below). On the other hand, the ligation step

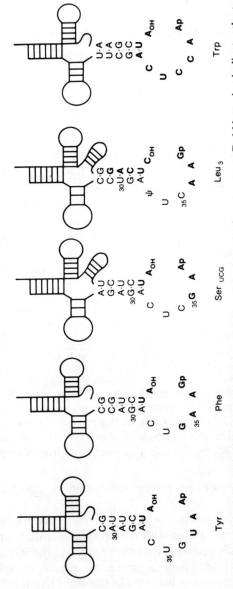

Figure 9 Half-tRNA molecules arising from excision of the intervening sequence. Bold lettering indicates the terminal oligonucleotides that were analyzed from RNase T1 or RNase A digestion of the separated halves. (Reprinted, with permission, from Knapp et al. 1979.)

of pre-tRNA splicing may be a common feature of all splicing reactions. If the 3′-phosphate requirement proves to be a feature of splicing reactions in general it may provide a useful means of identification and purification of this class of enzyme.

How does the tRNA splicing enzyme recognize the precursor and cleave it correctly to release the intervening sequence? This is perhaps a premature question, since we do not know that there is only one enzyme. However, preliminary enzyme purification has not revealed activities that distinguish among the nine precursors. Furthermore, mature tRNA selectively inhibits the ligation reaction (Peebles et al. 1979) but the inhibition is not tRNA-species-specific. For example, pure yeast tRNA^{Phe} inhibits equally the processing of pre-tRNA^{Phe} and pre-tRNA^{Tyr}. For these reasons and because RNA processing enzymes, in general, tend to have multiple roles (cf. RNase P, RNase III), we shall assume for the purposes of discussion that a single enzyme system recognizes all of the tRNA precursors shown in Figure 2. A composite structure, also shown in Figure 2, summarizes the similarities among the pre-tRNA sequences.

It is clear that the recognition cannot be sequence specific. The intervening sequences themselves are quite different among the tRNA species although they all have high A + U base compositions ([A + U]:[G + C] = [2:1]). There are two possibly significant sequence similarities among the precursors. First, the last two bases in the mature anticodon stem are the same in all five tRNAs. Second, there are similarities (see Fig. 2) in the base pairs in the stem of the D loop.

There are also some interesting structural similarities. The position of the intervening sequence, as determined by analysis of the excised intervening sequences, is the same in all five precursors—immediately adjacent to the nucleotide that usually is hypermodified in the mature tRNA. This means that the cleavage sites must occur in all precursors in exactly the same position relative to the mature portion of the molecule. It is also significant that in all five cases, the anticodon can always form base pairs with a complementary region in the intervening sequence. In some cases adjoining nucleotides are also base paired so that the anticodon helix may be extended almost to the helix of the anticodon stem. The variable regions of this structure are found in loops at either end of the anticodon helix. At one end of this helix, there is a hairpin loop varying from 5 to 13 nucleotides. At the other end, an interior loop is located between the helices of the anticodon and the anticodon stem. This loop can include a large helical stem and loop. It is not clear what a very large interior loop (e.g., pre-tRNA^{Trp} or pre-tRNA$^{Leu}_3$) would do to disrupt a continuous helix that might be formed by the anticodon helix and anticodon stem. However, in the case of the pre-tRNAs with shorter intervening sequences, the number of bases in the interior loop is small. For example, in pre-tRNA^{Tyr}, there are two unpaired nucleotides near the anticodon that are opposed by four nucleotides in the opposing strand. Thus, it appears that the helix could be fairly continuous. A preliminary model is proposed in Figure 10. From this structure it may be seen that the two cut sites, separated by a distance equivalent to 6 bp, would be situated

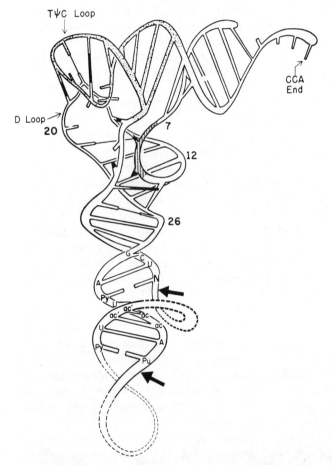

Figure 10 Proposed model for the tertiary structure of the tRNA precursors containing intervening sequences. This preliminary model has been drawn utilizing the conserved elements of secondary structure shown in Fig. 2. The tertiary structure of the tRNA has been retained except for the foreshortened anticodon stem. This stem has been extended in the form of the RNA double helix (A form). Splice points (→) are indicated. Invariant or semiinvariant positions are indicated in the anticodon stem and loop area and in the intervening sequence. The anticodon (ac-ac-A) and its complementary sequence (U-ac'-ac') in the intervening sequence are indicated. A position of variable nucleotide (N) is indicated. The numbers refer to the nucleotide positions in the mature tRNA sequence. The broken areas in the phosphate-ribose chain indicate the positions of variable size in the intervening sequence. (Adapted from Kang et al. 1979.)

conveniently on the same side of the helix. Also, the conserved region of the stem of the D loop is located on the same side of the precursor as the cuts. For the five precursors to all conform to this structure, the interior loop would have to fold out to one side of the helix or possibly be accommodated in the groove of the helix formed by the anticodon and its complement. These are, of course, speculations and knowledge of the true structure of the tRNA precursor will require further investigation using physical techniques.

CONCLUSION

The splicing of yeast tRNA precursors has been operationally separated into two distinct steps: an endonucleolytic cleavage of the pre-tRNA that produces the intervening sequence and two half-tRNA-sized molecules and a ligation step in which the halves are joined to produce the mature-sized tRNA. This separation of activities has allowed us to characterize the products and intermediates of the splicing reaction and thereby gain an insight into the mechanism of splicing. The cleavage of the precursor occurs by a simple scission of two phosphodiester bonds leaving 5'-OH and 3'-phosphate terminals on the intervening sequence and the halves. The 3'-phosphate is required for the subsequent ligation of the 5' half to its cognate 3' half. The probable participation of this 3'-phosphate may be a common feature of all eukaryotic splicing reactions.

In addition, the characterization of the products and intermediates of the splicing reaction and comparison of the known pre-tRNA sequences have shown common structural features among the precursors. It has also led to the proposal of a tertiary structure model.

ACKNOWLEDGMENT

These investigations were supported by grants from the National Institutes of Health (CA-10984, GM-05518, and GM-07199).

REFERENCES

Beckmann, J. S., P. F. Johnson, and J. Abelson. 1977. Cloning of yeast transfer RNA genes in *Escherichia coli*. *Science* **196**:205.

Blatt, B. and H. Feldmann. 1973. Characterization of precursors to tRNA in yeast. *FEBS Lett.* **37**:129.

Borer, P. N., B. Dengler, I. Tinoco, Jr., and O. C. Uhlenbeck. 1974. Stability of ribonucleic acid double-stranded helices. *J. Mol. Biol.* **86**:843.

Etcheverry, T., D. Colby, and C. Guthrie. 1979. A precursor to a minor species of yeast tRNA^{Ser} contains an intervening sequence. *Cell* **18**:11.

Fradin, A., H. Gruhl, and H. Feldmann. 1975. Mapping of yeast tRNAs by two-dimensional electrophoresis on polyacrylamide gels. *FEBS Lett.* **50**:185.

Goodman, H. M., M. V. Olson, and B. D. Hall. 1977. Nucleotide sequence of a mutant eukaryotic gene: The yeast tyrosine-inserting ochre suppressor *SUP4-O*. *Proc. Natl. Acad. Sci.* **74**:5453.

Hopper, A. K., F. Banks, and V. Evangelidis. 1978. A yeast mutant which accumulates precursor tRNAs. *Cell* **14**:211.

Hutchison, H. T., L. H. Hartwell, and C. S. McLaughlin. 1969. Temperature-sensitive yeast mutant defective in ribonucleic acid production. *J. Bacteriol.* **99**:807.

Kang, H. S., R. C. Ogden, G. Knapp, C. L. Peebles, and J. N. Abelson. 1979. Structure of yeast tRNA precursors containing intervening sequences. *ICN-UCLA Symp. Mol. Cell Biol.* **14**:69.

Knapp, G., R. C. Ogden, C. L. Peebles, and J. Abelson. 1979. Splicing of yeast tRNA precursors: Structure of the reaction intermediates. *Cell* **18**:37.

Knapp, G., J. S. Beckmann, P. F. Johnson, S. A. Fuhrman, and J. Abelson. 1978. Transcription and processing of intervening sequences in yeast tRNA genes. *Cell* **14**:221.

O'Farrell, P. Z., B. Cordell, P. Valenzuela, W. J. Rutter, and H. M. Goodman. 1978. Structure and processing of yeast precursor tRNAs containing intervening sequences. *Nature* **274**:438.

Ogden, R. C., J. S. Beckmann, H. S. Kang, J. Abelson, D. Söll, and O. Schmidt. 1979. *In vitro* transcription and processing of a yeast tRNA gene containing intervening sequences. *Cell* **17**:399.

Peebles, C. L., R. C. Ogden, G. Knapp, and J. Abelson. 1979. Splicing of yeast tRNA precursors: A two-stage reaction. *Cell* **18**:27.

Shiokawa, K. and A. O. Pogo. 1974. The role of cytoplasmic membranes in controlling the transport of nuclear messenger RNA and initiation of protein synthesis. *Proc. Natl. Acad. Sci.* **71**:2658.

Tinoco, I., Jr., P. N. Borer, B. Dengler, M. D. Levine, O. C. Uhlenbeck, D. M. Crothers, and J. Gralla. 1973. Improved estimation of secondary structure in ribonucleic acids. *Nat. New Biol.* **246**:40.

Valenzuela, P., A. Venegas, F. Weinberg, R. Bishop, and W. J. Rutter. 1978. Structure of yeast phenylalanine-tRNA genes: An intervening DNA segment within the region coding for the tRNA. *Proc. Natl. Acad. Sci.* **75**:190.

Yeast tRNA Precursors: Structure and Removal of Intervening Sequences by an Excision-Ligase Activity

Pablo Valenzuela, Patricia Z. O'Farrell, Barbara Cordell, Ted Maynard, Howard M. Goodman, and William J. Rutter
Department of Biochemistry and Biophysics
University of California, San Francisco
San Francisco, California 94143

The tRNA genes provide an attractive system for studying gene structure and expression in eukaryotes. Using recombinant DNA techniques, we have isolated and determined the structure of the yeast sequences coding for tRNA^{Tyr} (Goodman et al. 1977) and tRNA^{Phe} (Valenzuela et al. 1978). These studies resulted in the discovery that these DNAs contain additional sequences that are not used in encoding the final gene products. Immediately to the 3' side of their anticodon triplets, the yeast genes coding for tRNA^{Tyr} and tRNA^{Phe} contain an intervening DNA segment that is not present in the mature tRNAs.

There are eight unlinked genetic loci for tRNA^{Tyr} in yeast (Olson et al. 1977) and the nucleotide sequences of three have been determined (Goodman et al. 1977). All three genes contained a 14-bp intervening sequence, ATTTAYCACTACGA (Y is a pyrimidine). One of these was identified as coding for the suppressor tRNA associated with the genetic locus *sup4*. This demonstrated that the intervening sequence was present in an active gene, since the *sup4* locus for tRNA^{Tyr} containing the intervening sequence confers a dominant phenotype on *sup4-o* strains.

There are at least ten unlinked tRNA^{Phe} genes in a tetraploid yeast strain (Feldman 1977; Valenzuela et al. 1978). The nucleotide sequence of four alleles has been determined to contain an intervening 19- or 18-bp sequence, AAAAACTTCGGTCAAGTTA or AATACTTCGGTCAAGTTA (Valenzuela et al. 1978). Thus, the intervening sequences in the tRNA^{Phe} and tRNA^{Tyr} genes have no obvious sequence homology or structure, yet both occur in a similar position in the tRNAs.

The lack of colinearity between a gene and its active product has also been found in several other viral and cellular eukaryotic genes (for references, see O'Farrell et al. 1978; Knapp et al. 1978). The findings indicate that the biosynthetic pathway for some if not all tRNAs and mRNAs must involve elimination of the intervening sequence. Various possibilities for the removal of the intervening sequence include DNA splicing, RNA polymerase jumping, or cleavage and ligation of RNA precursors. We report here evidence for the latter mechanism. A precursor of yeast tRNA^{Tyr} has been isolated and shown to contain an intervening sequence identical to that found in the tRNA gene (O'Farrell et al. 1978; Knapp et al. 1978). The conformation of the pre-tRNA^{Tyr} appears to be similar to that of mature tRNA^{Tyr}, except for the anticodon loop. The loop is sensitive to endonucleolytic cleavage by nuclease S1 near the ends of the intervening sequence. This pre-tRNA is functionally inactive, as it cannot be aminoacylated and the anticodon is not accessible for hydrogen bonding (O'Farrell et al. 1978). Using crude nuclei preparations from yeast, pre-tRNA^{Tyr} and several other precursor tRNAs have been processed into tRNA^{Tyr} and 4S-sized molecules, respectively. We also report here the assay and partial purification from total yeast extracts of an excision-ligase system that specifically removes the intervening DNA sequence of several yeast tRNA precursors.

ISOLATION OF A PRECURSOR TO tRNA^{Tyr} THAT CONTAINS THE INTERVENING SEQUENCE

Following initial observations made by Hopper (Hopper et al. 1978; A. Hopper and F. Banks, pers. comm.), yeast tRNA precursors were isolated from the mutant strain ts136 (Hutchinson et al. 1969). Preparative separation and isolation of pre-tRNA^{Tyr} was achieved by polyacrylamide gel electrophoresis (Fig. 1). A comparison of the profile of the small RNAs isolated from wild-type (Fig. 1A) and ts136 (Fig. 1B) yeast strains shows an enhancement in the number of RNA species in the region of the gel between 5S RNA and tRNA in the mutant strain. Four of these species have been identified as precursors of tRNAs for tyrosine, phenylalanine, tryptophan, and serine. The tyrosine and phenylalanine precursors were identified by hybridization of the RNA eluted from the gel to filters containing cloned tRNA^{Tyr} or tRNA^{Phe} genes, respectively. The tryptophan and serine precursors were identified by direct RNA sequence analysis (Knapp et al. 1978; T. Etcheverry et al., pers. comm.).

For our own further studies on the putative tRNA precursor, we selected pre-tRNA^{Tyr} because it occurs in high yield and migrates in a relatively uncrowded region of the gel. Thus, it can be isolated in high purity by extraction from the 10% first-dimension gel (Fig. 1B, right).

pre-tRNA^{Tyr} isolated from the gel was further characterized by RNase T1 digestion (Fig. 2). The oligonucleotides were separated by electrophoresis on cellulose acetate paper at pH 3.5 in the first dimension and by homochromatography on DEAE-cellulose thin-layer chromatography plates in the second dimension (Sanger et al. 1965). The partial sequence

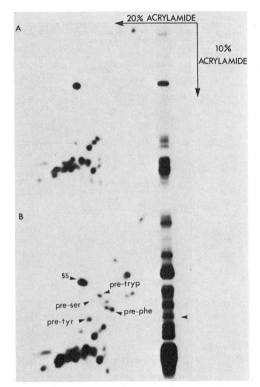

Figure 1 Two-dimensional gel electrophoresis of low-molecular-weight [^{32}P]-RNA. (*A*) Wild-type A364A strain. (*B*) *ts*136 strain. Cells were grown and labeled and gels were prepared and run as described previously (O'Farrell et al. 1978). pre-tRNA^{Tyr} and pre-tRNA^{Phe} were identified by hybridization to specific plasmid DNA (Goodman et al. 1977; Valenzuela et al. 1978). pre-tRNA^{Ser} and pre-tRNA were identified from data of T. Etcheverry et al. and J. Abelson et al. (both pers. comm.). (Reprinted, with permission, from O'Farrell et al. 1978.)

of each of the 18 T1 oligonucleotides was determined by subsequent digestion with RNase A (pancreatic RNase) and separation on DEAE paper at pH 3.5 (Brownlee 1972). These sequences were confirmed and the remaining T1 oligonucleotides identified by comparison of their positions on the fingerprint and the RNase A digestion products with the known sequences of tRNA^{Tyr} (Madison et al. 1966) and/or its gene (Goodman et al. 1977). Additional support for these assignments was obtained by the relative migration of the T1 products on cellulose acetate followed by DEAE paper in 7% formic acid in the second dimension (P. Valenzuela et al., unpubl.). The position of each T1 product in the sequence is shown in Figure 3.

Oligonucleotides 15, 16, and 17 occur in the precursor and are not

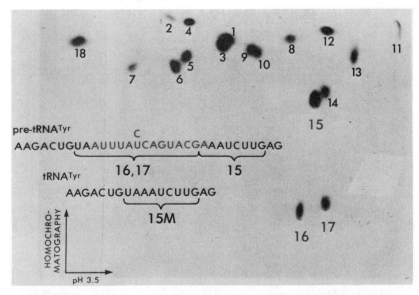

Figure 2 RNase T1 oligonucleotides from pre-tRNA^{Tyr}. The oligonucleotides were separated as described by Sanger et al. (1965) and are numbered 1–18, where 15M represents the oligonucleotide corresponding to the anticodon loop of tRNA^{Tyr}. The composition of each oligonucleotide was determined by hydrolysis with RNase A and subsequent electrophoresis at pH 3.5 in DE81 paper (Brownlee 1972). Their sequence is shown in Fig. 3.

present in the mature tRNA. They arise from the intervening segment (Fig. 3). Their sequence and position in the molecule proves that the intervening sequence in the gene is transcribed and contained in the precursor. Sequence analysis of the tyrosine genes shows that the intervening sequences differ by a single-base change (Goodman et al. 1977; and see Fig. 3). This heterogeneity is consistent with the sequence of the precursor tRNA^{Tyr}; T1 products 16 and 17 reflect the sequence change from a C to a U, respectively. The quantitation of these two oligonucleotides suggests that there are three genes with a C in the intervening sequence and five genes with a T (U in the RNA), if all eight tRNA genes have intervening sequences and all are transcribed with equal efficiency.

Although analysis of the minor bases occurring in pre-tRNA^{Tyr} is incomplete, evidence has been obtained for the following: D in T1 oligonucleotides numbers 9, 11, and 13; T in number 12; ψ in numbers 12, 15, 16, and 17; and m^{1}A in number 4. No 2'-O-methyl guanylic acid seems to occur, and hence oligonucleotide number 11 is isolated as DDGp rather than DDGmGp as predicted from the sequence of tRNA^{Tyr}. In general, pre-tRNA^{Tyr} apparently is undermethylated.

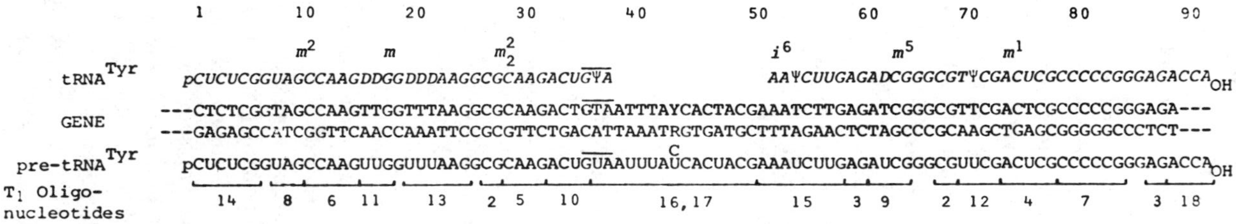

```
                 1           10              20              30              40              50              60              70              80             90
                                  m²               m                 m²₂                                     i⁶                      m⁵                    m¹
tRNA^Tyr         pCUCUCGGUAGCCAAGDGGDDDAAGGCGCAAGACUGᵁA         AAΨCUUGAGADCGGGCGTΨCGACUCGCCCCGGAGACCA_OH
GENE                              ---CTCTCGGTAGCCAAGTTGGTTTAAGGCGCAAGACTGTAATTTAYCACTACGAAATCTTGAGATCGGGCGTTCGACTCGCCCCGGAGA---
                                  ---GAGAGCCATCGGTTCAACCAAATTCCCGCGTTCTGACATTAAATRGTGATGCTTTAGAACTCTAGCCCGCAAGCTGAGCGGGGGCCTCT---
                                                                                    C
pre-tRNA^Tyr     pCUCUCGGUAGCCAAGUUGGUUUAAGGCGCAAGACUGUAAUUUAUCACUACGAAAUCUUGAGAUCGGGCGUUCGACUCGCCCCGGAGACCA_OH
T₁ Oligo-        └──14──┘ └8┘ └6┘└11┘└─13─┘ └─2─┘ └5┘ └10─┘       └──16,17──┘      └──15──┘  └3┘ └9┘  └2┘└12┘ └4┘  └─7─┘   └3┘└18┘
nucleotides
```

Figure 3 The nucleotide sequence of yeast tRNA^Tyr, the tRNA^Tyr gene, and pre-tRNA^Tyr. The anticodon and its coding triplet are overlined in both RNA and DNA sequences. The variable base pair within the intervening sequence is indicated (Y [pyrimidine], R [purine]). (Reprinted, with permission, from O'Farrell et al. 1978.)

Although pre-tRNA^{Tyr} is undermodified, both the 5' and 3' ends are already processed: The 5' end does not contain additional nucleotides and the 3'-terminal CCA$_{OH}$ has been added (Fig. 3). Presumably, the ends of the primary gene transcript are processed rapidly during synthesis of pre-tRNA^{Tyr} to remove an additional 5' sequence and to add the 3'-terminal CCA. Thus, the removal of the intervening sequence is apparently a late step in the processing pathway for tRNA, at least in the mutant *ts*136 strain.

It is not known what fraction of tRNA genes contain intervening sequences. We can detect about ten species of RNA that may be pre-tRNAs (see Fig. 1). The in vitro processing of all of these pre-tRNAs suggests that they all contain intervening sequences. Although some precursors may not be resolved from each other by the electrophoretic separation, it nevertheless seems that not all tRNAs can be synthesized from precursors that accumulate in the *ts*136 cells used in this study. The physiological reason some pre-tRNAs contain intervening sequences and others do not is not yet known. There may even be different routes for the synthesis of isoaccepting tRNAs, as it has been shown that tRNA^{Arg}, tRNA^{Asp}, and some genes for tRNA^{Ser} do not contain intervening sequences (J. Abelson; G. Page; both pers. comm.).

CONFORMATION OF YEAST pre-tRNA^{Tyr}

The susceptibility of an RNA molecule to nuclease S1 digestion has been a successful method for assessing its conformation. Nuclease S1 cleaves RNA predominantly in accessible single-strand regions. In all tRNAs studied, it hydrolyzes primarily at the anticodon loop and at the 3', single-stranded, CCA$_{OH}$ terminal producing oligonucleotides with 5'-phosphate terminals (Harada and Dahlberg 1975; Tal 1975).

We used this test on the pre-tRNA^{Tyr}. ^{32}P-labeled pre-tRNA^{Tyr} was digested with limiting levels of nuclease S1 so that less than 50% of the molecules were cleaved and purified by gel electrophoresis. The size of the two fragments produced from pre-tRNA^{Tyr} was estimated to be 40–42 nucleotides for the faster migrating fragment and 52–54 nucleotides for the more slowly migrating component (P. Z. O'Farrell and B. Cordell, unpubl.). The size of the fragments suggested that the site of cleavage of pre-tRNA^{Tyr} is within the intervening sequence.

This conclusion was substantiated by RNase T1 digestion and fingerprinting of the two fragments obtained after S1 digestion of pre-tRNA^{Tyr} (Fig. 4). The assignments of oligonucleotides were based on their position in the fingerprint (compare with Fig. 2), RNase A digestion (P. Z. O'Farrell and B. Cordell, unpubl.), and, in certain cases, minor base analysis (P. Z. O'Farrell and B. Cordell, unpubl.).

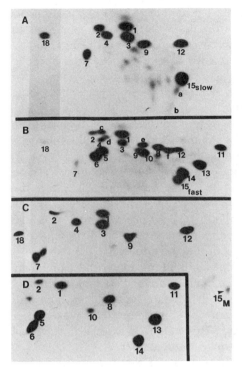

Figure 4 Analysis of pre-tRNA^{Tyr} and tRNA^{Tyr} treated with nuclease S1. The fragments were hydrolyzed with RNase T1 and the oligonucleotides were separated in two dimensions as in Fig. 2. Individual oligonucleotides were identified by digestion with RNase A and/or RNase T2 (Brownlee 1972) and are numbered in correspondence with the sequence map presented in Fig. 3 for pre-tRNA^{Tyr} and Fig. 5 for tRNA^{Tyr}. (*A*) The RNase T1 oligonucleotides present in the more slowly migrating or larger S1 fragment of pre-tRNA^{Tyr}. (*B*) The RNase T1 oligonucleotides present in the faster migrating or smaller S1 fragment of pre-tRNA^{Tyr}. New T1 oligonucleotide products in *A* and *B* are labeled a–f; a, b, e, and f are discussed in the text. Products c and d have not been analyzed in detail but are likely to be derived from S1 cleavage at the 3′ terminal in oligonucleotide 18. (15$_{slow}$) Oligonucleotide 15 derived from the more frequent cleavage by S1; (15$_{fast}$) oligonucleotide 15 derived from the minor cleavage by S1. (*C*) The RNase T1 oligonucleotides present in the larger S1 fragment of tRNA^{Tyr}. Oligonucleotide 7 migrated along the edge of the homochromatography plate and appears at two spots. (*D*) The RNase T1 oligonucleotides present in the smaller S1 fragment of tRNA^{Tyr}. (Reprinted, with permission, from O'Farrell et al. 1978.)

The fingerprint of the larger fragment (Fig. 4A) is unique in that only T1 oligonucleotides known to arise from the 3′ side of the molecule are seen (see Fig. 3 for sequences). Oligonucleotides 16 and 17, which result from the intervening sequence and occur in uncleaved pre-tRNA^{Tyr} (see

Fig. 2), are completely absent. In their place are a series of weaker spots in the area of spot 15. These arise from the 3' portion of oligonucleotides 16 and 17. All these results are consistent with major nuclease S1 cleavage sites in the intervening sequence 40 and 41 nucleotides from the 5' end of pre-tRNA^{Tyr}. These cleavages then produce a 3' fragment 51 or 52 nucleotides long (Fig. 4A) and a 5' fragment 40 or 41 nucleotides long. (pre-tRNA^{Tyr} is 92 nucleotides long.)

The fingerprint of the smaller fragment is consistent with its being from the 5' end of the molecule (Fig. 4B). All the T1 oligonucleotides (numbers 1, 2, 5, 6, 8, 10, 11, 13, and 14) from the 5' side are present in close to molar yield. Two new intense spots, e and f (Fig. 4B), also appear. These have been quantitated and analyzed by RNase A digestion. They have the sequences UAAUU$_{OH}$ (spot e; Up [1.16M], AAUp [1.0M]) and UAA-UUU$_{OH}$ (spot f; Up [2.04M], AAUp [1.0M]). These must arise by S1 cleavage of oligonucleotides 16 and 17 at the major sites at positions 40 (produces e) and 41 (produces f) from the 5' end of the precursor.

Also visible in Figure 4B is a minor set of T1 oligonucleotides (numbers 1, 2, 3, 4, 7, 9, 12, 15, and 18), which occur in about 20% molar yield and arise from the 3' side of the precursor. These data are only consistent with the occurrence of an additional minor (20%) nuclease S1 cleavage site at positions 50 or 51 from the 5' end. This minor cleavage produces a 3' fragment identical in size to the major 5' product.

pre-tRNA^{Tyr} was processed into mature tRNA^{Tyr} (e.g., Fig. 4B). After limited nuclease S1 cleavage of this RNA, the two fragments (35 and 43 nucleotides long) were isolated and digested with RNase T1 and fingerprinted (Fig. 4, C and D). Analysis of the fragments, as described above, shows that the larger fragments arise from the 3' side of the molecule and the smaller fragment from the 5' portion of the molecule. As the yield of oligonucleotides 10 and 15M (see Fig. 2) are reduced, nuclease S1 cleavage must occur at the anticodon.

These data are consistent with the secondary structure models of pre-tRNA^{Tyr} and tRNA^{Tyr} shown in Figure 5. Cleavage occurs in accessible single-strand regions. However, the anticodon is protected in a hydrogen-bonded structure in the precursor, but it is completely single-stranded in the mature tRNA. Previously, we had proposed that the anticodon triplet may be involved in base pairing with the intervening sequence in pre-tRNA^{Phe} (Valenzuela et al. 1978) and the present results on pre-tRNA^{Tyr} conform to this prediction.

pre-tRNA^{Tyr} CANNOT BE AMINOACYLATED IN VITRO

[^{32}P]RNA labeled to a low specific activity was prepared and pre-tRNA^{Tyr} was isolated in μg amounts after electrophoresis in a 10% gel. The purity of pre-tRNA^{Tyr} was assessed by fingerprinting (Fig. 2).

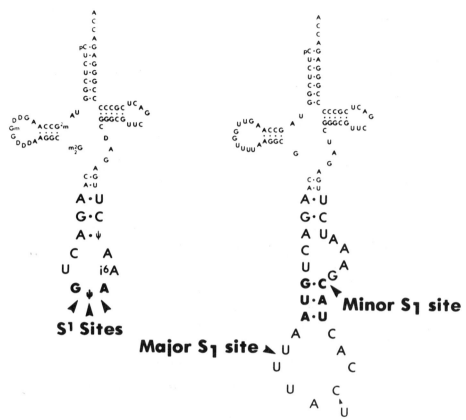

Figure 5 Secondary structure of tRNA^{Tyr} and pre-tRNA^{Tyr}. tRNA^{Tyr} is shown as previously described (Madison et al. 1966) and pre-tRNA is shown as predicted from the S1 cleavage data. (▶) Area of the major and minor sites of S1 cleavage.

The reaction conditions for aminoacylation of tRNA were optimized using highly purified tRNA^{Tyr} as substrate. Under these conditions, aminoacylation was complete. One mole of [^{3}H]tyrosine was added per mole of mature tRNA^{Tyr}; however, using identical conditions, pre-tRNA^{Tyr} was not aminoacylated (Table 1). The inability of pre-tRNA^{Tyr} to be aminoacylated was not due to the presence of an inhibitor because unpurified mature tRNA^{Tyr} isolated from the same gel as the precursor was a good substrate even in the presence of pre-tRNA^{Tyr} (Table 1).

The lack of biological activity could be due to a requirement for some base modification not yet completed in the isolated pre-tRNA^{Tyr}. Alternatively, it could reflect a requirement for a particular structure at the anticodon region of the molecule. The latter would provide a rationale for the existence, position, and sequence specificity of the intervening segments in tRNAs. According to this hypothesis, the intervening sequences must contain part or all of the coding sequence in position to base pair

Table 1 Lack of aminoacylation of pre-tRNA^Tyr

tRNA	[³H]Tyrosine (cpm incorporated)	
	experiment 1	experiment 2
No tRNA	2,300	1,714
tRNA^Tyr a	58,000	77,203
tRNA^Tyr b	11,299	10,349
pre-tRNA^Tyr	1,426	971
pre-tRNA^Phe c	1,601	1,083
tRNA^Tyr a + tRNA^Tyr b	68,768	50,487
tRNA^Tyr a + pre-tRNA^Tyr b	71,952	52,409
tRNA^Tyr a + pre-tRNA^Phe c	64,167	—
tRNA^Tyr b + pre-tRNA^Tyr	8,695	8,963

Aminoacyl-tRNA synthetases were prepared as described previously (O'Farrell et al. 1978). Aminoacylation of tRNAs was carried out in 10-μl reactions containing 0.04 M Tris-HCl (pH 7.9), 0.02 M $MgCl_2$, 0.04 M KCl, 5 mM ATP, 10 mM [³H]tyrosine (specific activity of 42 Ci/mmole) and 100 ng (or 200 ng) tRNA for reactions containing one (or two) species of tRNA (or pre-tRNA). Using these conditions, aminoacylation is not limited by substrate. After incubation at 37°C for 10 min, aliquots were removed, spotted on Whatman No. 3MM paper, precipitated with cold 10% trichloroacetic acid, washed, and counted.
ªPure tRNA^Tyr.
ᵇUnpurified tRNA^Tyr isolated from same 10% acrylamide gel as pre-tRNA^Tyr.
ᶜpre-tRNA^Phe isolated from same 10% acrylamide gel as pre-tRNA^Tyr and tRNA^Tyr.

with the anticodon triplet or with another sequence that base pairs in the vicinity of the anticodon so that this triplet is not available for its biological function. The nucleotides at the end of the intervening sequence are AT-rich. This favors the formation of loop structures that permit the protection of the anticodon and also provide a tertiary structure specific for the excision-ligase action.

PROCESSING OF YEAST PRECURSOR tRNAs BY AN ENZYMATIC ACTIVITY PRESENT IN YEAST CELLS

Previous experiments on in vitro transcription using yeast nuclei and endogenous or exogenous RNA polymerase III showed accumulation of tRNA-sized molecules (4S) (Tekamp et al. 1978). These results suggested that presumptive tRNA precursors of higher molecular weight were processed in this system. Accordingly, we tested these nuclear preparations as a source of tRNA processing activity.

Eight different pre-tRNAs were converted to 4S molecules by extracts of nuclei prepared from wild-type yeast. Figure 6 shows this activity using five different pre-tRNAs as substrates. The preparations of nuclei appear to be reasonably free of nonspecific nucleases, since no significant degradation of 5S RNA occurs during the incubation period (Fig. 6, lanes 2 and 3).

The optimal conditions for the cleavage-ligase reaction have been determined. Of particular interest is a requirement for ATP or GTP at approximately physiological concentrations (0.8 mM). As with other ATP-requiring reactions, Mg^{++} is also required (optimum is 8 mM). In addition, NaCl (160 mM; KCl, LiCl, NH_4Cl are equivalent), mercaptoethanol (10 mM), and slightly alkaline pH (50 mM Tris-HCl [pH 8.0]) are required for optimal activity. The reaction is inhibited by relatively high levels of yeast tRNA. Whether this inhibition is a general ionic strength effect or a specific inhibition by the mature tRNA product is not yet known.

The excision-ligase activity resembles other enzymatic reactions that make and break nucleotide bonds. The stimulation of the reaction by ATP suggests a requirement of phosphate-bond energy, as in the reactions of DNA ligase, RNA ligase, and DNA gyrase (Gellert et al. 1976; Sugino et al. 1977; Marians et al. 1977).

To be biologically effective, the specificity of the excision and ligation must be stringent. The fidelity of the in vitro reaction was characterized using pre-tRNA^{Tyr} as a defined substrate. ^{32}P-labeled pre-tRNA^{Tyr} was

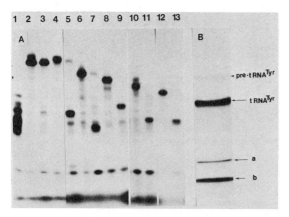

Figure 6 Processing of yeast tRNA precursors to molecules of 4S size by an enzyme system from yeast nuclei. (*A*) Assay of processing activity using several pre-tRNAs and 5S RNA as substrates: total yeast 4S RNA (*1*); 5S RNA (*2*); pre-tRNA^{Leu} (*4*) (Knapp et al. 1978; J. Abelson, pers. comm.); unidentified, putative pre-tRNA (*6*); unidentified, putative pre-tRNA (*8*); pre-tRNA^{Phe} (*10*); pre-tRNA^{Tyr} (*12*). Lanes *3, 5, 7, 9, 11,* and *13* correspond to the same RNAs in lanes *2, 4, 6, 8, 10,* and *12,* respectively, but after incubation with the nuclei system described above. (*B*) Preparative processing of pre-tRNA^{Tyr}. (Reprinted, with permission, from O'Farrell et al. 1978.)

treated with yeast nuclei and the resulting products were analyzed by electrophoresis (Fig. 6B). RNase T1 digestion and fingerprinting of the 4S RNA shows that it is authentic tRNA^{Tyr} (Fig. 7). All the T1 oligonucleotides of tRNA^{Tyr} are present. T1 oligonucleotides 16 and 17, which contain the intervening sequence of the precursor, are absent (compare Figs. 2 and 7). Oligonucleotide 15 is also absent and is replaced by 15M. The sequence of this T1 product is ψAAAψCUUG. (Preliminary analysis suggests there is also a modified A in this oligonucleotide.) This oligonucleotide proves the precise excision of the intervening sequence and religation to form mature tRNA^{Tyr}.

PARTIAL PURIFICATION OF THE YEAST EXCISION-LIGASE SYSTEM

The tRNA excision-ligase can be obtained in soluble form following centrifugation of yeast cell extracts at 100,000g in 0.5 M KCl or 0.15 M KCl in 0.02 M Tris-HCl (pH 8.0), 0.5 mM EDTA, 10% glycerol, 1 mM phenylmethylsulfonyl fluoride, 0.01 M 2-mercaptoethanol (Buffer A; Valenzuela et al. 1976). This crude enzyme fraction is stable for several weeks at $-70°C$, but is remarkably labile to certain standard protein purification procedures. For example, activity is frequently lost on precipitation with ammonium sulfate or on sedimentation in sucrose density gradients. Recombination of different fractions does not usually recon-

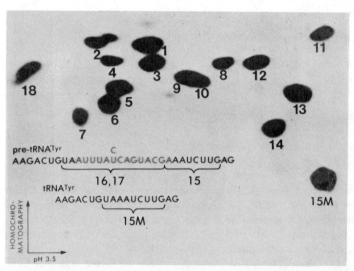

Figure 7 RNase T1 oligonucleotides from in-vitro-matured tRNA^{Tyr}. Oligonucleotides identical to those of Fig. 2 are labeled 1–14. (15M) New oligonucleotide spot corresponding to the anticodon loop of mature tRNA^{Tyr}.

stitute activity. (Occasionally, however, partial recovery of activity has been observed.) These results suggest that the cleavage-ligase activity is found in a complex that is, in part, stabilized by ionic bonds and hence dissociates under the high-ionic-strength conditions employed in ammonium-sulfate precipitation. The lack of our ability to reconstitute activity suggests that one or more of the dissociated components is labile or ineffective at high dilution. Whether the components of the complex include simple proteins, nucleic acids, or small-molecular-weight cofactors is conjectural.

In spite of the difficulties encountered with fractionation of the excision-ligase activity, we have achieved a 20-fold purification with good recovery of activity (about 50%) using the following three steps.

The first step requires Polymin P treatment. The 100,000g extract is adjusted to 0.15 M ammonium sulfate and precipitated with 0.07 volumes of a 5% Polymin P solution (pH 8.0). After centrifugation, the pellet is washed with 0.1 M ammonium sulfate in Buffer A and extracted with 0.3 M ammonium sulfate in Buffer A. The extract is then diluted to 0.2 M ammonium sulfate and filtered through a small column of phosphocellulose equilibrated with 0.2 M ammonium sulfate in Buffer A. This last step is necessary to completely remove the highly inhibitory Polymin P.

The second step employs DEAE-Sephadex chromatography. The enzyme fraction from Polymin P is loaded on a DEAE-Sephadex column equilibrated with 0.15 M ammonium sulfate in Buffer A and eluted with 0.3 M ammonium sulfate in the same buffer (Fig. 8). The excision-ligase activity elutes slightly behind the main peak of protein and before the main peak of nucleic acids.

The third step involves the processes of Biogel P-200 filtration and estimation of the molecular weight. Although filtration through Biogel P-200 does not increase the specific activity of the enzyme, it greatly increases its stability during storage at $-70°C$. The active fractions from DEAE-Sephadex were passed through a Biogel P-200 column equilibrated with 0.15 M ammonium sulfate in Buffer A. The results are shown in Figure 9. The fraction containing excision-ligase activity is able to process all yeast pre-tRNAs tested. The proteins in this fraction have an apparent m.w. of 75,000, assuming they are globular molecules.

THE CELLULAR LOCALIZATION
OF THE EXCISION-LIGASE ACTIVITY IS UNKNOWN

It seems reasonable that the excision-ligase function should be carried out close to the site of transcription and perhaps prior to the transport of the processed tRNA to the cytoplasm. Indeed, we first detected the activity in isolated nuclei; however, we also find it in cellular extracts, and Knapp and coworkers (1978) extracted the activity by salt washes of ribosomes.

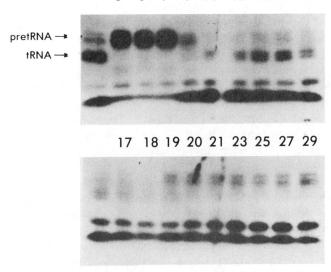

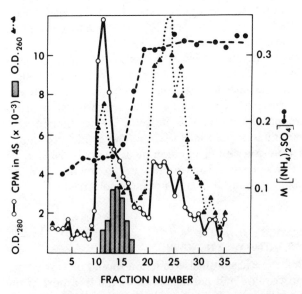

Figure 8 DEAE-Sephadex column chromatography of the yeast pre-tRNA excision-ligase activity. (*Top*) Assay of excision-ligase activity of different fractions by gel electrophoresis. (*Bottom*) Column elution profile.

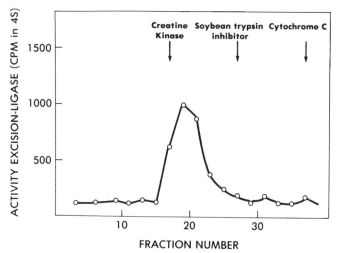

Figure 9 Filtration of yeast pre-tRNA excision-ligase activity on Biogel P-200.

Thus the enzyme is dispersed throughout various isolated cellular components. Whether this is due to adventitious absorption during the homogenization or reflects the specific cellular localization is unknown at present.

DISCUSSION

The availability of the mutant strain ts136 has substantially aided our studies on the tRNA precursors. The accumulation of pre-tRNAs in this strain might be due to a defect in the excision-ligase system itself. This appears unlikely, since the temperature sensitivity of the excision-ligase activity in extracts of both wild-type and ts136 yeast is similar (both are inactivated at temperatures higher than 30°C). Furthermore, it has been shown that other RNAs, for example, pre-rRNA and mRNA, accumulate in the nucleus of ts136 (Shiokawa and Pogo 1974; Hopper et al. 1978). Since the excision-ligase is unlikely to be the processing agent for all these RNA species, it seems that some other component must be involved in the ts136 mutation (e.g., a protein involved in the transport from the nucleus to the cytoplasm [Shiokawa and Pogo 1974]).

The substrate specificity for the excision ligase is yet unknown. Because there is considerable divergence in the intervening sequences, it is possible that separate cleavage ligases are involved. This seems unlikely to us for the following reasons:

1. A plethora of specific excision-ligase enzymes would be required, depending on the number of RNA precursors containing intervening sequences. Clearly in the limiting case, in which the genes for the

excision-ligase activity contained intervening sequences, separate cleavage-ligase enzymes for each RNA would not be allowed.
2. There is no evidence for resolution of different excision-ligase enzymes by the limited fractionation procedures thus far employed.
3. Yeast pre-tRNA^{Tyr} appears to be processed faithfully in heterologous hosts, such as in *Xenopus* (J. Gurdon, pers. comm.). It is unlikely that the sequences at junctions are strongly conserved in two species that are widely separated evolutionarily.
4. Because sequences at both junctions are different, if the recognition of a specific sequence were determining the reaction, two cleavage enzyme activities might be required for a single excision-ligase cycle. The available data, though limited, suggest that the process occurs in a concerted fashion (no intermediates accumulate). This is more consistent with a process that senses some other aspect of structure, for example, conformation.

It is already established that processing of phage T4 tRNA precursors proceeds precisely, but by a mechanism not involving sequence specificity (Smith 1976; and references therein). Thus, a precedent for processing by another aspect of structure is established. Our preliminary studies on the conformation of the pre-tRNA^{Tyr} suggest that substantial tertiary structural homology exists between this molecule and the mature tRNA^{Tyr}. However, there is a significant modification in the region of the anticodon where the intervening sequence is located. It is the altered conformation in this region that may be "sensed" by the excision-ligase system. Since the intervening sequence is located at the same site in all tRNA precursors thus far examined, the altered conformation in the anticodon region may be similar in all these precursors. Thus, one excision-ligase activity may suffice.

Because of redundancy of the bases at the junction, the precise site of cleavage ligation is uncertain in both the pre-tRNA^{Tyr} and the pre-tRNA^{Phe}, as it is in all other intervening sequences thus far examined (Breathnach et al. 1978). Thus, the cleavage ligation might occur at staggered sites, provided an intervening sequence of constant length is removed. This implies that either the length of the intervening sequence or some feature of the ends being ligated is "sensed" by the enzyme. In the case of tRNA precursors, the most reasonable feature of detection is the conformation of the molecule (Quigley and Rich 1976).

ACKNOWLEDGMENTS

We thank John Abelson (University of California, San Diego) for communication of results before publication. This research was supported by funds from the National Institutes of Health to H.M.G., W.J.R., and P.V.

REFERENCES

Breathnach, R., C. Benoit, K. O'Hare, F. Gannon, and P. Chambon. 1978. Ovalbumin gene: Evidence for a leader sequence in mRNA and DNA sequences at the exon-intron boundaries. *Proc. Natl. Acad. Sci.* **75:**4853.

Brownlee, G. G. 1972. A two-dimensional ionophoretic fractionation method for labeled oligonucleotides. In *Laboratory techniques in biochemistry and molecular biology* (eds. T. S. Work and E. Work), vol. 3, p. 67. Elsevier, New York.

Feldman, H. 1977. A comparison of transcriptional linkage of tRNA cistrons in yeast and *E. coli* by the ultraviolet light technique. *Nucleic Acids Res.* **4:**2831.

Gellert, M., K. Mizuichi, M. H. O'Dea, and H. Nash. 1976. DNA gyrase: An enzyme that introduces superhelical turns into DNA. *Proc. Natl. Acad. Sci.* **73:**3872.

Goodman, H. M., M. V. Olson, and B. D. Hall. 1977. Nucleotide sequence of a mutant eukaryotic gene: The yeast tyrosine-inserting ochre suppressor SUP 4-O. *Proc. Natl. Acad. Sci.* **74:**5453.

Harada, F. and J. E. Dahlberg. 1975. Specific cleavage of tRNA by nuclease S1. *Nucleic Acids Res.* **2:**865.

Hopper, A. K., F. Banks, and V. Evangelidis. 1978. A yeast mutant which accumulates precursor tRNAs. *Cell* **14:**211.

Hutchinson, H. T., L. H. Hartwell, and C. S. McLaughlin. 1969. Temperature-sensitive yeast mutant defective in ribonucleic acid production. *J. Bacteriol.* **99:**807.

Knapp, G., J. S. Beckman, P. F. Johnson, S. A. Fuhrman, and J. Abelson. 1978. Transcription and processing of intervening sequences in yeast tRNA genes. *Cell* **14:**221.

Madison, J. T., G. A. Everett, and H. Kung. 1966. Nucleotide sequence of a yeast tyrosine transfer RNA. *Science* **153:**531.

Marians, K. J., J. Ikeda, S. Schlagman, and J. Hurwitz. 1977. Role of DNA gyrase in ϕX replicative-form replication in vitro. *Proc. Natl. Acad. Sci.* **74:**1965.

O'Farrell, P. Z., B. Cordell, P. Valenzuela, W. J. Rutter, and H. M. Goodman. 1978. Structure and processing of yeast precursor tRNA containing intervening sequences. *Nature* **274:**438.

Olson, M. V., D. L. Montgomery, A. K. Hopper, G. S. Page, F. Horodystii, and B. D. Hall. 1977. Molecular characterization of the tyrosine tRNA genes of yeast. *Nature* **267:**639.

Quigley, G. J. and A. Rich. 1976. Structural domains of transfer RNA molecules. *Science* **194:**796.

Sanger, F., G. G. Brownlee, and B. G. Barrell. 1965. A two-dimensional fractionation procedure for radioactive nucleotides. *J. Mol. Biol.* **13:**373.

Shiokawa, K. and A. O. Pogo. 1974. The role of cytoplasmic membranes in controlling the transport of nuclear messenger RNA and initiation of protein synthesis. *Proc. Natl. Acad. Sci.* **71:**2658.

Smith, J. D. 1976. Transcription and processing of transfer RNA precursors. *Prog. Nucleic Acid Res. Mol. Biol.* **16:**25.

Sugino, A., H. M. Goodman, H. L. Heyneker, J. Shine, H. W. Boyer, and N. Cozzarelli. 1977. Interaction of bacteriophage T4 RNA and DNA ligases in joining of duplex DNA at base-paired ends. *J. Biol. Chem.* **252:**3987.

Tal, J. 1975. The cleavage of transfer RNA by a single strand specific endonuclease from *Neurospora crassa*. *Nucleic Acids Res.* **2:**1073.

Tekamp, P., P. Valenzuela, T. Maynard, G. Bell, and W. J. Rutter. 1978. Specific gene transcription in yeast nuclei and chromatin by added homologous RNA polymerases I and III. *J. Biol. Chem.* **254:**955.

Valenzuela, P., F. Weinberg, G. Bell, and W. J. Rutter. 1976. Yeast DNA-dependent RNA polymerase I. A rapid procedure for the large scale purification of homogeneous enzyme. *J. Biol. Chem.* **251:**1464.

Valenzuela, P., A. Venegas, F. Weinberg, R. Bishop, and W. J. Rutter. 1978. Structure of yeast phenylalanine tRNA genes: An intervening DNA segment within the region coding for the tRNA. *Proc. Natl. Acad. Sci.* **75:**190.

Gene Arrangement

The Organization of tRNA Genes

John Abelson
Department of Chemistry
University of California, San Diego
La Jolla, California 92093

Discussion of the organization of tRNA genes may be divided into two general topics. First, it is of interest to determine the individual primary transcription units for a number of tRNA genes in a variety of organisms. tRNA is an ideal transcription product to assay because it is stable in crude extracts. In addition, a good deal of information has already been acquired from studying tRNA transcription. As the following ten chapters show, much more information becomes available with continued investigation of tRNA transcription.

As discussed by Mazzara and McClain (this volume), all tRNA molecules are matured from the de novo transcript in a series of RNA processing reactions. The primary transcription unit for a particular tRNA must be known before it is possible to define all of the processing reactions that are involved in producing the mature tRNA. This information is available only for several tRNAs in *Escherichia coli*. In particular, the organization of the $tRNA_1^{Tyr}$ transcription unit has been studied in detail. Rossi et al. (this volume) review that work. Not only is the transcription unit known for this gene, but through the work of Khorana and his collaborators (Ryan et al., this volume) the gene has been chemically synthesized and shown to function in vivo. This work has raised a puzzling and very interesting question. Cells harboring a plasmid carrying the synthetic $tRNA_1^{Tyr}$ gene produce enhanced levels of $tRNA^{Tyr}$ (as expected, since there are a larger number of gene copies). In contrast, cells harboring a plasmid containing the natural gene do not show enhanced levels of expression. The most obvious difference between the two cloned genes is that the synthetic gene only contains 25 bp following the 3' end of the tRNA, whereas the natural gene contains a 178-bp repeating sequence that begins with the last 19 bp of the tRNA and extends a distance of 560 bp. It is a possibility that some feature of this sequence is involved in regulation of the level of $tRNA^{Tyr}$ in the cell.

Beyond the organization of individual transcription units, it is of interest to know what is the general arrangement of tRNA genes. How many tRNA genes are there? How do they map in the chromosome? Are tRNA genes clustered and, if so, are they contained in multifunctional transcription units? There are partial answers to these questions for several organisms and much of the available information is included in this volume.

tRNA GENES IN E. COLI

Information on the arrangement of tRNA genes in *E. coli* has come from several sources. The first information came through the genetic mapping of five nonsense and four missense suppressor genes. In general, this information led to the early impression that the tRNA genes in *E. coli* are distributed around the chromosome. Many of these suppressors have been isolated on transducing phages; through characterization of the genes in these phages it was discovered that other tRNA genes are linked to the suppressor gene. For example, Rossi et al. (this volume) describe the organization of the tRNA$_2^{Tyr}$ gene cluster, which also contains genes for two tRNAs^{Thr} and a tRNA^{Gly}. All of the tRNA gene locations obtained in this way are summarized in Appendix II (this volume).

The existence of transcription units containing more than one tRNA is indicated from studies of tRNA precursors. In mutants of *E. coli*, temperature sensitive for RNase P, a wide variety of precursors accumulate. Some of these are monomeric precursors containing extra nucleotides but only a single tRNA sequence, as in the case of pre-tRNA$_1^{Tyr}$. Others contain multiple tRNA sequences. A comprehensive survey of these precursors is presented by Shimura et al. (this volume). Some of the precursors contain different tRNAs; for example, there is a precursor that contains noninitiator tRNA^{Met}, tRNA$_1^{Gln}$, and an as yet unidentified tRNA. There are also precursors that contain multiple copies of the same tRNA. For example, tRNA$_1^{Leu}$ is found in a precursor that may contain as many as five copies of the tRNA sequence.

It has also been discovered that tRNA genes are contained in the rRNA transcription units. The tRNA genes are located in the spacer region between the 16S and 23S genes. In several cases the genes are located at the distal end of the operon. A comprehensive discussion of the gene organization of the spacer tRNAs and their synthesis can be found in several articles in this volume (Morgan et al.; Lund et al.; Young et al.).

It has been estimated that there are about 60 tRNA genes in *E. coli* (Brenner et al. 1970). This is about twice as many as would be required to translate the 61 codons if the wobble rules (Crick 1967) generally pertain. It is likely that all of these will eventually be mapped through application of the recombinant DNA technology. In the future, the study of the molecular biology of *E. coli* will move from the detailed characterization of the control of individual genes to the study of the overall control of the metabolism of the cell. An understanding of the control of tRNA biosynthesis will be important to this study. Why are there multiple tRNA genes for a particular tRNA? What is the function of tRNAs made in small amounts? Why are some tRNAs synthesized in ribosomal transcription units? It may be that answers to these questions will only be apparent in the context of an understanding of the control of cellular metabolism.

tRNA GENES IN BACTERIOPHAGE T4

Bacteriophage T4 carries the genes for eight tRNAs and two stable RNAs of unknown function (Hsu et al. 1967; Daniel et al. 1970; Abelson et al. 1975; Guthrie et al. 1975). Early genetic results showed that all of the tRNA genes are clustered between gene *e* (lysozyme) and gene *57* on the T4 genome. More recently a restriction endonuclease map of the region was constructed (Fig. 1); this clearly reveals two gene clusters separated by a distance of about 600 bp (Fukada 1979).

Six of the tRNAs are synthesized via dimeric precursors. These are the tRNA^{Gln}-tRNA^{Leu}, tRNA^{Pro}-tRNA^{Ser}, and tRNA^{Thr}-tRNA^{Ile} precursors, which have all been sequenced (Guthrie et al. 1975; McClain et al. 1972). The maturation of these precursors has been studied in detail so that a good deal is known about the terminal steps in the maturation of the tRNAs (see Mazzara and McClain, this volume). The other tRNAs are synthesized via monomeric precursors. None of the precursors, however, are primary transcription products and there must be an earlier set of processing reactions that (for reasons discussed below) release these five precursors from a larger RNA molecule.

The bacteriophage T4 tRNAs are synthesized early (Scherberg et al. 1970) and continuously through the lytic cycle (McClain et al. 1972). The early evidence for the clustering of the tRNA genes had suggested to us that these RNAs might be synthesized in a single transcription unit (Wilson et al. 1972). Evidence supporting this notion was obtained by in vitro transcription experiments (Kaplan and Nierlich 1975; Goldfarb et al. 1978). More recently we have obtained further evidence that the tRNA genes are included in a single transcription unit (Fukada 1979). The DNA fragment containing cluster 1 was cloned in a λgt vector. This fragment contains a region that appears to be deleterious to *E. coli*. Deletions of the deleterious region are easily isolated (the extent of one of the deletions, ΔλgtT411, is shown in Fig. 1). All of these deletions delete the T4 tRNA promoter. There is an RNA polymerase binding site in this region and it is indicated in Figure 1. Thus, there is good evidence that the promoter for this transcription unit is located approximately 1000 bp upstream from cluster 1. We have no evidence concerning how far this transcription unit extends, but experiments by Black and Gold (1971) suggest that it could extend as far as gene *e*, a distance of approximately 7000 bp.

Further understanding of the organization of the tRNA genes was obtained from the sequence of cluster 1 (Fukada 1979). This sequence, presented schematically in Figure 2, helps to answer the question of how the precursors that accumulate in the absence of RNase P are generated. The RNase P precursors are simply joined together in the DNA sequence suggesting that single endonucleolytic cleavages produce each dimeric or monomeric precursor. The positions of these cleavages (Fig. 1) imply

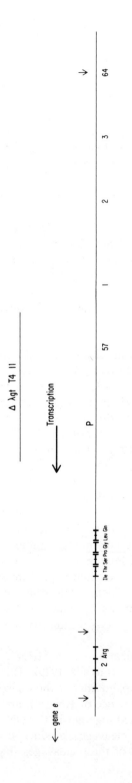

Figure 1 The map of the tRNA gene cluster on the T4 genome. The tRNA gene cluster for all the eight phage T4 tRNAs and two stable RNAs, species 1 and Species 2 RNAs is mapped between genes *e* and *57* (Wilson et al. 1972; Fukada 1979). The overall size of the tRNA gene cluster is about 1600 bp. There are two subclusters, clusters 1 and 2, which are separated by a distance of about 600 bp. (↓) *Eco*RI sites used for cloning these phage T4 tRNA gene clusters. The larger *Eco*RI fragment, which contains cluster 1, was cloned into a λ vector but turned out to contain some region "lethal" to the host. Some clones deleted the lethal region. The extent of one of the deletions, ΔλgtT411, is shown. The deletions also removed a T4 tRNA promoter. (P) An RNA polymerase binding site in the deleted region, which is about 1000 bp upstream from cluster 1; (→) direction of transcription of tRNA genes from the promoter. Genes *57*, *1*, *2*, *3*, and *64* are also in this fragment (J. Velten and J. Abelson, unpubl.).

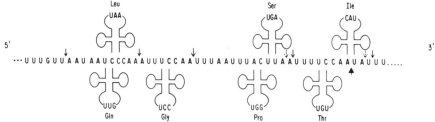

Figure 2 Schematic representation of cluster 1 containing seven of the T4 tRNA genes. The sequence of each tRNA is shown by the characteristic cloverleaf configuration with its anticodon sequence. All the interstitial sequences are arranged in a line to show the characteristic features of the sequence discussed in the text. (↓) Major terminals of tRNA precursors isolated in vivo (Abelson et al. 1974); (↓) minor terminals of tRNA precursors isolated in vivo. The terminal heterogeneities of the other dimeric precursors were also reported. (↑) Cleavage site that implies an alternative pathway in maturation of tRNA^{Thr} and tRNA^{Ile}. Precursors detected in vivo were connected side by side in the sequence. Single endonucleolytic cleavages generate each precursor.

the existence of an endonuclease that recognizes the tRNA and hydrolyzes a phosphodiester bond producing the CCA$_{OH}$ end where it is encoded (tRNA^{Leu} and tRNA^{Gly}) or producing a 3' end of equivalent length (e.g., UAA$_{OH}$ in tRNA^{Ser}). The two dimers, tRNA^{Pro}-tRNA^{Ser} and tRNA^{Gln}-tRNA^{Leu}, are not cleaved by this enzyme, but it is possible that the tRNA^{Thr}-tRNA^{Ile} dimer, with its larger interstitial sequence, including CCA$_{OH}$, is cleaved. This endonuclease is probably a host function, but such an activity has not yet been described.

All of the bacteriophage T4 tRNAs are synthesized in a single transcript. However, the amounts of different mature tRNA species can vary as much as fourfold. It appears that the precise arrangement of sequences surrounding the T4 tRNA genes serves to modulate the efficiency of RNA processing and thus control the level of the various tRNAs.

tRNA GENES IN EUKARYOTES

Although the complexity of tRNA populations in eukaryotes seems to be similar to that seen in bacteria, the study of the organization of the tRNA genes is inherently more complex. Hybridization studies indicate that the haploid yeast genome contains approximately 300 tRNA genes (Schweizer et al. 1969); the haploid *Drosophila* genome contains approximately 600 tRNA genes (Weber and Berger 1976); and the *Xenopus* genome may contain as many as 8000 genes (Clarkson and Birnstiel 1974).

Since the pattern of tRNA isoacceptors in eukaryotes is fairly simple, it is likely that in some cases there are a number of identical genes for the same tRNA. In yeast, the tRNA^{Tyr} genes have been studied extensively because they can be converted to nonsense suppressor genes (see Olson et al., this volume). There are eight identical tRNA^{Tyr} genes in yeast and they are located at eight unlinked genetic loci. This pattern of gene organization is probably generally true in yeast (Beckmann et al. 1977), but one case of gene clustering has been observed. Four separate recombinant clones were obtained in which a tRNA$_3^{Arg}$ gene was closely linked to a tRNA^{Asp} gene. DNA sequence analysis (H. Sakano et al., unpubl.) revealed that in one case the two genes are separated by only 9 bp.

The recombinant DNA technology is also now providing us with a detailed first glimpse of the new organization in *Drosophila* (Yen et al. 1977; Hovemann et al., this volume) and *Xenopus* (Clarkson et al. 1978). It will take some time before enough recombinant clones are analyzed to obtain a general picture. In the meantime, a graphic view of the overall organization of tRNA genes in *Drosophila* is being provided by the powerful technique of in situ hybridization to the polytene chromosomes (Tener et al.; Elder et al.; Kubli et al.; all this volume). This technique demonstrates that the tRNA genes are generally dispersed on most of the chromosomes but that some clusters are evident.

Relatively little is known about the transcription of tRNA genes in eukaryotes, but the prospects for increased understanding in this area are good. In vitro transcription of tRNA genes has been accomplished in several systems, notably in those utilizing *Xenopus* oocytes. It is possible to inject cloned tRNA genes directly into the large oocyte nucleus, as described by Cortese et al. (this volume), or to prepare a germinal vesicle extract (Birkenmeier et al. 1978). In both systems mature tRNA is produced in response to cloned tRNA genes. tRNA genes from *Xenopus*, nematode (Cortese et al., this volume), *Drosophila* (Schmidt et al. 1978; Hovemann et al., this volume), and yeast (de Robertis and Olson 1979; Ogden et al. 1979) have been shown to function. In the latter case, the tRNA^{Tyr} gene (de Robertis and Olson 1979) and tRNA^{Trp} gene (Ogden et al. 1979), both of which contain intervening sequences, are transcribed and the tRNA precursors are processed. This shows that the tRNA splicing enzyme system must be present in the oocyte. Evidence for specific initiation of transcription at sites close to the structural gene has been seen for *Drosophila* (Silverman et al. 1979) and yeast tRNA genes (de Robertis and Olson 1979; Ogden et al. 1979). It will be important to prove that the initiation sites are the same in yeast. It is already clear that the systems are capable of correct initiation and termination in the transcription of *Xenopus* 5S genes (Brown and Gurdon 1977; Birkenmeier et al. 1978).

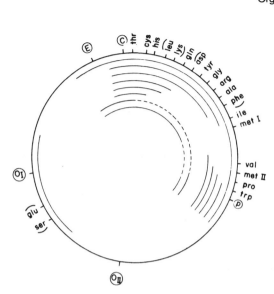

Figure 3 Deletion map of yeast mitochondrial tRNA genes.

ORGANELLE tRNA GENES

The yeast mitochondrial DNA contains at least 20 tRNAs (Martin et al. 1977). These genes have been mapped by petite-deletion mapping (Martin et al. 1977) and by restriction-endonuclease mapping (Van Ommen et al. 1977). The deletion map is reproduced in Figure 3. Most of the tRNA genes are clustered in 10% of the genome between the chloramphenicol- and erythromycin-resistance markers. More recently DNA sequences of the genes for two tRNAs^{Ser}, a tRNA^{Phe}, a tRNA^{Gly}, a tRNA^{Val}, a tRNA^{Arg} (Martin et al. 1979), and genes for tRNA^{Cys} and tRNA^{His} (Grivell et al. 1979) have been determined. None of the genes contain intervening sequences or coded CCA$_{OH}$ ends. It is interesting that the tRNA^{Arg} and tRNA$_2^{Ser}$ genes are separated by only 3 nucleotides, suggesting they are probably transcribed together. Mitochondrial tRNA genes from HeLa cells and *Xenopus* appear to be dispersed and are transcribed from both strands (Angerer et al. 1976; Dawid et al. 1976). The organization of tRNA genes in chloroplast DNA is reviewed by Steinmetz et al. (this volume).

REFERENCES

Abelson, J., K. Fukada, P. Johnson, H. Lamfrom, D. P. Nierlich, A. Otsuka, G. V. Paddock, T. C. Pinkerton, A. Sarabhai, S. Stahl, J. H. Wilson, and H. Yesian. 1975. Bacteriophage T4 tRNAs: Structure, genetics and biosynthesis. *Brookhaven Symp. Biol.* **26:**77.

Angerer, L., N. Davidson, W. Murphy. P. Lynch, and G. Attardi. 1976. An electron microscope study of the relative positions of the 4S and ribosomal RNA genes in Hela cell mitochondial DNA. *Cell* **9:**81.
Beckmann, J. S., P. F. Johnson, and J. Abelson. 1977. Cloning of yeast transfer RNA genes in *Escherichia coli. Science* **196:**205.
Birkenmeier, E. H., D. D. Brown, and E. Jordan. 1978. A nuclear extract of *Xenopus laevis* oocytes that accurately transcribes 5S RNA genes. *Cell* **15:**1077.
Black, L. W. and L. M. Gold. 1971. Pre-replicative development of the bacteriophage T4: RNA and protein synthesis *in vivo* and *in vitro. J. Mol. Biol.* **60:**365.
Brenner, D. J., M. J. Fournier, and B. P. Doctor. 1970. Isolation and partial characterization of the transfer ribonucleic acid cistrons from *Escherichia coli. Nature* **227:**448.
Brown, D. D. and J. B. Gurdon. 1977. High-fidelity transcription of 5S DNA injected into *Xenopus* oocytes. *Proc. Natl. Acad. Sci.* **74:**2064.
Clarkson, S. G. and M. L. Birnstiel. 1974. Clustered arrangement of tRNA genes of *Xenopus laevis. Cold Spring Harbor Symp. Quant. Biol.* **38:**451.
Clarkson, S. G., V. Kurer, and H. O. Smith. 1978. Sequences organization of a cloned tDNA$_1^{Met}$ fragment from *Xenopus laevis. Cell* **14:**713.
Crick, F. H. C. 1967. Codon-anticodon pairing: The wobble hypothesis. *J. Mol. Biol.* **19:**548.
Daniel, V., S. Sarid, and U. Z. Littauer. 1970. Bacteriophage induced transfer RNA in *Escherichia coli. Science* **167:**1682.
Dawid, I. B., C. Klukas, S. Ohi, J. L. Ramirez, and W. B. Upholt. 1976. Structure and evolution of animal mitochondrial DNA. In *The genetic function of mitochondrial DNA* (ed. C. Saccone and A. M. Kroon), p. 3. North-Holland, Amsterdam.
De Robertis, E. M. and M. V. Olson. 1979. Transcription and processing of yeast tyrosine tRNA genes microinjected into frog oocytes. *Nature* **278:**137.
Fukada, K. 1979. "Studies on the tRNA biosynthesis of bacteriophage T4." Ph.D. thesis, University of California, San Diego.
Goldfarb, A., E. Seeman, and V. Daniel. 1978. *In vitro* transcription and isolation of a polycistronic RNA product of the T4 tRNA operon. *Nature* **273:**562.
Grivell, L. A., A. C. Arnberg, P. H. Boer, P. Borst, J. L. Bos, E. F. J. van Bruggen, G. S. P. Groot, N. B. Hecht, L. A. M. Hensgens, G. J. B. van Ommen, and H. F. Tabak. 1979. Transcripts of yeast mitochondrial DNA and their processing. *ICN-UCLA Symp. Mol. Cell Biol.* **15:** (in press).
Guthrie, C., J. G. Seidman, M. M. Comer, R. M. Bock, F. J. Smith, B. G. Barrell, and W. H. McClain. 1975. The biology of bacteriophage T4 transfer RNAs. *Brookhaven Symp. Biol.* **26:**106.
Hsu, W. T., J. W. Foft, and S. B. Weiss. 1967. Effect of bacteriophage infection on the sulfur-labeling of sRNA. *Proc. Natl. Acad. Sci.* **58:**2028.
Kaplan, D. A. and D. P. Nierlich. 1975. Intiation and transcription of a set of transfer RNA genes *in vitro. J. Biol. Chem.* **250:**934.
Martin, N. C., M. Rabinowitz, and H. Fukuhara. 1977. Yeast mitochondrial DNA specifies tRNA for 19 amino acids. Deletion mapping of the tRNA genes. *Biochemistry* **16:**4672.
Martin, N. C., D. L. Miller, J. E. Donelson, C. Sigurdson, J. L. Hartley, P. S.

Moynihan, and H. D. Pham. 1979. Identification and sequencing of yeast mitochondrial tRNA genes in mitochondrial DNA-pBR 322 recombinants. *ICN-UCLA Symp. Mol. Cell Biol.* **15:** (in press).

McClain, W. H., C. Guthrie, and B. G. Barrell. 1972. Eight transfer RNAs induced by infection of *Escherichia coli* with bacteriophage T4. *Proc. Natl. Acad. Sci.* **69:** 3703.

Ogden, R. C., J. S. Beckmann, J. Abelson, H. S. Kang, D. Söll, and O. Schmidt. 1979. In vitro transcription and processing of a yeast tRNA gene containing an intervening sequence. *Cell* **17:** 399.

Scherberg, N. H., G. Arabinda, W. Hsu, and S. B. Weiss. 1970. Evidence for the early synthesis of T4 bacteriophage-coded transfer RNA. *Biochem. Biophys. Res. Commun.* **40:** 919.

Schmidt, O., J. Mao, S. Silverman, B. Hovemann, and D. Söll. 1978. Specific transcription of eukaryotic tRNA genes in *Xenopus* germinal vesicle extracts. *Proc. Natl. Acad. Sci.* **75:** 4819.

Schweizer, E., C. Mackechnie, and H. O. Halvorson. 1969. The redundancy of ribosomal and transfer RNA genes in *Saccharomyces Cerevisiae*. *J. Mol. Biol.* **40:** 261.

Silverman, S., O. Schmidt, D. Söll, and B. Hovemann. 1979. The nucleotide sequence of a cloned *Drosophila* arginine tRNA gene and its in vitro transcription in *Xenopus* germinal vesical extracts. *J. Biol. Chem.* **254:** 290.

Van Ommen, G.-J. B., G. S. P. Groot, and P. Borst. 1977. Fine structure physical mapping of 4S RNA genes on mitochondrial DNA of *Saccharomyces cerevisiae*. *Mol. Gen. Genet.* **154:** 255.

Weber, L. and E. Berger. 1976. Base sequence complexity of the stable RNA species of *Drosophila melanogaster*. *Biochemistry* **15:** 5511.

Wilson, J. H., J. S. Kim, and J. Abelson. 1972. Bacteriophage T4 transfer RNA III. Clustering of the genes for the T4 transfer RNAs. *J. Mol. Biol.* **71:** 547.

Yen, P. H., A. Sodja, M. Cohen, S. E. Conrad, M. Wu, and N. Davidson. 1977. Sequence arrangement of tRNA genes on a fragment of *Drosophila melanogaster* DNA cloned in *E. coli*. *Cell* **11:** 763.

Escherichia coli tRNA^{Tyr} Gene Clusters: Organization and Structure

John Rossi, James Egan,* Michael L. Berman, and Arthur Landy
Section of Microbiology and Molecular Biology
Division of Biology and Medicine
Brown University
Providence, Rhode Island 02912

Escherichia coli contains two species of tRNA^{Tyr} that differ from each other by only two bases in the variable loop region (Goodman et al. 1968, 1970). The locus *tyrT* at 27 minutes (Garen et al. 1965; Signer et al. 1965; Bachman et al. 1976) has been shown to contain duplicate copies of the tRNA$_1^{Tyr}$ sequence, since the bacteriophage ϕ80p$su^{+-}3$ stimulates the production of both su^+ and su^- tRNA$_1^{Tyr}$ (Russell et al. 1970). Derivatives of this bacteriophage, carrying only a single su^+ tRNA$_1^{Tyr}$ copy, have been isolated. These are presumed to arise by unequal recombination involving the two tRNA$_1^{Tyr}$ mature structural sequences (Russell et al. 1970). The *tyrU* locus, on the opposite side of the *E. coli* chromosome at 88 minutes (Orias et al. 1972), codes for tRNA$_2^{Tyr}$ (Eggerston and Adelberg 1965; Orias et al. 1972; Squires et al. 1973) and has been shown to contain a single tRNA$_2^{Tyr}$ mature structural sequence.

We are attempting to determine those aspects of tRNA gene structure and organization that are crucial to the initiation and termination of transcription and the regulation of gene expression. The tRNA^{Tyr} genes of *E. coli*, which have already figured prominently in this area, afford a model system well suited for studying this group of problems. In the experiments described here, the tRNA$_1^{Tyr}$ transducing phages ϕ80p$su^{+-}3$ (doublet) and ϕ80psu^+3 (singlet) (Russell et al. 1970), and the tRNA$_2^{Tyr}$ transducing phages λh80d$glyTsu^+36$ (Squires et al. 1973) and $\lambda rif^d 18$ (Kirschbaum and Konrad 1973) have been used to study the structural organizations of regulatory and coding sequences for these tRNA genes.

The results presented demonstrate that the tRNA$_1^{Tyr}$ doublet structure consists of two 85-bp mature structural sequences separated by a 200-bp intergenic spacer. Studies of the tRNA$_1^{Tyr}$ regulatory regions include DNA sequence analyses of tRNA$_1^{Tyr}$ promoter mutants, as well as an unusual 178-bp sequence adjoining the tRNA$_1^{Tyr}$ mature structural sequence that is

---

*Present address: Department of Pediatrics, University of Connecticut Health Center, Farmington, Connecticut 06032.

repeated 3.14 times, and contains an in vitro, rho-dependent termination site (Küpper et al. 1978). The regions surrounding and including the tRNA$_2^{Tyr}$ mature structural sequence have also been analyzed. The sequence immediately upstream of tRNA$_2^{Tyr}$ codes for a previously undescribed tRNA^{Thr} (tRNA$_4^{Thr}$). The sequences for tRNA$_2^{Gly}$-tRNA$_3^{Thr}$ are tightly clustered 115 bp downstream of the tRNA$_2^{Tyr}$ sequence.

Although a great deal of information has been obtained from specialized transducing phages carrying tRNA genes, there are serious questions, which are shown here to be well founded, about genetic alterations and rearrangements that may have taken place during the formation and/or growth of these phages. In experiments described here, the structural organizations of the tRNA^{Tyr} gene clusters have been analyzed directly on the *E. coli* chromosome in conjunction with the bacteriophage analyses.

STRUCTURAL FEATURES OF THE tRNA$_1^{Tyr}$ GENES IN THE ϕ80psu$^{+-}3$ AND ϕ80psu$^+3$ BACTERIOPHAGE

The identification and isolation of *Hin*II and *Hin*II + III restriction fragments carrying the tRNA$_1^{Tyr}$ gene from either ϕ80psu$^{+-}3$ or ϕ80psu$^+3$ transducing phages (Landy et al. 1974a,b) have facilitated detailed structural analyses of the tRNA$_1^{Tyr}$ regulatory and coding regions. The locations of sites within these fragments for several restriction endonucleases have been determined (Fig. 1) and have served to delineate the organizational pattern of the tRNA$_1^{Tyr}$ genes. From these analyses, as well as previously published sequences of the promoter (Sekiya and Khorana 1974; Sekiya et al. 1976), precursor (Altman and Smith 1971), and mature structural sequences (Goodman et al. 1968, 1970), the relative locations of the regulatory and coding sequences within the *Hin*II and *Hin*II + III fragments have been determined (Fig. 1).

As mentioned earlier, the arrangement of the tRNA$_1^{Tyr}$ genes undergoes an unequal recombination event whereby the duplicated tRNA$_1^{Tyr}$ genes give rise to a singlet tRNA$_1^{Tyr}$ (Russell et al. 1970). The results of digestions of the *Hin*II + III-1020 (B10a) and *Hin*II + III-730 (C1a) fragments with several restriction endonucleases demonstrate that the only restriction sites not common to both fragments are those known to derive from the intergenic spacer and second or *su*$^-$ tRNA$_1^{Tyr}$ sequences (Fig. 1). These results demonstrate that the only differences between the tRNA$_1^{Tyr}$ doublet and singlet configurations are the absence of the 200-bp intergenic spacer and the second or *su*$^-$ tRNA$_1^{Tyr}$ copy (Fig. 1). This result is important in reference to correlating transcriptional studies for the ϕ80psu$^+3$ bacteriophage to the naturally occurring doublet configuration (see below).

THE tRNA₁^Tyr PROMOTER

A tRNA₁^Tyr precursor of 129 nucleotides has been isolated from *E. coli* cells infected with $\phi 80psu^+3$ (Altman and Smith 1971). This precursor contained a 5'-pppG end group and was shown to contain 41 nucleotides preceding the mature tRNA₁^Tyr structural sequence. In addition, the nucleotide sequence extending 59 bp upstream of the transcriptional start point, including the presumptive promoter sequence, has been determined (Sekiya and Khorana 1974; Sekiya et al. 1976) (see Fig. 2). In vitro, promoter-dependent transcripts that initiated at the same site as the in vivo transcript have been demonstrated utilizing the *Hin*II + III-1020 (B10a) and *Hin*II + III-730 (C1a) restriction fragments (see Fig. 1) (Küpper et al. 1975). Several features of the tRNA₁^Tyr promoter set it apart from other promoters that have been studied. Among these are relatively unstable binding of RNA polymerase and high sensitivity of the RNA polymerase-promoter complex to salt, heparin, and rifampicin (Küpper et al. 1975). It had been postulated that the high G + C content of the tRNA₁^Tyr promoter around the region of initiation may be responsible for the above properties (Küpper et al. 1975).

Analyses of the promoter region have been extended with the isolation of several promoter mutants obtained using tRNA₁^Tyr-lac operon fusions. These mutants have been genetically recombined into the tRNA₁^Tyr gene in $\phi 80psu^+3$, and characterized by their ability to suppress bacteriophage T4 amber mutants (Berman and Beckwith 1979; see Table 1). DNA sequence analyses of the promoter region from several of these mutants have been carried out. The DNA source for this region is an *Hae*III 300-bp fragment that contains the tRNA₁^Tyr promoter sequence (see Fig. 1).

A common feature of *E. coli* promoters is a 7-bp sequence that is centered approximately 10 bp upstream from the start of transcription

Table 1 Amber suppressor activity of promoter mutants in the su^+3 allele of the *tyrT* gene

	phage T4					
Promoter allele	T4D	H39	am17	N133	N319	Mutant class
Parental su^+3	+	+	+	+	+	
74	+	+	very poor	+	poor	I
119	+	poor	0	0	poor	II
27	+	very poor	0	0	very poor	III

The T4 phage carrying amber mutations in various genes were tested as described previously (Berman and Beckwith 1979). T4D is wild type. (+) full growth; (0) no growth. The nucleotide changes associated with each class of mutants are presented in Fig. 2.

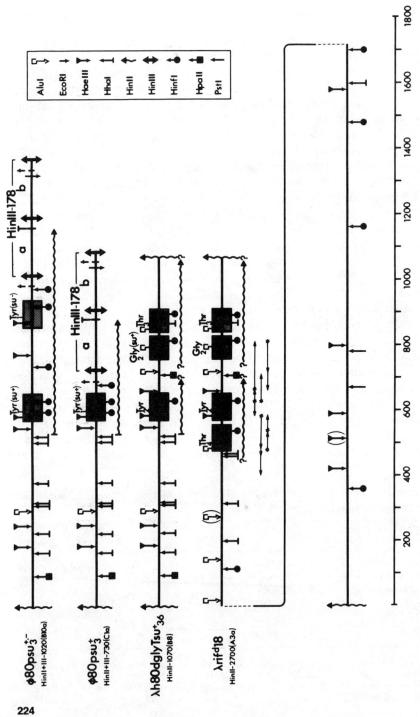

Figure 1 (See facing page for legend.)

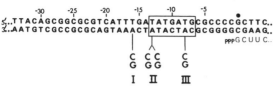

Figure 2 Base-pair alterations obtained from mutants in the tRNA$_1^{Tyr}$ (tyrT) gene promoter region. The nucleotide sequence extending 34 bp into the tRNA$_1^{Tyr}$ promoter region is presented (Sekiya and Khorana 1974; Sekiya et al. 1976). The alterations obtained from four independently isolated promoter mutants and their corresponding positions within the sequence are indicated. (I, II, and III) Phenotypic classes of the various mutants described in Table 1; (□) promoter heptamer sequence (Pribnow 1975; Gilbert 1976); (●) transcriptional initiation start point (Altman and Smith 1971; Küpper et al. 1975); (→) regions of inverted repeat.

(Pribnow 1975). The ideal sequence, defined by promoter mutants, is TATAATG, but this sequence differs by one or more positions in most promoters (for a review, see Gilbert 1976). In the tRNA$_1^{Tyr}$ gene, this sequence is TATGATG (Sekiya and Khorana 1974). Three of the four tRNA$_1^{Tyr}$ mutants described in Figure 2 are AT→GC transitions within this region. The fourth mutant, also an AT→GC transition, lies just outside this region (Fig. 2). It should be noted that a single-base-pair change at each of the sites indicated in Figure 2 is sufficient in itself to affect expression of the tRNA$_1^{Tyr}$ gene, since no other base changes or alterations were found in any of the mutant strains within a sequence extending over 100 bp upstream and 30 bp downstream from each of the mutant sites (M. Berman et al., unpubl.).

Figure 1 Structural organization of tRNA^{Tyr} gene clusters in the transducing bacteriophages $\phi 80psu^{+-}3$ (tRNA$_1^{Tyr}$), $\phi 80psu^+3$ (tRNA$_1^{Tyr}$), $\lambda h80dglyTsu^+36$ (tRNA$_2^{Tyr}$), and $\lambda rif^d 18$ (tRNA$_2^{Tyr}$). Restriction endonuclease sites within the HinII and HinII + III fragments were determined previously (Rossi et al. 1979). Sequences corresponding to (coding for) the mature tRNAs (▨) and direction of transcription (⤳) are indicated. The two HinIII 178-bp fragments adjoining HinII + III-1020 (B10a) and HinII + III-730 (C1a) (Landy et al. 1974; Egan and Landy 1978) contain an in vitro rho-dependent transcriptional termination site (T) and two potential termination sites (t) in the two flanking repeat units (Küpper et al. 1978). (←∗) DNA sequences obtained from $\lambda rif^d 18$ (by the technique of Maxam and Gilbert 1977) that are relevant to this study (asterisks indicate the labeled end of the restriction fragment used; arrowheads indicate the direction and extent of sequences obtained from the labeled end). The scale is given in bp. (Reprinted, with permission, from Rossi and Landy 1979.)

THE REGION ADJOINING THE 3' END
OF THE tRNA₁ᵀʸʳ MATURE STRUCTURAL SEQUENCE

Analyses of the region adjoining the tRNA₁ᵀʸʳ mature structural sequence in $\phi 80psu^{+-}3$ and $\phi 80psu^+3$ have revealed several unusual and interesting features (Landy et al. 1974a; Egan and Landy 1978). The major feature of this region is a 178-bp sequence that commences with the last 19 bases of the tRNA₁ᵀʸʳ mature structural sequence and is repeated tandemly 3.14 times (Egan and Landy 1978; Fig. 3). In the 3.14 repeat

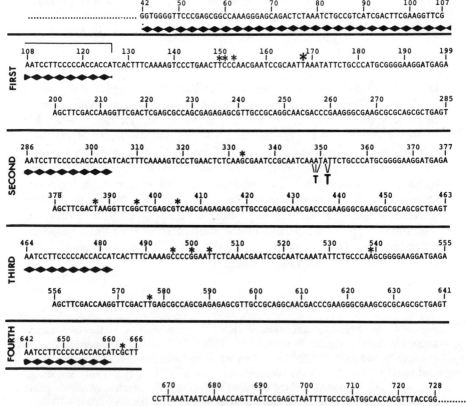

Figure 3 Sequence of the distal region of the tRNA₁ᵀʸʳ (*tyrT*) gene. The sequence presented is that determined previously (Egan and Landy 1978). The DNA strand having the same sequence as the tRNA is arranged to facilitate comparison of the 178-bp repeating units. The first base shown, at position 42, corresponds to the first base of the mature tRNA. Eighty-five contiguous bases correspond to the mature tRNA₁ᵀʸʳ (◆◆) and the last 19 bases of this sequence occur at the beginning of each repeat unit. (∗) Positions in one repeat that differ from the others; (T) locations of the rho-dependent termination of transcription (see text and Küpper et al. 1978).

units, comprising a total of 559 bp, there are only 14 single-base substitutions and no deletions or insertions (Fig. 3).

One particularly interesting site of nonidentity in the repeated sequences may be relevant to mechanisms controlling termination of transcription of the tRNA$_1^{Tyr}$ genes. Transcription studies with the 1536-bp HinII restriction fragment derived from ϕ80psu^+3 identified an in vitro, rho-dependent termination site in the second repeating unit, at the 3' end of sequence CAATCAAATAT (positions 342–352 in Fig. 3) (Küpper et al. 1978). Termination in the t_{R_1} region of bacteriophage λ takes place at the 3' end of sequence CAATCAAT (Rosenberg et al. 1978). There are seven nucleotides that are common to the λt_{R_1} and tRNA$_1^{Tyr}$ termination sequences. Some or all of these nucleotides may be required for rho-dependent termination in the tRNA$_1^{Tyr}$ gene, since rho-dependent termination was not observed in the first repeat unit, which has the altered sequence CAATTAAATAT (position 164–174, Fig. 3). This conclusion however, must be qualified for two reasons. First, there are four other single-base differences between these two repeat units (Fig. 3). Second, the RNA transcript is, of course, different by the time RNA polymerase reaches the second repeat. The sequence in the third repeat unit at positions 520–530 is identical to that found in the corresponding region of the second repeat, suggesting this sequence could also function as a rho-dependent termination site. In the in vitro studies reported, termination at the second repeat was very efficient and therefore the properties of the third repeat could not be analyzed (Küpper et al. 1978).

The following experiments demonstrate that the repeated structures are not peculiar to the su^+ transducing phage in which it was identified and analyzed, but it does exist in the chromosome of several E. coli strains. The ϕ80psu^+3-derived fragments HinIII-178 a and b, which are centrally derived from these repeats (Figs. 1 and 4), were labeled at their 5' terminals with [γ-^{32}P]ATP and polynucleotide kinase and used as probes for the repeated structures in HinII- and HinII + III-digested ϕ80p$su^{+-}3$ and E. coli strains CA274 (tRNA$_1^{Tyr}$ su^-) and CA275 (tRNA$_1^{Tyr}$ su^{+-}) (Russell et al. 1970). The results of these experiments (Fig. 4) demonstrate hybridization of the HinIII-178 a and b probes to the fragments expected from analyses utilizing the tRNA$_1^{Tyr}$ transducing phage (Egan and Landy 1978; Figs. 1 and 4). Similar results have been obtained using E. coli strain HB101, which is of an unrelated genetic background (data not shown). It should be noted that the tRNA$_1^{Tyr}$ transducing phage ϕ80sus_2psu^+3 (Kyoto University) (Andoh and Ozeki 1968), whose origin is different from that of ϕ80p$su^{+-}3$ (Molecular Research Council) (Russell et al. 1970) used in the present studies, does not contain the repeated sequences adjoining the tRNA$_1^{Tyr}$ mature structural sequence, although sequences upstream of the mature structural sequence are intact (Fig. 4). Thus, caution should be exercised when interpreting transcriptional studies involving this bacteriophage.

228

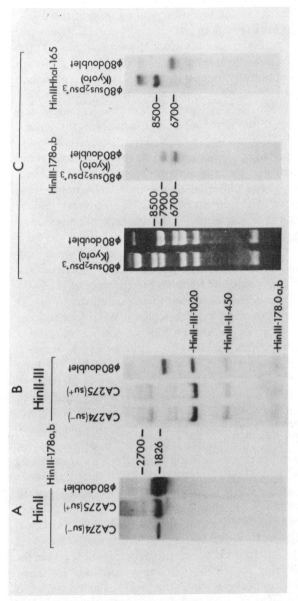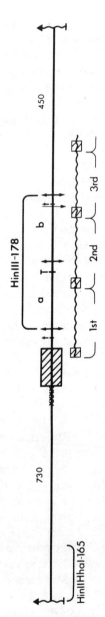

Figure 4 (See facing page for legend.)

STRUCTURAL ANALYSES OF THE tRNA$_2^{Tyr}$ GENE REGION

The tRNA$_2^{Tyr}$ transducing phage λh80d*glyTsu*$^+$*36* (Squires et al. 1973) and λ*rif*^{d}*18* (Kirschbaum and Konrad 1973) were the sources of *Hin*II fragments carrying the tRNA$_2^{Tyr}$ gene (see Fig. 1). Structural analyses of these *Hin*II fragments revealed several surprising features associated with the tRNA gene organization in these bacteriophages. It had been previously shown that λh80d*glyTsu*$^+$*36* carries the genes for tRNA$_2^{Tyr}$, tRNA$_2^{Gly}$, and tRNA$_3^{Thr}$ (Squires et al. 1973; Wu et al. 1973). Conflicting evidence concerning the relative order of these three genes (Wu et al. 1973; Chang and Carbon 1975) has now been rectified. The gene order (proceeding in the direction of transcription) is tRNA$_2^{Tyr}$-tRNA$_2^{Gly}$-tRNA$_3^{Thr}$ (Fig. 1). The identification of this same tRNA gene cluster on λ*rif*^{d}*18* (Nomura 1976) prompted a comparison of the gene organization in the λ*rif*^{d}*18* and λh80d*glyTsu*$^+$*36* bacteriophages. Although the region including and downstream of the tRNA$_2^{Tyr}$ gene is identical or near identical in the two bacteriophages, the region upstream of the tRNA$_2^{Tyr}$ gene is very different (Figs. 1, 5, and 7).

Figure 4 Identification of the tRNA$_1^{Tyr}$-gene-associated, 178-bp repeated sequence on the *E. coli* CA274 and CA275 chromosomes. 50 to 100 μg/lane of *Hin*II (*A*) or *Hin*II + III (*B*) digested *E. coli* CA274 or *E. coli* CA275 and 2 μg/lane φ80p*su*$^{+-}$*3* DNAs were electrophoresed in 2.5% agarose, transferred to nitrocellulose (Southern 1975) and hybridized with the 5'-^{32}P-labeled *Hin*III-178 probes depicted in the diagram (a and b). (*C*) 2 μg/lane of *Eco*RI digests of φ80*sus*$_2$p*su*$^+$*3* (Andoh and Ozeki 1968) and φ80p*su*$^{+-}$*3* (Russell et al. 1970) were electrophoresed in 1% agarose, transferred to nitrocellulose, and hybridized with either *Hin*III-178 a and b or *Hin*II*Hha*I-165. The schematic (*bottom*) represents the tRNA$_1^{Tyr}$ gene regions in φ80p*su*$^+$*3* from which the probes were derived. Symbols used are the same as those in Fig. 1 except: (↓) *Eco*RI site; (▨⌒) the repeat unit adjoining the tRNA$_1^{Tyr}$ mature structural sequence (Egan and Landy 1978) (see also Fig. 3); (xxx) known tRNA$_1^{Tyr}$ precursor and promoter sequences (Altman and Smith 1971; Sekiya and Khorana 1974; Sekiya et al. 1976). In φ80p*su*$^{+-}$*3*, the *Hin*II + III-730 fragment is replaced by the larger *Hin*II + III-1020 fragment (see Fig. 1). The sizes (in bp) of *Hin*II and *Hin*II + III hybridizing fragments were determined from previous estimates of φ80p*su*$^{+-}$*3* and φ80p*su*$^+$*3* fragment sizes (Landy et al. 1974a; Egan and Landy 1978; Rossi et al. 1979). The φ80*sus*$_2$p*su*$^+$*3* 8500-bp *Eco*RI fragment contains the tRNA$_1^{Tyr}$ mature structural sequences, as well as a φ80 cohesive end (J. Rossi and A. Landy, unpubl.). This fragment can anneal with the other φ80 *Eco*RI 8000-bp end, forming the larger of the two hybridizing fragments in the experiment with the *Hin*II*Hha*I-165 probe (*C*). Faint, unmarked bands of hybridization are due to incompletely digested fragments. The faint band at 2700 bp is due to hybridization between the 19 bp of tRNA$_1^{Tyr}$ mature structural sequence contained in the probes and the corresponding 19 bp in the tRNA$_2^{Tyr}$ mature structural sequence. The band of hybridization at 1826 bp in *B* is undigested *Hin*II-1826. (Reprinted, with permission, from Rossi and Landy 1979.)

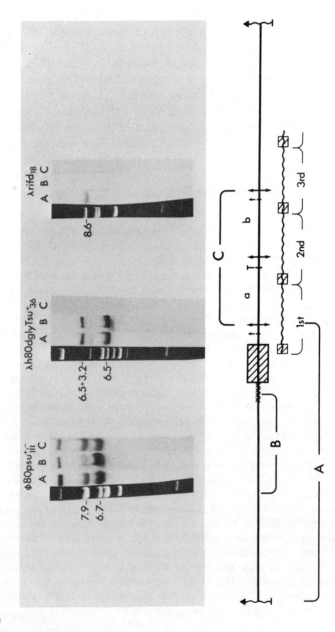

Figure 5 (See facing page for legend.)

The sequences upstream of tRNA$_2^{Tyr}$ in the λh80d*glyTsu$^+$36*-derived *Hin*II 1070-bp restriction fragment have a pattern of restriction endonuclease sites that are identical to those found on the tRNA$_1^{Tyr}$ transducing phage φ80p*su^{+-}3* and φ80p*su$^+$3*-derived *Hin*II + III fragments (Fig. 1). Such an identical pattern involving five different restriction endonucleases and encompassing 13 different sites cannot be simply fortuitous. The structural organization of the tRNA gene region on λh80d*glyTsu$^+$36* thus appears to be a hybrid of tRNA$_1^{Tyr}$ upstream and tRNA$_2^{Tyr}$-tRNA$_2^{Gly}$-tRNA$_3^{Thr}$ downstream sequences. This interpretation is verified by hybridization experiments (Fig. 5) and comparison of DNA sequences (Fig. 7). From the results presented in Figure 5, it can be seen that the restriction probe derived from the region upstream of tRNA$_1^{Tyr}$ in φ80p*su$^+$3*, which contains known tRNA$_1^{Tyr}$ promoter and precursor sequences, hybridizes with λh80d*glyTsu$^+$36* but not λ*rif^{d}18* sequences. Similar results were obtained using restriction fragments spanning the entire upstream sequence in the *Hin*II + III-730 (C1a) fragment as probes (data not shown). In addition, it should be noted that the repeated sequences adjoining the tRNA$_1^{Tyr}$ mature structural sequences (see above) are not present in either the λh80d*glyTsu$^+$36* or λ*rif^{d}18* genomes (Fig. 5).

The DNA sequences extending 59 bp into the tRNA^{Tyr} promoter in φ80p*su^{+-}3*, φ80p*su$^+$3*, and λh80d*glyTsu$^+$36* have previously been shown to be identical (Sekiya et al. 1976). The present results extend this homology for at least 500 bp (see Fig. 1).

The DNA sequences in the regions surrounding and including the tRNA$_2^{Tyr}$ mature structural sequence on λ*rif^{d}18* have also been determined (Figs. 6 and 7). The sequence immediately upstream of tRNA$_2^{Tyr}$ can be written in the cloverleaf structure of a tRNA (Fig. 6). When drawn in

Figure 5 Hybridization of specific DNA restriction fragment probes derived from *Hin*II + III-730 (C1a) with *Eco*RI digests of φ80p*su^{+-}3*, λh80d*glyTsu$^+$36*, and λ*rif^{d}18*. *Eco*RI digests of the bacteriophage DNA (2 μg/lane) were electrophoresed in 1% agarose. After staining with ethidium bromide, the fragments were transferred to nitrocellulose (Southern 1975) and hybridized to the ^{32}P-labeled probes depicted in the diagram. (*A*) Hybridization with *Hin*II + III-730 (C1a); (*B*) hybridization with *Hae*III-300; (*C*) hybridization with *Hin*III-178 a and b. The schematic represents the tRNA$_1^{Tyr}$ gene region on φ80p*su$^+$3* singlet, which is complementary to the threonine codon ACA. Comparison of this same as those described for Figs. 1 and 4. The numbers beside each digest depict (in 1000 bp) the *Eco*RI fragments that contain either the tRNA$_1^{Tyr}$ or the tRNA$_2^{Tyr}$-tRNA$_2^{Gly}$-tRNA$_3^{Thr}$ genes. In the case of λh80d*glyTsu$^+$36*, the 6.5k-bp fragment contains a φ80 "sticky end" that anneals with the λ 3.2k-bp end to form the larger fragment (6.5 and 3.2). In the φ80p*su^{+-}3* lanes (*left*), the top band of hybridization is undigested DNA. (Reprinted, with permission, from Rossi et al. 1979.)

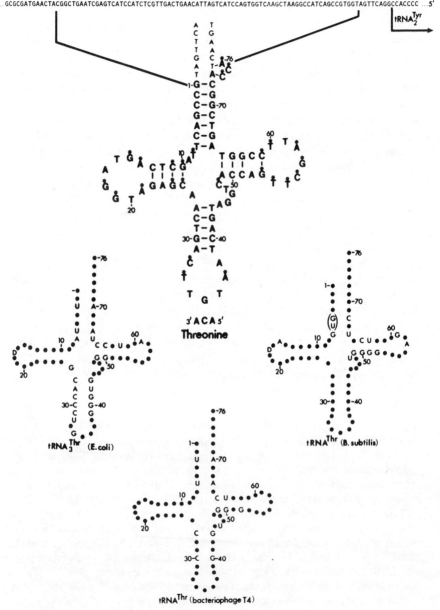

Figure 6 (See facing page for legend.)

such a configuration, this sequence contains the anticodon 5' TGT 3', which is complementary to the threonine codon ACA. Comparison of this sequence with those determined for other prokaryotic tRNA^{Thr} species demonstrates extensive regions of homology, especially with the tRNA^{Thr} species from *Bacillus subtilis* (Hasegawa and Ishikura 1978) and the bacteriophage-T4-coded species (Guthrie et al. 1978), both of which possess the anticodon 5' UGU 3' (Fig. 6).

Evidence to be presented below (see Discussion) strongly suggests that this sequence codes for a tRNA^{Thr}. We propose that this gene be called *thrU*, and the tRNA for which it codes be called tRNA$_4^{Thr}$.

An unusual feature of the sequence presented in Figure 6 is a mismatch in potential base pairing in the stem at positions 52 and 62. At present it cannot be determined whether this represents a mutant structure for this gene, or if, in fact, this gene codes for an unusual species of tRNA.

The nucleotide sequence for the region between the tRNA$_2^{Tyr}$ and tRNA$_2^{Gly}$ mature structural sequences in $\lambda rif^d 18$ has also been determined and is composed of 115 bp (J. Rossi and A. Landy, unpubl.). A portion of this sequence is presented in Figure 7 and is identical to the sequence determined for the analogous region in $\lambda h80dglyTsu^+36$ (Sekiya et al. 1976).

All of the above evidence strongly suggests that the tRNA$_2^{Tyr}$ gene region in bacteriophage $\lambda h80dglyTsu^+36$ is a hybrid structure comprised of tRNA$_1^{Tyr}$ upstream and tRNA$_2^{Tyr}$-tRNA$_2^{Gly}$-tRNA$_3^{Thr}$ downstream sequences. A possible mechanism for the origin of this unusual hybrid tRNA gene structure is presented and described in Figure 8.

ANALYSIS OF THE tRNA^{Tyr} GENE CLUSTERS ON THE E. COLI CHROMOSOME

As a means of probing the structural organizations of the tRNA^{Tyr} gene regions directly on the *E. coli* chromosome, advantage has been taken of restriction endonuclease fragments isolated from the bacteriophages $\phi 80psu^+3$ and $\lambda rif^d 18$, which carry specific information from the tRNA^{Tyr}

Figure 6 Sequence of the *thrU* gene and comparison of its structure with other known prokaryotic tRNA^{Thr} species. DNA sequences for this region were obtained from the $\lambda rif^d 18$-derived fragments indicated in Fig. 1. The DNA sequence of the noncoding strand is drawn in a tRNA cloverleaf configuration. (∗) Bases that are invariant or semiinvariant in most tRNAs; (—) inverted repeat flanking the 5' and 3' tRNA$_4^{Thr}$ mature structural sequences. The three other prokaryotic tRNA^{Thr} species for which complete sequences are known tRNA$_3^{Thr}$ (Chang and Carbon 1975); and tRNA^{Thr} *B. subtilis* (Hasegawa and Ishikura 1978); bacteriophage-T4-coded tRNA^{Thr} (Guthrie et al. 1978) are shown for comparison. (●) Bases in these tRNA^{Thr} species that are common to those presented for tRNA$_4^{Thr}$. (Reprinted, with permission, from Rossi and Landy 1979.)

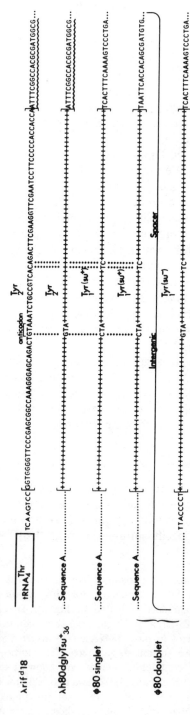

Figure 7 Comparison of tRNA^{Tyr} mature structural and adjoining sequences in the tRNA$_2^{Tyr}$ transducing phages, $\lambda rif^d 18$ and $\lambda h80dglyTsu^+36$, and the tRNA$_1^{Tyr}$ transducing phages $\phi 80psu^+3$ and $\phi 80psu^{+-}3$. (+) Residues common to both tRNA$_2^{Tyr}$ and tRNA$_1^{Tyr}$ mature structural sequences; ($\sim$) sequence adjoining the 3' end of the tRNA$_2^{Tyr}$ mature structural sequence is identical in the $\lambda rif^d 18$ and $\lambda h80dglyTsu^+3$ bacteriophages. The tRNA$_2^{Tyr}$ mature structural and adjoining sequences in $\lambda rif^d 18$ were determined from DNA sequence analyses of the restriction fragments illustrated in Fig. 1. The tRNA$_2^{Tyr}$ DNA sequences determined are in complete agreement with the previously published RNA sequences (Goodman et al. 1970). The tRNA$_2^{Tyr}$ mature structural sequence (as opposed to tRNA$_1^{Tyr}$) in $\lambda h80dglyTsu^+36$ is assumed from the identification of Squires et al. (1973) and V. Daniel (pers. comm.). DNA sequences adjoining the tRNA$_2^{Tyr}$ mature structural sequences in $\lambda h80dglyTsu^+36$ are from Sekiya et al. (1976). The tRNA$_1^{Tyr}$ mature structural DNA sequences are from Egan and Landy (1978) and J. Egan and A. Landy (unpubl.) and are entirely colinear with those determined for the RNA (Goodman et al. 1968, 1970). Sequences adjoining the tRNA$_1^{Tyr}$ mature structural sequences are from Altman and Smith (1971). Sekiya et al. (1976), and Egan and Landy (1978 and unpubl.). Sequence A contains the tRNA$_1^{Tyr}$ promoter and precursor regions, the sequences of which are identical or near identical in bacteriophages $\phi 80psu^+3$, $\phi 80psu^{+-}3$, and $\lambda h80dglyTsu^+36$ (Sekiya et al. 1976). (Reprinted, with permission, from Rossi and Landy 1979.)

gene regions. HinII + III-digested DNAs from *E. coli* strain MB93 (tRNA$_1^{Tyr}$ *su*$^+$-singlet; Smith et al. 1970) and CA274 (tRNA$_1^{Tyr}$ *su*$^-$-doublet; Russell et al. 1970) were probed with the radioactively labeled restriction fragments described in Figure 9. In each case, one of the hybridizing *E. coli* DNA fragments corresponded in size to the tRNA$_1^{Tyr}$-containing fragment from either φ80p*su*$^{+-}$*3* or φ80p*su*$^+$*3*. The other hybridizing *E. coli* DNA fragment always corresponded in size to the fragment for λ*rif*^{d}*18* containing tRNA$_4^{Thr}$-tRNA$_2^{Tyr}$-tRNA$_2^{Gly}$-tRNA$_3^{Thr}$.

An additional experiment utilized a ^{32}P-labeled, λ*rif*^{d}*18*-derived, *Hae*III-74.0 fragment, which contains tRNA$_2^{Tyr}$ structural information, to probe *Eco*RI-digested DNA from *E. coli* strains CA274 and MB93 (Fig. 9D). These results show that the *E. coli* CA274 and MB93 8.6k-bp fragments that hybridize with this probe are analogous to the λ*rif*^{d}*18*-derived fragment, and therefore contain the tRNA$_2^{Tyr}$ gene region.

A smaller hybridizing fragment was also observed in each of the *E. coli* digests (6.1k bp in CA274 and 5.8k bp in MB93) (Fig. 9D). These fragments carry the doublet and singlet, respectively, tRNA$_1^{Tyr}$ genes. Since the φ80p*su*$^{+-}$*3* *Eco*RI fragment carrying the tRNA$_1^{Tyr}$ genes contains both *E. coli* and phage DNA (J. Rossi and A. Landy, unpubl.), a direct size correlation between the phage- and *E. coli*-derived fragments cannot be made, but the differences in size between the CA274 (doublet) and MB93 (singlet) hybridizing fragments (6.1k vs 5.8k bp) (Fig. 9D) is consistent with our structural analyses of the tRNA$_1^{Tyr}$ genes on the singlet and doublet bacteriophages.

DISCUSSION

The tRNA$_1^{Tyr}$ and tRNA$_2^{Tyr}$ gene clusters are found on opposite sides of the *E. coli* chromosome. Despite the fact that the tRNA^{Tyr} mature structural sequences differ by only two nucleotide pairs (Goodman et al. 1968, 1970; Fig. 7), the structural organization of the regions surrounding these sequences are very different (Figs. 1, 5, and 7).

The tRNA$_1^{Tyr}$ genes are characterized by having two tRNA$_1^{Tyr}$ mature structural sequences, separated by a 200-bp intergenic spacer sequence (Fig. 1). In contrast, there is a single-sequence coding for tRNA$_2^{Tyr}$. The organization of genes in this region is tRNA$_4^{Thr}$-8 bp-tRNA$_2^{Tyr}$-115 bp-tRNA$_2^{Gly}$-6 bp-tRNA$_3^{Thr}$.

A major feature of the tRNA$_1^{Tyr}$ genes that is not found in the tRNA$_2^{Tyr}$ gene region (Fig. 5) is the occurrence of a 178-bp sequence that commences with the last 19 bp of the tRNA$_1^{Tyr}$ mature structural sequence and is tandemly repeated 3.14 times (Egan and Landy 1978; Figs. 1, 3, and 4). In addition to the unusual structure of this region, an in vitro, rho-dependent transcriptional termination site has been identified in the

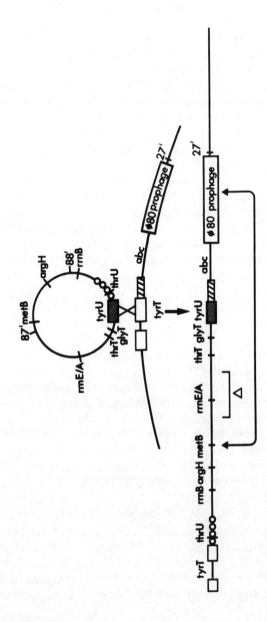

Figure 8 (See facing page for legend.)

second repeating unit (Küpper et al. 1978; Figs. 1, 3, and 4). The DNA sequence is reproduced in each of the 178-bp repeat units with very few changes (Fig. 3), leaving open the possibility that there are potential termination sites in the first and third repeating units (Figs. 1, 3, and 4). Such a pattern of repeats could provide an in vivo mechanism for modulation of transcriptional termination.

An important question concerns the origin of these repeat units. Are they present in the *E. coli* CA275 chromosome, from which the transducing phages were derived, or were they generated during the formation of the su^+3 transducing phages? The experiments presented demonstrate that the repeat structures are present in both the *E. coli* CA275 and CA274 chromosomes (Fig. 4) and in the unrelated strain HB101 (data not shown) with the structural organization predicted from bacteriophage analyses (Landy et al. 1974a; Egan and Landy 1978). This result is especially important, since another tRNA$_1^{Tyr}$ transducing phage analyzed, $\phi 80sus_2psu^+3$ (Kyoto University) (Andoh and Ozeki 1968) does not contain the repeated sequences. Tandemly duplicated sequences are known to occur frequently in the *E. coli* genome, but most of these, unless maintained by selection, are unstable and readily revert to the nonduplicated structure (for a review, see Anderson and Roth 1977). The stable maintenance of the tRNA$_1^{Tyr}$-associated repeat units, even under nonselective conditions in the

Figure 8 Proposed mechanism for the formation of the tRNA gene hybrid structure in $\lambda h80dglyTsu^+36$. The construction of $\lambda h80dglyTsu^+36$ started with an unusual $\phi 80dmetB$ transducing phage (Squires et al. 1973). The latter had been isolated from an *E. coli* strain in which the *metB* region of the chromosome had been transposed near the bacteriophage $\phi 80$ attachment site (Konrad 1969). At the time of its isolation there was no apparent mechanism or genetic pathway for this unusual transposition; however, the results reported here do seem to provide one. Tandem duplications between the multiple rRNA cistrons in *E. coli* occur with high frequency (Hill et al. 1969, 1977; Hill and Cambriato 1973). A large fraction of these *rrn* duplications includes the *metB* and *glyT* loci. Hill et al. (1977) have isolated and characterized covalently closed supercoiled circles generated by recombination between these duplicated segments. This model postulates the formation of such a circle (class II from Hill et al. 1977), which then reintegrates into the chromosome near $\phi 80att$ by the only known homologies in this region, the *tyrT* and *tyrU* loci. This recombination event would thus generate the hybrid structure between *tyrT* and the gene cluster containing *tyrU*. A deletion of chromosomal material between *thrT* and *metB*, including the hybrid *rrn* E/A genes, is proposed to accommodate *metB* and the tRNA gene cluster in the transducing phage and coincide with evidence (data not presented) that demonstrates $\lambda h80dglyTsu^+36$ does not carry rRNA genes (J. Rossi and A. Landy, unpubl.). The relative locations of genetic markers are from Bachman et al. (1976). (▨) tRNA$_2^{Tyr}$ mature structural sequences; (□) tRNA$_1^{Tyr}$ mature structural sequences; (OOOO) sequences upstream of tRNA$_2^{Tyr}$; (▨) sequences upstream of tRNA$_1^{Tyr}$. (Reprinted, with permission, from Rossi and Landy 1979.)

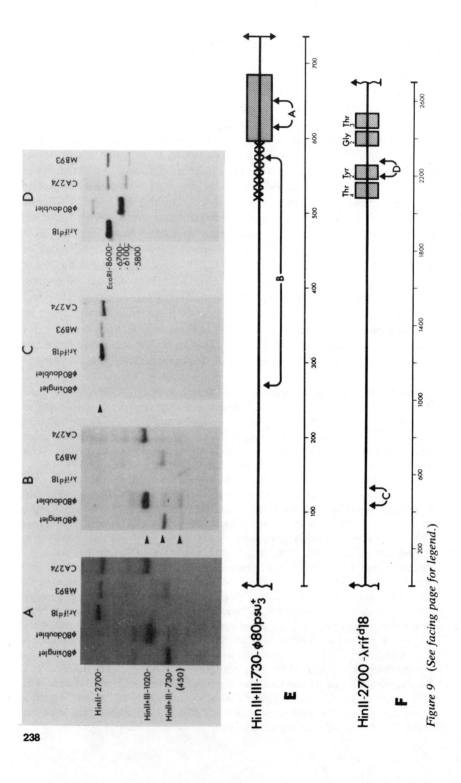

Figure 9 (See facing page for legend.)

su^- CA274 strain (Fig. 4), strongly suggests that these structures have an essential function in vivo. The longest in vivo tRNA$_1^{Tyr}$ transcripts described to date extend only a few nucleotides beyond the CCA terminal (Altman and Smith 1971; Ghysen and Cellis 1974). The data presented here suggest that these transcripts represent partially processed products of an as yet unidentified primary transcript.

The isolation of several tRNA$_1^{Tyr}$ promoter mutants (Berman and Beckwith 1979) represents the first such mutants characterized for a stable RNA gene (M. Berman and A. Landy, in prep.). All of the mutants analyzed thus far consist of single-base-pair changes occurring either within or just outside the promoter heptamer sequence (Fig. 2). These results further support the proposed importance of this region in RNA polymerase-promoter recognition (Pribnow 1975; Gilbert 1976). The

Figure 9 Analysis of the tRNA^{Tyr} gene regions on the *E. coli* MB93 and CA274 chromosomes. (*A–C*) 2–3 μg/lane of either *Hin*II + III-digested ϕ80psu^+3, ϕ80p$su^{+-}3$, or λrif^d18 DNAs, and approximately 200 μg/lane of *Hin*II + III-digested *E. coli* MB93 or 300 μg/lane of *Hin*II + III-digested CA274 DNAs were electrophoresed in 2.5% agarose, transferred to nitrocellulose (Southern 1975), and hybridized to the ^{32}P-labeled, nick-translated probes depicted (*E, F*). (*D*) Either 2 μg/lane of *Eco*RI-digested ϕ80p$su^{+-}3$ or λrif^d18, or 25 μg/lane of *E. coli* MB93 or CA274 DNAs were electrophoresed in 1% agarose and treated as described above. The origins of the radioactive restriction fragment probes were *Hin*fI-35.0 from ϕ80psu^+3 (*A*), *Hae*III-300 from ϕ80psu^+3 (*B*), *Hae*III-93 from λrif^d18 (*C*), and *Hae*III-74.0 from λrif^d18 (*D*). (*E*) Regions in the ϕ80psu^+3 *Hin*II + III-730 (C1a) fragment from which *A* and *B* were derived; (*F*) regions from the λrif^d18 *Hin*II-2700 fragment from which *C* and *D* were derived. Symbols used are the same as those described for Figs. 1 and 4. The scales given are in bp. Sizes (in bp) of *Hin*II and *Hin*II + III hybridizing fragments are from previously determined estimates for ϕ80psu^+3, ϕ80p$su^{+-}3$, and λrif^d18 tRNA^{Tyr}-carrying fragments (Rossi et al. 1979). *Eco*RI hybridizing fragment sizes (in bp) are from previously determined estimates for ϕ80p$su^{+-}3$ (Rossi et al. 1979) and λrif^d18 (Nomura 1976). The *Eco*RI *E. coli* hybridizing fragment sizes were estimated from their electrophoretic mobility relative to *Eco*RI-digested λ DNA fragments electrophoresed in an adjacent lane (not shown). Faint bands of hybridization that are not indicated are due to partial endonuclease digestion in most cases. A singlet-size tRNA$_1^{Tyr}$-containing fragment can also be seen in the *Hin*II + III digest for ϕ80p$su^{+-}3$ and *E. coli* CA274. These minor bands represent the conversion, by unequal recombination, of the doublet to singlet structures (Russell et al. 1970). In addition, a 450-bp fragment that hybridizes with the *Hae*III-300 probe (*B*), has been identified as *Hin*III-II-450 and maps downstream from the tRNA$_1^{Tyr}$ mature structural sequences in ϕ80p$su^{+-}3$ and ϕ80psu^+3 (Landy et al. 1974a; Egan and Landy 1978). The homologous sequences in the probe and the *Hin*III-II-450 fragment are not known. The small discrepancies between electrophoretic mobilities of the bacteriophage vs *E. coli Hin*II + III hybridizing fragments are due to differences in the amount of DNA electrophoresed into each of the lanes.

tRNA$_1^{Tyr}$ promoter exhibits relatively poor in vitro RNA polymerase binding (Küpper et al. 1975). This and other unusual properties of this promoter have been attributed to the high G+C content of this region (Küpper et al. 1975). Each of the four mutants described here represent AT→GC transitions, which would tend to further stabilize this region, thus lending support to this hypothesis.

The results of in vivo and in vitro tRNA$_1^{Tyr}$ transcriptional studies suggest that both tRNA$_1^{Tyr}$ copies are cotranscribed in the same transcriptional unit (Ghysen and Cellis 1974; Daniel et al. 1975; Küpper et al. 1975). Direct comparisons of in vitro transcripts from the restriction fragments *Hin*II + III-1020 (B10a) and *Hin*II + III-730 (C1a) demonstrated that tRNA$_1^{Tyr}$ transcription initiated at the same start signal in both fragments; also, the doublet transcript contained equimolar amounts of tRNA$_1^{Tyr}$ su^+ and su^- sequences (Küpper et al. 1975). The transcript produced from the doublet restriction fragment was about 235 bases longer than that produced from the singlet fragment, which is in close agreement with the differences we have described for the doublet and singlet structures (Fig. 1).

Taking into account the start and stop points of in vivo and in vitro transcription (Altman and Smith 1971; Küpper et al. 1975, 1978), the total length of the tRNA$_1^{Tyr}$ doublet primary transcript should be approximately 640 nucleotides or 3.7 times larger than the two 85-base tRNA$_1^{Tyr}$ mature structural sequences (see Fig. 1). Whether or not any of the 470 extra nucleotides that are processed from the primary transcript have any cellular function is an interesting but unanswered question that is currently under investigation.

The organization of the tRNA$_2^{Tyr}$ gene region is quite different from that determined for the tRNA$_1^{Tyr}$ genes (Fig. 1). This region includes a cluster of four different tRNA genes within 450 bp.

The following evidence strongly suggests that the sequence upstream of tRNA$_2^{Tyr}$ codes for a tRNA^{Thr} species (tRNA$_4^{Thr}$) (Fig. 6). M. Yamamoto and M. Nomura (pers. comm.), using total cellular tRNA, found a previously unidentified tRNA species that hybridizes to the $\lambda rif^d 18$ genome. We presume this tRNA to be the newly identified tRNA$_4^{Thr}$ shown in Figure 6. E. Lund and J. Dahlberg (pers. comm.) have obtained an RNAse T1 fingerprint of a previously unidentified tRNA. This fingerprint pattern is consistent with the predicted set of oligonucleotides from the known DNA sequence (Fig. 6). Ilgen et al. (1976) and Shimura and Sakano (1977), working with strains temperature sensitive for the processing enzyme RNAse P, have isolated precursors (approximately 200 nucleotides long) containing tRNA$_2^{Tyr}$ and an unidentified tRNA. It is quite likely that the unidentified tRNA is tRNA$_4^{Thr}$. In addition to the 200-nucleotide precursor, Ilgen et al. (1976) isolated a separate tRNA$_2^{Gly}$-tRNA$_3^{Thr}$ precursor. Thus, it still remains to be determined whether all four tRNAs are contained in a single primary transcript. Experiments already in progress that are designed to study the transcription of the tRNA$_2^{Tyr}$ gene regions in $\lambda rif^d 18$, should provide answers to these problems. It should be noted that unlike other tRNA genes located distal to rRNA operons (Morgan et al. 1978), the tRNA$_4^{Thr}$-tRNA$_2^{Tyr}$-tRNA$_2^{Gly}$-tRNA$_3^{Thr}$ gene cluster is not transcribed from the *rrnB* promoter (M. Yamamoto and M. Nomura, pers. comm.).

An interesting feature of the tRNA$_4^{Thr}$ gene sequence presented in Figure 6 is the small inverted repeat flanking the 5' and 3' ends of the mature tRNA structure, which thus extends the base-pairing possibilities. At present, it is not known whether this extended base pairing plays a role in processing of the tRNA-containing transcripts. A possible analogy for such interactions is the base pairing between sequences at the 5' and 3' ends of 16S rRNA, which does appear to be important for processing by the endonuclease RNase III (Young and Steitz 1978).

From the standpoint of using specialized transducing bacteriophage to study *E. coli* gene organization and regulation, it is important to note that out of a total of four independently isolated transducing phages, two carrying the tRNA$_1^{Tyr}$ genes, $\phi 80psu^{+-}3$ and $\phi 80sus_2psu^+3$, and two carrying the tRNA$_2^{Tyr}$ gene, $\lambda rif^d 18$ and $\lambda h80dglyTsu^+36$, only one phage from each group ($\phi 80psu^{+-}3$ for tRNA$_1^{Tyr}$ and $\lambda rif^d 18$ for tRNA$_2^{Tyr}$) actually reflects the structure of the tRNA^{Tyr} gene regions on the *E. coli* chromosome (Figs. 4 and 9). The differences between sequences carried in the $\phi 80sus_2psu^+3$ and $\lambda h80dglyTsu^+36$ bacteriophages and those found on the *E. coli* chromosome are in regions known to be involved in transcriptional initiation or termination and presumably transcript processing.

Finally, it should be noted that the DNA sequences obtained for each of the six tRNAs in this study are entirely colinear with the known RNA sequences, thus eliminating the possibility of genomic sequences, whose transcripts must be spliced to form the mature tRNA, as has been found for several yeast tRNA genes (Goodman et al. 1977; Valenzuela et al. 1978 and this volume).

REFERENCES

Altman, S. and J. D. Smith. 1971. Tyrosine tRNA precursor molecule polynucleotide sequence. *Nat. New Biol.* **233**:35.

Anderson, P. and J. R. Roth. 1977. Tandem genetic duplications in phage and bacteria. *Annu. Rev. Microbiol.* **31**:473.

Andoh, T. and H. Ozeki. 1968. Suppressor gene su3$^+$ of *E. coli*, a structural gene for tyrosine tRNA. *Proc. Natl. Acad. Sci.* **59**:792.

Bachman, B. J., K. B. Low, and A. L. Taylor. 1976. Recalibrated linkage map of *Escherichia coli* K-12. *Bacteriol. Rev.* **40**:116.

Berman, M. L. and I. Beckwith. 1979. Use of gene fusions to isolate promoter mutants in the transfer RNA gene *tyrT* of *Escherichia coli*. *J. Mol. Biol.* **130**:303.

Chang, S. and J. Carbon. 1975. The nucleotide sequence of a precursor to the glycine- and threonine-specific transfer ribonucleic acids of *Escherichia coli*. *J. Biol. Chem.* **250**:5542.

Daniel, V., J. I. Grimberg, and M. Zeevi. 1975. In vitro synthesis of tRNA precursors and their conversions to mature size tRNA. *Nature* **257**:193.

Egan, J. and A. Landy. 1978. Structural analysis of the tRNA$_1^{Tyr}$ gene of *Escherichia coli*. *J. Biol. Chem.* **253**:3607.

Eggerston, G. and E. A. Adelberg. 1965. Map positions and specificities of suppressor mutations in *Escherichia coli* K-12. *Genetics* **52**:319.

Garen, A., S. Garen, and R. C. Wilhelm. 1965. Suppressor genes for nonsense mutations. I. The su^{-1}, su^{-2}, and su^{-3} genes of *Escherichia coli*. *J. Mol. Biol.* **14**:167.

Ghysen, A. and J. E. Cellis. 1974. Joint transcription of two tRNA$_1^{Tyr}$ genes from *Escherichia coli*. *Nature* **249**:418.

Gilbert, W. 1976. Starting and stopping sequences for the RNA polymerase. In *RNA polymerase* (ed. R. Losick and M. Chamberlin), p. 193. Cold Spring Harbor Laboratory, Cold Spring Harbor, New York.

Goodman, H. M., M. V. Olson, and B. D. Hall. 1977. Nucleotide sequence of a mutant eukaryotic gene: The yeast tyrosine-inserting ochre suppressor sup4-0. *Proc. Natl. Acad. Sci.* **74**:5453.

Goodman, H. M., J. Abelson, A. Landy, S. Brenner, and J. D. Smith. 1968. Amber-suppression: A nucleotide change in the anticodon of a tyrosine transfer RNA. *Nature* **217**:1019.

Goodman, H. M., J. N. Abelson, A. Landy, S. Zadrazil, and J. D. Smith. 1970. The nucleotide sequences of tyrosine transfer RNAs of *Escherichia coli*: Sequences of the amber suppressor su$_{III}^+$ transfer RNA, the wild type su$_{III}^-$ transfer RNA and tyrosine transfer RNAs species I and II. *Eur. J. Biochem.* **13**:461.

Guthrie, C., C. A. Scholla, H. Yesian, and J. Abelson. 1978. The nucleotide sequence of threonine transfer RNA coded by bacteriophage T4. *Nucleic Acids Res.* **5**:1833.

Hasegawa, T. and H. Ishikura. 1978. Nucleotide sequence of threonine tRNA from *Bacillus subtilis*. *Nucleic Acids Res.* **5**:537.

Hill, C. W. and G. Cambriato. 1973. Genetic duplication induced at very high frequency by ultraviolet irradiation in *E. coli*. *Mol. Gen. Genet.* **127**:197.

Hill, C. W., J. Foulds, L. Soll, and P. Berg. 1969. Instability of a missense suppressor resulting from a duplication of genetic material. *J. Mol. Biol.* **39**:563.

Hill, C. W., R. H. Graftstrom, B. W. Harnish, and B. S. Hillman. 1977. Tandem duplications resulting from recombination between ribosomal RNA genes in *Escherichia coli*. *J. Mol. Biol.* **116**:407.

Ilgen, C., L. L. Kirk, and J. Carbon. 1976. Isolation and characterization of large transfer ribonucleic acid precursors from *Escherichia coli*. *J. Biol. Chem.* **251**:922.

Kirschbaum, J. B. and E. B. Konrad. 1973. Isolation of a specialized lambda transducing bacteriophage carrying the Beta subunit gene for *Escherichia coli* ribonucleic acid polymerase. *J. Bacteriol.* **116**:517.

Konrad, B. 1969. "The genetics of chromosomal duplications." Ph.D. thesis, Harvard Medical School, Cambridge, Massachusetts.

Küpper, H., R. Contreras, A. Landy, and H. G. Khorana. 1975. Promoter-dependent transcription of tRNA$_1^{Tyr}$ genes using DNA fragments produced by restriction enzymes. *Proc. Natl. Acad. Sci.* **72:**4754.

Küpper, H., T. Sekiya, M. Rosenberg, J. Egan, and A. Landy. 1978. A ρ-dependent termination site in the gene coding for tyrosine tRNA su$_3$ of *Escherichia coli. Nature* **272:**423.

Landy, A., C. Foeller, and W. Ross. 1974a. DNA fragments carrying genes for tRNA$_1^{Tyr}$. *Nature* **249:**738.

Landy, A., E. Ruedisueli, L. Robinson, C. Foeller and W. Ross. 1974b. Digestion of deoxyribonucleic acids from bacteriophage T7, λ, and ϕ80h with site-specific nucleases from *Hemophilus influenzae* strain Rc and strain Rd. *Biochemistry* **13:**2134.

Maxam, A. M. and W. Gilbert. 1977. A new method for sequencing DNA. *Proc. Natl. Acad. Sci.* **74:**560.

Morgan, E. A., T. Ikemura, L. Lindahl, A. M. Fallon, and M. Nomura. 1978. Some rRNA operons in *E. coli* have tRNA genes at their distal ends. *Cell* **13:**335.

Nomura, M. 1976. Organization of bacterial genes for ribosomal components: Studies using novel approaches. *Cell* **9:**633.

Orias, E., T. K. Gartner, J. E. Lannan, and M. Betlach. 1972. Close linkage between ochre and missense suppressors in *Escherichia coli. J. Bacteriol.* **109:**1125.

Pribnow, D. 1975. Nucleotide sequence of an RNA polymerase binding site at an early T7 promoter. *Proc. Natl. Acad. Sci.* **72:**784.

Rosenberg, M., D. Court, H. Shimatake, C. Brady, and D. L. Wulff. 1978. The relationship between function and the DNA sequence in an intercistronic regulatory region in phage λ. *Nature* **272:**414.

Rossi, J. J. and A. Landy. 1979. Structure and organization of the two tRNA^{Tyr} gene clusters on the *E. coli* chromosome. *Cell* **16:**523.

Rossi, J. J., W. Ross, J. Egan, D. J. Lipman, and A. Landy. 1979. Structural organization of *Escherichia coli* tRNA^{Tyr} gene clusters in four different transducing bacteriophages. *J. Mol. Biol.* **128:**21.

Russell, R. L., J. N. Abelson, A. Landy, M. L. Gefter, S. Brenner, and J. D. Smith. 1970. Duplicate genes for tyrosine transfer RNA in *Escherichia coli. J. Mol. Biol.* **47:**1.

Sekiya, T. and H. G. Khorana. 1974. Nucleotide sequence in the promoter region of the *Escherichia coli* tyrosine tRNA gene. *Proc. Natl. Acad. Sci.* **71:**2978.

Sekiya, T., R. Contreras, H. Küpper, A. Landy, and H. G. Khorana. 1976. *Escherichia coli* tyrosine transfer ribonucleic acid genes. *J. Biol. Chem.* **251:**5124.

Shimura, Y. and H. Sakano. 1977. Processing of tRNA precursors in *Escherichia coli*. In *Nucleic acid-protein recognition* (ed. H. J. Vogel), p. 293. Academic Press, New York.

Signer, E. R., J. Beckwith, and S. Brenner. 1965. Mapping of suppressor loci in *Escherichia coli. J. Mol. Biol.* **14:**153.

Smith, J. D., L. Barnett, S. Brenner, and R. L. Russell. 1970. More mutant transfer ribonucleic acids. *J. Mol. Biol.* **54:**1.

Southern, E. M. 1975. Detection of specific sequences among DNA fragments separated by gel electrophoresis. *J. Mol. Biol.* **98**:503.

Squires, C., B. Konrad, J. Kirschbaum, and J. Carbon. 1973. Three adjacent transfer RNA genes in *Escherichia coli*. *Proc. Natl. Acad. Sci.* **70**:438.

Valenzuela, P., A. Venegas, F. Weinberg, R. Bishop, and W. J. Rutter. 1978. Structure of yeast phenylalanine-tRNA genes: An intervening segment within the region coding for the tRNA. *Proc. Natl. Acad. Sci.* **75**:190.

Wu, M., N. Davidson, and J. Carbon. 1973. Physical mapping of the transfer RNA genes on $\lambda h80dglyTsu_{36}^{+}$. *J. Mol. Biol.* **78**:23.

Young, R. A. and J. A. Steitz. 1978. Complementary sequences 1700 nucleotides apart form a ribonuclease III cleavage site in *E. coli* ribosomal precursor RNA. *Proc. Natl. Acad. Sci.* **75**:3593.

Cloning of Two Chemically Synthesized Genes for a Precursor to the su^+3 Suppressor tRNA^{Tyr}

**Michael J. Ryan,* Eugene L. Brown,
Ramamoorthy Belagaje, and H. Gobind Khorana**
Departments of Biology and Chemistry
Massachusetts Institute of Technology
Cambridge, Massachusetts 02139

Hans-Joachim Fritz
Institut für Genetik der Universität Köln
Köln, Federal Republic of Germany

There are three tRNA^{Tyr} genes in *Escherichia coli*. One of these genes is found near 88 minutes on the *E. coli* genetic map; the other two are present as a tandem duplication near the bacteriophage ϕ80 attachment site at 27 minutes. The structures and organization of these three genes are described in detail by Rossi et al. (this volume). The su^+3 amber suppressor tRNA^{Tyr} has arisen as a result of an anticodon mutation in one of the tandem, duplicate *tyr* tRNA genes. This su^+3 tRNA gene was an ideal goal for the chemical synthesis of a gene because its promoter sequence could be determined (Sekiya et al. 1975, 1976a,b), it is a relatively small gene, and it has an in vivo function that can be detected easily. This paper describes the cloning and characterization of both the chemically synthesized gene for a precursor to the su^+3 tRNA as well as a second, modified derivative of this gene.

THE CHEMICALLY SYNTHESIZED su^+3 *tyr* tRNA GENE

Figure 1 represents the total synthetic gene for a precursor to the su^+3 tRNA. This synthetic gene includes 56 bp in the promoter region, 126 bp in the structural gene that codes for the su^+3 *tyr* tRNA precursor sequenced previously (Altman and Smith 1971), and 25 bp beyond the mature 3' end of this tRNA. Originally, it was thought that this region might contain a transcription termination signal. However, recent data have shown that transcription in vitro proceeds for approximately 225 nucleotides before a rho-dependent termination site was encountered (Küpper et al. 1978). Therefore, the synthetic gene differs from the

---

*Present address: Microbiological Sciences, Schering Corporation, Bloomfield, New Jersey 07003.

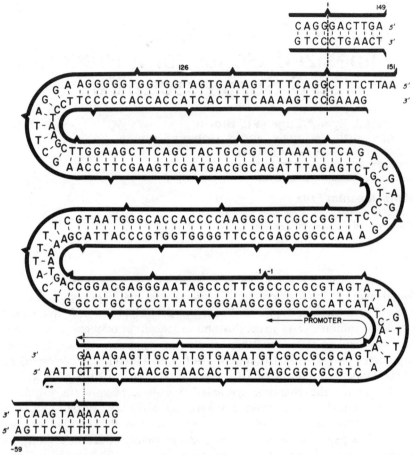

Figure 1 Chemically synthesized gene for a precursor to the suppressor tRNA^{Tyr}. (Λ) Points at which T4 polynucleotide ligase was used to join oligodeoxynucleotides to one another. (Reprinted, with permission, from Ryan et al. 1979.)

naturally occurring analog because it lacks this transcription termination signal contained within the remarkable 178-bp sequence repeated 3.14 times in the region downstream of the naturally occurring su^+3 gene (Egan and Landy 1978).

T4 polynucleotide ligase was used to join the chemically synthesized oligonucleotides to one another to produce a number of double-stranded DNA segments with protruding single-stranded ends; these were, in turn, joined to one another by T4 ligase to give the complete gene. In the upper-right and lower-left corners of Figure 1 are the naturally occurring nucleotide sequences that were modified to allow the inclusion of the protruding single-stranded AATT 5' ends produced by the EcoRI restriction endonuclease.

Cloning the Synthetic su^+3 tRNA Gene

The presence of these AATT sticky ends on the synthetic gene has allowed it to be joined enzymatically to EcoRI-digested vector chromosomes. The vectors and procedures used are shown schematically in Figure 2. The first vector, Charon 3A (Blattner et al. 1977), is a derivative of bacteriophage λ, which has only two sites for digestion by EcoRI on either side of a nonessential region of the phage chromosome. In addition, this phage carried a small deletion in its immunity region, resulting in an essentially virulent phenotype and amber mutations in genes A and B, which are required for capsid formation. With this phage vector, the successful cloning and expression of the synthetic suppressor tRNA gene is indicated by the presence of plaque-forming phage particles after transformation of nonsuppressing bacteria with recombinant DNA carrying this synthetic gene. In the second procedure, this synthetic gene was joined to the EcoRI-digested, ampicillin-resistant ColE1 plasmid (So et al. 1975) to form recombinant DNAs that were used to transform a strain of E. coli carrying amber mutations in genes required for the biosynthesis of histidine and tryptophan. Here, the ability to isolate ampicillin-resistant, prototrophic bacteria indicated that the synthetic suppressor tRNA gene was functioning correctly in vivo. The results of these experiments (Table 1) demonstrated that both bacterial and bacteriophage amber mutations were suppressed in vivo after transformation of nonsuppressing bacteria with recombinant DNA presumed to carry the synthetic suppressor gene.

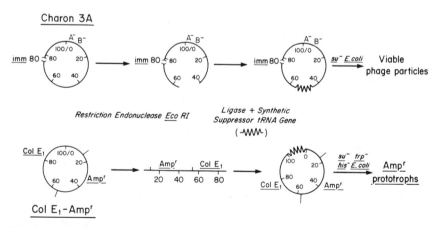

Figure 2 Outline of procedures and relevant structural features of the plasmid and bacteriophage DNAs used to prepare recombinant DNAs containing the chemically synthesized suppressor tRNA gene. The mutations in the bacteriophage vector, Charon 3A (A^-B^-), as well as those in the su^- host ($his^- trp^-$), are all amber mutations. (Reprinted, with permission, from Ryan et al. 1979.)

Table 1 In vivo suppression of phage and bacterial amber mutations by the synthetic suppressor tRNA gene

Experiment I

Vector DNA (0.8 µg)	Synthetic suppressor DNA	Plaque-forming units/ml after transformation of	
		CA274	LS340
Charon 3A	−	0	0
Charon 3A	+	1.6×10^6	—
Charon 3A	+	—	1.8×10^7

Experiment II

Vector DNA (0.5 µg)	Synthetic suppressor DNA	Transformants/ml of LS340 found to be:	
		amp^r	amp^r prototrophs
ColE1-Ap^r [a]	−	398	0
ColE1-Ap^r [a]	+	297	47

In each experiment the vector DNA was digested with the restriction endonuclease *Eco*RI and aliquots were treated with T4 polynucleotide ligase in the presence or absence of 0.01 µg of the chemically synthesized suppressor tRNA^{Tyr} gene. The amount of each vector DNA shown below in parentheses is the level that was present in the ligase reaction mixture used for the transformation assay. In all cases, phages were titered using *E. coli* CA274 (*lacZ* amber, *trp* amber) as the host strain. Ampicillin-resistant derivatives of LS340 (*his* amber, *trp* amber) were detected by growth on LB broth plates containing 35 µg/ml of the antibiotic. The same level of ampicillin was used when scoring ampicillin-resistant prototrophs on M9 minimal plates containing 40 µg/ml L-methionine and 0.2% glucose.
[a] Ampicillin-resistant ColE1.

Characterization of Recombinant DNA

Although the suppression of amber mutations indicated that the cloned synthetic gene was able to function in vivo, it was necessary to show that the suppressor function was part of the presumed recombinant DNA and that the cloned synthetic DNA could be reisolated from the amplified recombinant DNA and shown to be the correct size and to code for suppressor activity in further transformation experiments. Plasmid DNA was therefore isolated from the ampicillin-resistant prototrophs obtained after transformation with recombinant DNAs believed to carry the synthetic suppressor gene. This plasmid was then used in other transformation experiments and found to confer both antibiotic resistance and suppressor activity on the host bacteria (Table 2).

*Eco*RI restriction endonuclease was used to digest the chromosomal

Table 2 Transformation of *E. coli* LS340 with purified cloned recombinant plasmid DNA carrying the synthetic suppressor tRNA gene

Transforming DNA	amp^r transformants/ml	amp^r prototrophs/ml
ColE1-Ap^r [a]		
(0.7 μg)	2.4×10^4	0
(0.07 μg)	0.8×10^4	0
pSSU101		
(0.9 μg)	6.2×10^4	5.1×10^4
(0.09 μg)	1.9×10^4	1.3×10^4

The presumed pSSU101 plasmid DNA was isolated from an ampicillin-resistant, prototrophic transformant of LS340 constructed as shown in Table 1. Transformants were scored as in Table 1.
[a] Ampicillin-resistant ColE1.

DNA of a Charon 3A derivative presumed to carry the cloned synthetic suppressor gene. This reaction was analyzed by electrophoresis on a polyacrylamide gel (Fig. 3B), which revealed the presence of a new, low-molecular-weight fragment of DNA present in the phage genome assumed to carry suppressor activity. This fragment was isolated by chromatography on Sepharose 4B (Fig. 3A) and shown to be the same size as the original, chemically synthesized gene, as judged by its electrophoretic mobility (Fig. 3C). Another sample of this DNA fragment was isolated by preparative electrophoresis and shown to function as a source of suppressor activity when recloned in the phage vector Charon 3A, as well as two plasmid vectors, ampicillin-resistant ColE1 and pMB9 (Bolivar et al. 1977). The recombinant plasmids formed by incorporating this synthetic suppressor gene into pMB9 and ampicillin-resistant ColE1 have been designated pSSU1 and pSSU101, respectively.

In all of the situations described to this point, the ability of this cloned synthetic gene to suppress amber mutations was examined under conditions in which there were many copies of this gene per bacterial chromosome. Therefore, the virulent Charon 3A derivative carrying the cloned synthetic su^+3 gene was crossed with a temperate λ phage to produce recombinant, lysogenizing phage carrying the synthetic suppressor gene. This phage, designated λssu1, was used to demonstrate that this synthetic gene can still suppress amber mutations when integrated into the bacterial chromosome as part of a λ prophage. The presence of the cloned synthetic su^+3 gene in these recombinant phages and plasmids, selected by their suppressor activity, has been confirmed by digestions of each of these DNAs with *Eco*RI followed by electrophoretic analysis on polyacrylamide gels (Fig. 4).

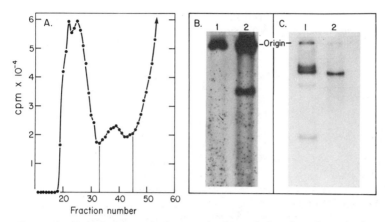

Figure 3 Reisolation and characterization of the cloned synthetic suppressor tRNA gene from Charon 3A *ssu1*. DNA samples from Charon 3A and Charon 3A*ssu1* were digested with the restriction endonuclease *Eco*RI and the 5′ ends were first dephosphorylated with bacterial alkaline phosphatase and then rephosphorylated with [γ-^{32}P]ATP and polynucleotide kinase. The end-labeled DNA fragments were then analyzed on a 7.5% polyacrylamide gel shown in *B*: Charon 3A (*lane 1*), Charon 3A*ssu1* (*lane 2*). A sample of Charon 3A*ssu1* (410 μg), after *Eco*RI digestion and end-labeling, was chromatographed on a Sepharose 4B column and, as indicated in *A*, fractions 33–45 were pooled. After concentration and dialysis, approximately 1 pmole of this cloned, low-molecular-weight restriction fragment was electrophoresed on a 10% polyacrylamide gel (*C, lane 2*); the adjacent track (*C, lane 1*) contained 0.1 pmole of the synthetic suppressor tRNA^{Tyr} gene, which had been phosphorylated with [γ-^{32}P]ATP (90,000 cpm/pmole) and polynucleotide kinase. The extra bands seen in this synthetic marker DNA sample are the result of a single nick in one strand of a fraction of these DNA molecules. (Reprinted, with permission, from Ryan et al. 1979.)

In Vitro Transcription Studies

Transcription of the recombinant plasmid pSSU1 with purified *E. coli* RNA polymerase holoenzyme yielded a heterogeneous set of products, most of which were too large to enter the polyacrylamide gel used for their analysis (Fig. 5A). This would be expected, since these transcription experiments were done in the absence of rho factor and, as described above, the synthetic su^+3 gene lacks the termination site contained within the 178-bp repeat that occurs downstream of the naturally occurring su^+3 gene. Therefore, under these conditions there is probably a good deal of read-through transcription across the plasmid genome starting from any active promoter. However, when this primary transcript was exposed to a crude extract of *E. coli* for increasing periods of time, almost all of the high-molecular-weight material was degraded (Fig. 5, B–D). Simul-

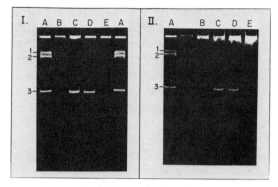

Figure 4 EcoRI restriction endonuclease digestion products of recombinant DNAs carrying the cloned synthetic suppressor tRNA^{Tyr} gene. After digestion with the restriction endonuclease EcoRI, each of the DNA samples was precipitated with ethanol, redissolved, and analyzed on a 10% polyacrylamide gel. The molecular weight markers (*A* in *I* and *II*), 1 (630 bp), 2 (520 bp), and 3 (205 bp) were derived from pKB252 (Backman et al. 1976) by digestion with both the EcoRI and HindIII restriction endonucleases. (*I*) EcoRI restriction pattern of ampicillin-resistant ColE1 (*B*), pSSU101 (*C*), pSSU1 (*D*), and pMB9 (*E*). (*II*) Restriction patterns of Charon 3A (*B*), Charon 3A*ssu1* (*C*), λ*ssu1* (*D*), and λp*lac5Sam7c*I857 (*E*). (Reprinted, with permission, from Ryan et al. 1979.)

taneously, two other bands appeared. The slower moving of the two, which decreased with time, was shown by two-dimensional fingerprint analysis to correspond to the 126-nucleotide-long Altman and Smith *tyr* tRNA precursor. The faster moving band was similarly identified as the

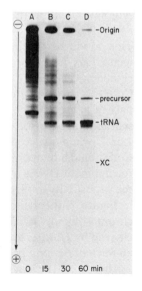

Figure 5 Autoradiograph of polyacrylamide gel separation of products formed in the in vitro transcription of plasmid pSSU1 and their processing with *E. coli* MRE600 S100 extract. The plasmid was transcribed using [α-^{32}P]UTP as one of the four nucleoside triphosphates. The transcripts and cleavage products were fractionated by electrophoresis on a 10% polyacrylamide gel containing 7 M urea. (*A*) Primary transcripts; (*B–D*) cleavage products obtained after processing with the S100 extract for the indicated times. ^{32}P-labeled 5S RNA was electrophoresed in parallel with the cleavage products and found to comigrate with the precursor (not shown). (Reprinted, with permission, from Ryan et al. 1979.)

85-nucleotide-long mature *tyr* tRNA. Under these conditions the removal of the first 41 nucleotides by an endonucleolytic cleavage carried out by RNase P (Robertson et al. 1972) appears to be rate limiting. These findings were expanded by treating the primary transcript with purified RNase P alone and with a crude extract prepared from *E. coli* A49, which carries a temperature-sensitive RNase P (Schedl and Primakoff 1973). In the former case, the first 41 nucleotides corresponding to the 5' end of the Altman and Smith precursor were released intact and possessed a GTP as the first nucleotide residue. In the latter case, the 126-nucleotide-long precursor was formed. This molecule was purified and then cleaved by purified RNase P to yield both the 85-nucleotide-long mature tRNA and the 41-nucleotide-long 5'-end fragment. The results of these experiments demonstrate that the cloned synthetic su^+3 gene is transcribed from its own promoter. Furthermore, they indicate that the nucleotide sequence of the structural gene has been conserved and is accurate.

CLONING AND CHARACTERIZATION OF THE MODIFIED SYNTHETIC su^+3 GENE

The first sequence alteration introduced into the synthetic su^+3 gene involved the substitution of four AT base pairs for the four GC base pairs that immediately precede the point of transcription initiation (Fig. 6). There are three features of the new sequence that should be noted. First, the rather high GC content of this promoter was decreased and, therefore, might permit a higher rate of transcription of this gene by facilitating strand separation by RNA polymerase. Second, some of the twofold symmetry within this promoter region was destroyed. And third, a recognition sequence for the *Hin*dIII restriction endonuclease was created. This modified synthetic suppressor gene was cloned using the same techniques described above and found to suppress amber mutations when cloned in Charon 3A, as well as the plasmids pMB9 and ampicillin-resistant ColE1. It also functioned when integrated into the bacterial chromosome as part of a λ prophage. The recombinant molecules carrying this modified synthetic suppressor gene have been designated pSSU2, pSSU102, and λ*ssu2*, when the vector chromosomes were pMB9, ampicillin-resistant ColE1, and a temperate λ phage, respectively. It was possible to check the recombinant DNAs for the presence of this altered sequence by examining their sensitivity to digestion by *Hin*dIII restriction endonuclease. As shown in Figure 7, the synthetic su^+3 genes could be excised from their vector chromosomes by *Eco*RI digestion, but only the modified synthetic gene was cleaved by *Hin*dIII to yield the 151-nucleotide-long structural gene and the 56-nucleotide-long promoter.

The *Hin*dIII cleavage site in this gene, at the juncture between the

Modified Promoter Sequence

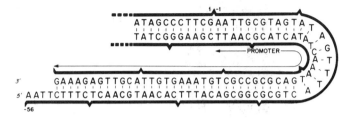

Promoter Sequence

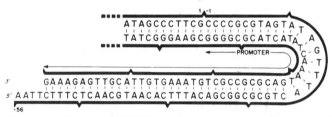

Figure 6 The partial DNA sequence (*top*) includes the modified promoter in which the sequence of nucleotides from −4 to −1, TTAA, has replaced the CCCC sequence found in the naturally occurring gene (*bottom*).

promoter and structural gene, also allowed an in vitro test of the ability of the *E. coli* RNA polymerase to recognize this region. As seen in Figure 8, RNA polymerase inhibited the digestion of the modified synthetic su^+3 gene by *Hin*dIII restriction endonuclease. This figure also shows that RNA polymerase inhibited cleavage at the other *Hin*dIII site on this plasmid, which occurs within the promoter region of the gene coding for tetracycline resistance (Boyer et al. 1977).

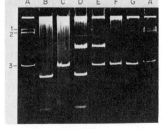

Figure 7 Polyacrylamide gel electrophoresis of restriction-enzyme-digested recombinant plasmid DNAs carrying the synthetic su^+3 genes. (*A*) pKB252 (Backman et al. 1976) + *Eco*RI + *Hin*dIII; (*B*) pSSU102 + *Eco*RI + *Hin*dIII; (*C*) pSSU102 + *Eco*RI; (*D*) pSSU2 + *Eco*RI + *Hin*dIII; (*E*) pSSU1 + *Eco*RI + *Hin*dIII; (*F*) pSSU2 + *Eco*RI; (*G*) pSSU1 + *Eco*RI. (Reprinted, with permission, from Ryan et al. 1979.)

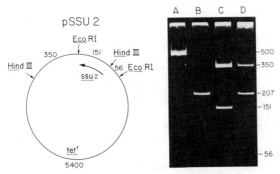

Figure 8 RNA polymerase inhibition of cleavage of pSSU2 by HindIII restriction endonuclease. Each 0.1-ml reaction contained 11 µg (approximately 3 pmole) of pSSU2 in 50 mM Tris-HCl (pH 7.6), 7 mM MgCl$_2$, 1 mM by GTP, 0.265 mM CTP, 1.15 mM dithiothreitol, 10 mM NaCl, 30 mM KCl, 0.015 mM EDTA, 6 mM KPO$_4$ (pH 8), and 7.5% glycerol. The reactions were preincubated at 37°C for 18 min before adding the following restriction enzymes: HindIII (*A*), EcoRI (*B*), HindIII (*C*), and EcoRI (*D*). The *E. coli* RNA polymerase holoenzyme (27 pmole), present in *D* only, was added prior to the preincubation step. The enzymes were destroyed by heating at 65°C for 10 min. The digested DNAs were precipitated with ethanol, redissolved, and analyzed on a 10% polyacrylamide gel followed by staining in ethidium bromide.

CLONING OF THE NATURALLY OCCURRING su⁺3 GENE

An *Eco*RI restriction fragment having a molecular weight of approximately 4.3 × 10⁶ and carrying the naturally occurring *su⁺3* gene was initially cloned from ɸ80*su⁺3* (Russell et al. 1970) into Charon 3A. The recombinant phage formed in this experiment has been designated Charon 3A *su⁺3*. Because direct DNA sequence analysis (Egan and Landy 1978) has discovered a recognition sequence for the *Eco*RI restriction endonuclease 500 nucleotides from the start of the *su⁺3* structural gene, this cloned *Eco*RI restriction fragment should contain the rho-dependent termination site detected in in vitro transcription studies (Küpper et al. 1978). The virulent Charon 3A *su⁺3* was later crossed with a temperate λ phage to yield a derivative, designated λ*su⁺3*, which has suppressor activity and can lysogenize *E. coli*. The *Eco*RI digestion patterns of each of these phages is shown in Figure 9. From this data, it appears that the suppressor gene in λ*su⁺3* is now present on the much smaller (approximately 1200 bp) *Eco*RI restriction fragment seen in lane 5. This observation was tested by first joining *Eco*RI-digested λ*su⁺3* and pMB9 DNAs with T4 polynucleotide ligase and then transforming cells carrying *his* and *trp* amber mutations with this mixture of recombinant molecules. Plasmid DNA, designated pMR243, isolated from a tetra-

Figure 9 EcoRI digestion patterns of recombinant DNAs. Each DNA sample was digested with EcoRI, precipitated with ethanol, redissolved, and electrophoresed on a horizontal 1% agarose gel followed by staining with ethidium bromide. The resulting patterns correspond to the following EcoRI-digested chromosomes: $\phi 80 su^+3$ (*1*), Charon 3A (*2*), Charon 3A su^+3 (*3*), $\lambda plac5$ (*4*), λsu^+3 (*5*), pMR243 (*6*), Charon 3A *ssu1* (*7*), $\lambda ssu1$ (*8*), λsu^+3 (*9*).

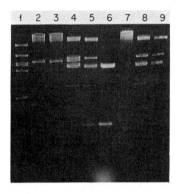

cycline-resistant, prototrophic transformant was shown to have incorporated this small fragment of DNA, indicating that the su^+3 gene is contained within it (lane 6). Both the phage λsu^+3 and the plasmid pMR243 have been used as controls for the experiments described below.

RELATIVE EXPRESSION OF THE CLONED SUPPRESSOR GENES IN VIVO

The level of expression of these cloned suppressor genes in vivo was measured by determining the amount of chargeable $tRNA^{Tyr}$ in the host cells relative to the level of chargeable $tRNA^{Phe}$. The data are presented as a ratio to obviate the problem of differential recoveries. As shown in Table 3, the presence of the cloned synthetic genes results in a very substantial increase in the level of total $tRNA^{Tyr}$ in these cells (experiments 1 and 2). As the level of *tyr* tRNA increases, the rate of growth decreases, suggesting that the excess tRNA, which should all be suppressor tRNA, is harming the cell in some way, perhaps by causing the readthrough of normal translation termination amber codons.

The greater increases were seen with both synthetic genes when incorporated into the pMB9 plasmid vector, which is expected to be present in approximately twice as many copies per cell as the ampicillin-resistant ColE1 plasmid vector (So et al. 1975). When excess L-valine was added to cultures of MR161 carrying the recombinant plasmid pSSU2 to repress the biosynthesis of isoleucine and thereby inhibit protein synthesis, this ratio increased to twelve, indicating that about one-third of the total cellular tRNA should be su^+3 tRNA. This larger increase probably results from a combination of the stringent response depressing the rate of expression of the bacterial tRNA genes while the recombinant plasmid is amplified. This increase in the copy number of the synthetic su^+3 gene might allow it to escape regulation through the stringent response, since this phenomenon has been observed in cells infected with $\phi 80 su^+3$ (Primakoff and Berg 1971). In those cases where the suppressor genes are

Table 3 Relative expression of cloned suppressor gene in vivo

Experiment	Host strain	Plasmid/ phage	Generation time (min)	Ratio tRNA^{Tyr}:tRNA^{Phe}
1	MR209	pMB9	60	0.84
1	MR161	pSSU1	93	3.52
1	MR161	pSSU2	101	4.87
2	MR209	ColE1-Ap^r [a]	71	0.89
2	MR161	pSSU101	70	1.20
2	MR161	pSSU102	84	1.65
3	MR209	—	120	0.75
3	MR161	λsu^+3	118	1.12
3	MR161	λssu1	115	1.03
3	MR161	λssu2	123	1.13
4	MR209	pMB9	62	0.77
4	MR161	pSSU2	81	4.64
4	MR161	pMR243	95	0.86

The host strains used were MR209 (his amber trp amber tyrT [=su^+3]) and MR161 (his amber trp amber). Within each experiment, the cultures were grown in M9 minimal medium plus glucose to a turbidity of either 100 or 150 Klett units and then centrifuged and washed. The bulk tRNA was extracted with phenol and precipitated with ethanol. The redissolved material was charged with [^{3}H]tyrosine and [^{3}H]phenylalanine using a crude E. coli extract as a source of synthetases. The growth temperature was 37°C for experiments 1, 2, and 4; it was 30°C in experiment 3, since each phage is thermoinducible. In experiment 4, the medium was supplemented with alanine, arginine, asparagine, aspartic acid, glutamine, glutamic acid, glycine, lysine, phenylalanine, proline, serine, threonine, and tyrosine.
[a] Ampicillin-resistant ColE1.

carried on lysogenizing phages (Table 3, experiment 3), the increase observed is about that expected for a change from three to four tyr tRNA genes per bacterial chromosome.

All the data obtained so far indicate that the modified synthetic suppressor gene is transcribed to a greater degree than the unmodified gene. However, this enhancement is less than twofold, suggesting that, although the four nucleotides changed do affect the expression of this gene, they are not of critical importance. In contrast with these results is the observation that the portion of the naturally occurring su^+3 gene within the 1200 bp cloned in plasmid vector pMB9 is expressed poorly in vivo (Table 3, experiment 4). The generation time and tyr tRNA levels measured suggest the possibility that there is too little su^+3 tRNA in these cells for normal growth under these selective conditions. The reasons for this low level of expression are not at all clear. The most obvious structural difference between these cloned synthetic genes and the naturally occurring su^+3 gene lies in the repeated sequences found after the latter gene. It is possible that this unique repeated sequence, which has been conserved, plays an important role in the regulation and/or processing of the *tyr* tRNA precursor in vivo. The synthetic su^+3 genes might escape this control, since they lack these sequences. Alternatively, the naturally occurring su^+3 gene might normally be expressed more efficiently than these synthetic genes in vivo. In this case, there would be a great deal of selective pressure for any mutation that could depress the level of expression of the naturally occurring su^+3 gene when it was cloned on a multicopy plasmid.

CONCLUDING REMARKS

The data presented in this report indicate that the synthetic genes for a precursor to the su^+3 suppressor tRNA^{Tyr} are expressed and do function in vivo. The availability of these synthetic genes and the procedures for chemical synthesis of DNA will allow an unrestricted, systematic study of structure-function relationships within this suppressor tRNA gene and its promoter region. Studies of this nature are especially well suited to the testing of hypotheses that have been proposed to explain the influence of sequence alterations detected through genetic studies. The advantage of chemical synthesis is that it gives complete flexibility in the design of predetermined mutations—any single nucleotide or nucleotide sequence(s) can be changed at will. The enormous range of potential applications of this technology will undoubtedly lead to the introduction of evermore rapid synthetic procedures, and, consequently it is possible that the chemical synthesis and cloning of functional genes will become almost routine.

ACKNOWLEDGMENTS

We wish to thank F. Blattner (University of Wisconsin, Madison), S. Altman (Yale University), R. D. Wells (University of Wisconsin, Madison), D. Helinski (University of California, San Diego), M. Ptashne (Harvard University), and J. D. Smith (Medical Research Council) for bacterial and bacteriophage strains and U. L. RajBhandary (Massachusetts Institute of Technology) and S. Altman for providing enzymes. We would also like to thank B. M. Livesey for typing this manuscript. This work has been supported by grant CA-11981-07 from the National Cancer Institute; grant PCM73-06757, awarded by the National Science Foundation; grant NP-140 from the American Cancer Society; and by funds made available to the Massachusetts Institute of Technology by the Sloan Foundation. E.L.B. was a Postdoctoral Fellow (1974–1976) of the National Institutes of Health (fellowship no. CA-01599). M.H.R. was supported by a National Institutes of Health traineeship (no. T32 CA-09112). H.-J.F. was the recipient of postdoctoral fellowships from NATO (1974–1975) and from Deutsche Forschungsgemeinschaft (1976).

REFERENCES

Altman, S. and J. D. Smith. 1971. Tyrosine tRNA precursor molecule polynucleotide sequence. *Nat. New Biol.* **233:**35.

Backman, K., M. Ptashne, and W. Gilbert. 1976. Construction of plasmids carrying the cI gene of bacteriophage λ. *Proc. Natl. Acad. Sci.* **73:**4174.

Blattner, F. R., B. G. Williams, A. E. Blechl, K. Denniston-Thompson, H. E. Faber, L.-A. Furlong, D. J. Grunwald, D. O. Kieter, D. D. Moore, J. W. Schumm, E. L. Sheldon, and O. Smithies. 1977. Charon phages: Safer derivatives of phage lambda for DNA cloning. *Science* **196:**161.

Bolivar, F., R. L. Rodriguez, M. C. Betlach, and H. W. Boyer. 1977. Construction and characterization of new cloning vehicles. I. Ampicillin-resistant derivatives of the plasmid pMB9. *Gene* **2:**75.

Boyer, H. W., M. Betlach, F. Bolivar, R. L. Rodriguez, H. L. Heyneker, J. Shine, and H. M. Goodman. 1977. The construction of molecular cloning vehicles. In *Recombinant molecules: Impact on science and society* (ed. R. F. Beers, Jr. and E. G. Bassett), p. 9. Raven Press, New York.

Egan, J. and A. Landy. 1978. Structural analysis of the $tRNA_1^{Tyr}$ gene of *Escherichia coli*. A 178-base-pair sequence that is repeated 3.14 times. *J. Biol. Chem.* **253:**3607.

Küpper, H., T. Sekiya, M. Rosenberg, J. Egan, and A. Landy. 1978. A ρ-dependent termination site in the gene coding for tyrosine $tRNAsu_3^+$ of *Escherichia coli*. *Nature* **272:**423.

Primakoff, P. and P. Berg. 1971. Stringent control of transcription of phage $\phi 80su_3^+$. *Cold Spring Harbor Symp. Quant. Biol.* **35:**391.

Robertson, H. D., S. Altman, and J. D. Smith. 1972. Purification and properties of a specific *Escherichia coli* ribonuclease which cleaves a tyrosine transfer ribonucleic acid precursor. *J. Biol. Chem.* **247:**5243.

Russell, R. L., J. N. Abelson, A. Landy, M. L. Gefter, S. Brenner, and J. D. Smith. 1970. Duplicate genes for tyrosine transfer RNA in *Escherichia coli*. *J. Mol. Biol.* **47:**1.

Ryan, M. J., E. L. Brown, T. Sekiya, H. Küpper, and H. G. Khorana. 1979. Total synthesis of a tyrosine suppressor tRNA gene (18): Biological activity and transcription, *in vitro*, of the cloned gene. *J. Biol. Chem.* **254:**5817.

Schedl, P. and P. Primakoff. 1973. Mutants of *Escherichia coli* for the synthesis of transfer RNA. *Proc. Natl. Acad. Sci.* **70:**2091.

Sekiya, T., H. van Ormondt, and H. G. Khorana. 1975. The nucleotide sequence in the promoter region of the gene for an *Escherichia coli* tyrosine transfer ribonucleic acid. *J. Biol. Chem.* **250:**1087.

Sekiya, T., R. Contreras, H. Küpper, A. Landy, and H. G. Khorana. 1976a. *Escherichia coli* tyrosine transfer ribonucleic acid genes, nucleotide sequence of their promoters and of the region adjoining CCA ends. *J. Biol. Chem.* **251:**5124.

Sekiya, T., M. J. Gait, K. Norris, B. Ramamoorthy, and H. G. Khorana. 1976b. The nucleotide sequence in the promoter region of the gene for an *Escherichia coli* tyrosine transfer ribonucleic acid. *J. Biol. Chem.* **251:**4481.

So, M., R. Gill, and S. Falkow. 1975. The generation of a ColE1-Ap^r cloning vehicle which allows detection of inserted DNA. *Mol. Gen. Genet.* **142:**239.

tRNA Genes in rRNA Operons of *Escherichia coli*

Edward A. Morgan,* Toshimichi Ikemura,† Leonard E. Post, and Masayasu Nomura
Institute for Enzyme Research
Departments of Genetics and Biochemistry
The University of Wisconsin
Madison, Wisconsin 53706

In *Escherichia coli*, there are at least seven (Kenerley et al. 1977), and probably only seven (Kiss et al. 1977), rRNA operons located at different sites in the chromosome. Seven different rRNA operons have been isolated on hybrid plasmids (Kenerley et al. 1977) or on λ specialized transducing phages (Deonier et al. 1974; Lindahl et al. 1975; Jørgensen 1976; Jørgensen and Fiil 1976; Yamamoto and Nomura 1976). Early kinetic studies and investigations with the 30S precursor had indicated that each operon consisted of genes for all three rRNA species, transcribed in the order 16S-23S-5S (for review, see Nomura et al. 1977). In the course of examining the individual operons isolated in this laboratory and elsewhere, we have shown that there are tRNA genes located in all seven *E. coli* rRNA transcription units (Lund et al. 1976; Ikemura and Nomura 1977; Morgan et al. 1977). We have found that all rRNA operons have tRNA genes located between 16S and 23S rRNA genes. In addition, some operons have tRNA genes located at their distal ends (Morgan et al. 1978). In this discussion, we will summarize the evidence that tRNA genes are located in rRNA operons and comment on the significance of this discovery.

ISOLATION OF rRNA TRANSCRIPTION UNITS

Several *E. coli* rRNA operons were isolated on λ specialized transducing phages (Deonier et al. 1974; Lindahl et al. 1975; Jørgensen 1976; Jørgensen and Fiil 1976; Yamamoto and Nomura 1976). These phages contain rRNA operons from four regions of the chromosome (*rrnB*, *rrnC*, *rrnD*, and *rrnE*). In addition, rRNA operons from six regions of the chromosome were isolated by Kenerley et al. (1977) by screening the

---
*Present address: Department of Biochemistry, University of Illinois, Urbana, Illinois 61801.
†On leave from the Department of Biophysics, Faculty of Science, Kyoto University, Kyoto, Japan.

hybrid plasmid bank made by Clarke and Carbon (1976). Altogether, seven different rRNA operons have been isolated.

HETERODUPLEX ANALYSIS AND SPACER tRNA GENES

When heteroduplex molecules are formed between two different rRNA operons and examined by electron microscopy, they are not always found to be homologous throughout the entire operon. In some heteroduplex structures, there is a nonhomology region approximately 250 bp long located between the 16S and 23S rRNA genes (the spacer region) (Deonier et al. 1974; Ohtsubo et al. 1974; Lund et al. 1976; Kenerley et al. 1977). Heteroduplex analysis of all seven rRNA operons has revealed that there are two classes of spacer regions as determined by this method (Kenerley et al. 1977). Heteroduplex structures formed between rRNA operons within a class show no observable nonhomology in the spacer region, whereas heteroduplexes formed between members of the different classes reveal a nonhomologous spacer region. The experiments described below show that the spacer regions of all rRNA operons contain genes for tRNA, and that the two different classes of operons, as revealed by heteroduplex analysis, are due to two different arrangements of tRNA genes in the rRNA operon.

tRNA genes in rRNA operons present on specialized transducing phages were analyzed by infecting UV-irradiated cells with the phages and then adding [^{32}P]orthophosphate (Lund et al. 1976; Yamamoto et al. 1976; E. A. Morgan and M. Nomura, in prep.). In these experiments, radioactive RNA is synthesized only under the direction of genes carried by the infecting phage. The RNA species produced were separated by two-dimensional gel electrophoresis and identified by fingerprinting after RNase T1 digestion.

The expression of tRNA genes isolated on ColE1 hybrid plasmids was also analyzed (Ikemura and Nomura 1977; Morgan et al. 1978). In these experiments, chloramphenicol (or amino acid starvation) was used to increase the copy number of the plasmid. [^{32}P]Orthophosphate was then added. If a tRNA gene and its promoter are present, the tRNA species synthesized from the plasmid are greatly overproduced compared to synthesis of the tRNAs whose genes are located only on the chromosome. The tRNAs that were overproduced were then separated by two-dimensional gel electrophoresis and identified by fingerprinting.

The location and identity of these tRNA genes were also demonstrated by RNA-DNA filter hybridization techniques (Lund et al. 1976; Morgan et al. 1977, 1978). Radioactive tRNAs were prepared by several methods. [^{32}P]tRNAs were isolated from *E. coli* cells grown in the presence of [^{32}P]orthophosphate and separated by two-dimensional gel electrophoresis. Individual tRNAs were eluted from gels and used directly for hybridization in some cases. Another method was to charge a mixture of unlabeled tRNAs with individual ^{3}H-labeled amino acids and then use the charged tRNAs for hybridization (Morgan et al. 1977). The radioactive tRNAs were hybridized to whole plasmid or phage DNA or to purified restriction nuclease fragments generated by digestion of a plasmid or phage DNA containing an rRNA operon. Another technique used as a general method to identify tRNA genes was to hybridize a mixture of [^{32}P]orthophosphate-labeled tRNA to purified DNA fragments produced by restriction nuclease digestion of a plasmid or phage DNA containing an rRNA operon. The tRNAs that hybridized were then eluted from the hybridization filter, separated by two-dimensional gel electrophoresis, and identified by fingerprinting after RNase T1 digestion (Morgan et al. 1977, 1978).

The results of these experiments proved that there are two types of spacer tRNA gene arrangements in *E. coli*. Four rRNA operons have tRNA$_2^{Glu}$ genes in the spacer region, whereas three tRNA operons have genes for both tRNA$_1^{Ile}$ and tRNA$_{1B}^{Ala}$ in the spacer region. This information, together with the chromosomal locations of these genes, is summarized in Table 1.

We have sequenced the spacer region of *rrnE*, which is carried by λ*metA20* (Fig. 1). The sequence of this region confirms the presence of one tRNA$_2^{Glu}$ gene at this location. Two spacer regions, which each have tRNA$_1^{Ile}$ and tRNA$_{1B}^{Ala}$ genes, have been sequenced by J. A. Steitz and her coworkers (Young et al. 1978). Young and Steitz (1978) have proposed that

Table 1 tRNA genes associated with known rRNA operons in *E. coli* K12

rRNA Operon	Chromosomal location	Spacer region tRNA	Distal tRNA
rrnA	85	tRNA$_1^{Ile}$, tRNA$_{1B}^{Ala}$	
rrnB	88	tRNA$_2^{Glu}$	
rrnC	83	tRNA$_2^{Glu}$	tRNA$_1^{Asp}$, tRNA^{Trp}
rrnD	71	tRNA$_1^{Ile}$, tRNA$_{1B}^{Ala}$	tRNA$^{Thr\ a,b}$
rrnE	89	tRNA$_2^{Glu}$	
rrnF	74	unknown	unknown
Group I	unmapped	tRNA$_1^{Ile}$, tRNA$_{1B}^{Ala}$	tRNA$_1^{Asp\ b}$
Group VI	unmapped	tRNA$_2^{Glu}$	

A summary of the map locations of rRNA operons and their associated tRNA genes. The locations and designations of rRNA operons have been summarized and discussed previously (Nomura et al. 1977). Group I and group VI refer to unmapped rRNA operons isolated on plasmids by Kenerley et al. (1977). One of these groups may correspond to *rrnF*.
[a] A tentative assignment of this tRNA as a threonine-accepting species (see Ikemura and Nomura 1977; Lund and Dahlberg 1977).
[b] These tRNA genes are believed to be cotranscribed with rRNA genes, although this has not been rigorously shown.

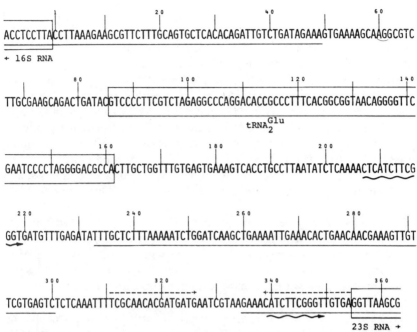

Figure 1 The sequence of the spacer region of *rrnE*. Only the coding strand is shown. The structural genes are boxed and identified. Regions of almost perfect homology to the $tRNA_1^{Ile}$-$tRNA_{1B}^{Ala}$ spacers sequenced by Young et al. (1978) are underlined with a solid line. The remaining sequences have no detectable homology between spacer sequences. The sequences showing homology immediately adjacent to 16S and 23S RNA sequences are probably involved in RNase III recognition in precursor rRNA molecules (Young et al. 1978). Sequences distal to the 3' end of 23S RNA could conceivably also pair with sequences 209–219, which is an 11-nucleotide direct repeat of sequences 339–349 (∽). By this mechanism, two structures similar to RNase III cleavage sites may be formed by interactions between spacer sequences and sequences at the 3' end of 23S RNA. It is also interesting to note that sequence 339–354, which involves the above 11-nucleotide repeat, can make an almost perfect base pairing with sequence 311–326 (-----). There are other stem and loop structures outside the structural genes that might be important for rRNA processing; for example, sequence 195–205 could pair with sequence 224–234. Details of the methods used to obtain the sequence are described elsewhere (Post 1979).

sequences in the spacer region base pair with sequences near the 5' terminal of the 16S RNA gene and with sequences near the 3' terminal of 23S RNA. These base-paired regions form cleavage sites for RNase III in vivo. A comparison of the $tRNA_1^{Ile}$-$tRNA_{1B}^{Ala}$ spacer sequence (Young et al. 1978) with that presented in Figure 1 reveals that the spacer sequences that are adjacent to 16S and 23S sequences and involved in forming the RNase III cleavage sites are conserved in both types of spacer. In addition, an 85-bp stretch of high AU (55/85) content is conserved in both spacer regions. This sequence may also be involved in forming some unidentified RNase cleavage site. Further analysis of the tRNA$_2^{Glu}$ spacer reveals several interesting features noted in Figure 1. The conserved 85-bp sequence, and other features of secondary structure, may serve an important, as yet unknown, function in rRNA processing. Analysis of RNA processing events, and further DNA sequencing, particularly in the region of 5S RNA, will be required to clarify the importance of these sequences of the 30S rRNA precursor. For discussion of the RNA processing, see Lund and Dahlberg (this volume).

tRNA GENES AT THE DISTAL END OF rRNA OPERONS

The techniques discussed above revealed that, in addition to tRNA genes between 16S and 23S rRNA, three rRNA operons had tRNA genes located distal to 5S rRNA (Ikemura and Nomura 1977; Morgan et al. 1977, 1978). These include genes for a tRNA tentatively identified as a tRNA^{Thr} species, two genes for tRNA$_1^{Asp}$, and a gene for tRNA^{Trp}. The location of these genes is summarized in Table 1.

We have shown that the tRNA$_1^{Asp}$ and tRNA^{Trp} genes located at the end of the *rrnC* operon are cotranscribed with rRNA genes and spacer tRNA genes. There are three lines of evidence for this conclusion. First, in experiments where chloramphenicol is used to increase the copy number of plasmids carrying tRNA$_1^{Asp}$ and tRNA^{Trp}, these tRNAs are overproduced only if the rRNA promoter is present (Morgan et al. 1978). Secondly, in normal *E. coli*, the amount of tRNA^{Trp} synthesized after rifampicin addition is about fourfold higher than those of spacer tRNAs, indicating that this tRNA is located approximately 6000 bp from its promoter (Morgan et al. 1978). The third line of evidence concerns deletion analysis of λ*ilv5su7*, a specialized transducing phage carrying a complete rRNA operon and a mutant form of the tRNA^{Trp} gene, which suppresses amber mutations. A deletion of this phage that removes the rRNA promoter but leaves part of 16S rRNA and all other rRNA genes and all tRNA genes intact results in loss of the ability of this phage to suppress amber mutations in lysogens. In addition, this deletion, called C116, and other similar deletions of the rRNA promoter are unable to stimulate synthesis of any rRNA or tRNA species in UV-irradiated cells (E. A. Morgan and M. Nomura, in prep.). It is therefore concluded that these distal tRNA genes are cotranscribed with the rRNA genes and spacer tRNA genes of *rrnC*.

The genes for tRNA$_1^{Asp}$ at the distal end of an unmapped rRNA operon (see Table 1) and a gene for a tRNA^{Thr} species at the end of *rrnD* have not been rigorously shown to be cotranscribed with rRNA genes. However, the tRNA^{Thr} species can be isolated after in vitro processing of the 30S rRNA precursor (Lund and Dahlberg 1977). Therefore, it is likely that this tRNA species, and perhaps also tRNA$_1^{Asp}$ at the distal end of an unmapped rRNA operon, are cotranscribed with rRNA genes. The tRNA^{Tyr}, tRNA^{Gly}, and two tRNA^{Thr} genes located near the distal end of *rrnB* (Orias et al. 1972; Squires et al. 1973; Nomura 1976; J. Rossi and A. Landy, pers. comm.; M. Yamamoto and M. Nomura, in prep.) are not cotranscribed with rRNA genes (M. Yamamoto and M. Nomura, in prep.).

ROLE OF tRNA GENES IN rRNA OPERONS

At present, there are no experiments to support or disprove the idea that it is necessary or advantageous for *E. coli* to have certain tRNA genes located in rRNA operons rather than elsewhere on the chromosome. Perhaps cotranscription of these genes may aid RNA maturational events or may be important in the assembly of ribosomes. Possibly the presence of tRNA genes in rRNA operons plays a role in the regulation of their own expression, as discussed elsewhere (Nomura et al. 1977). However these possibilities remain mostly speculative.

The presence of tRNA in rRNA transcripts indicates that tRNA-processing enzymes may play an important, as yet unconfirmed, role in the processing of the 30S rRNA precursor. It has been suggested that, in RNase III⁻ mutants, the 30S rRNA precursor is processed by different enzymes, resulting in precursors that are intermediate in the processing pathway and differ in size from the intermediate precursors pre16S and pre23S RNA, which occur in normal *E. coli* (Gegenheimer et al. 1977). It is possible that tRNA-processing enzymes cleaving near spacer tRNAs and perhaps near distal tRNAs are responsible for the observed cleavages of the 30S precursor found in RNase III⁻ cells. These may, in fact, represent previously undetected processing events occurring in normal cells.

Although the significance of the presence of tRNA genes in rRNA operons is not clear, these genes may provide a means to analyze rRNA operons by genetic techniques. In particular, the amber-suppressing form of tRNA^{Trp} may allow isolation of additional types of mutations, such as mutations affecting the promoter function of rRNA operon.

ACKNOWLEDGMENTS

Work described in this article (paper number 2296 from the Laboratory of Genetics) was supported in part by the College of Agriculture and Life Sciences, University of Wisconsin, and by grants from the National Science Foundation (GB-31086) and the National Institutes of Health (GM-20427). E.A.M. and L.E.P. were supported by a National Institutes of Health postdoctoral fellowship (GM-06033) and by a National Science Foundation graduate fellowship, respectively.

REFERENCES

Clarke, L. and J. Carbon. 1976. A colony bank containing synthetic Col E1 hybrid plasmids representative of the entire *E. coli* genome. *Cell* **9**:91.

Deonier, R. C., E. Ohtsubo, H. J. Lee, and N. Davidson. 1974. Electron microscope heteroduplex studies of sequence relations among plasmids of *Escherichia coli*. VII. Mapping of ribosomal RNA genes of plasmid F14. *J. Mol. Biol.* **89**:619.

Gegenheimer, P., N. Watson, and D. Apirion. 1977. Multiple pathways for the primary processing of ribosomal RNA in *Escherichia coli*. *J. Biol. Chem.* **252**:3064.

Ikemura, T. and M. Nomura. 1977. Expression of spacer tRNA genes in ribosomal RNA transcription units carried by hybrid Col E1 plasmids in *E. coli*. *Cell* **11**:779.

Jørgensen, P. 1976. A ribosomal RNA gene of *Escherichia coli* (*rrnD*) on λd*aroE* specialized transducing phages. *Mol. Gen. Genet.* **146**:303.

Jørgensen, P. and N. P. Fiil. 1976. Ribosomal RNA synthesis in vitro. In *Alfred Benzon Symposium IX: Control of Ribosome Synthesis* (ed. N. O. Kjeldgaard and O. Maaløe), p. 370. Academic Press, New York.

Kenerley, M. E., E. A. Morgan, L. Post, L. Lindahl, and M. Nomura. 1977. Characterization of hybrid plasmids carrying individual ribosomal ribonucleic acid transcription units of *Escherichia coli*. *J. Bacteriol.* **132**:931.

Kiss, A., B. Sain, and P. Venetianer. 1977. The number of rRNA genes in *Escherichia coli*. *FEBS Lett.* **79**:77.

Lindahl, L., S. R. Jaskunas, P. P. Dennis, and M. Nomura. 1975. Cluster of genes in *Escherichia coli* for ribosomal proteins, ribosomal RNA, and RNA polymerase subunits. *Proc. Natl. Acad. Sci.* **72**:2743.

Lund, E. and J. E. Dahlberg. 1977. Spacer transfer RNAs in ribosomal RNA transcripts of *E. coli*: Processing of 30S ribosomal RNA in vitro. *Cell* **11**:247.

Lund, E., J. E. Dahlberg, L. Lindahl, S. R. Jaskunas, P. P. Dennis, and M. Nomura. 1976. Transfer RNA genes between 16S and 23S rRNA genes in rRNA transcription units of *E. coli*. *Cell* **7**:165.

Morgan, E. A., T. Ikemura and M. Nomura. 1977. Identification of spacer tRNA genes in individual ribosomal RNA transcription units of *Escherichia coli*. *Proc. Natl. Acad. Sci.* **74**:2710.

Morgan, E. A., T. Ikemura, L. Lindahl, A. M. Fallon, and M. Nomura. 1978. Some rRNA operons in *E. coli* have tRNA genes at their distal ends. *Cell* **13**:335.

Nomura, M. 1976. Organization of bacterial genes for ribosomal components: Studies using novel approaches. *Cell* **9**:633.

Nomura, M., E. A. Morgan, and S. R. Jaskunas. 1977. Genetics of bacterial ribosomes. *Annu. Rev. Genet.* **11**:297.

Ohtsubo, E., L. Soll, R. C. Deonier, H. J. Lee, and N. Davidson. 1974. Electron microscope heteroduplex studies of sequence relations among plasmids of

Escherichia coli. VIII. The structure of bacteriophage $\phi 80d_3ilv^+su^+7$, including the mapping of the ribosomal RNA genes. *J. Mol. Biol.* **89**:631.

Orias, E., T. K. Gartner, J. E. Lannan, and M. Betlach. 1972. Close linkage between *ochre* and missense suppressors in *Escherichia coli. J. Bacteriol.* **109**:1125.

Post, L. E. 1979. "DNA sequences from ribosomal protein operons of *Escherichia coli.*" Ph.D. thesis, University of Wisconsin, Madison.

Squires, C., B. Konrad, J. Kirschbaum, and J. Carbon. 1973. Three adjacent transfer RNA genes in *Escherichia coli. Proc. Natl. Acad. Sci.* **70**:438.

Yamamoto, M. and M. Nomura. 1976. Isolation of λ transducing phages carrying rRNA genes at the *metA-purD* region of the *Escherichia coli* chromosome. *FEBS Lett.* **72**:256.

Yamamoto, M., L. Lindhal, and M. Nomura. 1976. Synthesis of ribosomal RNA in *E. coli*: Analysis using deletion mutants of a λ transducing phage carrying ribosomal RNA genes. *Cell* **7**:179.

Young, R. A., R. Macklis, and J. A. Steitz. 1979. Sequence of the 16S–23S spacer region in two ribosomal operons of *E. coli. J. Biol. Chem.* **254**:3264.

Young, R. A. and J. A. Steitz. 1978. Complementary sequences 1700 nucleotides apart form a ribonuclease III cleavage site in *Escherichia coli* ribosomal precursor RNA. *Proc. Natl. Acad. Sci.* **75**:3593.

Yeast Suppressor tRNA Genes

**Maynard V. Olson, Guy S. Page,* André Sentenac,† Kate Loughney,‡
Janet Kurjan, Joshua Benditt, and Benjamin D. Hall**
Genetics Department SK-50
University of Washington
Seattle, Washington 98195

The advent of rapid techniques for isolating and characterizing specific eukaryotic DNA sequences offers the promise of studying the structure and function of eukaryotic tRNA genes at the level of individual nucleotides. tRNA genes in any eukaryotic organism are readily accessible by these techniques; both their small size and the availability of purified tRNAs as RNA-DNA hybridization probes facilitate applications of molecular cloning and DNA sequencing to these genes. It has been our view, however, that the complex series of steps by which the coding sequence of a tRNA gene is expressed as a mature, functional tRNA molecule is unlikely to be decipherable by structural studies alone. We have chosen, therefore, to focus our efforts on the molecular cloning and physical analysis of a small set of yeast tRNA genes that have been genetically identified as nonsense suppressors (Hawthorne and Leupold 1974). Twelve genetically mapped loci in yeast correspond to nonsense suppressors of known amino-acid-insertion specificity (Table 1). In only two cases, $SUP5_{UAG}$ and $SUPRL1_{UAG}$ (Piper et al. 1976; Piper 1978), have these nonsense suppressor mutations been shown to affect the primary structure of a tRNA, but there is now little doubt that all the loci in Table 1 correspond to tRNA genes. The tyrosine-inserting loci have been particularly intensively studied at the DNA level and a $SUP4_{UAA}$ gene was shown to be mutant in the DNA sequence specifying a $tRNA^{Tyr}$ anticodon (Goodman et al. 1977; M. V. Olson et al. 1977, 1979, in prep.).

HYBRIDIZATION TO RESTRICTION DIGESTS

We have studied the DNA coding sequences for $tRNA^{Tyr}$ and $tRNA^{Ser}$ using three different hybridization probes: the single known $tRNA^{Tyr}$ species (Madison et al. 1966), the major serine species $tRNA_2^{Ser}$ (Zachau et al. 1966), and a mixture of highly homologous minor serine isoac-

---

Present addresses: *Department of Biochemistry and Biophysics, University of California, San Francisco Medical Center, San Francisco, California 94143: ‡Department of Physiological Chemistry, University of Wisconsin, Madison, Wisconsin 53706. †Permanent address: Service de Biochimie, Centre D'Études Nucléaires de Saclay, 91190 Gif-sur-Yvette, France.

Table 1 Yeast nonsense suppressor loci of known amino acid insertion specificities

Locus	Insertion specificity	Map position	Known alleles	References
SUP2	tyrosine	IV	UAA, UAG	Hawthorne and Leupold (1974)
SUP3	tyrosine	XV	UAA, UAG	
SUP4	tyrosine	X	UAA, UAG	Gilmore et al. (1971)
SUP5	tyrosine	F8	UAA, UAG, UGA	Hawthorne (1976)
SUP6	tyrosine	VI	UAA, UAG	Liebman et al. (1976)
SUP7	tyrosine	X	UAA, UAG	
SUP8	tyrosine	XIII	UAA, UAG	
SUP11	tyrosine	VI	UAA, UAG	
SUP16	serine	XVI	UAA, UAG	Cox (1965); Ono et al. (1979); Liebman et al. (1976)
SUP17	serine	IX	UAA	Ono et al. (1979)
SUPRL1	serine	III	UAG, UAA	Brandriss et al. (1975, 1976)
SUP52	leucine	X	UAG	Liebman et al. (1977)

ceptors, which we will refer to as tRNA$_{min}^{Ser}$ (Piper 1978). Our studies began with experiments in which a size-fractionated set of yeast EcoRI restriction fragments was hybridized to an individual purified tRNA (Southern 1975). Figure 1 shows that, in such experiments, eight hybridization bands are seen for tRNA^{Tyr} (Olson et al. 1977), eleven for tRNA$_2^{Ser}$, and four for the tRNA$_{min}^{Ser}$ mixed probe. The hybridization probes used in these experiments were labeled with Na-^{125}I in the presence of H$_2$O$_2$ and horseradish peroxidase. These gentle iodination conditions lead to i^{6}A-specific labeling; hence, the Southern autoradiographs (Fig. 1) are relatively free of the numerous impurity bands obtained with uniformly labeled tRNA probes.

We use the designations *tyrA–tyrH*, *ser$_2$A–ser$_2$K*, and *ser$_{min}$A–ser$_{min}$D* to characterize each of the tRNA genes according to the size of the EcoRI fragment on which it occurs. The fragment size provides a reliable identifying criterion for a specific tRNA gene in any given haploid yeast strain, but there is considerable strain-to-strain variability. The reference strains for the labeling of the tRNA^{Tyr}, tRNA$_2^{Ser}$, and tRNA$_{min}^{Ser}$ genes are specified in Figure 1.

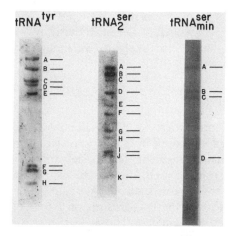

Figure 1 Hybridization of tRNA^{Tyr}, tRNA$_2^{Ser}$, and tRNA$_{min}^{Ser}$ to yeast DNA. The yeast DNA was cleaved with EcoRI, fractionated on a 0.7% agarose gel, and transferred to nitrocellulose for hybridization. The yeast DNA for the tRNA^{Tyr} and tRNA$_2^{Ser}$ hybridizations was extracted from the strain B596, whereas strain Y4A DNA was used for the tRNA$_{min}^{Ser}$ hybridizations. The sizes of the best characterized of these fragments (in 1000 bp) are as follows: tRNA^{Tyr} 14 (A), 9.9 (B), 6.7 (C), 6.3 (D), 5.2 (E), 1.3 (F), 1.2 (G), 0.9 (H); tRNA$_2^{Ser}$ 9.7 (A), 9.2 (B), 5.1 (F), 4.1 (H), 1.6 (J); tRNA$_{min}^{Ser}$ 4.7 (B), 4.3 (C).

MOLECULAR CLONING

We used ^{125}I-labeled tRNA^{Tyr}, tRNA$^{Ser}_{2}$, and tRNA$^{Ser}_{min}$ to screen pools of λ-yeast hybrid phage for clones containing these tRNA genes. A phage DNA preparation was made from each positive clone; then the phage DNA was EcoRI cut, gel electrophoresed, transferred to nitrocellulose, and hybridized to ^{125}I-labeled tRNA (Olson et al. 1979). The identity of the yeast tRNA gene in each clone was determined from the size of the EcoRI fragment that hybridized. By using this procedure, we have cloned seven of the eight tRNA^{Tyr} genes, four out of eleven tRNA$^{Ser}_{2}$ genes, and all four of the genes for tRNA$^{Ser}_{min}$ (Olson et al. 1979; Page 1978).

ASSIGNMENT OF tRNA HYBRIDIZATION BANDS TO *SUP* LOCI

As a first step towards relating mutations at suppressor loci to DNA sequence changes in the corresponding tRNA genes, we determined the approximate chromosomal map locations for many of the tRNA genes displayed in Figure 1. Three types of experiments were carried out.

Disome Analysis

We compared the Southern hybridization patterns for two related yeast strains, one haploid and the other differing from it by having a second copy of one single chromosome. Bands that differ quantitatively or qualitatively between haploid and disome are likely to correspond to tRNA genes on the disomic chromosome.

Meiotic Linkage Analysis

Both this method and the related mitotic procedure discussed below take advantage of the frequently observed variation, from one yeast strain to another, in the size of a given restriction fragment. These size variations behave as Mendelian traits and can thus be used to mark the restriction fragment bearing a given tRNA gene. Suppressor alleles that map within the gene will exhibit close linkage to their parental restriction fragment variant when progeny from crosses are examined. The meiotic test for linkage requires the construction of diploids heterozygous both for a particular suppressor and for one or more restriction-fragment-size variations. These diploids can be used to make direct tests for linkage between the two traits.

Mitotic Recombination

In the mitotic linkage test, a diploid of the same genotype as that employed for meiotic analysis is used. In this case, however, we screen for those rare diploid cells that have become homozygous either for *SUP* or *sup*$^+$ by mitotic recombination. DNA is extracted from these recombinant clones, *Eco*RI restricted, and analyzed by Southern hybridization with the appropriate tRNA. For restriction fragments of only a few thousand base pairs, the suppressor gene is expected to be closely linked to the locus determining the size of the restriction fragment; therefore, homozygosis for *SUP* or *sup*$^+$ will cause a corresponding homozygosis for one of the variant restriction fragments.

The effects of restriction fragment variation upon the tRNA^{Tyr} hybridization pattern are seen in Figure 2, where bands *tyrA* and *tyrF* differ in size between the two haploid strains. In the disome for chromosome X, two variant forms of band *tyrA* are present, indicating that *tyrA* is either *SUP4* or *SUP7*, both of which occupy genetic map positions on chromosome X. The tRNA^{Tyr} hybridization pattern of the chromosome VI disome shows no variant bands or other qualitative differences from the patterns for euploid strains; however, bands *tyrE* and *tyrG* are increased in intensity relative to the corresponding bands in haploid strains, suggesting that *SUP6* and *SUP11* (which map to chromosome VI) correspond to the pair of bands *tyrE* and *tyrG*. As is apparent in Figure 2, this increase in the intensity of particular bands in disomic strains is often only marginally detectable because of the considerable random variability of band intensities. A less ambiguous result was obtained, however, when ^{125}I-labeled tRNA$^{Ser}_{min}$ was hybridized to DNA from a strain disomic for chromosome III: Band C is clearly more intense than the three remaining bands (Fig. 3a). In the haploid control, four bands of equal intensity were

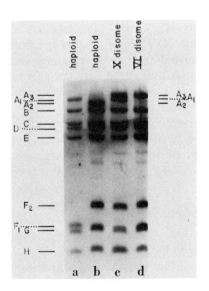

Figure 2 Hybridization of tRNA^{Tyr} to yeast DNA. The two haploid strains are D311-3A (*a*) and S288C (*b*). The disomic (n + 1) strains are DH3787-5D (*c*) and X4001-39A (*d*). (Reprinted, with permission, from Olson et al. 1979.)

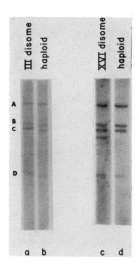

Figure 3 Hybridization of tRNA$^{Ser}_{min}$ to yeast DNA. The four yeast strains are 96C, a chromosome III disome (*a*); 96CH, a haploid derived from 96C by mitotic chromosome loss (*b*); 4093-5D, a chromosome XVI disome (*c*); and 4093-5DH, a haploid derived from 4093-5D by mitotic chromosome loss (*d*).

observed (Fig. 3b). On the basis of this result, the tRNA$^{Ser}_{min}$ gene of band C is tentatively identified as *SUPRL1*.

Figure 3c shows the pattern of bands obtained by hybridization of tRNA$^{Ser}_{min}$ to a strain disomic for chromosome XVI, the chromosome to which *SUP16* (*SUQ5*) maps. The aberrant pattern of five bands reflects a restriction-fragment-size heterozygosity of the type discussed above in connection with the presence of two *tyrA* bands in a chromosome X disome. When the tRNA$^{Ser}_{min}$ probe is hybridized to DNA from a haploid derivative of the chromosome XVI disome strain, the standard four-band pattern was once again observed; one of the restriction-fragment alleles was lost (Fig. 3d).

To determine which of the three bands, -A, B, or D-, is allelic to the variant band observed in the chromosome XVI disomic pattern, DNA from clones of these fragments was labeled by nick translation and annealed to the disome DNA. Band B DNA hybridized very strongly to the variant fragment, indicating that *ser*$_{min}$*B* is on chromosome XVI and thus is a likely candidate for the tRNA^{Ser} gene corresponding to *SUP16*.

In the case of the tRNA^{Tyr} genes, the results of the disome hybridizations were complemented by a variety of more direct linkage tests. For example, both meiotic and mitotic tests of linkage were carried out between the *tyrA* variation and *SUP4* and *SUP7*. The diploid strains used had the following genetic constitutions.

4H: $sup4^+$ $tyrA_1$ $tyrF_1$ $ade2_{UAA}$ × *SUP4* $tyrA_2$ $tyrF_2$ $ade2_{UAA}$ (1)

7H: $sup7^+$ $tyrA_1$ $tyrF_1$ $ade2_{UAA}$ × *SUP7* $tyrA_2$ $tyrF_2$ $ade2_{UAA}$ (2)

Sporulation of 4H yielded four-spored asci that showed 2:2 segregation of the suppressor. The tRNA hybridization patterns exhibited by meiotic segregants of the diploid 4H are shown in Figure 4. The lack of recombination between *SUP4* and the *tyrA*$_2$ variant suggests meiotic linkage. In contrast, four out of seven meiotic segregants from diploid 7H are recombinant for the markers *SUP7* and A$_2$. Confirmation for the linkage of *tyrA* to *SUP4* and not to *SUP7* was obtained by hybridizing tRNA^{Tyr} to *SUP4/SUP4* and *sup4$^+$/sup4$^+$* homozygote clones that arose from diploid 4H as a result of mitotic crossing over between *SUP4* and the centromere of chromosome X. Such mitotic recombinant clones can be visually distinguished from *SUP4/sup4$^+$* diploid clones on an agar growth plate containing low levels of adenine. Heterozygotes (*SUP4/sup4$^+$*) are pink because of incomplete suppression of *ade2*$_{UAA}$; homozygous *SUP4/SUP4* colonies are white because of complete suppression; and homozygous *sup4/sup4$^+$* colonies are red because they lack suppression (Cox 1965).

The results of tRNA^{Tyr} hybridization to DNA from the two halves of such a sectored colony show disappearance of *tyrA*$_2$/*tyrA*$_1$ heterozygosity accompanying homozygosis for the *SUP4* or *sup4$^+$* alleles (Fig. 5a). The white half of the colony yields only fragment *tyrA*$_2$ in its DNA and the red half only fragment *tyrA*$_1$. An analogous experiment showing the cosectoring of *SUP2* with *tyrF*$_2$ and *sup2$^+$* with *tyrF*$_1$ is shown in Figure 5b. Further applications of this technique have allowed us to associate *tyrB* with *SUP5*, *tyrC* with *SUP8*, *tyrD* with *SUP3*, and *tyrH* with *SUP7* (M. V. Olson et al., in prep.). P. Philippsen et al. (pers. comm.) have associated *tyrE* with *SUP6* by finding that an *SUP6* suppressor mutation alters 1 bp within the *tyrE* tRNA gene. By elimination, we can then infer that the remaining unassigned band, *tyrG*, corresponds to *SUP11*, a result that is consistent with the chromosome VI disome hybridization results (Fig. 2).

The tRNA^{Ser} bands that we have tentatively identified with genes include *ser*$_{min}$*B* and *ser*$_{min}$*C*, which map, respectively, to chromosomes XVI and III and are therefore likely to correspond to the suppressors *SUP16* (Ono et al. 1979) and *SUPRL1* (Brandriss et al. 1976). In addition, one of the genes for tRNA$_2^{Ser}$ has been localized on the yeast chromosomal map at a position where no suppressor has yet been found. The *ser*$_2$*B* EcoRI fragment cloned from strain B596 was found to contain the *cyc1* gene (Montgomery et al. 1978). Consequently, this potential suppressor gene, *SUP100*, is precisely mapped to a location on the right arm of chromosome X, 2100 bp toward the centromere from *cyc1*.

THE SEQUENCE CHANGE ACCOMPANYING A SUPPRESSOR MUTATION

The genetics of tyrosine-inserting yeast suppressors is most easily understood by postulating that each suppressor consists of a point mutation in

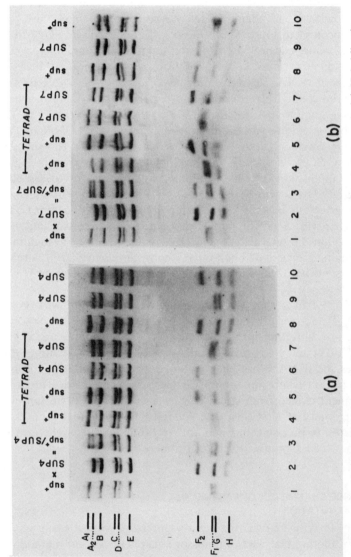

Figure 4 Demonstration of meiotic linkage between *SUP4* and the *tyrA$_2$* fragment. (*a*) Hybridization patterns of the *sup4$^+$* haploid parent (*1*), the *SUP4* haploid parent (*2*), and the *sup4$^+$/SUP4* diploid (*3*); hybridization patterns of four haploid segregants derived from a single ascus after sporulation of the diploid (*4–7*); hybridization patterns of three random spores from the same cross (*8–10*). (*b*) This experiment is analogous to that in *a*, except that in this case *SUP7* is segregating. (Reprinted, with permission, from Olson et al. 1979.)

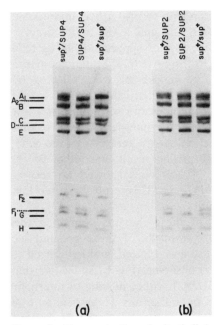

Figure 5 Demonstration of mitotic linkage between *SUP4* and the *tyrA*$_2$ fragment and *SUP2* and the *tyrF*$_2$ fragment. (*a*) The *sup*$^+$/*SUP4* diploid is heterozygous for *sup4*$^+$/*SUP4*, *tyrA*$_1$/*tyrA*$_2$, and *tyrF*$_1$/*tyrF*$_2$. The *SUP4*/*SUP4* and *sup4*$^+$/*sup4*$^+$ homozygotes were isolated from the two halves of a single symmetrically sectored colony. (*b*) This experiment is analogous to that in *a*, but in this case the original diploid is heterozygous for *SUP2*. (Reprinted, with permission, from Olson et al. 1979.)

one tRNA gene out of the set of eight. Having identified tRNA genes on specific cloned restriction fragments with particular suppressor loci, we were able to test this postulate directly by DNA sequencing (Goodman et al. 1977). Within the coding sequence, the *tyrA* tRNA^{Tyr} genes from *sup4*$^+$ and *SUP4* strains differ by a single-base-pair change, GC→TA at the position coding for the wobble-position base of the tRNA anticodon. This change allows the *SUP4* suppressor tRNA to have the anticodon 3'-AψU* (where U* is a modified U). These results rule out the possibility that the ocher-suppressing anticodon is 3'-AψI, as has been suggested to explain their ocher specificity. It appears, instead, that the *SUP4* suppressor tRNA contains as its wobble-position base a modified U residue that cannot pair with G, preventing translation of UAG codons (Hawthorne and Leupold 1974; Piper et al. 1976).

MUTATIONS INACTIVATING *SUP4* FUNCTION

A variety of mutational events in a suppressor tRNA gene can cause loss of suppressor activity. We have isolated such second-site mutations at the

SUP4 locus in an *SUP4*$_{UAA}$ haploid strain with several ocher suppressible markers, *can1-100, ade2-1, lys2-1, met4-1*, and *trp5-2*. The *SUP4* strain is white and canavanine-sensitive, whereas loss of suppression leads to a red, canavanine-resistant phenotype (Rothstein 1977). Mutations at the *SUP4* locus that abolish activity of the suppressor were selected by plating on synthetic media containing canavanine. Mutant *SUP4* genes bearing defects at different sites within the gene can recombine at a low frequency to give *SUP4* recombinants. The frequencies of these recombination events can be used to construct a fine-structure map of the gene. Such an analysis provided a rather surprising result: 18 out of 98 independently isolated mutants at the *SUP4* locus showed no recombination with any other mutant within the gene. This result suggested that these mutants were deletions of the entire gene.

A biochemical analysis of several of these mutants confirmed this conclusion. *Eco*RI-cleaved yeast DNA from six of these presumptive deletion strains was run on 0.7% agarose gels, blotted, and hybridized to ^{125}I-labeled tRNA^{Tyr}. No *tyrA* band was observed (Fig. 6), indicating that this fragment, which corresponds to *SUP4*, is either missing or internally deleted for its tRNA^{Tyr} gene in all six mutant strains. Hybridization to a ^{32}P-labeled, nick-translated, cloned DNA fragment of the *tyrA* band indicates that not all the *tyrA* DNA has been deleted; this probe hybridizes to an *Eco*RI fragment in the deletion strains that is smaller than the parental *tyrA* fragment.

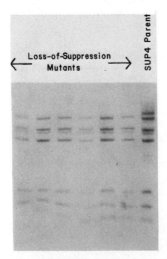

Figure 6 Demonstration that six loss-of-suppression mutations at the *SUP4* locus involve deletions of the entire tRNA^{Tyr} coding region. The *SUP4* parent is the haploid suppressor strain in which the six independent loss-of-suppression mutations were selected.

INTERVENING SEQUENCES

The genes for several tRNA species contain intervening sequences; for yeast tRNA^{Tyr}, yeast tRNA^{Phe}, and *Xenopus laevis* tRNA^{Tyr} genes, such sequences have been directly demonstrated by DNA sequencing of the cloned genes (Goodman et al. 1977; Valenzuela et al. 1978; S. Clarkson, pers. comm.). In addition, the tRNA$^{Ser}_{min}$ gene that we have placed on chromosome III (presumably *SUPRL1*) has an intervening sequence (M. Olson et al., unpubl.). Unlike these tRNA genes, the DNA of three genes coding for yeast tRNA$^{Ser}_{2}$ contains no intervening sequences (Page 1978). This structural variability within the population of yeast tRNA genes suggests that whatever the function of the intervening sequence, it is not intrinsically required for the expression of eukaryotic tRNA genes.

APPLICATIONS

The nine yeast tRNA genes that we have cloned and identified with nonsense suppressor loci provide excellent subjects for combined biochemical and genetic studies of tRNA gene function. Demonstration of the sequence changes accompanying $sup^+ \to SUP_{UAA}$ mutation at the *SUP4* and *SUP6* loci (Goodman et al. 1977; P. Philippsen et al., pers. comm.) are initial examples of such investigations. The analysis of these mutant genes, which, like the wild type, contain intervening sequences, led to the significant conclusion that an actively expressed tRNA gene could contain an intervening sequence. This conclusion followed from the fact that the suppressor genes are necessarily active, since they confer a dominant phenotype on suppressor-containing cells.

1. The analysis of intragenic loss-of-suppression mutations. Such studies should allow the identification of all functional regions within a tRNA gene, including sequences involved in the initiation and termination of transcription and sequences essential for RNA processing. A single *SUP4-o* gene containing a second-site mutation that causes loss of suppression has been cloned and sequenced (J. Kurjan et al., unpubl.); it proved to have a G residue deleted from the tract of five G residues present near the 3' end of the wild-type gene. Of particular interest will be efforts to relate the position of loss-of-suppression mutations within the gene to the particular nature of the functional defect they cause.

2. Introduction of cloned suppressor tRNA genes (with or without further sequence alterations) back into living cells to test their function. The most promising systems are yeast cells themselves, which can be transformed with naked DNA (Hinnen et al. 1978) and *X. laevis* oocytes, which can be injected with recombinant plasmid DNA. Following the introduction of yeast tRNA^{Tyr} genes into *Xenopus* oocytes,

both mature and precursor forms of the yeast tRNA are observed (deRobertis and Olson 1979).

It is hoped that by bringing a wide variety of methods—cloning, DNA sequencing, yeast genetics, yeast transformation, and the use of in vitro transcription, and RNA processing systems—to bear on yeast suppressor genes, it will prove possible to relate, in detail, the structure of these genes to their in vivo function.

ACKNOWLEDGMENTS

We would like to acknowledge the assistance of Peter W. Piper (Imperial Cancer Research Fund Laboratories), who supplied us with a sample of tRNA$^{Ser}_{min}$, and the assistance of Donald C. Hawthorne (University of Washington, Seattle), who provided several unpublished yeast strains.

REFERENCES

Brandriss, M. C., L. Soll, and D. Botstein. 1975. Recessive lethal amber suppressors in yeast. *Genetics* **79**:551.

Brandriss, M. C., J. W. Stewart, F. Sherman, and D. Botstein. 1976. Substitution of serine caused by a recessive lethal suppressor in yeast. *J. Mol. Biol.* **102**:467.

Cox, B. S. 1965. ψ, A cytoplasmic suppressor of super-suppressor in yeast. *Heredity* **20**:505.

deRobertis, E. M. and M. V. Olson. 1979. Transcription and processing of cloned yeast tyrosine tRNA genes microinjected into frog oocytes. *Nature* **278**:137.

Gilmore, R. A., J. W. Stewart, and F. Sherman. 1971. Amino acid replacements resulting from super-suppression of nonsense mutants of iso-l-cytochrome c from yeast. *J. Mol. Biol.* **61**:157.

Goodman, H. M., M. V. Olson, and B. D. Hall. 1977. Nucleotide sequence of a mutant eukaryotic gene: The yeast tyrosine-inserting ochre suppressor SUP4-o. *Proc. Natl. Acad. Sci.* **74**:5453.

Hawthorne, D. C. 1976. UGA mutatious and UGA suppressors in yeast. *Biochimie* **58**:179.

Hawthorne, D. C. and U. Leupold. 1974. Suppressor mutations in yeast. *Curr. Top. Microbiol. Immunol.* **64**:1.

Hinnen, A., J. B. Hicks, and G. R. Fink. 1978. Transformation of yeast. *Proc. Natl. Acad. Sci.* **75**:1929.

Liebman, S. W., F. Sherman, and J. W. Stewart. 1976. Isolation and characterization of amber suppressors in yeast. *Genetics* **82**:251.

Liebman, S. W., J. W. Stewart, J. H. Parker, and F. Sherman. 1977. Leucine insertion caused by a yeast amber suppressor. *J. Mol. Biol.* **109**:13.

Madison, J. T., G. A. Everett, and H. Kung. 1966. Nucleotide sequence of a yeast tyrosine transfer RNA. *Science* **153**:531.

Montgomery, D. L., B. D. Hall, S. Gillam, and M. Smith. 1978. Identification and isolation of the yeast cytochrome c gene. *Cell* **14**:673.

Olson, M. V., K. Loughney, and B. D. Hall. 1979. Identification of the yeast DNA sequences that correspond to specific tyrosine-inserting nonsense suppressor loci. *J. Mol. Biol.* **131:**(in press).

Olson, M. V., B. D. Hall, J. R. Cameron, and R. W. Davis. 1979. Cloning of the yeast tyrosine transfer RNA genes in bacteriophage lambda. *J. Mol. Biol.* **127:**285.

Olson, M. V., D. L. Montgomery, A. K. Hopper, G. S. Page, F. Horodyski, and B. D. Hall. 1977. Molecular characterisation of the tyrosine tRNA genes of yeast. *Nature* **267:**639.

Ono, B.-I., J. W. Stewart, and F. Sherman. 1979. Yeast UAA suppressors effective in ψ+ strains. Serine-inserting suppressors. *J. Mol. Biol.* **128:**81.

Page, G. S. 1978. "The major seryl tRNA genes of *Saccharomyces cerevisiae*." PhD. thesis, University of Washington, Seattle.

Piper, P. W. 1978. A correlation between a recessive lethal amber suppressor mutation in *Saccharomyces cerevisiae* and an anticodon change in a minor serine transfer RNA. *J. Mol. Biol.* **122:**217.

Piper, P. W., M. Wasserstein, F. Engbaek, K. Kaltoft, J. E. Celis, J. Zeuthen, S. Liebman, and F. Sherman. 1976. Nonsense suppressors of *Saccharomyces cerevisiae* can be generated by mutation of the tyrosine tRNA anticodon. *Nature* **262:**757.

Rothstein, R. J. 1977. A genetic fine structure analysis of the suppressor 3 locus in *Saccharomyces. Genetics* **85:**55.

Southern, E. M. 1975. Detection of specific sequences among DNA fragments separated by gel electrophoresis. *J. Mol. Biol.* **98:**503.

Valenzuela, P., A. Venegas, F. Weinberg, R. Bishop, and W. J. Rutter. 1978. Structure of yeast phenylalanine-tRNA genes: An intervening DNA segment within the region coding for the tRNA. *Proc. Natl. Acad. Sci.* **75:**190.

Zachau, H. G., D. Dütting, and H. Feldman. 1966. The structures of two serine transfer ribonucleic acids. *Z. Physiol. Chem.* **347:**212.

Mapping of tRNA Genes on the Circular DNA Molecule of *Spinacia oleracea* Chloroplasts

André Steinmetz, Mfika Mubumbila, Mario Keller, Gérard Burkard, and Jacques H. Weil
Institut de Biologie Moléculaire et Cellulaire
67084 Strasbourg, France

Albert J. Driesel, Edwin J. Crouse, Karl Gordon, Hans-Jürgen Bohnert, and Reinhold G. Herrmann
Botanisches Institut der Universität Düsseldorf
D-4000 Düsseldorf, Federal Republic of Germany

Chloroplasts contain their own complement of tRNAs, which are different from those of the cytoplasm and those of the mitochondria (Weil et al. 1977). The chloroplast tRNA structure is similar to that of prokaryotic tRNAs, at least for the two chloroplast tRNAs whose nucleotide sequences have been determined (Chang et al. 1976; Guillemaut and Keith 1977), and they are coded for by chloroplast DNA (Tewari and Wildman 1970; Gruol and Haselkorn 1976; Haff and Bogorad 1976; McCrea and Hershberger 1976; Schwartzbach et al. 1976).

The genes coding for chloroplast rRNAs have been mapped on higher plant and algal chloroplast DNAs. In maize and spinach, the chloroplast rRNA genes are located on two inverted repeats (Bedbrook et al. 1977; Whitfeld et al. 1978); in *Euglena*, the chloroplast rRNA genes exist as three clustered tandem repeats (Gray and Hallick 1978; Rawson et al. 1978). But until now, the chloroplast tRNA genes have not been localized. In this paper we describe the fractionation and identification of spinach (*Spinacia oleracea*) chloroplast tRNAs and the first results of our efforts to map the tRNA genes on spinach chloroplast DNA (a circular molecule of $\sim 90 \times 10^6$ daltons).

Unbroken chloroplasts (Herrmann et al. 1975) were used in the preparation of tRNAs, aminoacyl-tRNA synthetases, and high-molecular-weight DNA. The tRNAs were prepared by phenol treatment of detergent-lysed chloroplasts, precipitated by ethanol, dissolved in 1 M NaCl, diluted to 0.2 M NaCl, incubated with DNase, and purified on DEAE-cellulose columns (Burkard et al. 1970). Aminoacyl-tRNA synthetases were obtained by Sephadex G-75 chromatography of the 105,000g supernatant of sonicated or mechanically broken (Dounce homogenizer) chloroplasts. Chloroplast DNA was isolated as described previously (Herrmann et al. 1975).

FRACTIONATION OF ISOACCEPTING tRNAs

Chloroplast tRNAs were electrophoretically separated on two-dimensional polyacrylamide gels (Fradin et al. 1975). The individual isoaccepting tRNAs were visualized by methylene-blue staining, eluted from the minced gel by phenol extraction, and chromatographed on Sephadex G-25. Fractionation of spinach chloroplast tRNAs yields about 35 spots (Fig. 1), whereas two-dimensional gel electrophoresis of spinach-leaf tRNAs yields about 80 spots. Most of the chloroplast tRNA spots have been identified by

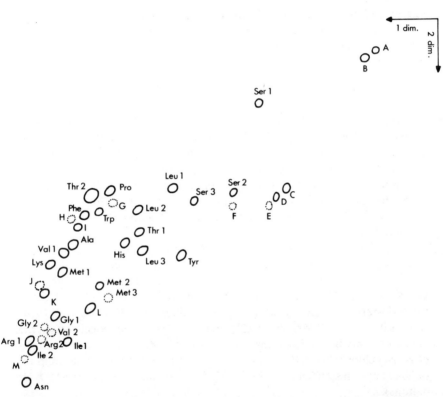

Figure 1 Two-dimensional gel electrophoresis of S. oleracea chloroplast tRNAs. For experimental conditions, see Fradin et al. (1975). Unlabeled chloroplast tRNA was run on a 10% polyacrylamide gel in the first dimension for 40 hr at 450 V, and then on a 20% polyacrylamide gel in the second dimension for 140 hr at 350 V. The individual spots were eluted from the gel (see text) and either aminoacylated to allow identification of the tRNA or labeled with ^{125}I and used in hybridization studies. Spots A–E are probably not tRNAs; spots F–M are probably tRNAs that have not yet been identified. The tRNAs that have been identified are indicated by the amino acid that they accept.

aminoacylation with both chloroplast and *Escherichia coli* aminoacyl-tRNA synthetases. To date, single tRNA species have been demonstrated for alanine, asparagine, histidine, lysine, phenylalanine, proline, tryptophan, and tyrosine. Multiple spots were found for eight amino acids: two isoaccepting tRNA species for threonine, valine, isoleucine, arginine, and glycine and three for leucine, serine, and methionine.

MAPPING OF tRNA GENES

The tRNA genes were localized on a physical map of the circular DNA molecule, which was constructed using the cleavage sites of the restriction endonucleases *Sal*I, *Pst*I, *Kpn*I, and *Xma*I. After digestion of plastid DNA with one enzyme, the resulting fragments were isolated and redigested with another enzyme, and vice versa. Analysis of common end pieces allowed the determination of the relative order of the cleavage sites (Fig. 2). An outstanding characteristic of the circular molecule is the duplication of a region of about 15×10^6 daltons in inverted orientation. The duplicate regions are separated from each other by single-copy regions of 50×10^6 daltons and 10×10^6 daltons, respectively (Herrmann et al. 1976).

For the localization of chloroplast tRNA genes, the individual tRNA species were radioactively labeled with ^{125}I (Commerford 1971). Chloroplast DNA fragments produced by digestion with one enzyme, by double digestion, and by redigestion of individual DNA fragments with a second enzyme were transferred from gels to nitrocellulose filters (Southern 1975). The labeled tRNAs were then hybridized to these DNA fragments. Autoradiography revealed that most of the tRNA genes are present in the large single-copy region (Fig. 2).

DISCUSSION

On *E. coli* DNA, tRNA genes have been located in the spacer region between the 16S and the 23S rRNA genes (Lund et al. 1976) and at the distal ends of some rRNA operons (Morgan et al. 1978). In the case of spinach chloroplast DNA, only a tRNA^{Ile} gene has been found in the intercistronic spacer between the 16S and 23S rRNA genes; two of the three chloroplast leucine-isoaccepting tRNAs hybridize to the inverted repeat regions, but they are separated by a distance of at least 2.5×10^6 daltons from the clustered rRNA cistrons and therefore cannot be part of the polycistronic rDNA transcript (Bohnert et al. 1977).

The fact that we obtain 35 chloroplast tRNA spots upon two-dimensional gel electrophoresis (Fig. 1) is in good agreement with the reported figures concerning the number of tRNA genes on chloroplast DNA: 26 tRNA genes

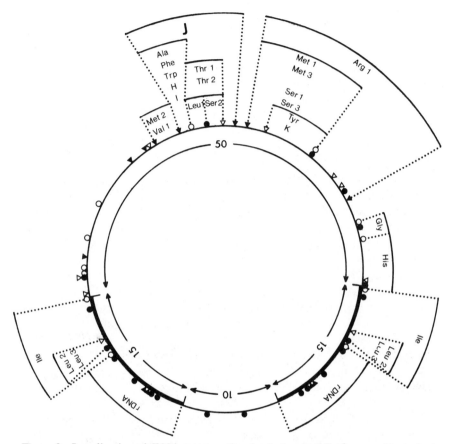

Figure 2 Localization of tRNA genes on the physical map of *S. oleracea* chloroplast DNA. ^{125}I-labeled individual tRNAs were hybridized to chloroplast DNA fragments (obtained upon action of a restriction endonuclease, separated by electrophoresis on agarose gel, and transferred to nitrocellulose strip) in 2 × standard saline citrate, 50% formamide at 37°C for 24 hr, and the strip was washed with RNase T1 or pancreatic RNase before autoradiography. Cleavage sites of the four restriction endonucleases used: (○) *Sal*I; (▽) *Pst*I; (▼) *Kpn*I; (●) *Xma*I. The smallest DNA region to which hybridization was observed is indicated for each chloroplast tRNA tested. The heavy line corresponds to the two inverted repeat regions containing the two known polycistronic rDNA sites (the gene arrangement is 16S-spacer-23S-4.5/5S, with the transcription polarity toward the small single-copy region).

on tobacco chloroplast DNA (Tewari and Wildman 1970), 20-26 on *Euglena* chloroplast DNA (Gruol and Haselkorn 1976; McCrea and Hershberger 1976; Schwartzbach et al. 1976), and 20-26 on maize chloroplast DNA (Haff and Bogorad 1976).

When there are several chloroplast isoaccepting tRNAs for the same amino acid, they seem to be encoded by different genes, as shown by the fact that serine isoacceptors hybridize to different DNA fragments; this is also the case for leucine and methionine isoacceptors (Fig. 2).

ACKNOWLEDGMENTS

These studies were aided by grants from the Deutsche Forschungsgemeinschaft, Heinrich-Hertz-Stiftung, Deutscher Akademiker Austauch Dienst, European Molecular Biology Organization, the Commissariat à l'Energie Atomique, and the Fondation pour la Recherche Médicale Française. The expert technical assistance of Barbara Schiller is gratefully acknowledged.

REFERENCES

Bedbrook, J. R., R. Kolodner, and L. Bogorad. 1977. Zea mays chloroplast ribosomal RNA genes are part of a 22,000 base pair inverted repeat. *Cell* **11**:739.
Bohnert, H. J., A. J. Driesel, and R. G. Herrmann. 1977. Transcription and processing of transcripts in isolated, unbroken chloroplasts. In *Acides nucléiques et synthèse des protéines chez les végétaux. Colloques Internationaux du Centré National de la Recherche Scientifique* (ed. L. Bogorad and J. H. Weil), no. 261, p. 213. CNRS, Paris.
Burkard, G., P. Guillemaut, and J. H. Weil. 1970. Comparative studies on the tRNAs and the aminoacyl-tRNA synthetases from the cytoplasm and the chloroplasts of *Phaseolus vulgaris*. *Biochim. Biophys. Acta* **224**:184.
Chang, S. H., C. K. Brum, M. Silberklang, U. L. RajBhandary, L. I. Hecker, and W. E. Barnett. 1976. The first nucleotide sequence of an organelle transfer RNA: Chloroplast tRNA^{Phe}. *Cell* **9**:717.
Commerford, S. L. 1971. Iodination of nucleic acids *in vitro*. *Biochemistry* **10**:1993.
Fradin, A., H. Gruhl, and H. Feldmann. 1975. Mapping of yeast tRNAs by two dimensional electrophoresis on polyacrylamide gels. *FEBS Lett.* **50**:185.
Gray, P. W. and R. B. Hallick. 1978. Physical mapping of the *Euglena gracilis* chloroplast DNA and ribosomal RNA gene region. *Biochemistry* **17**:284.
Gruol, D. J. and R. Haselkorn. 1976. Counting the genes for stable RNA in the nucleus and chloroplasts of Euglena. *Biochim. Biophys. Acta* **447**:82.
Guillemaut, P. and G. Keith. 1977. Primary structure of bean chloroplastic tRNA^{Phe} *FEBS Lett.* **84**:351.
Haff, L. A. and L. Bogorad. 1976. Hybridization of maize chloroplast DNA with transfer ribonucleic acids. *Biochemistry* **15**:4105.
Herrmann, R. G., H. J. Bohnert, A. Driesel, and G. Hobom. 1976. The location of rRNA genes on the restriction endonuclease map of *Spinacia oleracea* chloroplast DNA. In *Genetics and biogenesis of chloroplasts and mitochondria: Interdisciplinary Conference on the Genetics and Biogenesis of Chloroplasts and Mitochondria* (ed. T. Bucher et al.), p. 351. North-Holland, Amsterdam.
Herrmann, R. G., H. J. Bohnert, K. V. Kowallik, and J. M. Schmitt. 1975. Size, conformation and purity of chloroplast DNA of some higher plants. *Biochim. Biophys. Acta* **378**:305.

Lund, E., J. E. Dahlberg, L. Lindahl, S. R. Jaskunas, P. P. Dennis, and M. Nomura. 1976. Transfer RNA genes between 16S and 23S rRNA genes in rRNA transcription units of *E. coli. Cell* **7**:165.

McCrea, J. M. and C. L. Hershberger. 1976. Chloroplast DNA codes for transfer RNA. *Nucleic Acids Res.* **3**:2005.

Morgan, E. A., T. Ikemura, L. Lindahl, A. M. Fallon, and M. Nomura. 1978. Some rRNA operons in *E. coli* have tRNA genes at their distal ends. *Cell* **13**:335.

Rawson, J. R., S. R. Kushner, D. Vapnek, N. K. Alton, and C. L. Boerma. 1978. Chloroplast ribosomal RNA genes in *Euglena gracilis* exist as three clustered tandem repeats. *Gene* **3**:191.

Schwartzbach, S. D., L. I. Hecker, and W. E. Barnett. 1976. Transcriptional origin of Euglena chloroplast tRNAs. *Proc. Natl. Acad. Sci.* **73**:1984.

Southern, E. M. 1975. Detection of specific sequences among DNA fragments separated by gel electrophoresis. *J. Mol. Biol.* **98**:503.

Tewari, K. K. and S. G. Wildman. 1970. Information content in the chloroplast DNA. In *Control of organelle development. Symposia of the Society for Experimental Biology* (ed. P. L. Miller), vol. 24, p. 147. Academic Press, New York.

Weil, J. H., G. Burkard, P. Guillemaut, G. Jeannin, R. Martin, and A. Steinmetz. 1977. tRNAs and aminoacyl-tRNA synthetases in plant cytoplasm, chloroplasts and mitochondria. In *Nucleic acids and protein synthesis in plants. NATO Advanced Study Institute Series A* (ed. L. Bogorad and J. H. Weil), Life Sciences vol. 12, p. 97. Plenum Press, New York.

Whitfeld, P. R., R. G. Herrmann, and W. Bottomley. 1978. Mapping of the ribosomal RNA genes on spinach chloroplast DNA. *Nucleic Acids Res.* **5**:1741.

Transcription and Processing of Nematode tRNA Genes Microinjected into the Frog Oocyte

Riccardo Cortese,* Douglas Melton, Theresa Tranquilla, and John D. Smith
MRC, Laboratory of Molecular Biology
Hills Road, Cambridge, England

With the development of cloning techniques, large amounts of pure segments of DNA, which may include pure single genes, can be obtained and their nucleotide sequences determined. The purified segments can be used to obtain general information on the organization of the genes in the total genome and also to study gene expression. To this end, in vitro systems have been successfully employed for eukaryotic genes (Birkenmeier et al. 1978; Schmidt et al. 1978; Ng et al. 1979). An alternative in vivo system based on the special properties of the frog oocyte has been developed by Gurdon and his colleagues (Mertz and Gurdon 1977; Brown and Gurdon 1977, 1978; Gurdon et al. 1979). The oocyte is large enough to allow purified DNA segments to be injected directly into its nucleus. This system has the advantage that one may reasonably expect that all the transcription and processing components will be intact. Here we present a summary of our results on cloned tRNA genes of the nematode *Caenorhabditis elegans* and their expression in *Xenopus laevis* oocytes.

The tRNA genes are particularly suitable for studies aimed at correlating structure with function because the tRNAs themselves are easily characterized. They constitute a group of genes of related function that are expressed coordinately in many physiological conditions. On the other hand, the expression of some tRNA genes is subject to an individual mode of regulation, particularly evident in some extreme cases, such as in the silk gland of *Bombyx mori* (Garel et al. 1970) where some tRNA genes are disproportionately expressed. Some tRNA genes in eukaryotes differ strikingly from the known prokaryotic tRNA genes in having intervening sequences (Goodman et al. 1977; Valenzuela et al. 1978; Knapp et al. 1978; O'Farrell et al. 1978). Furthermore, the arrangement of the genes on the chromosome varies in different eukaryotes, for example, yeast (Beckmann et al. 1977) and *Xenopus* (Clarkson et al. 1973). Consequently, it is interesting to compare the tRNA genes from many different eukaryotes. We

---

*Present address: I Istituto di Chimica Biologica, II Facolta di Medicina e Chirurgia, University of Naples, Naples, Italy.

chose the nematode *C. elegans* because of the advanced genetic study of this organism (Brenner 1974) and the possibility that some of the suppressors isolated may be altered tRNAs (Waterson and Brenner 1978).

CLONING OF NEMATODE tRNA GENES

C. elegans has about 300 tRNA genes and 55 genes for 18S and 28S rRNA (Sulston and Brenner 1974). An analysis of the organization of RNA genes by *Eco*RI restriction digests of genomic DNA showed that the rRNA genes occur in a unique 6500-bp fragment containing both 18S and 28S sequences. At least 40 different-sized *Eco*RI fragments of genomic nematode DNA hybridize to nematode tRNA, indicating that the tRNA genes are rather dispersed in the genome (Cortese et al. 1978).

To clone tRNA genes, nematode DNA was digested with *Eco*RI endonuclease and the fragments were ligated into the unique *Eco*RI site of the plasmid ColE1. Clones carrying tRNA and rRNA genes were identified by the colony hybridization procedure of Grunstein and Hogness (1975) using 4S nematode RNA labeled with [^{32}P]ATP at the 5' end using T4 polynucleotide kinase (Chang et al. 1976) as a probe. Plasmid DNA was extracted from the positive clones and hybridized separately to preparations of 4S RNA labeled in two different manners. In one hybridization, the 4S RNA was labeled with ^{32}P at the 5' end (containing labeled tRNA but also rRNA fragments) and the second group had 4S RNA labeled with ^{32}P at the 3'-terminal A residue using tRNA nucleotidyl transferase, which specifically labels tRNA. Clones that hybridized only to the 5'-labeled probe contained rRNA genes; those hybridizing to both probes contained tRNA genes (Cortese et al. 1978).

We concentrated our further studies on three clones, Cet 1, Cet 7, and Cet 18, carrying nematode DNA inserts of 5500, 5000, and 2500 bp, respectively. These clones have the following properties. They code for, or at least hybridize to, a single tRNA species, indicating they each contain only one type of gene. Quantitative hybridization experiments indicate that each contains just one tRNA gene of this type. The three cloned DNA segments do not cross-hybridize and therefore do not have common sequences (homologies as small as 20–30 base sequences would have been detected). They do not contain repeated sequences. The DNA segments did not undergo detectable rearrangements during the cloning procedure. Work is now in progress to determine sequences of the DNA and the corresponding tRNAs.

AN ASSAY FOR GENE ACTIVITY: THE FROG OOCYTE

Largely due to the pioneering work of Gurdon and his colleagues (Mertz and Gurdon 1977; Brown and Gurdon 1977, 1978; Laskey et al. 1978; Gurdon et al. 1979), it is now well established that naked DNA becomes active following microinjection into the nucleus of frog oocytes. This means that it is capable of all reactions necessary for correct transcription. The physiological accuracy of such transcription has been well documented in several instances using defined, *X. laevis* DNAs (Brown and Gurdon 1978; Kressman et al. 1978). We have used this heterologous system to investigate the functional properties of cloned nematode tRNA genes (Cortese et al. 1978).

The general protocol of this type of injection experiment is to inject cloned DNA together with α-^{32}P-labeled nucleoside triphosphate ([α-^{32}P]NTP) RNA precursors into the oocyte nucleus and incubate the oocytes at room temperature for a given time. The labeled RNA is then extracted and analyzed by gel electrophoresis. We injected a number of plasmid cloned DNAs and found that clones containing tRNA genes promote the synthesis of specific 4S RNA species, whereas other plasmid cloned DNAs, for example, those carrying rRNA genes, do not. This shows that the *Xenopus* oocyte injection system can be used to analyze cloned DNAs from a heterologous source.

The frog oocyte can perform all the biosynthetic steps leading from an injected tRNA gene to its gene product. The first step is the interaction of the injected DNA with the transcriptional machinery of the cell. Injected DNA is now known to be immediately assembled into chromatin (Laskey et al. 1978), so that this event may precede the synthesis of the immediate transcription product. In the case of tRNA, the primary transcriptional product is a long precursor RNA. The maturation of the precursor tRNA involves polynucleotide chain cleavages and nucleoside modifications. At some stage during this process, the tRNA must be transported to the cytoplasm.

Transcription

The synthesis of tRNA following injection of the cloned tDNA into oocyte nuclei shows that the cloned segments contain sufficient information for the accurate expression of a tRNA gene. Obviously, the essential sequences are only small parts of the large fragments of cloned nematode DNAs. To identify the essential sequences we have injected smaller fragments derived from restriction endonuclease digests and measured their capacity to provide tRNA synthesis. In one case, the Cet-1 clone, our data indicate that a 300-bp DNA segment contains all the information necessary for making tRNA.

tRNA genes are known to be transcribed by RNA polymerase III (Roeder 1976). This polymerase, which is also responsible for 5S RNA transcription, is distinguished from the other polymerases by its intermediate sensitivity to α-amanitin (Roeder 1976). This inhibitor can be injected into the oocyte nucleus, along with DNA, and its effect on in vivo transcription investigated (Gurdon and Brown 1978). In such experiments, tRNA synthesis is reduced to 50% at α-amanitin concentrations of 1–10 μg/ml. This agrees with in vitro results and confirms that RNA polymerase III is responsible for transcription of nematode tRNA genes in injected oocytes.

Different tRNA genes and 5S RNA gene compete directly for the same transcriptional machinery and RNA polymerase III. This was shown by competition experiments where two different cloned DNAs were coinjected into the same nucleus at various concentrations. Measuring separately the amounts of each tRNA and 5S RNA synthesized, we could evaluate the extent to which each gene had been transcribed. We concluded that the different tRNA genes and the 5S RNA genes compete with each other for transcriptional enzymes and precursors. SV40 DNA, transcribed by RNA polymerase II (D. Melton and R. Cortese, unpubl.), does not compete with tRNA synthesis at any injected concentration. We are investigating whether there is a difference in the affinity of RNA polymerase III for different genes. This possibility is suggested by the fact that several yeast tRNA genes injected under similar conditions are transcribed more slowly than nematode tRNA genes. In similar experiments, *Escherichia coli* tRNA genes and human mitochondrial tRNA genes are not expressed (R. Cortese et al., unpubl.).

The synthesis of tRNA is dependent on the amount of tRNA genes injected, but the relationship is not linear. Below 10×10^6 gene copies per oocyte, little or no tRNA is synthesized. Above that amount, tRNA synthesis is linearly proportional to the number of gene copies injected. This lag effect can be eliminated so that tRNA synthesis becomes linearly proportional to the number of injected genes at any concentration by coinjecting 2 μg of carrier DNA (e.g., ColE1 DNA) together with the DNA containing tRNA genes. We do not know, at this stage, whether the lag at low concentration of DNA is physiologically significant, perhaps reflecting a negative control that can be overcome by high DNA concentrations.

Maturation

tRNA genes are transcribed first into precursor molecules of varying lengths usually containing 5' or 3' extensions and, occasionally, small intervening sequences. The precursor molecules are matured to produce tRNA by nucleolytic cleavage and base modifications. Our results and those of others show that all these reactions occur following gene injections, producing a mature tRNA.

The maturation intermediates have, in general, short half-lives reaching low steady-state concentrations and are therefore hardly detectable in both prokaryotes and eukaryotes. In some instances, however, tRNA precursors are clearly detectable. Following injection of Cet-7 DNA, we observed a precursor molecule, Cet-7 pre-RNA, made in relatively large amounts. Fingerprint analysis shows that Cet-7 pre-RNA differs from Cet-7 tRNA at the 5' terminal, but has the same 3' end. It appears to be a precursor with extra nucleotides at the 5' end similar to the precursor molecules identified and characterized in prokaryotes, which are converted to mature tRNA by action of RNase P (Smith 1976).

In the experiments described so far [α-^{32}P]NTP was used to label the RNA. By using [γ-^{32}P]NTP only those RNAs that preserve an intact 5'-triphosphate end will be detected; the γ label is not internally incorporated into the RNA. Cet-7-cloned DNA, injected with either [γ-^{32}P]ATP or [γ-^{32}P]GTP, directs the synthesis of labeled RNA species of about the same size as the Cet-7 pre-RNA produced with α labels. This suggests that the Cet-7 pre-RNA may be the intact primary transcriptional product, assuming no changes at the 3' end. We observed both γ-ATP-labeled and γ-GTP-labeled precursors of slightly different mobilities, indicating that perhaps RNA polymerase III starts at least at two different points, probably adjacent on the gene. Such alternative starting positions of the RNA polymerase have been observed also for other tRNA genes (R. Cortese et al., unpubl.), in other systems involving RNA polymerase III (Vennström et al. 1978), and in vitro, in some instances (Debenham 1979).

The large size of the frog oocyte allows for the manual separation of the nucleus from the cytoplasm, thereby offering an opportunity to study the intracellular localization of molecules (Merriam and Hill 1976). Dissection experiments of this type show that after injection of Cet-7 DNA, the Cet-7 pre-RNA is found exclusively in the nucleus, whereas the mature tRNA is present both in the nucleus and in the cytoplasm. With increasing time after injection, the tRNA accumulates in the cytoplasm. Since the intranuclear concentration of Cet-7 pre-RNA is quite high, there may be a barrier preventing its transport into the cytoplasm. Cet-7 tRNA, only slightly smaller, readily goes across the nuclear membranes in both directions. It will be interesting to establish the molecular basis of this selectivity.

We have injected purified Cet-7 pre-RNA, previously extracted from oocytes and separated by gel electrophoresis, directly into the nucleus or into the cytoplasm. Only the precursor injected into the nucleus is processed to mature tRNA. Injection of the precursor into the cytoplasm results in its slow degradation, without the appearance of a mature tRNA. These experiments suggest that the Cet-7 pre-RNA processing enzyme—possibly the eukaryotic RNase P—is located in the nucleus. In addition, since the Cet-7 pre-RNA exists exclusively in the nucleus, the modified bases present in this molecule, including ψ and T, must also be synthesized in the nucleus. This indicates the intranuclear localization of the corresponding enzymes.

REFERENCES

Beckmann, J. S., P. F. Johnson, and J. Abelson. 1977. Cloning of yeast tRNA genes in *E. coli. Science* 196:205.

Birkenmeier, E. H., D. D. Brown, and E. Jordan. 1978. A nuclear extract of *Xenopus laevis* oocytes that accurately transcribes 5S RNA genes. *Cell* 15:1077.

Brenner, S. 1974. The genetics of *Caenorhabditis elegans. Genetics* 77:71.

Brown, D. D. and J. B. Gurdon. 1977. High fidelity transcription of 5S DNA injected into *Xenopus* oocyte. *Proc. Natl. Acad. Sci.* 74:2064.

―――. 1978. Cloned single repeating units of 5S DNA direct accurate transcription of 5S RNA when injected into *Xenopus* oocytes. *Proc. Natl. Acad. Sci.* 75:2849.

Chang, S. H., C. K. Brum, M. Silberklang, U. L. RajBhandary, L. I. Hecker, and W. E. Barnett. 1976. The first nucleotide sequence of an organelle tRNA: chloroplastic tRNA^{Phe}. *Cell* 9:717.

Clarkson, S. G., M. L. Birnstiel, and V. Serra. 1973. Reiterated tRNA genes of *Xenopus laevis. J. Mol. Biol.* 79:391.

Cortese, R., D. Melton, T. Tranquilla, and J. D. Smith. 1978. Clonng of nematode tRNA genes and their expression in the frog oocyte. *Nucleic Acids Res.* 5:4593.

Debenham, P. 1979. The influence of ribonucleoside triphosphates, and other factors, on the formation of very salt stable RNA polymerase $-su^+_{III}$ tRNA (tRNA^{tyr}) promoter complexes. *Eur. J. Biochem.* 96:535.

Garel, J. P., P. Mandel, G. Chavancy, and J. Dallie. 1970. Functional adaptation of tRNA's to fibroin biosynthesis in the silk gland of *Bombyx mori. FEBS Lett.* 7:327.

Goodman, H. M., M. V. Olson, and B. P. Hall. 1977. Nucleotide sequence of a mutant eukaryotic gene: The yeast tyrosine-inserting ochre suppressor SUP4-o. *Proc. Natl. Acad. Sci.* 74:5453.

Grunstein, M. and D. S. Hogness. 1975. Colony hybridization: A method for the isolation of cloned DNAs that contain a specific gene. *Proc. Natl. Acad. Sci.* 72:3961.

Gurdon, J. B. and D. D. Brown. 1978. The transcription of 5S DNA injected into *Xenopus* oocytes. *Dev. Biol.* 67:346.

Gurdon, J. B., D. Melton, and E. DeRobertis. 1979. Genetics in an oocyte. *Ciba Symp.* (in press).

Knapp, G., J. S. Beckmann, P. F. Johnson, S. A. Fuhrman, and J. Abelson. 1978. Transcription and processing of intervening sequences in yeast tRNA genes. *Cell* 14:221.

Kressman, A., S. G. Clarkson, V. Pirrotta, and M. L. Birnstiel. 1978. Transcription of cloned tRNA gene fragments and subfragments injected into the oocyte of *Xenopus laevis. Proc. Natl. Acad. Sci.* 75:1176.

Laskey, R. A., B. M. Honda, A. D. Mills, H. R. Morris, A. H. Wyllie, J. E. Mertz, E. De Robertis, and J. B. Gurdon. 1978. Chromatin assembly and transcription in eggs and oocytes of *Xenopus laevis. Cold Spring Harbor Symp. Quant. Biol.* 42:171.

Merriam, R. W. and R. J. Hill. 1976. The germinal vesicle nucleus of *Xenopus laevis* oocytes as a selective storage receptacle for proteins. *J. Cell Biol.* 69:659.

Mertz, J. and J. B. Gurdon. 1977. Purified DNAs are transcribed after microinjection into *Xenopus* oocytes. *Proc. Natl. Acad. Sci.* 74:1502.

Ng, S. Y., C. S. Parker, and R. G. Roeder. 1979. Transcription of cloned *Xenopus* 5S

RNA genes by *Xenopus laevis* RNA polymerase III in reconstituted systems. *Proc. Natl. Acad. Sci.* **76:**136.

O'Farrell, P. Z., B. Cordell, P. Valenzuela, W. J. Rutter, and H. M. Goodman. 1978. Structure and processing of yeast precursor tRNAs containing intervening sequences. *Nature* **274:**438.

Roeder, R. G. 1976. Eukaryotic RNA polymerases. In *RNA polymerase* (ed. R. Losick and M. Chamberlin), p. 285. Cold Spring Harbor Laboratory, Cold Spring Harbor, New York.

Schmidt, O., J. I. Mao, S. Silverman, B. Hovemann, and D. Söll. 1978. Specific transcription of eukaryotic tRNA genes in *Xenopus* germinal vesicle extracts. *Proc. Natl. Acad. Sci.* **75:**4819.

Smith, J. D. 1976. Transcription and processing of tRNA precursors. *Prog. Nucleic Acid Res. Mol. Biol.* **16:**25.

Sulston, J. and S. Brenner. 1974. The DNA of *Caenorhabditis elegans*. *Genetics* **77:**95.

Valenzuela, P., A. Venegas, F. Weinberg, R. Bishop, and W. J. Rutter. 1978. Structure of yeast phenylalanine tRNA genes: An intervening DNA segment within the region coding for the tRNA. *Proc. Natl. Acad. Sci.* **75:**190.

Vennström, B., U. Petterson, and L. Philipson. 1978. Two initiation sites for adenovirus 5.5S RNA. *Nucleic Acids Res.* **5:**195.

Waterson, R. H. and S. Brenner. 1978. A suppressor mutation in the nematode acting on specific alleles of many genes. *Nature* **275:**715.

tRNA Genes of Drosophila melanogaster

Gordon M. Tener, Shizu Hayashi, Robert Dunn,
Allen Delaney, Ian C. Gillam, Thomas A. Grigliatti,
Thomas C. Kaufman, and David T. Suzuki
Departments of Biochemistry and Zoology
University of British Columbia
Vancouver, British Columbia, Canada V6T 1W5

Although the organization of tRNA genes in *Drosophila* has not been studied as extensively as those of the genes in *Xenopus* (Birnstiel et al. 1972; Clarkson et al. 1973a,b, 1978; Clarkson and Kurer 1976) and *Saccharomyces* (Olson et al. 1977; Goodman et al. 1977; Valenzuela et al. 1978; O'Farrell et al. 1978), we are nevertheless beginning to get a much clearer picture of this organization. The reason for this is that in addition to the techniques of genetic analysis, in vitro hybridization, and incorporation of DNA fragments into plasmids, it is possible with *Drosophila* to localize genes by in situ hybridization of radioactive tRNA to its polytene chromosomes. After autoradiography, tRNA genes may be assigned visually to specific sites on the chromosomes.

THE NUMBER OF tRNA GENES DETERMINED BY HYBRIDIZATION

One of the first questions asked about tRNA genes is how many of them exist in the haploid genome for each tRNA. Ritossa et al. (1966) estimated that there are 750 genes for all species of tRNA by observing the saturation level on hybridizing labeled tRNA to *Drosophila* DNA. More recently, the hybridization results of Weber and Berger (1976) suggested that *Drosophila* 4S RNA is composed of approximately 59 families of sequences encoded for by 590 genes. As they pointed out, the proposed number of different sets of tRNA genes is based on several assumptions and, at best, is an average value. The 4S RNA fraction contains not only tRNAs but also fragments of rRNA and other RNAs. Because of these impurities and large differences in concentration of individual tRNAs, the interpretation of hybridization studies with the total mixture is subject to errors.

A number of years ago, we showed by RPC-5 column chromatography that the crude 4S RNA fraction of *Drosophila* contained about 100 different species of tRNA, some of which are homogenic species resulting from posttranscriptional modification (Grigliatti et al. 1974b). Thus we expected to find less than 100 sets of tRNA genes.

To obtain meaningful hybridization results, it was considered essential to carry out hybridizations using highly purified tRNAs. The purification was done by successive column chromatography using BD-cellulose, Sepharose 6B, and RPC-5. Most products rechromatographed as single components and showed an uptake of better than 1700 pmoles of their specific amino acid per A_{260} unit. Details of a typical procedure, the purification of tRNAs^{Val}, are presented elsewhere (Dunn et al. 1978).

For subsequent hybridization studies, these purified tRNAs were converted into radioactive species using the method of Commerford (1971). This involves the reaction of the tRNA with ^{125}I label in the presence of thallium trichloride. The iodine was introduced at the C5 residues and, under standard conditions, it was possible to introduce about one I atom per tRNA molecule. This ^{125}I-labeled tRNA has a specific activity of 1.6×10^8 dpm/μg. For the in situ experiments to be discussed later, it is necessary to have labeled tRNA with a specific activity of this order.

The ^{125}I-labeled tRNAs were also used for in vitro hybridization with DNA, both in solution and on cellulose nitrate discs. We had expected the results to be straightforward, but this was not the case. Figure 1 (top) shows the time course of hybridization when the molar ratio of tRNA$_2^{Lys}$ to DNA was about 260:1. This represents a large excess of tRNA$_2^{Lys}$. With time, the amount of hybrid formed reached a plateau value and from this value it is possible to calculate that about 9 genes on the DNA have hybridized with the tRNA. However, as shown in Figure 1 (bottom), when the concentration of tRNA in solution was altered, different plateau values of between 4 and 18 genes were reached, even though in all cases there was an excess of tRNA over the sites on the DNA. Thus, the plateau level attained appears to be an equilibrium value. One should achieve complete saturation at very high concentrations of tRNA, but such experiments are limited in practice by traces of contaminating tRNAs reaching concentration levels that also give hybrids.

The hybrids that form have variable stabilities. Figure 2 shows the rate of release of tRNA$_2^{Lys}$ from the isolated hybrid under hybridization conditions. About 7 moles of tRNA are released very rapidly, whereas the remaining hybrids are about an order of magnitude more stable. We cannot yet interpret this experiment, but it may reflect the structure of the genes. For example, analogous to yeast tRNA genes (Olson et al. 1977; Valenzuela et al. 1978), some genes may have inserts that would be expected to labilize the hybrid or there may be pseudogenes or segments of genes for tRNA present (Jacq et al. 1977; Botchan et al. 1977; Egan and Landy 1978). This hybridization problem appears to be confined to tRNAs and tRNA genes because when this method is applied to 5S RNA, the extent of hybridization is not dependent on concentration, provided sufficient 5S RNA is initially present to hybridize with all 168 genes. Because of this problem, we cannot determine the exact number of tRNA genes present in *Drosophila* by using this hybridization method.

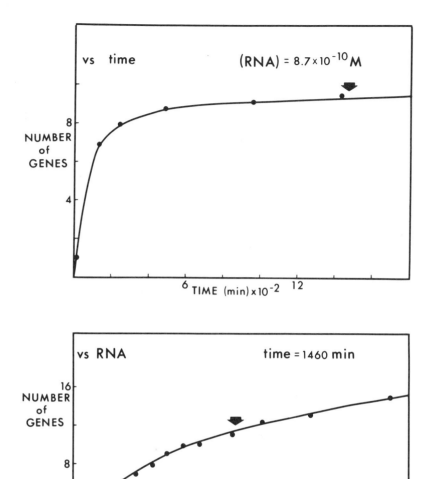

Figure 1 Hybridization of tRNA$_2^{Lys}$ to *Drosophila* DNA. ^{125}I-labeled tRNA$_2^{Lys}$ was hybridized with denatured *Drosophila* DNA in 70% formamide containing 80 mM K$_2$HPO$_4$, 10 mM Na$_2$ EDTA, and 0.5 M KCl (adjusted to pH 7.0 with H$_3$PO$_4$). Incubation temperature was 50°C. At the end of the incubation period, the samples were chilled, diluted with two volumes of 5 M NaCl, and run onto a column (0.7 × 45 cm) of Bio-Gel A-5m at 50°C and eluted with 150 mM NaCl, 10 mM EDTA, 15 mM sodium phosphate buffer (pH 7) to separate hybrid from nonhybridized tRNA. (*Top*) tRNA concentration 8.7 × 10^{-10} M, DNA concentration 2.2 × 10^{-12} M, aliquots taken at the times shown; (*Bottom*) hybridization plateau levels at 1460 min and DNA concentration 3.8 × 10^{-12} M with different concentrations of tRNA.

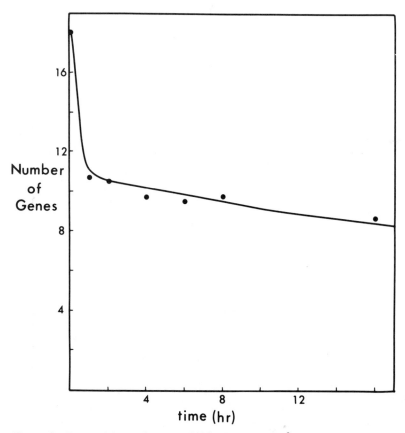

Figure 2 Rate of breakdown at 50°C of the tRNA$_2^{Lys}$ DNA hybrid; prepared under hybridization conditions listed in Fig. 1. Time, 960 min; RNA concentration, 3.63×10^{-9} M; DNA concentration, 3.3×10^{-12} M.

THE NUMBER OF tRNA GENES DETERMINED BY RESTRICTION ENDONUCLEASE CLEAVAGE

Further information about the number of tRNA genes can be obtained by cleaving *Drosophila* DNA with restriction endonucleases and separating the resulting fragments by agarose gel electrophoresis. The fragments bearing the tRNA genes are localized by the Southern transfer techniques (Southern 1975) using the ^{125}I-labeled tRNA. Autoradiography shows that both *Eco*RI and *Hin*dIII cleave *Drosophila* DNA into fragments, and approximately 18 of these fragments hybridize with ^{125}I-labeled tRNA$_2^{Lys}$, as shown in Figure 3. We do not know if any of these fragments contain more than one gene, but it appears that there are at least 18 genes. Since

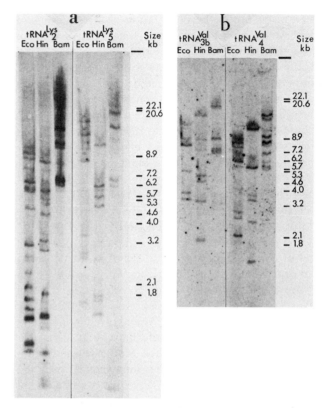

Figure 3 Autoradiographs of ^{125}I-labeled tRNAs hybridized to Drosophila DNA restricted with EcoRI, HindIII, or BamI and separated by agarose gel electrophoresis by the procedure of Southern (1975). (a) tRNA$_2^{Lys}$ (left) and tRNA$_5^{Lys}$ (right); (b) tRNA$_{3b}^{Val}$ (left) and tRNA$_4^{Val}$ (right).

at least some occur on rather large fragments, the genes must be well dispersed along the Drosophila DNA. Also, as can be seen in Figure 3, the genes for tRNA$_2^{Lys}$ appear to be well separated from those for tRNA$_5^{Lys}$. With tRNA$_{3b}^{Val}$ as the probe, one sees six bands after EcoRI cleavage and seven after HindIII cleavage. These genes occur on fragments with sizes different from those bearing the tRNA$_4^{Val}$ genes.

tRNA GENE LOCALIZATION

By in situ hybridization of ^{125}I-labeled tRNAs to polytene chromosomes followed by autoradiography using the techniques described by Gall and Pardue (1971), the sites for genes may be visualized. We have incorporated two major modifications in this technique. First, the hybridization time is reduced to 2 hours or less to reduce the extent of hybridization by contaminants and, second, in the preparation of the squashes of polytene chromosomes, we employ an acetylation step that results in much cleaner backgrounds on autoradiography (Hayashi et al. 1978). Under these conditions, multiple silver grains are found only over a few bands on the chromosomes. Some of the autoradiographs are shown in Figure 4 and the results are plotted on the composite photographs of polytene chromosomes prepared by Lefevre (1976) in Figure 5.

One generalization based on these results is that tRNA genes are found scattered along the chromosomes, for example, clusters of genes for $tRNA_2^{Met}$ occur on the 2R, 3L, and 3R chromosomal arms. Genes for $tRNA_{3b}^{Val}$ are found at two major and one minor site on the right arm of the third chromosome. The major sites are at 84D and 92B, and the minor site occurs at 90BC. From grain counts, $tRNA_{3b}^{Val}$ hybridizes to these sites in the ratio of 5:4:1 or 6:3:1. When $tRNA_{3a}^{Val}$ and $tRNA_4^{Val}$ are used as probes, a few grains are also found at these sites. We think this reflects a low level of contamination of these tRNAs with $tRNA_{3b}^{Val}$, but it could also reflect some sequence homology.

Genes for $tRNA_2^{Lys}$ are found at two major sites, 42A and 42E, and three minor sites, 50B, 62A, and 63B. The grain counts over these sites were approximately in the ratio of 5:3:1:1:1. When considered in light of the earlier data, this result indicates that there may be 22 genes for $tRNA_2^{Lys}$ in the haploid genome. In earlier studies, Steffensen and Wimber (1971), using ^{3}H-labeled 4S RNA as a probe, found grains over the 42A region but none over 42E. Yen et al. (1977) studied a recombinant plasmid carrying genes for *Drosophila* tRNA and found that it hybridized to chromosomes in the 42A regions.

The lines above the chromosomes in Figure 5 cover regions where *Minute* mutants map. K. C. Atwood (referred to in Lindsley and Grell 1968) suggested that mutations of the *Minute* class result from mutations in tRNA genes. However, there is no apparent relationship between the sites of tRNA genes and sites of known *Minute* mutants, nor is there any apparent relationship between tRNA genes and regions of puffing.

To date, only about 15% of tRNA genes have been localized. None has been definitely localized on the X chromosome or on the fourth chromosome. However, when labeled total 4S RNA is hybridized to polytene chromosomes, at least two sites on the X chromosome become labeled. One of these, site 12E, may be a possible site for $tRNA_5^{Lys}$. Localizations of tRNA genes are listed in Table 1.

tRNA IN *DROSOPHILA* MUTANTS

Another approach for examining gene number and gene localization uses mutant flies that have duplications or deletions (deficiencies) at sites on the chromosome carrying tRNA genes. Although the homozygous state is lethal for most deletions and duplications, the heterozygote can be

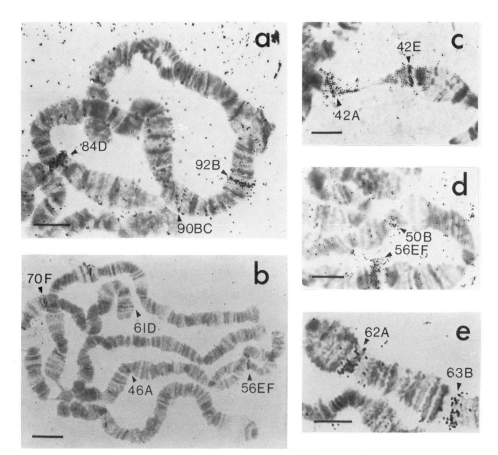

Figure 4 (a) Hybridization of ^{125}I-labeled tRNA$_{3b}^{Val}$ (1.9×10^{18} cpm/mole) in 2× standard saline citrate (SSC) to an unacetylated preparation of polytene chromosomes at 65°C for 8 hr. Sites 84D, 90BC, and 92B are labeled. Exposure was for 12 days. (Reprinted, with permission, from Dunn et al. 1979.) (b) Hybridization of ^{125}I-labeled tRNA$_3^{Met}$ (1.41×10^{18} cpm/mole) in 0.12 M sodium phosphate buffer (pH 7.0), 0.01 M EDTA to an acetylated preparation of polytene chromosomes at 65°C for 2 hr. Sites 46A, 56EF, 61D, and 70F are labeled. Exposure was for 30 days. (c) Hybridization of ^{125}I-labeled tRNA$_2^{Lys}$ (9.5×10^{18} cpm/mole) in 2× SSC to an unacetylated preparation of polytene chromosomes at 65°C for 4 hr. Sites 42A and 42E1,2 are labeled. Exposure was for 28 days. (d) Hybridization of ^{125}I-labeled tRNA$_2^{Lys}$ (9.15×10^{18} cpm/mole) in 2× SSC to an acetylated preparation of polytene chromosomes at 65°C for 4 hr. Sites 50B and 56EF are labeled. Exposure was for 16 days. (e) Same as d, except sites 62A and 63B are labeled and exposure was for 16 days. Bars in lower-left corners represent 10 µm.

raised in sufficient numbers for analysis. Two stocks of mutant flies were studied. One, a deficiency mutant designated $Df(3R)Antp^{Ns+R17}$, has material missing from 84B1,2 through 84D12, inclusive. This gap includes

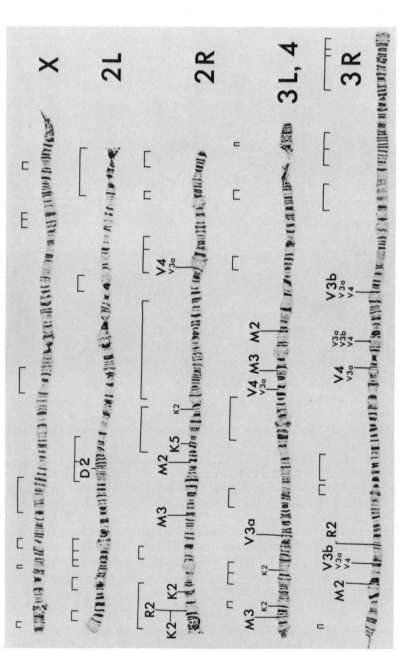

Figure 5 Localization of tRNA genes on the polytene chromosome of *Drosophila* (composite photographs by Lefevre [1976]). Amino acid abbreviations are used to designate locations of the corresponding tRNA genes: D, tRNA[Asp]; K, tRNA[Lys]; M, tRNA[Met]; R, tRNA[Arg]; V, tRNA[Val]. (Reprinted, with permission, from Lefevre 1976.)

Table 1 Localization of tRNA genes

tRNA	Approximate no. of genes/haploid genome	Location
$tRNA_2^{Arg}$	—	42A; 84F
$tRNA_2^{Asp}$	—	29DE
$tRNA_2^{Lys}$	18	42A; 42E; 50B; 62A; 63B
$tRNA_2^{Met}$	7	48B; 72F; 83F–84A
$tRNA_3^{Met}$	6	46A; 61D; 70F
$tRNA_{3a}^{Val}$	2	64DE
$tRNA_{3b}^{Val}$	10	84D3,4; 92B; 90BC
$tRNA_4^{Val}$	—	56D; 70BC; 89B
$tRNA_{3a}^{Gly}$	12	56E–57A[a]
$tRNA_5^{Lys}$	—	48F–49A [84A; 87B; 29DE; 12E][a]
$tRNA_2^{Phe}$	—	56EF[a]
5S RNA	168	56F

[a]Tentative assignments.

the $tRNA_{3b}^{Val}$ gene cluster at 84D3,4 (Duncan and Kaufman 1975). The origin and cytology of the second stock, carrying the duplication $In\,(3R)\,dsx^{C+R3L}Antp^{BR}$, is complicated (T. C. Kaufman et al., unpubl.), and for present purposes it is sufficient to indicate that the material from 84B2 through 84D12 is duplicated.

The acceptances of valine by 4S RNAs from these mutants as well as from the wild-type flies were compared using acceptance of lysine and alanine as internal controls. The results are shown in Table 2. The deficiency mutant showed essentially wild-type levels of valine acceptance. That is, Drosophila cells compensate for the gene deficiency by producing more tRNAs^{Val}. The duplication mutant showed an increased level of valine acceptance, the level being somewhat greater than that expected if the amount of gene product were directly related to the number of genes present.

More detailed analyses of these 4S RNA fractions showed that the ratio of $tRNA_{3b}^{Val}$ to $tRNA_4^{Val}$ was increased 30% in the duplication mutant and was decreased 31% in the deficiency mutant (Table 3). Grain counts on autoradiographs described above support a ratio of $tRNA_{3b}^{Val}$ genes at 84D, 90BC, and 92B of approximately 5:1:4 or 6:1:3. If the first ratio is correct, we might expect a change in the $tRNA_{3b}^{Val}/tRNA_4^{Val}$ ratio of ±25% in the heterozygous mutants in comparison to the ratio with wild-type tRNA and if the second ratio is correct the value would be 30%. We find a value of about 30%. However, grain counts are subject to considerable error, as are analyses of the amounts of various isoaccepting species of tRNA after several chromatographic steps. Nevertheless, this

Table 2 Valine acceptance of tRNA prepared from *Drosophila* stocks with a deletion or duplication around site 84D

Strain	Ratio of amino acid acceptance	% of control	Calculated[a]
Valine/lysine[b]			
Control	0.96	100	100
Deficiency	1.01	105	90
Duplication	1.14	119	110
Valine/alanine[c]			
Control	0.89	100	100
Deficiency	0.92	103	90
Duplication	1.02	115	110

[a]These numbers assume tRNA$_{3b}^{Val}$ to be 35% of the total tRNAs^{Val} in the control (from column chromatography data) with changes in the mutants proportional to the change in the number of genes.
[b]These values are the average of two experiments each.
[c]These values are the average of four experiments each.

Table 3 Amounts of tRNA$_{3b}^{Val}$ isoacceptors measured by chromatography in the RPC-5 system

Drosophila strain	tRNA$_{3a}^{Val}$/tRNA$_{4\&5}^{Val}$	tRNA$_{3b}^{Val}$/tRNA$_{4\&5}^{Val}$	tRNA$_{3b}^{Val}$ as % of control
Control	0.29	0.73	100
Duplication	0.28	0.90	123
Deficiency	0.29	0.52	71

Chromatography in a buffer system containing 1 mM EDTA, 1 mM 2-mercaptoethanol, and 10 mM sodium formate (pH 3.8) with an increasing concentration of NaCl.

Drosophila strain	tRNA$_{3a\&3b}^{Val}$/ tRNA$_4^{Val}$	tRNA$_{3a}^{Val}$/ tRNA$_4^{Val}$	tRNA$_{3b}^{Val}$/ tRNA$_4^{Val}$	tRNA$_{3b}^{Val}$ as % of control
Control	1.14	0.31	0.83	100
Duplication	1.39	0.31	1.08	130
Deficiency	0.88	0.31	0.57	69

Chromatography in a buffer system containing 10 mM MgCl$_2$, 1 mM 2-mercaptoethanol, and 10 mM sodium acetate (pH 4.5) with an increasing concentration of NaCl.

type of analysis does support the idea that we are dealing with tRNA genes at these sites and provides an indication of the number of genes present.

CONCLUSIONS

A number of different studies are providing results that help to clarify the organization of tRNA genes on *Drosophila* DNA. In situ hybridizations reported here and elsewhere (Grigliatti et al. 1974a; Kubli and Schmidt 1978; Schmidt et al. 1978; Elder 1978) have clearly demonstrated that multiple copies of genes for many individual tRNAs occur in two or more clusters at widely separated sites on the chromosomes. At the level of the DNA, even those genes within a cluster are apparently separated by large spacer regions. The analysis of a plasmid containing *Drosophila* DNA that hybridizes with *Drosophila* 4S RNA showed that the 4S RNA genes were separated by spacers of several kilobases (Yen et al. 1977). This observation is supported by studies (Fig. 3) on the distribution of tRNA genes on fragments of DNA resulting from digestion by restriction nucleases. Many of the genes occur apparently as single gene copies on fragments of DNA containing more than three kilobases. As yet, we have no understanding of the biological significance of the distribution of the genes, nor do we understand how the cell maintains the fidelity of sequence for genes that are so widely dispersed. These problems are subjects for future investigation.

ACKNOWLEDGMENTS

This research was supported by the Medical Research Council of Canada (MT-1279), the National Research Council of Canada (A-1764), and the National Cancer Institute (contract 6051).

REFERENCES

Birnstiel, M. L., B. H. Sells, and I. F. Purdom. 1972. Kinetic complexity of RNA molecules. *J. Mol. Biol.* **63:**21.
Botchan, P., R. H. Reeder, and I. B. Dawid. 1977. Restriction analysis of the nontranscribed spacers of *Xenopus laevis* ribosomal DNA. *Cell* **11:**599.
Clarkson, S. G. and V. Kurer. 1976. Isolation and some properties of DNA coding for tRNA^{Met} from *Xenopus laevis*. *Cell* **8:**183.
Clarkson, S. G., M. L. Birnstiel, and I. F. Purdom. 1973a. Clustering of transfer RNA genes of *Xenopus laevis*. *J. Mol. Biol.* **79:**411.
Clarkson, S. G., M. L. Birnstiel, and V. Serra. 1973b. Reiterated transfer RNA genes of *Xenopus laevis*. *J. Mol. Biol.* **79:**391.
Clarkson, S. G., V. Kurer, and H. O. Smith. 1978. Sequence organization of a cloned tDNA$_1^{Met}$ fragment from *Xenopus laevis*. *Cell* **14:**713.

Commerford, S. L. 1971. Iodination of nucleic acids *in vitro*. *Biochemistry* **10:** 1993.

Dunn, R., W. R. Addison, I. C. Gillam, and G. M. Tener. 1978. The purification and properties of valine tRNAs of *Drosophila melanogaster*. *Can. J. Biochem.* **56:** 618.

Dunn, R., S. Hayashi, I. C. Gillam, A. D. Delaney, G. M. Terrer, T. A. Grigliatti, T. C. Kaufman, and D. T. Suzuki. 1979. Genes coding for valine transfer ribonucleic acid-3b in *Drosophila melanogaster*. *J. Mol. Biol.* **128:** 277.

Egan, J. and A. Landy. 1978. Structural analysis of the $tRNA_1^{Tyr}$ gene of *Escherichia coli*. *J. Biol. Chem.* **253:** 3607.

Elder, R. T. 1978. Genes for a single transfer RNA are present at two chromosomal sites in *Drosophila melanogaster*. *Fed. Proc.* **37:** 1732.

Goodman, H. M., M. W. Olson, and B. D. Hall. 1977. Nucleotide sequence of a mutant eukaryotic gene: The yeast tyrosine-inserting ochre suppressor SUP4-O. *Proc. Natl. Acad. Sci.* **74:** 5453.

Grigliatti, T. A., B. N. White, G. M. Tener, T. C. Kaufman, and D. T. Suzuki. 1974a. The localization of transfer RNA_5^{Lys} genes in *Drosophila melanogaster*. *Proc. Natl. Acad. Sci.* **71:** 3527.

Grigliatti, T. A., B. N. White, G. M. Tener, T. C. Kaufman, J. J. Holden, and D. T. Suzuki. 1974b. Studies on the transfer RNA genes of *Drosophila*. *Cold Spring Harbor Symp. Quant. Biol.* **38:** 461.

Hayashi, S., I. C. Gillam, A. D. Delaney, and G. M. Tener. 1978. Acetylation of chromosome squashes of *Drosophila melanogaster* decreases the background in autoradiographs from hybridization with [^{125}I]-labeled tRNA. *J. Histochem. Cytochem.* **26:** 677.

Jacq, C., J. R. Miller, and G. G. Brownlee. 1977. A pseudogene structure in 5S DNA of *Xenopus laevis*. *Cell* **12:** 109.

Kubli, E. and T. Schmidt. 1978. The localization of $tRNA_4^{Glu}$ genes from *Drosophila melanogaster* by in situ hybridization. *Nucleic Acids Res.* **5:** 1465.

Lefevre, G., Jr. 1976. A photographic representation and interpretation of the polytene chromosomes of *Drosophila melanogaster* salivary glands. In *The genetics and biology of* Drosophila (ed. M. Ashburner and E. Novitski), vol. 1a, p. 31. Academic Press, New York.

Lindsley, D. and E. H. Grell. 1968. Genetic variations of *Drosophila melanogaster*. *Carnegie Inst. Wash. Publ.* **627:** 152.

O'Farrell, P. Z., B. Cordell, P. Valenzuela, W. J. Rutter, and H. M. Goodman. 1978. Structure and processing of yeast precursor tRNAs containing intervening sequences. *Nature* **274:** 438.

Olson, M. V., D. L. Montgomery, A. K. Hopper, G. S. Page, F. Horodyski, and B. D. Hall. 1977. Molecular characterization of the tyrosine tRNA genes of yeast. *Nature* **267:** 639.

Ritossa, F. M., K. C. Atwood, and S. Spiegelman. 1966. On the redundancy of DNA complementary to amino acid transfer RNA and its absence from the nucleolar organizer region of *Drosophila melanogaster*. *Genetics* **54:** 663.

Schmidt, T., A. H. Egg, and E. Kubli. 1978. The localization of $tRNA_{2\gamma}^{Asp}$ genes from *Drosophila melanogaster* by *in situ* hybridization. *Mol. Gen. Genet.* **164:** 249.

Southern, E. M. 1975. Detection of specific sequences among DNA fragments separated by gel electrophoresis. *J. Mol. Biol.* **98**:503.

Steffensen, D. M. and D. E. Wimber. 1971. Localization of tRNA genes in the salivary chromosomes of *Drosophila* by RNA:DNA hybridization. *Genetics* **69**:163.

Valenzuela, P., A. Venegas, F. Weinberg, R. Bishop, and W. J. Rutter. 1978. Structure of yeast phenylalanine-tRNA genes: An intervening DNA segment within the region coding for the tRNA. *Proc. Natl. Acad. Sci.* **75**:190.

Weber, L. and E. Berger. 1976. Base sequence complexity of the stable RNA species of *Drosophila melanogaster*. *Biochemistry* **15**:5511.

Yen, P. H., A. Sodja, M. Cohen, S. E. Conrad, M. Wu, N. Davidson, and C. Ilgen. 1977. Sequence arrangement of tRNA genes on a fragment of *Drosophila melanogaster* DNA cloned in *E. coli*. *Cell* **11**:763.

The Localization of the Genes for tRNA$_4^{Glu}$ and tRNA$_2^{Asp}$ in *Drosophila melanogaster* by In Situ Hybridization

Eric Kubli, Thomas Schmidt, and Albert H. Egg
Zoologisches Institut der Universität Zürich
CH-8057 Zürich, Switzerland

The localization of the genes coding for tRNAs has been greatly simplified by the introduction of the method of in situ hybridization (Gall and Pardue 1969; John et al. 1969). Early attempts to localize the tRNA genes in *Drosophila* by hybridizing ^{3}H-labeled total tRNA to polytene salivary gland chromosomes were hampered by the low specific activity of the RNA obtained after feeding ^{3}H-labeled RNA precursors to *Drosophila* larvae (Steffensen and Wimber 1971). This difficulty was overcome by using the in vitro iodination procedure of RNA developed by Commerford (1971). The first successful attempt to localize the genes for a purified *Drosophila* tRNA was made by Grigliatti et al. (1974). These authors localized the genes for tRNA$_5^{Lys}$ at the 48F-49A region. Some label, however, was also found at 56EF, the locus of the 5S RNA genes. The purification procedures used (RPC-5 chromatography) obviously did not completely eliminate traces of the 5S RNA from this tRNA sample. Therefore, we decided to use two more specific isolation procedures for the purification of *Drosophila* tRNA isoacceptors (Kubli and Schmidt 1978; Schmidt et al. 1978).

THE LOCALIZATION OF THE tRNA$_4^{Glu}$ GENES

Grosjean et al. (1973) developed an anticodon-anticodon affinity chromatography for the isolation of tRNA^{Glu} and tRNA^{Phe} isoacceptors. This method has been applied to the isolation of a tRNA^{Glu} isoacceptor from Sepharose 4B prefractionated *Drosophila* tRNA. Double-label experiments have shown that this isoacceptor corresponds to tRNA$_4^{Glu}$ of Grigliatti et al. (1974).

After in vitro iodination, the purified tRNA$_4^{Glu}$ was hybridized to salivary gland chromosomes of the mutant *giant* (*gt*, 1–0.9). Three regions were found to be labeled after an exposure time of 55 days: 52F, 56EF, and 62A (Fig. 1). Since 56EF is the locus for the 5S RNA genes and it was known from Grigliatti et al. (1974) that 5S RNA can be a contaminant of tRNA preparations, cold 5S RNA was added as a competitor. The grain counts at

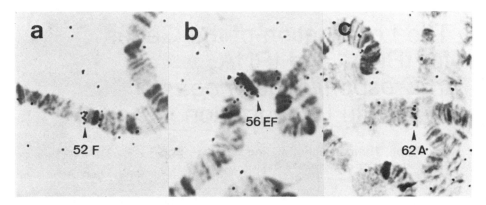

Figure 1 Hybridization of ^{125}I-labeled tRNA$_4^{Glu}$ to salivary gland chromosomes of the mutant *giant* at regions 52F (*a*), 56EF (*b*), and 62A (*c*). (Reprinted, with permission, from Kubli and Schmidt 1978.)

56EF, however, showed no reduction. Indeed, an increase in the grain numbers was observed at the regions 52F and 56EF. This was interpreted as an effect of reduced RNase action on the iodinated tRNA$_4^{Glu}$ by the addition of cold 5S RNA. Hence, the genes for the isoacceptor tRNA$_4^{Glu}$ are localized at three regions of the *Drosophila* genome.

The sensitivity of the in situ hybridization technique does not permit determination of whether the 5S RNA and the tRNA$_4^{Glu}$ genes at the region 56EF are intermingled or merely in close proximity. Although tRNA genes have not been found on 5S RNA genes containing plasmids (Artavanis-Tsakonas et al. 1977), there is more than enough DNA available in this region to code for 160 5S RNA genes and a few tRNA genes. Analysis of mutants located at this site (Nix 1973; Procunier and Tartof 1975) and a combined restriction-hybridization approach may give further insights into this problem.

Four mutants are of interest in the region 52F: *M(2)d*, *M(2)S7*, *l(2)me*, and *Su(f)* (Lindsley and Grell 1968). The mutants *M(2)d* and *Su(f)* have been lost. The tRNA isoacceptor pattern for 20 amino acids has been analyzed by White (1974) for *M(2)S7*. No differences in comparison with the wild-type patterns could be demonstrated, however. Lethal-meander [*l(2)me*, 2–72±] is a recessive lethal mutant that stops growth 2 days after hatching from the egg (Schmid 1949). Most of the homozygous animals never pupate, dying as 7- to 8-day-old larvae (Dübendorfer et al. 1974). The hypothesis has been proposed by Züst et al. (1972) that a deletion for stage-specific activated tRNA genes might be responsible for the lethal effect. Although no exact correlation has been established between genetic loci and the cytogenetic map at this region, we decided to analyze the tRNA^{Glu} isoacceptor pattern of this mutant by RPC-5 chromatography (Kubli 1978). Double-label experiments with wild-type and *l(2)me/l(2)me* tRNA charged with glutamic acid revealed a change in the concentration of the tRNA^{Glu} isoacceptors (Fig. 2). The concentration of the tRNA$_4^{Glu}$ is clearly reduced in the mutant. The total acceptance for glutamic acid, however, is only slightly lowered in the tRNA isolated from homozygous *l(2)me* larvae (Kubli 1978). This indicates that not only the concentration of isoacceptor tRNA$_4^{Glu}$ is reduced, but at the same

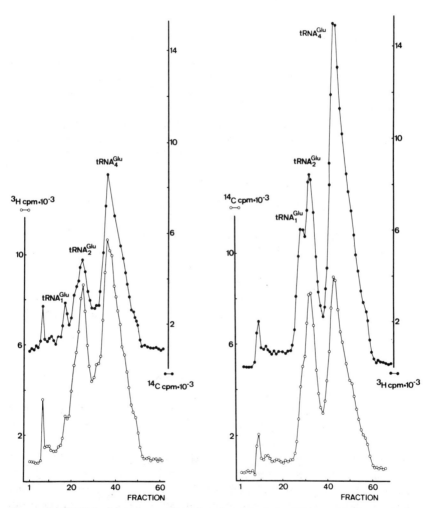

Figure 2 RPC-5 chromatography of unfractionated tRNA aminoacylated with glutamate isolated from +/+ and *l(2)me/l(2)me* flies. (*Left*) ^{14}C-labeled glutamyl-tRNA from 4-day-old wild-type larvae (●); ^{3}H-labeled glutamyl-tRNA from 4-day-old *l(2)me* homozygotes (○). (*Right*) ^{3}H-labeled glutamyl-tRNA from 4-day-old wild-type larvae (●); ^{14}C-labeled glutamyl-tRNA from 4-day-old *l(2)me* homozygotes (○). Modified from Kubli (1978).

time the concentration of tRNA$_2^{Glu}$ is increased. If the assumption that *l(2)me* is a deletion of tRNA^{Glu} genes is correct, it follows that the cells possess a mechanism to keep the tRNA concentration at a constant level, even though a certain number of genes are missing. This is in accordance with the findings of R. Dunn et al. (in prep.), who demonstrated that the amount of tRNA^{Val} is the same in wild-type flies and in flies carrying a deletion for the tRNA$_{3b}^{Val}$ genes (see also, Tener et al., this volume).

THE LOCALIZATION OF tRNA$_{2\delta}^{Asp}$ GENES

In *Drosophila* and other organisms, a modified nucleoside Q is found in the first position of the anticodon in tRNA^{Asn}, tRNA^{Asp}, tRNA^{His}, and tRNA^{Tyr} (Harada and Nishimura 1972; White et al. 1973; Kasai et al. 1975). In mammalian tRNA^{Asp}, mannose is covalently linked to position 4 of the cyclopentene diol moiety of Q. Since concanavalin A (Con A) binds specifically α-D-glucosyl and sterically related residues, this isoacceptor can be isolated with a Con-A-Sepharose column (Okada et al. 1977). We have applied the same procedure to *Drosophila* tRNA. A pure tRNA^{Asp} isoacceptor can be isolated. Double-label experiments show that this isoacceptor corresponds to tRNA$_{2\delta}^{Asp}$ of Grigliatti et al. (1974).

The results of the in situ hybridization experiments are shown in Figure 3. Statistically significant labeling was found at two regions: 29D and 29E. Sometimes grains could also be observed at 25D. Since the only difference between the two major tRNA^{Asp} isoacceptors is the Q base modification in the anticodon, they are transcribed from the same cistrons. When transcription of the tRNA^{Asp} genes is proportional to the isoacceptor concentrations, we have determined the prevalent gene clusters for tRNA^{Asp}.

GENERAL COMMENTS

Our results clearly show that the genes for a specific tRNA isoacceptor can be distributed over the *Drosophila* genome. This raises interesting questions about the regulation and possible stage-specific activation of these genes. We have not calculated the number of genes from the grain counts, since too many uncertainties are involved.

Assuming an average of 13 genes per tRNA species (Ritossa et al. 1966), each cluster for tRNA$_4^{Glu}$ and tRNA$_{2\delta}^{Asp}$ probably contains about 5 genes. However, it has to be emphasized that the in situ hybridization procedure allows the detection of complementary sequences in the DNA, but does not give any information about the function of these genes during the development of an organism. Therefore, analyses of mutants such as *l(2)me* have to complement these studies.

Atwood has proposed the hypothesis that *Minutes* are sites of tDNA (Ritossa et al. 1966). The following *Minute* loci coincide with the sites for

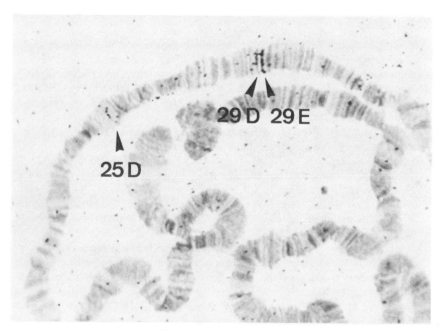

Figure 3 Hybridization of ^{125}I-labeled tRNA$_{2\delta}^{Asp}$ to salivary gland chromosomes of the mutant *giant*. Regions 29D, 29E, and 25D are labeled. (Reprinted, with permission, from Schmidt et al. 1978.)

tRNA$_4^{Glu}$ and tRNA$_2^{Asp}$: *M(2)d, M(2)S7, M(2)b, M(2)173,* and *M(2)e*. Lindsley et al. (1972) have produced a set of aneuploids by combining elements of two Y-autosome translocations with displaced autosomal breakpoints. Forty-one *Minute* phenotypes could be found by analyzing deficiencies over the whole *Drosophila* genome. Among them are deficiencies covering the regions 56EF and 62A (tRNA$_4^{Glu}$) and 29D and 29E (tRNA$_2^{Asp}$). Although these findings do not prove that *Minutes* are tDNA deletions, they support Atwood's hypothesis (Ritossa et al. 1966).

Finally, we would like to mention that all the tRNA loci determined in our work have been confirmed by hybridizing total ^{125}I-labeled tRNA to *Drosophila* salivary gland chromosomes (see Elder et al., this volume).

ACKNOWLEDGMENTS

These investigations were supported by grant number 3.024.76 from the Swiss National Science Foundation, the Hescheler- and the Julius-Klaus Stiftung. We wish to express our gratitude to M. Bienz and M. Milner for critically reading the manuscript.

REFERENCES

Artavanis-Tsakonas, S., P. Schedl, C. Tschudi, V. Pirotta, and W. J. Gehring. 1977. The 5 S genes of *Drosophila melanogaster*. *Cell* **12**:1057.

Commerford, S. L. 1971. Iodination of nucleic acids *in vitro*. *Biochemistry* **10**:1993.

Dübendorfer, K., R. Nöthiger, and E. Kubli. 1974. A selective system for a biochemical analysis of the lethal mutation *l(2)me* of *Drosophila melanogaster*. *Biochem. Genet.* **12**:203.

Gall, J. G. and M. L. Pardue. 1969. Formation and detection of RNA-DNA hybrid molecules in cytological preparations. *Proc. Natl. Acad. Sci.* **63**:378.

Grigliatti, T. A., B. N. White, G. M. Tener, T. C. Kaufman, J. J. Holden, and D. T. Suzuki. 1974. Studies on the transfer RNA genes of *Drosophila*. *Cold Spring Harbor Symp. Quant. Biol.* **38**:461.

Grosjean, H., C. Takada, and J. Petre. 1973. Complex formation between transfer RNAs with complementary anticodons: Use of matrix bound tRNA. *Biochem. Biophys. Res. Commun.* **53**:882.

Harada, F. and S. Nishimura. 1972. Possible anticodon sequences of tRNA^{His}, tRNA^{Asn} and tRNA^{Asp} from *Escherichia coli* B. Universal presence of nucleoside Q in the first position of the anticodons of these transfer ribonucleic acids. *Biochemistry* **11**:301.

John, H. A., M. L. Birnstiel, and K. W. Jones. 1969. RNA-DNA hybrids at the cytological level. *Nature* **223**:582.

Kasai, H., Y. Kuchino, K. Nihei, and S. Nishimura. 1975. Distribution of the modified nucleoside Q and its derivatives in animal and plant transfer RNA's. *Nucleic Acids Res.* **2**:1931.

Kubli, E. 1978. Der Letalfaktor *l(2)me* von *Drosophila melanogaster*: Eine Deletion für tRNA Gene? *Rev. Suisse Zool.* **85**:790.

Kubli, E. and T. Schmidt. 1978. The localization of tRNA$_4^{Glu}$ genes from *Drosophila melanogaster* by "in situ" hybridization. *Nucleic Acids Res.* **5**:1465.

Lindsley, D. L. and E. H. Grell. 1968. Genetic variations of *Drosophila melanogaster*. *Carnegie Inst. Wash. Publ.* **627**:151.

Lindsley, D. L., L. Sandler, B. S. Baker, A. T. C. Carpenter, R. E. Denell, J. C. Hall, P. A. Jacobs, G. L. G. Miklos, B. K. Davies, R. C. Gethmann, R. W. Hardy, A. Hessler, S. M. Miller, H. Nozawa, D. M. Parry, and M. Gould-Somero. 1972. Segmental aneuploidy and the genetic gross structure of the *Drosophila* genome. *Genetics* **71**:157.

Nix, C. E. 1973. Molecular studies on the 5 S RNA genes of *Drosophila melanogaster*. *Mol. Gen. Genet.* **120**:309.

Okada, N., N. Shinodo-Okada, and S. Nishimura. 1977. Isolation of mammalian tRNA^{Asp} and tRNA^{Tyr} by lectin-sepharose affinity column chromatography. *Nucleic Acids Res.* **4**:415.

Procunier, J. D. and K. D. Tartof. 1975. Genetic analysis of the 5 S RNA genes in *Drosophila melanogaster*. *Genetics* **81**:515.

Ritossa, F. M., K. C. Atwood, D. L. Lindsley, and S. Spiegelman. 1966. On the chromosomal distribution of DNA complementary to ribosomal and soluble RNA. *Natl. Cancer Inst. Monogr.* **23**:449.

Schmid, W. 1949. Analyse der letalen Wirkung des Faktors *l(2)me* (*letal-meander*) von *Drosophila melanogaster*. *Z. Indukt. Abstammungs-Vererbungsl.* **83**:220.

Schmidt, T., A. H. Egg, and E. Kubli. 1978. The localization of tRNA$_2^{Asp}$ genes from *Drosophila melanogaster* by "in situ" hybridization. *Mol. Gen. Genet.* **164**: 249.

Steffensen, D. M. and D. E. Wimber. 1971. Localization of tRNA genes in the salivary gland chromosomes of *Drosophila* by RNA-DNA hybridization. *Genetics* **69**: 163.

White, B. N. 1974. An analysis of tRNAs in five *Minutes* and two suppressors. *Drosophila Inform. Serv.* **51**: 58.

White, B. N., G. M. Tener, J. J. Holden, and D. T. Suzuki. 1973. Activity of a transfer RNA modifying enzyme during the development of *Drosophila* and its relationship to the *su(s)* locus. *J. Mol. Biol.* **74**: 635.

Züst, H., A. H. Egg, H. A. Hosbach, and E. Kubli. 1972. Regulation von Darmenzym-aktivitäten und Isozymmuster der Hexokinase bei der Letalmutante *l(2)me* von *Drosophila melanogaster*. *Verh. Schweiz. Naturforsch. Ges.* **152**: 172.

4S RNA Gene Organization in *Drosophila melanogaster*

Robert T. Elder and Olke C. Uhlenbeck
Department of Biochemistry
University of Illinois
Urbana, Illinois 61801

Paul Szabo
Sloan-Kettering Institute for Cancer Research
New York, New York 10021

Drosophila melanogaster has 300–750 genes for tRNAs per haploid genome (Ritossa et al. 1966; Tartof and Perry 1970). Since the kinetic complexity of the mixture suggests 60 different species (Weber and Berger 1976), each tRNA sequence is present an average of 5–12 times per haploid genome. We have studied the organization of the genes for several purified tRNAs. When combined with experiments using the unfractionated 4S RNA mixture, we can obtain an overall view of the organization of tRNA genes in *Drosophila*.

A major advantage of studying *Drosophila* is that with the high specific activities obtained by iodination of tRNA with ^{125}I (Prensky 1975), the tRNA genes can be located directly on the salivary polytene chromosomes by in situ hybridization. This technique is carried out by treating chromosomes fixed to a microscope slide under conditions that will denature some of their DNA without greatly disrupting their cytological structure. Radioactive RNA is then hybridized to the chromosomes and unhybridized RNA is removed by ribonuclease treatment and extensive washing. After autoradiography, the silver grains resulting from radioactive decay localize the hybridized tRNA molecules on the cytogenetic map (Gall and Pardue 1969). By counting the number of grains at a chromosomal site on many chromosomes, one can determine the amount of RNA-DNA hybrid that is formed at that site. In this way, the rate of the in situ hybridization reaction and the amount of hybrid at saturation can be determined. Such a quantitative approach has been used successfully for the analysis of the in situ hybridization reaction of 5S RNA and 18 + 28S RNA to *Drosophila* polytene chromosomes (Szabo et al. 1977). Although the precision is rather low, it is our opinion that quantitation is essential for the reliable localization of tRNA genes.

STUDIES ON PURIFIED tRNAs

Purification and Iodination

We have purified five RNA species from the 4S RNA mixture obtained from *Drosophila* larvae by two procedures. The first is a direct method, involving separating the 4S RNAs into several classes on DEAE-Sephadex A-25 (Szabo et al. 1977), fractionating each class on a low-pressure RPC-5 column, and pooling peaks that showed a promising degree of purification and rerunning them on a high-pressure RPC-5 column with a shallow salt gradient. Three 4S RNAs, RNA 1, RNA 2, and RNA 3 were purified by this procedure.

The second purification procedure involved aminoacylating the unfractionated RNA with a single amino acid. After deproteinizing, the reaction mixture is reacted with activated CH-Sepharose (the N-hydroxy-succinimide ester of 6-aminohexanoic acid coupled to Sepharose 4B), which results in the covalent attachment of the aminoacylated tRNAs through the α amino group of the amino acid. Extensive washing removes all unreacted tRNA and the group of isoacceptors for the chosen amino acid is eluted from the Sepharose by raising the pH. The isoacceptors are then fractionated on a high-pressure RPC-5 column. This procedure was used to obtain $tRNA_2^{Met}$ and $tRNA_2^{Arg}$.

All five 4S RNAs were reasonably pure by several criteria. Each was predominately a single peak on high-resolution RPC-5 chromatography and each was predominately a single band on denaturing polyacrylamide gels. The 4S RNAs were iodinated to a specific activity of 1×10^8 to 4×10^8 dpm/μg using a variation (Szabo et al. 1977) of the procedure of Prensky (1975). Total base hydrolysis of the iodinated RNAs indicated that greater than 90% of the ^{125}I was in the form of 5-iodocytidine, except for RNA 1 in which 35% of the ^{125}I was incorporated into an unidentified modified base. A two-dimensional fractionation of an RNase T1 digest of the iodinated RNA gave a small number (6–10) of heavily labeled spots, indicating fairly uniform labeling and providing more evidence for the purity of the samples.

Rate of In Situ Hybridization

An analysis of the kinetics of the in situ hybridization reaction is important in interpreting a tRNA gene localization experiment. The extent of the in situ hybridization reaction is most conveniently expressed in terms of $C_r t$, the product of the RNA concentration and the time of reaction. Thus, at each value of $C_r t$, the number of grains over a given site is determined for a number of chromosomes and a mean is determined. We have shown previously for 5S RNA that the data follow a pseudo first-order rate law. The data can be fit to this rate law by a computer program to give the grains at saturation and the $C_r t_{1/2}$, the $C_r t$ at which the amount of hybrid is one half of the amount at saturation (Szabo et al. 1977). A similar behavior is expected for a purified tRNA. Since there are 20–50 times fewer gene copies for an average tRNA than for 5S RNA, there will be far fewer grains. However, since the complexity of tRNA is quite similar to 5S RNA, the rate of the in situ hybridization of a tRNA should be similar to the rate of 5S RNA. Any impurity present in the tRNA preparation should hybridize at a slower rate because of its lower concentration.

In Figure 1, the data for the in situ hybridization of tRNA$_2^{Met}$ illustrate this concept. This tRNA hybridizes to the regions 48AB on chromosome II and 72F–73A on chromosome III at essentially the same rate and at a rate very similar to that obtained previously for 5S RNA (Table 1). This is good evidence that both 48AB and 72F–73A contain genes for tRNA$_2^{Met}$. In addition, sites at 84 and 63AB are also labeled with the same ^{125}I-labeled tRNA$_2^{Met}$ probe, but the $C_r t_{1/2}$ of these sites is nearly a factor of 5 higher. We interpret the labeling at 84 and 63AB to be the result of a small amount of one or more other tRNAs that are present as impurities in the tRNA$_2^{Met}$ preparation. Due to their lower concentration, they hybridize at a lower

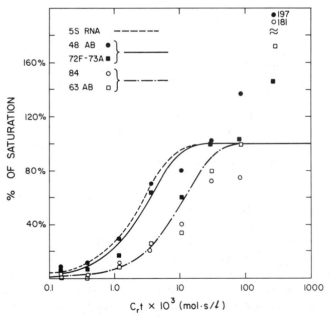

Figure 1 Rate of in situ hybridization of tRNA$_2^{Met}$ at four chromosomal sites. The sp. act. of tRNA$_2^{Met}$ was 1.8×10^8 dpm/μg. Hybridization was at different concentrations for 12 hr. Exposure time was 31 days. Saturation was 7.1, 7.7, 3.8, and 5.3 grains for 48AB, 72F–73A, 63AB, and 84, respectively.

Table 1 In situ hybridization of purified RNAs

Type of RNA	Site	$C_r t_{1/2} \times 10^3$	Gene copies
18 + 28S rRNA	nucleolus	50	100
5S rRNA	56F	2.2	240
4S RNA 1	12E	1.4	5
	23EF	0.9	2
4S RNA 2	22DE	3.6	6
4S RNA 3	90BC	3.8	9
$tRNA_2^{Met}$	48AB	2.2	2
	72F-73A	3.0	2
$tRNA_2^{Arg}$	42A	2.7	8
	84EF	2.1	5

apparent rate. By carefully analyzing the kinetics of the in situ hybridization reaction we can distinguish between the correct sites for a particular tRNA and sites that code for impurities in the tRNA preparation.

Number of tRNA Genes

Hybridization experiments of ^{125}I-labeled $tRNA_2^{Met}$ to *Drosophila* embryo DNA immobilized on nitrocellulose filter discs indicated about three gene copies per haploid genome. An independent determination of four or five gene copies was obtained by observing that number of *Eco*RI or *Hin*dIII restriction enzyme fragments present in *Drosophila* DNA when ^{125}I-labeled $tRNA_2^{Met}$ was used as a probe in a Southern blot hybridization experiment (Southern 1975). Since an equal number of grains appear at the 48AB and 72F-73A sites, there are probably two genes for $tRNA_2^{Met}$ at 72F-73A and two at 48AB.

The hybridization efficiency can be calculated from the number of genes for $tRNA_2^{Met}$. The hybridization efficiency is the fraction of complementary DNA in the chromosome that has formed hybrid when the in situ hybridization reaction has reached saturation. Based on two gene copies at each site, an autoradiographic efficiency of 13%, and an estimate of 500 for the degree of polytenization of the salivary chromosomes, the hybridization efficiency at both 48AB and 72F-73A is around 20%. This is close to the value of the hybridization efficiency observed for 5S RNA hybridizing to 56F under identical conditions. Measurements with other RNAs also indicate that the hybridization efficiency does not depend upon the chromosomal site so that an estimate of the number of genes can be made from the number of grains at saturation.

Results with Single Species of tRNA

In Table 1, the hybridization data for all 5 RNAs are summarized. Only those sites that have a sufficiently low value of $C_r t_{1/2}$ are assigned to that tRNA. Although the analysis is not as complete as with $tRNA_2^{Met}$, the combination of filter disc hybridization experiments and the number of grains at each site at saturation gives an estimate of the number of gene copies at that site. Table 1 shows that the genes for each Drosophila tRNA we have examined are present in 4–13 copies in the haploid genome and that they are grouped in clusters at one or two sites on the polytene chromosomes.

UNFRACTIONATED 4S RNA

With this quantitative understanding of the in situ hybridization reaction for several purified 4S RNAs in hand, we examined the in situ hybridization reaction of unfractionated 4S RNA. Since the kinetic complexity of this mixture is about 60 times greater than an individual tRNA (Weber and Berger 1976), much higher $C_r t$ values should be necessary to achieve saturation of a site. This was directly confirmed by scoring the number of grains over many chromosomes at one site at several values of $C_r t$. As an example, Figure 2 shows the kinetics of in situ hybridization at the 90C site

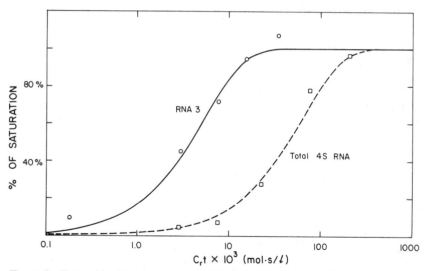

Figure 2 Rate of in situ hybridization at 90C with RNA 3 and with unfractionated 4S RNA. The sp. act. of RNA 3 was 1.1×10^8 dpm/μg. Hybridization was at different concentrations for 5 hr. Exposure was 21 days. Saturation was 11 grains. The sp. act. of unfractionated 4S RNA was 10^8 dpm/μg. Hybridization was at different concentrations for 9.2 hr. Exposure was 32 days. Saturation was 15 grains.

using both unfractionated 4S RNA and RNA 3 as probes. The $C_r t_{1/2}$ for the 90C site is 20 times higher with unfractionated 4S RNA than with RNA. This indicates that RNA 3 is present as 5% of the unfractionated 4S RNA, a number only slightly higher than expected for a single RNA species.

To identify many of the sites for 4S RNA on the *Drosophila* chromosomes, in situ hybridization of unfractionated 4S RNA was carried to high $C_r t$ values on Canton S polytene chromosomes and the even larger polytene chromosomes of the *giant* mutation (Kaufman 1972). About 10 sets of chromosomes of each type were examined at the highest $C_r t$ value (0.77 moles · sec/liter) to identify the 4S RNA sites. The list of the 54 4S RNA sites is given in Table 2. We distinguish 2 classes of sites, 26 strong sites with 4 or more gene copies (underlined) and 28 weak sites with 1–3 gene copies. In addition, there are 9 sites not listed in Table 1, which are possible weak sites (35DE, 36EF, 38AB, 40, 43A, 45DE, 56E, 62EF, and 94BC). All of the purified tRNA sites in Table 1 are included in Table 2, as well as most of the sites identified by others (Grigliatti et al. 1974; Kubli and Schmidt 1978; Schmidt et al. 1978). The distribution of the sites over the chromosomes is quite random except that significantly fewer 4S RNA sites are on the X chromosome. It is likely that Table 2 contains the majority of the sites for tRNA genes in *Drosophila*. The only tRNA genes that may not be localized are any in the unpolytenized or underpolytenized regions of the

Table 2   *Drosophila* 4S RNA Sites

X	2L	2R	3L	3R	4
3D	22DE	41CD	61D	84A	none
3F	23EF	42A	62A	84D	
5F6A	24DE	42E	63A	84F	
11A	28D	44EF	64DE	87BC	
12E	28F-29A	48AB	66B	87F-88A	
	29D	48CD	67B	89BC	
	29E	49AB	69F-70A	90C	
	35AB	49F-50A	70DE	90DE	
		50BC	72F-73A	92AB	
		52F-53A	79F	93A	
		54A		95F-96A	
		54D		97CD	
		55EF		99EF	
		56D			
		56F			
		57D			
		58AB			
		60E			

Underlining indicates 4 or more genes for 4S RNA; sites not underlined have 1–3 genes.

genome, such as the Y chromosome or nucleolus, and any genes present in only a few gene copies whose tRNAs are present as less than 0.3% of the unfractionated 4S RNA.

Atwood has proposed that *Minutes*, a class of mutations in *Drosophila*, correspond to deficiencies of tRNA genes (Lindsley and Grell 1968). The correlation between the locations of reasonably well mapped *Minutes* and the sites in Table 2 is no better than that expected from chance alone. In addition, several well-mapped *Minutes* occur at sites where no 4S RNA was found to hybridize. Thus, we suggest that, in general, *Minutes* are not deficiencies of the structural genes for tRNAs.

ACKNOWLEDGMENTS

We thank Dale Steffensen, Marcello Siniscalco, and Wolf Prensky for use of facilities and interest in this work. This research was supported by grants from the National Institutes of Health (GM-19059) and the National Cancer Institute (CA-17085) and a biomedical research grant from the University of Illinois. R. T. E. was a National Science Foundation predoctoral fellow.

REFERENCES

Gall, J. G. and M. L. Pardue. 1969. Formation and detection of RNA-DNA hybrid molecules in cytological preparations. *Proc. Natl. Acad. Sci.* **63**:378.

Grigliatti, T. A., B. N. White, G. M. Tener, T. C. Kaufman, and D. T. Suzuki. 1974. The localization of transfer RNA_3^{lys} genes in *Drosophila melanogaster*. *Proc. Natl. Acad. Sci.* **71**:3527.

Kaufman, T. C. 1972. Characterization of three new alleles of the *giant* locus of *Drosophila melanogaster*. *Genetics* **71**:S28.

Kubli, E. and T. Schmidt. 1978. Localization of transfer RNA_4^{glu} genes from *Drosophila melanogaster* by *in situ* hybridization. *Nucl. Acid Res.* **5**:1465.

Lindsley, D. L. and E. H. Grell. 1968. Genetic variations of *Drosophila melanogaster*. *Carnegie Inst. Wash. Publ.* **627**:151.

Prensky, W. 1975. The radioiodination of RNA and DNA to high specific activities. *Methods Cell Biol.* **13**:121.

Ritossa, F. M., K. C. Atwood, and S. Spiegelmann. 1966. On the redundancy of DNA complementary to amino acid transfer RNA and its absence from the nucleolar organizer region of *Drosophila melanogaster*. *Genetics* **54**:663.

Schmidt, T., A. H. Egg, and E. Kubli. 1978. The localization of $tRNA_{2B}^{asp}$ genes from *Drosophila melanogaster* by *in situ* hybridization. *Mol. Gen. Genet.* **164**:249.

Southern, E. M. 1975. Detection of specific sequences among DNA fragments separated by gel electrophoresis. *J. Mol. Biol.* **98**:503.

Szabo, P., R. Elder, D. M. Steffensen, and O. C. Uhlenbeck. 1977. Quantitative *in situ* hybridization of ribosomal RNA species to polytene chromosomes of *Drosophila melanogaster*. *J. Mol. Biol.* **77**:539.

Tartof, K. D. and R. P. Perry. 1970. The 5S RNA genes of *Drosophila melanogaster*. *J. Mol. Biol.* **51**:171.

Weber, L. and E. Berger. 1976. Base sequence complexity of the stable RNA species of *Drosophila melanogaster*. *Biochemistry* **15**:5511.

Arrangement and Transcription of *Drosophila* tRNA Genes

Bernd Hovemann, Otto Schmidt,
Hirotomo Yamada, Sanford Silverman, Jen-i Mao,
Donald DeFranco, and Dieter Söll
Department of Molecular Biophysics and Biochemistry
Yale University
New Haven, Connecticut 06520

Our knowledge of the details of tRNA biosynthesis in eukaryotic organisms is much more limited than that of prokaryotes (see Mazzara and McClain, this volume). The main stumbling block in unraveling the sequence of the complex enzymatic steps that lead from tRNA genes to mature tRNA has been the lack of well-characterized tRNA genes and of defined tRNA precursors. Many of the questions asked in an earlier review (Schaefer and Söll 1974) are still open. What is the arrangement of tRNA genes in the chromosomes? How is a tRNA gene organized; what is the nature of promoter, terminator, and operatorlike sites? What factors control the rate and extent of tRNA biosynthesis?

Progress in recent years was stimulated by the development of techniques for molecular cloning of DNA (Sinsheimer 1977), which allowed the isolation and study of single eukaryotic tRNA genes. We now have some insight into the organization of tRNA genes in the genome of yeast (Beckmann et al. 1977; Olson et al. 1979), *Drosophila* (Yen et al. 1977; Schmidt et al. 1978; Dunn et al. 1979), the nematode *Caenorhabditis elegans* (Cortese et al. 1978), and *Xenopus laevis* (Kressmann et al. 1978). In addition, the DNA sequences of a number of yeast tRNA genes have been elucidated (Goodman et al. 1977; Valenzuela et al. 1978; Ogden et al. 1979). These tRNA genes were shown to be noncolinear with their products; they contained intervening nucleotide sequences that are removed at the RNA level by novel enzymes of RNA metabolism (Knapp et al. 1978; O'Farrell et al. 1978; DeRobertis and Olson 1979; Ogden et al. 1979). However, a great deal of unpublished data indicate that the majority of eukaryotic tRNA genes does not contain intervening sequences. The lack of recognized mutants affecting tRNA biosynthesis in eukaryotic organisms (Hopper et al. 1978) has limited the availability of tRNA precursors. Such molecules, obtained by faithful in vitro transcription of tRNA genes (Schmidt et al. 1978; Ogden et al. 1979), would provide the substrates needed to engage in the search and characterization of the enzymes involved in tRNA maturation.

Studies on the enzymology of transcription in eukaryotic cells are progressing rapidly (see Roeder 1976). The nuclear RNA polymerase III is responsible for the transcription of 5S and tRNA genes. The enzyme has been purified from a number of eukaryotic sources and shown to contain ten different proteins. However, the purified polymerase (e.g., from *Xenopus*) is not specific in initiating RNA formation from pure DNA templates; both strands of the DNA are transcribed (Parker et al. 1976). More recent experiments make it likely that in addition to the polymerase, factors present in crude extracts are required for selective and asymmetric transcription of *Xenopus* 5S genes in vitro (Ng et al. 1979). Unfortunately the various components of such a system have not yet been purified. Our knowledge of the nature of the initiating nucleotide and of the promoter sequences recognized by this enzyme is very scant. *Xenopus* RNA polymerase initiates 5S RNA transcription with pppG at the position that corresponds to the 5' nucleotide of the mature RNA (Birkenmeier et al. 1978; Ng et al. 1979). For tRNA gene transcripts only one isoacceptor family has been investigated: Yeast tRNA^{Tyr} genes injected into *Xenopus* oocytes can be transcribed in vivo to form tRNA precursors with pppA at their 5' terminals (DeRobertis and Olson 1979). The length of the leader sequences is variable and appears to be a property of the particular gene. In the case of tRNAs, no regions of homology resembling a promoter (Goodman et al. 1977; Valenzuela et al. 1978) were found in the 5' flanking sequences of the yeast tRNA^{Tyr} and tRNA^{Phe} genes, whereas such sequences were observed in a comparison of several *Xenopus* 5S RNA genes (Korn and Brown 1978). However, the significance of such homologous sequences in the 5' flanking region is not clear (Federoff 1979; Telford et al. 1979).

Gene localization in *Drosophila* is greatly aided by in situ hybridization experiments with salivary gland polytene chromosomes (see Tener et al.; Kubli et al.; Elder et al.; all this volume). In this way some clustering of tRNA genes was observed in a few chromosomal regions (Steffensen and Wimber 1971; Kubli 1980; Elder et al., this volume). In addition, hybridization with many purified *Drosophila* tRNA species revealed their location to be widely scattered over the genome (see Tener et al., this volume).

In this discussion we summarize the arrangement of tRNA genes in a 9.3-kilobase (9.3-kb) fragment of *Drosophila* DNA as determined by extensive DNA sequence analysis and their in vitro transcription in *Xenopus* germinal vesicle extracts.

THE tRNA GENES ON pCIT12, A CLONED FRAGMENT OF *DROSOPHILA* DNA

Cytological hybridization studies had revealed that several tRNA genes appear to be clustered in the 42A region of the *Drosophila* genome

(Steffensen and Wimber 1971). A few years ago, Yen et al. (1977) constructed pCIT12, a plasmid containing a 9.3-kb *Drosophila* DNA fragment from that region. Using their techniques for localizing tRNA genes by electron microscopy (Angerer et al. 1976), they showed the DNA to contain three tRNA genes. A fourth one was detected by hybridization of radioactive tRNA to restriction enzyme fragments. These four tRNA genes lie approximately 2 kb apart. As suggested by Yen et al. (1977), we have designated these tRNA regions as regions 1–4 in Figure 1. The availability of several pure *Drosophila* tRNA species allowed the demonstration (Schmidt et al. 1978) that tRNA region 1 coded for tRNA$_2^{Arg}$, the major arginine isoacceptor in *Drosophila* (White et al. 1973); but tRNA regions 2–4 coded for tRNA$_2^{Lys}$, the major lysine isoacceptor in this organism (Silverman et al. 1979a). To our surprise, tRNA^{Asn} also hybridized, to both tRNA regions 2 and 4 (Schmidt et al. 1978). Thus, there were at least six tRNA genes contained in this plasmid. After extensive DNA sequence studies, the arrangement displayed in Figure 1 emerged. Region 1 contains a single tRNA$_2^{Arg}$ gene, whereas region 3 contains a single tRNA$_2^{Lys}$ gene. However, tRNA region 4 contains four tRNA genes, one for tRNA$_2^{Lys}$ and three for the same species of tRNA^{Asn}. Our DNA sequencing work of region 2 is not yet complete. So far we have shown a tRNA$_2^{Lys}$ gene, a tRNA^{Ile} gene, and at least one other tRNA gene to be present. This latter tRNA gene hybridizes to our *Drosophila* tRNA^{Asn} preparation, yet restriction endonuclease analysis on hybridization properties of transcription products shows it to be different from the tRNA^{Asn} genes in region 4. Thus, there are at least nine tRNA genes contained in the 9.3-kb DNA fragment of pCIT12.

A major result of the DNA sequence analysis is the elucidation of the arrangement of the individual tRNA genes. As can be seen from Figure 1, the direction of transcription is different for the various genes of the same isoacceptor. Thus, they are able to form inverted-repeat structures in which the homology extends over the entire coding region of the tRNA genes. Such structures were seen by Yen et al. (1977) in their electron microscopic analysis of heteroduplexes of pCIT12 with ColE1 DNA. As a matter of fact, we can now explain the majority of their observed inverted-repeat structures as being formed by homologous tRNA genes of opposite polarity. As indicated in Figure 1, we propose that b′ is the tRNA^{Asn} gene located at 9.1 kb, which can pair either with b$_1$, the tRNA^{Asn} gene at 9.0 kb (to form the foldback structure b$_1$b′, which contains no observable loop), or with b$_2$, the tRNA^{Asn} gene at 8.7 kb. This also explains why the electron microscopic study never detected the two inverted repeats b$_1$b′ and b$_2$b′ in the same molecule (Yen et al. 1977). The other inverted repeats are: a, the tRNA^{Lys} gene at 4.9 kb, and a′, the tRNA^{Lys} gene located at 6 kb. The inverted repeat cc′ with a loop length

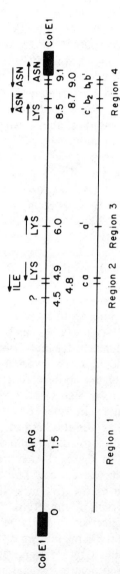

Figure 1 Scheme of tRNA gene arrangement on the 9.3-kb *Drosophila* DNA fragment in pCIT12. ColE1 marks the ends of the plasmid DNA; the numbers refer to the positions (in kb) of the *Drosophila* DNA; (?) denotes still unknown tRNA gene in region 2; (→) denotes transcription direction of the tRNA genes where known. The tRNA regions and the inverted repeat regions (a, a', b', b₁, b₂, c, c') correspond to those designated by Yen et al. (1977).

of 3.4 kb, which was seen by Yen et al. (1977) in the same molecule with the aa' repeat, is most interesting. The sequence c' must correspond to the tRNA^{Lys} gene at 8.5 kb, which pairs with 75% homology with c, the tRNA^{Ile} gene located at 4.8 kb. Of course, a pair ac' would also be possible, but then either the tRNA^{Ile} gene or a still unknown tRNA gene in this region should pair with the tRNA^{Lys} gene a'. Our DNA sequence studies have ruled out the presence of another tRNA^{Ile} gene in tRNA region 4. So far, we have not extended our sequence studies to the very short foldback structure observed by Yen et al. (1977).

The occurrence of inverted-repeat structures, especially with short loops, explains the difficulty or inability to detect the tRNA genes involved in their formation by hybridization with labeled tRNA probes (Yen et al. 1977; Schmidt et al. 1978). If such tRNA gene arrangements are common throughout the *Drosophila* (or any other) genome, then one would expect the gene numbers detected by hybridization with tRNA probes to be underestimates. We would not be surprised if the number of *Drosophila* tRNA genes is greater than 590, the figure determined by Weber and Berger (1976). For the same reason it may not be possible to detect each tRNA gene by in situ hybridization of salivary gland polytene chromosomes. Careful hybridization to small DNA fragments or transcription of tRNA genes (see Transcription of *Drosophila* tRNA Genes) may be the only methods short of DNA sequence analysis to establish a complete tRNA gene catalog.

What did our studies tell about DNA regions involved in regulation of tRNA gene transcription? We thought that the sequence analysis of a few *Drosophila* tRNA genes, including several examples of genes coding for the same isoacceptor species, may reveal promoter or terminator regions common to all tRNA genes or specific for a certain isoacceptor family.

The 3' flanking sequences of all the tRNA genes in pCIT12 are very AT rich. As in other RNA polymerase III genes (e.g., Korn and Brown 1978; Tekamp et al. 1979), these regions may represent terminator sequences. As shown below for the tRNA$_2^{Arg}$ gene from region 1 of pCIT12, transcription termination occurs over a stretch of thymidylate residues.

Such generalizations could not be made for possible promoter sequences in the 5' flanking regions, since a search did not reveal significant sequence homologies among the various tRNA genes. Comparison of the nucleotide sequences of the three tRNA$_2^{Lys}$ genes coded by pCIT12 and another tRNA$_2^{Lys}$ gene (not present in pCIT12 and probably from a different region of the *Drosophila* chromosome [J. Lowenberg and P. Wensink, unpubl.]) made it obvious that their 5' flanking sequences contained the undecanucleotide G*GC*AG*TTTT*TA, which was fairly well conserved (each individual sequence differs from the given one in maximally three bases in the positions indicated in italics; see Fig. 2 for more

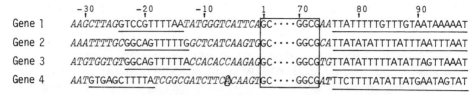

Figure 2 Flanking regions of four *Drosophila* tRNA$_2^{Lys}$ genes. The location of gene *4* in the *Drosophila* genome is unknown. Genes *1*, *2*, and *3* are located in pCIT12 in regions 2, 3, and 4, respectively. The boxed region between residues 1 and 73 is the same for all genes and corresponds to the mature tRNA sequence (Fig. 3). The circled nucleotide at position −6 in gene *4* denotes the RNA initiation site. The underlined nucleotides in the 5′ and 3′ flanking regions represent large homologies among the four genes and may be promoter and terminator regions.

details). Its 5′ end was located about 25 nucleotides before the start of the coding region for the mature tRNA, but it was not located at exactly the same position in all four genes. This sequence was observed only in the flanking regions of the tRNA$_2^{Lys}$ genes; the other tRNA genes in pCIT12 did not contain it. Whether this nucleotide sequence represents a promoter for RNA polymerase III has to await the results of transcription studies with cloned tRNA genes chemically mutagenized in this region or with deletion mutants of this region made by recombinant DNA technology. A similar search did not show significant homologies in the tRNA^{Asn} genes.

Sequence analysis of eight tRNA genes in pCIT12 and of the additional tRNA^{Lys} gene from another region revealed that none of these genes contain an intervening DNA sequence. They also do not code for the trinucleotide CCA, the 3′-terminal sequence of mature tRNA. This oligonucleotide is added by the tRNA nucleotidyl transferase during the maturation process (Schmidt et al. 1978).

Figure 3 shows the cloverleaf structures of either the transcribed tRNA species or of the tRNA genes for the four different amino acid acceptor RNA species coded on pCIT12. Like other tRNAs, they can be arranged into a cloverleaf model of secondary structure. In addition, the *Drosophila* tRNAs show great similarity to the corresponding mammalian tRNA sequences. The tRNA$_2^{Lys}$ shows exactly the same sequence as the corresponding rabbit liver tRNA (Raba et al. 1979). The tRNA^{Asn} differs in six positions from human tRNA^{Asn} (Chen and Roe 1978) and the tRNA$_2^{Arg}$ has four nucleotide substitutions from mouse tRNA^{Arg} (Harada 1978). No comparison is possible for tRNA^{Ile} because the mammalian isoacceptor has not yet been sequenced.

These results provide the first example of a detailed examination of a

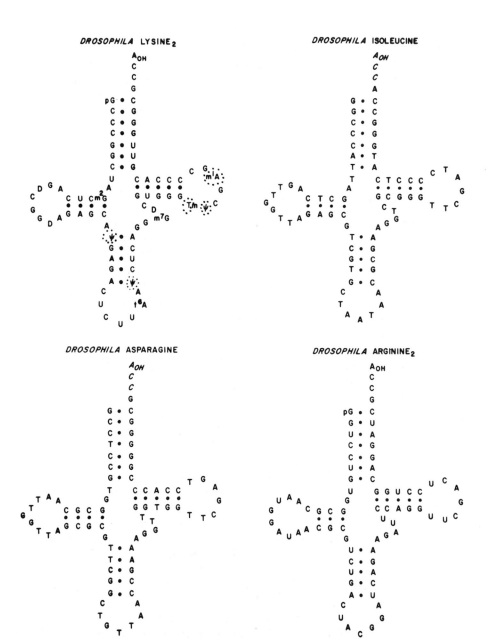

Figure 3 Cloverleaf model of four *Drosophila* tRNA species coded on pCIT12. For the tRNA^{Lys}, the sequence of the mature in-vivo-formed tRNA was determined (Silverman et al. 1979a); for tRNA$_2^{Arg}$, the mature in vitro transcription product was sequenced (Silverman et al. 1979b). For tRNA^{Asn} and tRNA^{Ile}, the DNA sequence is folded into a cloverleaf structure. The 3′-terminal CCA sequence is given in italics, since it is not coded in the gene.

tRNA gene cluster in a eukaryotic genome. The finding of genes for different tRNA species randomly intermingled and in random orientation is puzzling. Further studies will undoubtedly lead to the elucidation of transcription regulatory signals. However, determination of both the relative efficiency of their in vivo expression and the reason for gene clustering of seemingly unrelated tRNA species might not occur in the near future.

TRANSCRIPTION OF DROSOPHILA tRNA GENES

The availability of tRNA genes emphasized the need for routine tests of their biological function. Especially desirable would be an in vitro transcription system that would help to define the sites of initiation and termination of RNA formation and provide a test in the search for transcription regulatory DNA sequences of tRNA genes. The *Xenopus* oocyte is an excellent system for such studies. Cloned *Xenopus* genes for 5S RNA (Brown and Gurdon 1978) and for initiator tRNA (Kressmann et al. 1978) and cloned heterologous tRNA genes (Cortese et al. 1978; DeRobertis and Olson 1979) have been faithfully transcribed in vivo into their mature product by injection into the *Xenopus* oocyte nucleus. This procedure has been greatly simplified by the use of extracts prepared from germinal vesicles that allow efficient in vitro transcription of *Xenopus* 5S RNA (Birkenmeier et al. 1978) and *Drosophila* and yeast tRNA genes (Schmidt et al. 1978; Ogden et al. 1979) into their product. This crude polymerase system contains all components necessary for specific transcription of isolated tRNA genes.

We have transcribed a variety of plasmids containing *Drosophila* tRNA genes. In every case we obtained RNAs of tRNA and precursor tRNA length. As an example we present the transcription of pCIT12 DNA. Separation of the product by two-dimensional gel electrophoresis gave at least seven distinct RNA species (Fig. 4B). Species 2, 4, 5, 6, and 7 have tRNA length, whereas RNA-1 and RNA-3 migrate like 4.5S RNA. Hybridization experiments of these RNA species to separated pCIT12 DNA fragments made it likely that RNA-4 is tRNA^{Asn}, RNA-5 is tRNA^{Lys}, and RNA-6 and RNA-7 are tRNA^{Arg}. This was confirmed by sequence analysis that showed RNA-5 to be mature tRNA^{Lys} and RNA-7 to be mature tRNA^{Arg} (Fig. 3). These results represented the first in vitro synthesis of eukaryotic tRNA. They also indicated that only tRNA genes are transcribed, as judged by our finding that all the RNA species examined are either precursor tRNAs or mature tRNAs. The larger RNA species 1 and 3 are very likely precursors of the shorter RNAs. The fact that the mature tRNAs (e.g., RNA-4 and RNA-5) are formed from larger precursor RNAs is evident when the products of a short-term

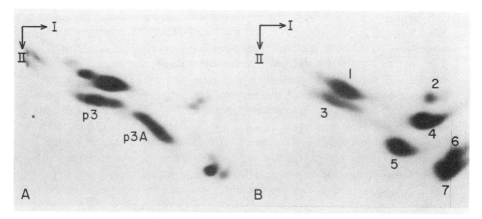

Figure 4 Separation of [α-^{32}P]CTP-labeled transcription products of pCIT12 DNA by two-dimensional polyacrylamide gel electrophoresis. RNA was extracted from a 20-min reaction (*A*) or 130-min reaction (*B*). For details, see Schmidt et al. (1978).

transcription (20 min) were analyzed by two-dimensional gel electrophoresis (Fig. 4A). Very few 4S-sized RNA molecules were observed. Hybridization and sequence data indicated that p1 is a precursor to RNA-4 and p3 is a precursor to RNA-5. These precursor RNAs contained very few modified nucleosides; however, the detailed analysis of RNA-4 and RNA-5 showed them to contain more modified nucleosides (the ones found in RNA-5, i.e., tRNA^{Lys}, are circled in Fig. 3; for tRNA^{Arg}, see Fig. 5). These species also contain the 3'-terminal CCA$_{OH}$ sequence of mature tRNA that is absent in the gene sequence. This trinucleotide is most likely added by tRNA nucleotidyl transferase. Thus, it appears that many of the enzymes responsible for tRNA maturation are localized in the germinal vesicle.

Since the DNA sequences of most of the tRNA genes on pCIT12 have been determined, the major obstacles in defining the RNA initiation and termination sites were the isolation and sequence analysis of the primary transcripts. These studies would also reveal the nature of the initiating nucleotide. Although the *Xenopus* germinal vesicle extract contained processing enzymes that convert the primary transcripts into mature tRNA species, we continued to use this crude extract because of the inability of pure RNA polymerase to initiate specific asymmetric transcription. Because of the variable yield of the primary transcript in different reactions, we decided to isolate these RNA molecules specifically by affinity chromatography according to Smith et al. (1978). Their method is based on the ability of nucleoside triphosphates containing a γ-phosphorothioate to be substrates for RNA polymerase. In the transcription product only the 5'-terminal nucleoside bears the phosphorothioate group. This RNA can then be isolated by selective binding to a mercury-agarose column from which it is eluted by dithiothreitol. This method gave the desired results when RNA that was transcribed from single tRNA genes and then recloned from pCIT12 was analyzed. The nucleotide sequence of the primary transcript of the tRNA^{Arg} gene (see Fig. 1) is given in Figure 5 together with the sequences of the gene and of the 4S-sized processed transcript (Silverman et al. 1979b). RNA initiation occurs with pppG. The 5'-terminal leader sequence is only 7 nucleotides long and is readily processed in the *Xenopus* extract. Although this nucleotide sequence defines the transcription initiation site on the DNA sequence, no obvious promoter sites can be recognized in the 5' flanking region. The 3'-terminal sequence of the precursor tRNA reveals a set of related oligonucleotides ending in U_n ($n = 3-6$, only the longest sequence is shown in Fig. 5), which indicates that termination occurred at different positions in an oligothymidylate stretch. Such a "stuttering" transcription termination is also found in *Escherichia coli* (Gilbert 1976). As expected, the primary transcript is colinear with the DNA sequence. This RNA is further processed in the *Xenopus* extracts to mature-sized tRNA^{Arg} containing 5'-pGp and 3'-CCA$_{OH}$ terminals. Since this trinucleotide is not coded in the gene, it must have been added by the tRNA nucleotidyl transferase present in the germinal vesicle extract. As mentioned, a ψ synthetase is also contained in the germinal vesicle as indicated by the presence of one ψ in the primary and processed transcript.

At present the primary transcript of only one other tRNA gene has been analyzed. In the case of the *Drosophila* tRNA$_2^{Lys}$ gene *4*, the transcript is initiated with pppA at position -6 (indicated in Fig. 2). Thus *Xenopus* RNA polymerase III initiates tRNA gene transcription with purine nucleosides (DeRobertis and Olson 1979; Silverman et al. 1979b). The RNA initiation site of the tRNA$_2^{Lys}$ gene *4* is only 13 nucleotides away from the putative promoter region. However, the significance of this sequence for transcription is not clear at present, as they may not be required for transcription initiation at the correct point (Federoff 1979).

Transcription of the tRNA$_2^{Lys}$ gene *1* (from a recloned *Hin*dIII fragment [either as linear DNA or inserted into pBR322] in which the *Drosophila* DNA extended only 3 nucleotides beyond this common possible promoter sequence [see Fig. 2]) proceeded very efficiently and correctly. Thus, the required length of 5' flanking DNA (if any) is very short. The knowledge of the RNA initiation sites in several *Drosophila* tRNA genes sets the stage for chemical mutagenesis and other alterations of the cloned genes with the aim of defining promoter regions within them.

Since we have used a heterologous system involving *Xenopus* poly-

```
Precursor 5'                              pppGUCAAGCCGGUCCUGGGCGCAAUGGAUAACGCGΨCUGACUACGGAUCAGAAGAUUCCAGGUUCGACUCCUGGCAGGAUCG———AAUUUUUU    3'
                  -20         -10         -1  1        10        20        30        40        50        60        70        80
GENE    5'..CTGTTACACTCGCACGTCAAGCGGTTCCTGTGGCGCAATGGATAACGCGTCTGACTACGGATCAGAAGATTCCAGGTTCGACTCCTGGCAGGATCG———AATTTTTTGGCGTT..3'

tRNA    5'                                pGGUCCUGGGCGCAAUGGAUAACGCGΨCUGACUACGGAUCAGAAGAUUCCAGGUUCGACUCCUGGCAGGAUCGCCA_{OH}   3'
```

Figure 5 Nucleotide sequences of the *Drosophila* tRNA$_2^{Arg}$ gene (from tRNA region 1 of pCIT12), its precursor RNA formed by transcription, and the in-vitro-processed tRNA.

merase and *Drosophila* DNA, a question may be raised whether the RNA initiation sites determined in this way reflects the sites used in *Drosophila*. Until we have isolated the primary, preferably in vivo, transcript from the homologous system, we can only speculate. However, the generality of the *Xenopus* enzyme for transcribing heterologous tRNA genes, such as yeast (Ogden et al. 1979; DeRobertis and Olson 1979; J. Mao and O. Schmidt, unpubl.) and *Bombyx mori* (O. Hagebüchle and K. Sprague, unpubl.), and the conservation of tRNA sequences between *Drosophila*, *Xenopus*, and mammals may argue for related RNA initiation signals in these organisms.

In addition to their significance in understanding eukaryotic gene transcription, these studies also provide precursors that are ideal substrates that are useful for detection and isolation of the complex nucleases and tRNA modifying enzymes involved in eukaryotic tRNA biosynthesis.

ACKNOWLEDGMENTS

We are grateful to P. Yen and N. Davidson for the gift of pCIT12 and for their continued interest. This work was supported by grants from the National Institutes of Health and the National Science Foundation. B. H. and O. S. were recipients of postdoctoral fellowships from the Deutsche Forschungsgemeinschaft.

REFERENCES

Angerer, L., N. Davidson, W. Murphy, D. Lynch, and G. Attardi. 1976. An electron microscopy study of the relative positions of the 4S and ribosomal RNA genes in HeLa cell mitochondrial DNA. *Cell* **9**:81.

Beckmann, J. S., P. F. Johnson, and J. Abelson. 1977. Cloning of yeast transfer RNA genes in *Escherichia coli*. *Science* **196**:205.

Birkenmeier, E. H., D. D. Brown, and E. Jordan. 1978. A nuclear extract of *Xenopus laevis* oocytes that accurately transcribes 5S RNA genes. *Cell* **15**:1077.

Brown, D. D. and J. B. Gurdon. 1978. Cloned single repeating units of 5S RNA direct accurate transcription of 5S RNA when injected into *Xenopus* oocytes. *Proc. Natl. Acad. Sci.* **75**:2849.

Chen, E. Y. and B. A. Roe. 1978. The nucleotide sequence of rat liver tRNA^{Asn}. *Biochem. Biophys. Res. Commun.* **82**:235.

Cortese, R., D. Melton, T. Tranquilla, and J. D. Smith. 1978. Cloning of nematode tRNA genes and their expression in the frog oocyte. *Nucleic Acids Res.* **5**:4593.

DeRobertis, E. M. and M. V. Olson. 1979. Transcription and processing of yeast tyrosine tRNA genes microinjected into frog oocytes. *Nature* **278**:137.

Dunn, R., A. D. Delaney, I. C. Gillam, S. Hayashi, G. M. Tener, T. Grigliatti, V. Misra, M. G. Spurr, D. M. Taylor, and R. C. Miller, Jr. 1979. Isolation and characterization of recombinant DNA plasmids carrying *Drosophila* tRNA genes. *Gene* **7**:197.

Federoff, N. V. 1979. Deletion mutants of *Xenopus laevis* 5S ribosomal DNA. *Cell* **16**:551.
Gilbert, W. 1976. Starting and stopping sequences for RNA polymerase. In *RNA polymerase* (ed. R. Losick and M. Chamberlin), p. 193. Cold Spring Harbor Laboratory, Cold Spring Harbor, New York.
Goodman, H. M., M. V. Olson, and B. D. Hall. 1977. Nucleotide sequence of a mutant eukaryotic gene: The yeast tyrosine-inserting ochre suppressor *sup4-0*. *Proc. Natl. Acad. Sci.* **75**:5453.
Harada, F. 1978. Primer tRNAs for DNA synthesis in RNA tumor viruses. *Seikagaku* **50**:397.
Hopper, A. K., F. Banks, and V. Evangelidis. 1978. A yeast mutant which accumulates precursor tRNAs. *Cell* **14**:211.
Knapp, G., J. S. Beckmann, P. F. Johnson, S. A. Fuhrman, and J. Abelson. 1978. Transcription and processing of intervening sequences in yeast tRNA genes. *Cell* **14**:221.
Korn, L. J. and D. D. Brown. 1978. Nucleotide sequence of *Xenopus borealis* oocyte 5S DNA: Comparison of sequences that flank several related eukaryotic genes. *Cell* **15**:1145.
Kressmann, A., S. G. Clarkson, V. Pirotta, and M. L. Birnstiel. 1978. Transcription of cloned tRNA gene fragments and subfragments injected into the oocyte nucleus of *Xenopus laevis*. *Proc. Natl. Acad. Sci.* **75**:1186.
Kubli, E. 1980. The genetics of transfer RNA in *Drosophila*. *Adv. Genet.* (in press).
Ng, S. Y., C. S. Parker, and R. G. Roeder. 1979. Transcription of cloned *Xenopus* 5S RNA genes by *X. laevis* RNA polymerase III in reconstituted systems. *Proc. Natl. Acad. Sci.* **76**:136.
O'Farrell, P. Z., B. Cordell, P. Valenzuela, W. J. Rutter, and H. M. Goodman. 1978. Structure and processing of yeast precursor tRNAs containing intervening sequences. *Nature* **274**:438.
Ogden, R. C., J. S. Beckmann, J. Abelson, H. S. Kang, D. Söll, and O. Schmidt. 1979. *In vitro* transcription and processing of a yeast tRNA gene containing an intervening sequence. *Cell* **17**:399.
Olson, M. V., B. D. Hall, J. R. Cameron, and R. W. Davis. 1979. Cloning of the yeast tyrosine transfer RNA genes in bacteriophage lambda. *J. Mol. Biol.* **127**:285.
Parker, C. S., Y.-Y. Ng, and R. D. Roeder. 1976. Selective transcription of the 5S RNA genes in isolated chromatin by RNA polymerase III. In *Molecular mechanisms in the control of gene expression* (ed. D. P. Nierlich et al.), vol. 5, p. 223. Academic Press, New York.
Raba, M., K. Limburg, M. Burghagen, J. R. Katze, M. Simsek, J. E. Heckman, U. L. RajBhandary, and H. J. Gross. 1979. Nucleotide sequence of three isoaccepting lysine tRNAs from rabbit liver and SV40 transformed mouse fibroblasts. *Eur. J. Biochem.* **97**:305.
Roeder, R. G. 1976. Eukaryotic nuclear RNA polymerases. In *RNA polymerase* (ed. R. Losick and M. Chamberlin), p. 285. Cold Spring Harbor Laboratory, Cold Spring Harbor, New York.
Schaefer, K. and D. Söll. 1974. New aspects in tRNA biosynthesis. *Biochimie* **56**:795.

Schmidt, O., J. Mao, S. Silverman, B. Hovemann, and D. Söll. 1978. Specific transcription of eukaryotic tRNA genes in *Xenopus* germinal vesicle extracts. *Proc. Natl. Acad. Sci.* **75**:4819.

Silverman, S., I. C. Gillam, G. M. Tener, and D. Söll. 1979a. The nucleotide sequence of lysine tRNA$_2$ from *Drosophila. Nucleic Acids Res.* **6**:435.

Silverman, S., O. Schmidt, D. Söll, and B. Hovemann. 1979b. The nucleotide sequence of a cloned *Drosophila* arginine tRNA gene and its *in vitro* transcription in *Xenopus* germinal vesicle extracts. *J. Biol. Chem.* **254**:10,290.

Sinsheimer, R. L. 1977. Recombinant DNA. *Annu. Rev. Biochem.* **46**:415.

Smith, M., A. E. Reeve, and R. C. C. Huang. 1978. Transcription of bacteriophage λ DNA *in vitro* using purine nucleoside 5′-(γ-S)triphosphates as affinity probes for RNA chain initiation. *Biochemistry* **17**:493.

Steffensen, D. M. and D. E. Wimber. 1971. Localization of tRNA genes in the salivary chromosomes of *Drosophila* by RNA-DNA hybridization. *Genetics* **69**:163.

Tekamp, P. A., P. Valenzuela, T. Maynard, G. I. Bell, and W. J. Rutter. 1979. Specific gene transcription in yeast nuclei and chromatin by added homologous RNA polymerases I and III. *J. Biol. Chem.* **254**:955.

Telford, J. L., A. Kressmann, R. A. Koski, R. Grosschedl, F. Müller, S. G. Clarkson, and M. L. Birnstiel. 1979. Delimitation of a promoter for RNA polymerase III by means of a functional test. *Proc. Natl. Acad. Sci.* **76**:2590.

Valenzuela, P., A. Venegas, F. Weinburg, R. Bishop, and W. J. Rutter. 1978. Structure of yeast phenylalanine-tRNA genes: An intervening DNA segment within the region coding for the tRNA. *Proc. Natl. Acad. Sci.* **75**:190.

Weber, L. and E. Berger. 1976. Base sequence complexity of the stable RNA species of *Drosophila melanogaster. Biochemistry* **15**:5511.

White, B. N., G. M. Tener, J. Holden, and D. T. Suzuki. 1973. Analysis of tRNAs during the development of *Drosophila. Dev. Biol.* **33**:185.

Yen, P. H., A. Sodja, M. Cohen, S. E. Conrad, M. Wu, N. Davidson, and C. Ilgen. 1977. Sequence arrangement of tRNA genes on a fragment of *Drosophila melanogaster* DNA cloned in *E. coli. Cell* **11**:763.

Suppression and Coding

Genetics of Nonsense Suppressor tRNAs in *Escherichia coli*

Haruo Ozeki, Hachiro Inokuchi, Fumiaki Yamao,
Mieko Kodaira, Hitoshi Sakano,*
Toshimichi Ikemura, and Yoshiro Shimura
Department of Biophysics, Faculty of Science
Kyoto University
Kyoto 606, Japan

The genetic approach to tRNA originated in the study of nonsense suppressors in *Escherichia coli* that could specifically restore nonsense mutations in bacteria or bacteriophages. Three codons, UAG (amber), UAA (ocher), and UGA (opal), do not code for an amino acid due to the lack of tRNAs with anticodons corresponding to these codons, called nonsense codons. They are not normally present within the coding regions of genes, but can be generated by nonsense mutations, resulting in the termination of peptide chain elongation at the site. Accordingly, if a mutation occurs in a tRNA gene that alters the coding specificity to recognize a nonsense codon, the altered tRNA may suppress the nonsense mutations in other genes by inserting a specific amino acid during the process of translation. Although the suppression may take place in various ways (for reviews, see Gorini 1971; Steege and Söll 1979), nonsense suppression has been generally attributed to the tRNA suppression. Missense or frameshift mutations are also suppressed by altered tRNAs specific to each mutation (for a review, see Steege and Söll 1979).

In the case of suppressor gene mutations, a mutant normally shows positive character (su^+) differing from ordinary genes and is dominant over its allelic wild-type gene (su^-). Accordingly, the detection of strains carrying nonsense suppressors or the isolation of new suppressor mutants from su^- strains is relatively easy in genetic operation, and, in fact, many different nonsense suppressors (e.g., su^+1, su^+2, su^+B, su^+C, etc.) have been detected before they were fully understood. They can be distinguished by their positions on the bacterial chromosome and by the amino acids they insert. Nonsense suppressors are classified according to their codon specificity, as amber (or UAG), ocher (or UAA), and opal (or UGA) suppressor.

The first crucial proof that a nonsense suppressor gene is the structural gene for a tRNA came from the studies on su^+3, which inserts tyrosine in

*Present address: Basel Institute for Immunology, Basel 5, Switzerland.

response to a UAG codon. Since the su^+3 gene is mapped very close to the attachment site of phage $\phi 80$, the transducing phages of $\phi 80$ carrying su^+3 are readily obtainable (Smith et al. 1966; Andoh and Ozeki 1968). Isolation of the transducing phages, $\phi 80 dsu^+3$ or $\phi 80 psu^+3$, made it possible to perform DNA-tRNA hybridization experiments with the phage DNA, or to analyze the tRNA that is specifically enriched upon phage infection. The su^+3 suppressor was thus identified as a mutant of the structural gene for tRNA$_1^{Tyr}$, in which a single-base substitution was detected in the anticodon (GUA→CUA). This enables the su^+3 tRNA to pair with the UAG codon, thus suppressing this type of nonsense mutation in various genes of bacteria or bacteriophages. This in turn indicated that the transducing phages carry a tRNA gene that is genetically marked. Therefore, such phages can be utilized to isolate a variety of mutants important for studies elucidating the function, structure, or biosynthesis of tRNAs. The earlier works on these aspects of tRNA genetics have been reviewed by Smith (1972) and by Shimura and Ozeki (1973).

The availability of transducing phages carrying suppressor genes offered a convenient method to identify a particular suppressor tRNA among the bulk of other cellular tRNAs. Infecting *E. coli* with an su^+ transducing phage leads to an enrichment of the suppressor tRNA, which then can be purified by gel electrophoresis and analyzed by fingerprinting. Appropriate UV irradiation of the host cells prior to infection provides almost exclusive integration of ^{32}P into the phage-specified tRNA molecules, thus helping the analysis. For this reason, the isolation of su^+ transducing phages was essential. To date, various transducing phages carrying different suppressor genes have been isolated and their tRNAs have been identified. These are summarized in the first section of this paper.

The second section is concerned with the mutants of suppressor tRNAs, particularly those of altered amino acid specificity (mischarging mutants). The genetic methods devised to select various mutants of su^+3 tRNA$_1^{Tyr}$, such as defective, temperature-sensitive, or mischarging suppressors, are basically applicable to other suppressors (for a review, see Shimura and Ozeki 1973). We have recently isolated the mutants of su^+2 tRNA$_2^{Gln}$ from a transducing phage λpsu^+2; they are described below. So far, the only extensive approach to this type of genetics has been with su^+3. The features of defective or temperature-sensitive mutants are more or less comparable between these two suppressor tRNAs, but the sites in the tRNA of the mischarging mutations are entirely different in each case. Based on these results, the recognition of tRNA by aminoacyl-tRNA synthetases is discussed.

Transducing phages often carry genes for other tRNA species in addition to the suppressor tRNA gene. An su^+ gene is often accompanied by its su^- wild-type allele in a transducing phage. In some cases, nonsuppressor tRNA genes are also carried. These tRNA genes are presumably located close to each other on the bacterial chromosome. Since the locations of suppressor genes on the *E. coli* chromosome are known, the map positions of such nonsuppressor tRNA genes are also determined, and detailed mapping of them may be worked out by using the transducing phage. In the third section of this paper, we discuss the clusters of tRNA genes elucidated in this way. The analysis of multimeric tRNA precursors provides the crucial evidence for clustered tRNA genes that form a transcription unit. Such tRNA precursors are efficiently accumulated in the temperature-sensitive RNase P mutants of *E. coli*, which were isolated with the aid of the su^+3 transducing phages (Schedl and Primakoff 1973; Sakano et al. 1974; Ozeki et al. 1974). The genetic studies on tRNA precursors are reviewed separately in this volume by Shimura et al.

NONSENSE SUPPRESSOR tRNAs

Conclusive identification of a suppressor with a tRNA structural gene can be made by showing a nucleotide change in suppressor tRNA. Among the known *E. coli* nonsense suppressors the following have been identified in this way: amber suppressors su^+2, su^+3, and su^+7; ocher suppressors su^+4, su^+8, su^+B, and $su^+\beta$; and opal suppressor su^+9. In all cases, except su^+9, the suppressor mutations are detected as a single-base change at the anticodon. The well-known amber suppressors su^+1 (inserts serine) and su^+6 (inserts leucine) have not yet been characterized in this fashion.

su^+3 Amber and su^+4 Ocher Suppressors
Inserting Tyrosine (*supF* or *tyrT* at 27 Minutes)

The su^+3 suppressor is a mutant of the structural gene for $tRNA_1^{Tyr}$, carrying a single-base substitution of Q→C in its anticodon (Q is a derivative of G). Thus, the tRNA is capable of inserting tyrosine at UAG codon (Fig. 1). The ocher suppressor su^+4 carries a Q→U mutation at the same position in the anticodon of $tRNA_1^{Tyr}$, thus inserting tyrosine at UAA or UAG. The su^+3 was the first identified tRNA suppressor and has been well documented (for reviews, see Smith 1972; Shimura and Ozeki 1973). In Figure 1, other base substitutions occurring in various mutants of the suppressor, such as temperature sensitives or defectives, are also indicated (see below). The detection of base changes in mutant suppressor tRNAs is taken as a good indication that the suppressor mutation is in a tRNA structural gene.

Russell et al. (1970) showed that $tRNA_1^{Tyr}$ is coded by two adjacent identical genes, one of which has mutated in su^+3, leaving the other intact

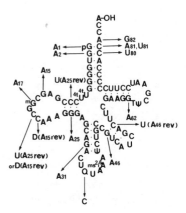

Figure 1 tRNA$_1^{Tyr}$: $su^-3 \to su^+3$ and other mutants (see text).

(su^-3). Accordingly, bacterial strains of different origin carrying su^+3-type suppressors may differ as to which tRNA$_1^{Tyr}$ gene has mutated. By unequal crossing-over involving these two tRNA$_1^{Tyr}$ genes, one-gene or three-gene derivatives are formed. We shall discuss this point further, together with other cases of tRNA gene duplication.

su^+7 Amber and su^+8 Ocher Suppressors Inserting Glutamine or Tryptophan (supU or trpT at 83 Minutes)

su^+7 is an anticodon mutant of a structural gene for tRNA^{Trp} (Fig. 2). This suppressor inserts either glutamine or tryptophan at a UAG codon. su^+7 was isolated from a merodiploid strain carrying F′ilv by Soll and Berg (1969). Since the su^+7 mutation is lethal without a second copy of wild-type allele (su^-7), presumably only one copy of this gene is normally present in E. coli. In agreement with this, only one species of tRNA^{Trp} has been detected so far. Detailed analysis of su^+7 tRNA was carried out by

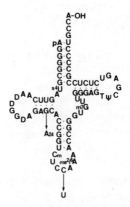

Figure 2 tRNA^{Trp}: $su^-7 \to su^+7$ and su^+9 (A24).

using $\phi 80$ transducing phages, and the anticodon mutation CCA → CUA in tRNA^{Trp} was identified (Soll 1974; Yaniv et al. 1974). An ocher suppressor su^+8 was also identified with the anticodon UUA, which may be derived from su^{-7} (CCA) by two steps of mutation via su^+7 (CUA). It is very interesting to note that in tRNA^{Trp}, a single mutation to su^+7 in the anticodon at the same time causes misacylation by glutaminyl-tRNA synthetase (Yaniv et al. 1974). This will be discussed below, together with the mischarging mutants of other tRNAs.

su^+9 Opal (UGA) Suppressor Inserting Tryptophan ($supU$ or $trpT$ at 83 Minutes)

su^+9 is another mutant of tRNA^{Trp} gene and, therefore, allelic to the su^+7 suppressor. The mutation site is not in the anticodon, but at position 24, changing A → G in the D stem, as indicated in Figure 2. So far, this is the only exceptional case in nonsense suppressors in which the mutation is detected outside the anticodon. The altered codon recognition to UGA is presumably due to a conformational change in the mutant tRNA, altering the coding properties in a quantitative way (Hirsh 1971; Hirsh and Gold 1971; Chan et al. 1971; see also Buckingham and Kurland, this volume).

su^+2 Amber Suppressor Inserting Glutamine ($supE$ at 15 Minutes)

Recently, su^+2 suppressor was identified as an anticodon mutant of tRNA$_2^{Gln}$ gene (Inokuchi et al. 1975; 1979b). The site of mutation is the third nucleoside in the anticodon, G → U (Fig. 3). su^+2 has been mapped at 15 minutes on the chromosome of E. coli. This is near to the attachment site for λ but too far to be picked by this phage. Therefore, a chromosome portion between att^λ and su^+2 was deleted first on an F' factor (F' $su^+2gal\lambda c$I857); then the transducing phages λdsu^+2 or λpsu^+2 were isolated. Upon infection of E. coli with λpsu^+2, the production of glutamine and methionine tRNAs was markedly stimulated. Separation of these tRNAs in polyacrylamide gel electrophoresis, followed by fingerprinting analysis, revealed three types of tRNAs: tRNA$_2^{Gln}$ (Yaniv and Folk 1975), a mutant of tRNA$_2^{Gln}$ carrying G → A substitution in the anticodon, and noninitiator tRNA^{Met}; tRNA$_1^{Gln}$ (Yaniv and Folk 1975) was not detected. Thus, it was concluded that the su^+2 was a structural gene for tRNA$_2^{Gln}$ and derived from one of the duplicated genes for this tRNA. A noninitiator tRNA^{Met} gene is closely linked to the paired tRNA$_2^{Gln}$ genes. As expected from the anticodon CUG, the su^-2 cannot mutate to an ocher suppressor by a single-base change, but becomes su^+2 ocher from su^+2 amber suppressor by an additional mutation, i.e., CUG → CUA → UUA (Inokuchi et al. 1979a). The first U is modified to N, like the corresponding nucleotide in tRNA$_1^{Gln}$ (Yaniv and Folk 1975).

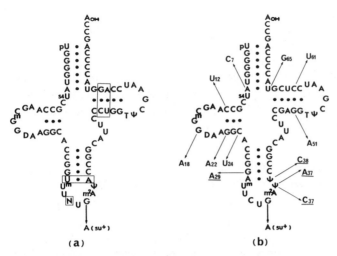

Figure 3 Two tRNAs^{Gln} (see text). (a) tRNA$_1^{Gln}$: $su^-B \to su^+B$; (b) tRNA$_2^{Gln}$: $su^-2 \to su^+2$ and other mutants.

su^+B Ocher Suppressor Inserting Glutamine (*supB* at 15 Minutes)

su^+B and su^+2 have been mapped at the same position on the *E. coli* chromosome and both are known to insert glutamine. The only difference observed was that su^+B can suppress ocher and amber mutations, but su^+2 suppresses only amber ones. su^+B has been identified in our laboratory as an anticodon mutant of tRNA$_1^{Gln}$, and therefore su^+B is not an ocher suppressor mutant of su^+2 (see above) but derived from a different gene. λpsu^+B was obtained from λpsu^+2 by marker exchange in an su^+B strain through integration and excision in the transducing region. The analysis of tRNAs encoded by λpsu^+B leads to the identification of the su^+B gene product (Inokuchi and Ozeki 1976; H. Inokuchi et al., in prep.). The base change caused by this suppressor mutation is indicated in Figure 3. Since the su^-B tRNA$_1^{Gln}$ is also enriched upon λpsu^+B infection, the genes for this tRNA are again duplicated. Besides tRNA$_1^{Gln}$, three more species of tRNA, tRNA$_2^{Gln}$, noninitiator tRNA^{Met}, and tRNA$_{36}^X$ (unidentified tRNA), are encoded by λpsu^+B. The arrangement of these tRNA genes is discussed below.

The first anticodon base in tRNA$_1^{Gln}$ is modified; N is a derivative of s^2U. This modification has been known to prevent base pairing with G, but not with A (Yaniv and Folk 1975). If this rule is strict, the anticodon NUA of su^+B tRNA should only recognize UAA and not UAG. Therefore, the amber suppression that occurs with an ocher suppressor (su^+B or su^+2 ocher) might be due to a fraction of unmodified suppressor tRNA in the cells that can suppress UAG by wobble pairing.

$su^+\beta$ Ocher Suppressor Inserting Lysine ($supL$ at 16 Minutes)

$su^+\beta$ (or sup-273) has been described as an amber suppressor that inserts lysine (Kaplan 1971). Recently we isolated from *E. coli* CA273, λ transducing phages carrying this suppressor and identified the $su^+\beta$ as an anticodon mutant (UUU → UUA) of tRNA^{Lys} (Kimura et al. 1976; M. Kimura et al., in prep.). Transducing phages are readily obtainable from ordinary lysogens carrying prophage λ at att^λ. A defective transducing phage λd$su^+\beta$ thus obtained carried *gal*, $su^+\beta$, and *gltA*, whereas a plaque-forming transducing phage λp$su^+\beta$ made from λd$su^+\beta$ had lost *gal*. One of the tRNAs enriched upon λp$su^+\beta$ infection showed a fingerprint identical to tRNA^{Lys} of *E. coli* B (Chakraburtty 1975). In addition, a mutant of tRNA^{Lys} was detected, carrying a U → A change in the anticodon, as expected for the suppressor tRNA of $su^+\beta$ (Fig. 4). Also, tRNA$_1^{Val}$ was produced. In contrast to its original classification, the $su^+\beta$ is now identified as an ocher suppressor with a UUA anticodon. When we tested the suppressor activity of $su^+\beta$ with various known ocher and amber mutants, it suppressed both types of nonsense mutations. At the same time, however, we also noticed that the suppression is not detected in many strains regardless of the type of nonsense codon. It appears that in many genes the lysine replacement by $su^+\beta$ does not permit restoration of function of proteins. If the ocher mutant used was not suppressed for this reason, but the amber one was, the suppressor tested may be misassigned as an amber suppressor.

Another lysine ocher-suppressor, su^+5, is known to be mapped near the *gal* region in *E. coli* (Garen 1968). The structure of su^+5 suppressor tRNA may or may not be the same as that of $su^+\beta$. Two loci have been allocated for lysine suppressors on the bacterial chromosome, *supG* (15.7 min) and *supL* (16.3 min) (Bachman et al. 1976). Although the $su^+\beta$ has not yet been mapped, *supL* is more likely the locus for this suppressor, as judged from the markers on the transducing phages described above. *supG* for su^+5 appears to be a little too far away to be picked by phage λ at att^λ, whereas the transduction of *supL* locus by phage λ has been reported (Eggertsson and Adelberg 1965). Since the λp$su^+\beta$ used does not seem to carry *supG*, the genes for tRNA^{Lys} in CA273 are duplicated at the position of *supL*, namely $su^+\beta$ and $su^-\beta$.

Nonsense Suppressors Yet to Be Identified (su^+1 and su^+6)

su^+1 (*supD* at 43 min) is one of the classical amber suppressors, inserting serine at a UAG codon. We have isolated λpsu^+1 from *E. coli* CR63 (su^+1) carrying prophage λ near *his*, but no evident stimulation of the production of tRNA^{Ser} has been observed so far on its infection (Yamao et al. 1975). D. A. Steege and D. Söll (unpubl.) also obtained similar results

Figure 4 tRNA^{Lys}: $su^-\beta \to su^+\beta$.

by using su^+1 transducing phages isolated by them (Steege and Low 1975). The infection of λpsu^+1 stimulates the production of tRNAs accepting asparagine and aspartic acid, but no nucleotide alteration on the tRNAs has been detected by comparing the fingerprints with those obtained from a corresponding strain of λpsu^-1.

Leucine suppressor su^+6 (near ilv) also remains to be identified. We have recently confirmed the map position of su^+6 by marker exchange with F'14 carrying the ilv region, and we are now attempting to construct su^+6 transducing phage to analyze the suppressor tRNA.

In all the instances analyzed, except su^+9, nonsense suppressors are identified by a single-base substitution in the anticodon region of tRNA structural genes. It is very reasonable that the creation of a new anticodon by mutation is the most efficient and satisfying way for changing specificity in codon recognition of suppressor tRNAs. In this way, however, the tRNA species for suppressors must be limited within those whose anticodons could mutate to a nonsense anticodon by single-nucleotide substitution. For instance, amber suppressors may arise only from the tRNAs having anticodons (in parentheses) for tyrosine (AUA or GUA), leucine (CAA), tryptophan (CCA), serine (CGA), glutamic acid (CUC), glutamine (CUG), and lysine (CUU); or for ocher suppressors, tyrosine (AUA or GUA), leucine (UAA), serine (UGA), glutamic acid (UUC), glutamine (UUG), and lysine (UUU). The underlined bases must be altered to CUA for amber or to UUA for ocher suppressors. All the amino acids detected so far at the insertion sites of suppressed nonsense mutations are within this range. Among those amino acids, glutamic acid insertion has not been observed. Recently, the genes for tRNA$_2^{Glu}$ (MUC anticodon, M is a modified U) have been detected at the spacer regions of several rRNA genes (rrn) (Ikemura and Nomura 1977). This tRNA is expected to become an ocher suppressor, possibly inserting glutamic acid. Our attempt, however, has been unsuccessful in isolating an ocher suppressor of tRNA$_2^{Glu}$ from a transducing phage carrying one of those rrn genes.

MUTANTS OF SUPPRESSOR tRNAs

Taking advantage of suppressor transducing phages, various mutants of su^+3 tRNA$_1^{Tyr}$ and su^+2 tRNA$_2^{Gln}$ have been isolated. An amber suppressor itself is already a mutant tRNA in which the anticodon is changed to CUA, but the general properties of the tRNA are apparently not affected, except the coding specificity. Studies of mutant suppressor tRNAs have provided pertinent general information on the correlations of in vivo function and structure of tRNA. They are summarized below.

Defective or Temperature-sensitive Mutants

To obtain su^- defective mutants or temperature-sensitive mutants that function at low temperature (e.g., 32°C) but not at high temperature (e.g., 42°C), certain methods must be devised to select rare mutants of negative character from an su^+ population. For this purpose, we used *E. coli* strains carrying amber mutations in two genes specifying the receptor sites for the virulent phages T6 and T5 (or BF23), for example, CA85 $lac_{am}^- T6_{am}^r T5_{am}^r$; whereas in Cambridge, strains carrying an amber mutation in the galactokinase gene (*galK*) and a nonsuppressible mutation in the galactose epimerase gene (*galE*) were used, for example, MB100 $ara_{am}^- trp_{am}^- galK_{am}^- galE^-$. When the former strain is lysogenized with an su^+ transducing phage, the suppressed bacteria become susceptible to the virulent phages T6 and T5, and thus su^- mutants (or su^{ts} at high temperature) may be selected from an su^+ population by treatment with a mixture of two virulent phages (phage selection); a combination of two virulent phages sufficiently reduces the chance to select nonsuppressible mutants in those phage-receptor genes. In the latter strain, MB100, the suppression of $galK_{am}^-$ leads to the accumulation of UDP-galactose due to the defect in $galE^-$, which causes the cell lysis; thus, su^- mutants can be selected from an su^+ population (suicide selection). su^- mutants of su^+ transducing phages may also be detected by plating the mutagenized phages on a lac_{am}^- strain in the presence of 5-bromo-4-chloro-3-indolyl-β-D-galactoside with inducer for β-galactosidase production (e.g., isopropyl-β-D-thiogalactoside): su^+ phages give blue plaques, whereas su^- mutants form colorless ones (dye indicator test) (Abelson et al. 1970).

Earlier work on the analysis of such mutants has been extensively carried out with the su^+3 amber suppressor (for a review, see Smith 1972; Shimura and Ozeki 1973). Various single-base substitutions have been identified along the su^+3 tRNA molecule (see Fig. 1); many caused defective or temperature-sensitive suppression, some affected the synthesis or processing of the tRNA precursor, and some exhibited an altered specificity toward aminoacyl-tRNA synthetases. Isolation of the revertants of defective (or temperature-sensitive) mutants often has resulted in double mutants, in which the effect of the original mutation is compensated by the second mutation in the same tRNA molecule. In this way, for instance, a GC pair at one of the double-helical regions in the cloverleaf arrangement could be replaced by an AU pair in two mutational steps. This implies that tRNA function requires the formation of a base pair itself and that the nature of bases at these positions is not crucial. The pairs of nucleotide residues that form hydrogen bonds in the tertiary structure of a tRNA function in vivo may be elucidated by such genetic studies.

Recently we isolated a number of mutants from su^+2 tRNA$_2^{Gln}$; they are summarized in Figure 3 (Yamao et al. 1977). The single-base mutants C7, A51, and G65 are the defectives, and A18, A22, and U24 are the temperature-sensitive suppressors. Two independently isolated A22 mutants showed very weak suppression even at low temperature. The mutation U12 causes no appreciable production of su^+2 tRNA. The double mutants C7-G65, U12-A22, and A51-U61, which show normal activity of su^+2 suppression, were isolated as the revertants of G65, A22, and A51, respectively. Single mutants of C7 and U12 were obtained from C7-G65 and U12-A22, respectively, by backcrossing these double mutants with the original su^+2. General features of these defective and temperature-sensitive mutants of su^+2 are quite comparable to those of su^+3 (cf. Fig. 1 and 3). Other mutations A29, C37, A37, and C38 are related to the mischarging of su^+2 tRNA$_2^{Gln}$ that is described below.

Mischarging Mutants

It has been shown in su^+3 tRNA$_1^{Tyr}$ that amino acids specificity can be altered by a single-base mutation in the tRNA. The mutants of altered amino acid acceptor specificity have been designated as mischarging mutants. In these tRNA mischarging mutants an actual alteration in specificity towards the aminoacyl-tRNA synthetase is accomplished by a minimal change in the tRNA. Therefore, these mutants may provide the pertinent information on the specific sites in the tRNA molecule recognized by the synthetase.

Such mutants have been isolated by using amber suppressor tRNAs in the following way. Among amber mutants of bacteria or phages, some may be found that can be suppressed by su^+1 (serine) or su^+2 (glutamine), but not by su^+3 (tyrosine). In such an amber mutant, the amino acid corresponding to the UAG mutation site can be replaced by either serine or glutamine, but not by tyrosine, to restore the function of the protein coded by the gene. Accordingly, the mutants of su^+3 that can suppress such nonsense mutations may be expected to be mischarging mutants of su^+3 tRNA^{Tyr}. In this way, many mutants of su^+3 were isolated and analyzed. It was found that the mutation sites were exclusively localized within the terminal region of the amino acid acceptor stem, such as A1, A2, G82, A81, U81, and U80 (see Fig. 1). In these mutants, a single-base substitution was sufficient to allow the misacylation of the tRNA^{Tyr} by glutaminyl-tRNA synthetase, although the efficiency of misacylation differed in the various mutations. A double mutant A1-G82, which was constructed by recombination between two single mutants A1 and G82, exhibited more efficient mischarging of glutamine. Thus, it was strongly suggested that the terminal region of the amino acid acceptor stem is involved in the synthetase recognition (Aono et al. 1969; Smith et al. 1970; Hooper et al. 1972; Shimura et al. 1972; Celis et al. 1973; Ghysen and Celis 1974; Inokuchi et al. 1974). These results are also in good agreement with a hypothesis that the fourth base from the 3' terminal may serve as a first discriminator for synthetase recognition (Crothers et al. 1972).

It should be noted that every mischarging mutant of su^+3 tRNA$_1^{Tyr}$ carries another mutation at the anticodon as an amber suppressor. Obviously, this anticodon mutation alone is not sufficient to cause mischarging in su^+3 tRNA, but it might be a necessary alteration required to change the synthetase specificity. Indeed, in the case of tRNA^{Trp}, its mutation to su^+7 amber suppressor at the middle base of the anticodon is already sufficient to cause misacylation by glutaminyl-tRNA synthetase, as mentioned above (Yaniv et al. 1974). Thus, the su^+7 itself is a mischarging mutant of tRNA^{Trp}.

Since with both su^+3 and su^+7 glutaminyl-tRNA synthetase is involved in mischarging, it was interesting to isolate the mischarging mutants of the tRNA^{Gln} suppressor. By using a one-gene derivative of λpsu^+2, which had lost the su^-2 gene by unequal crossing over, we isolated the mischarging mutants of su^+2 tRNA$_2^{Gln}$ (Yamao et al. 1978; F. Yamao et al., in prep.). About 30 independent mutants were isolated and analyzed. The altered residues were found exclusively at ψ37 residue, changing it to either A37 or C37 (Fig. 3). Thus, this residue plays a key role in the synthetase recognition of su^+2 tRNA$_2^{Gln}$. The suppressor mutation at the anticodon may or may not be involved. The suppression patterns of these mischarging mutants on *E. coli* and T4 strains are summarized in Table 1. As judged from the suppression patterns on defined amber markers in *E. coli* or T4, the altered amino acid in A37 (or C37) is not serine, tyrosine, lysine, leucine, and tryptophan, although the actual amino acid has not yet been determined. The mischarging mutant of su^+2 either with a mutation at the terminal region of the amino acid acceptor stem, as in su^+3 mutants, or with tyrosine as the altered amino acid, as a reverse situation of su^+3 ones, has not been obtained so far.

From these single-base mischarging mutants of su^+2, we have attempted to isolate secondary mischarging mutants in which the amino acid specificity has been further altered by an additional mutation in the tRNA. *E. coli* strains carrying met_{am3} were used for such a selection, since

Table 1 Suppression pattern

	su^+1 Ser	su^+2 Gln	su^+3 Tyr	$su^+\beta$ Lys	su^+6 Leu	su^+7 Trp·Gln	su^+3 A1G82 Gln	su^+2 mischarging mutants			
								A37	C37	A37C38	A37 A29
E. coli											
lac_{am1000}	−	+	−	−	−	+	+	+	+	+	+
met_{am3}	−	−	+	−	−	+	−	−	−	+	+
cys_{am235}	−	−	+	+	+	+	−	+	+	+	+
lac_{am125}	+	+	+	+	+	+	+	+	+	+	+
Phage T4											
N133	+	−	+	−	−	−	−	−	−	±	±
NG187	+	−	+	−	+	−	−	−	−	+	+
NG60	+	−	+	−	+	−	−	+	+	+	+
S116	−	−	−	−	−	−	−	−	−	−	−
Bu33	+	−	−	−	−	−	−	−	−	−	−
C266	+	+	−	−	−	+	+	+	+	+	+

this gene cannot be suppressed by either A37, C37, or su^+2 (Table 1). From the A37 mutant of su^+2, two types of double mutants that can suppress met_{am3} were obtained, namely, A37-A29 and A37-C38 (see Fig. 3). No such mutant capable of suppressing met_{am3} was obtained either from the C37 mutant or directly from the original su^+2. Two particular mutations, therefore, may be required on su^+2 tRNA, such as A37 and A29 (or C38). The suppression patterns of these double mutants suggest that the mischarged amino acid could be tryptophan (Table 1). This possibility is further supported by the fact that mutations in the $trpS$ gene (for tryptophanyl-tRNA synthetase) of the host bacteria specifically abolish the mischarging suppression on met_{am3} by A37-A29 and A37-C38 (F. Yamao et al., in prep.). These $trpS$ mutants were selected from $trpS^{ts}$ strains, such as those that showed no mischarging suppression on met_{am3} at low temperature. Some temperature-sensitive $trpS$ mutants that are known to increase the expression of the trp operon (Kano et al. 1968; Ito 1972) also showed a similar effect at 37°C, either abolishing or reducing the suppression on met_{am3} by A37-A29 and A37-C38. These results strongly suggest that by two steps of mutations su^+2 tRNA$_2^{Gln}$ becomes capable of accepting tryptophan in vivo.

In Figure 5, the sites of base changes in the mischarging mutations of su^+2, su^+3, and su^+7 are indicated on a folding pattern of yeast tRNA^{Phe} of Kim et al. (1974), showing their corresponding positions by arrows. Although these sites are different in each tRNA, they appear to be located on the diagonal side of the folded tRNA molecules. These sites are located at positions whose alterations do not seem to affect the total folding structure. Rich and Schimmel (1977) proposed that a major part

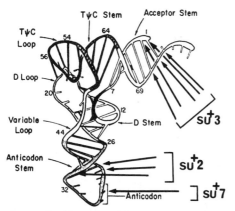

Figure 5 Altered residues in the mischarging mutants of su^+2, su^+3, and su^+7. Altered residues are indicated (→) showing the corresponding positions on the tertiary structure diagram of yeast tRNA^{Phe} (Kim et al. 1974). The anticodon mutations in su^+2 and su^+3 are not indicated.

of the binding site for sythetases is along the diagonal side of the tRNA structure. In general, tRNAs may interact with synthetases, even with noncognate ones to a certain extent, at their diagonal side, and the residues at both terminal regions of the arms of L-shaped tRNA molecules appear to provide additional fine interactions that account for the specificity of aminoacylation in vivo. The alteration of these residues by mutation could provide incremental interaction with certain noncognate synthetases, resulting in the mischarging suppression in vivo.

The results obtained with *trpS* mutants described above suggests that an incremental interaction provided by the A29 mutation in su^+2 A37-A29 (or by C38 in A37-C38) may be compensated by a mutation in tryptophanyl-tRNA synthetase, which still retains appreciable activity toward the cognate tRNA. Incidentally, the A29 mutation in su^+2 tRNA$_2^{Gln}$ corresponds to the A31 mutation in su^+3 tRNA$_1^{Tyr}$, which is known to alter the kinetics of aminoacylation by increasing the K_m 10–20 times (Abelson et al. 1970). Thus, this residue in tRNA$_1^{Tyr}$ is also involved in the synthetase recognition.

Recently we isolated a mutant of *E. coli* glutaminyl-tRNA synthetase that causes the mischarging suppression by normal su^+3 (H. Inokuchi et al., in prep.). The mutants of *glnS* (*glnS**) were isolated by using a transducing phage λp*glnS*$^+$ such as those capable of suppressing *E. coli* *lac*$_{am1000}$ under the presence of su^+3. As is seen in Table 1, the *lac*$_{am1000}$ can be suppressed by glutamine insertion but not by tyrosine insertion at the amber mutation site. The *glnS** itself cannot suppress *lac*$_{am1000}$ without su^+3. Other amber suppressors su^+1 and su^+6 cannot replace the position of su^+3. These results suggest that the *glnS** charges glutamine to su^+3 tRNA$_1^{Tyr}$ specifically. Thus, the alteration of synthetases by mutation could also provide the incremental interactions with noncognate tRNAs in vivo, resulting in mischarging suppression. This implies that missense suppressions could occur by the mutations not only in tRNAs but also in aminoacyl-tRNA synthetases. In the case of nonsense mutations, the presence of mutant tRNAs having nonsense anticodons is required for suppression. However, in the case of missense mutations, tRNA changes are not necessarily required, since tRNAs with matching anticodons are always available in the cells.

DUPLICATE tRNA GENES

As mentioned briefly above, suppressor transducing phages, in almost every case, stimulate the production of not only the su^+ tRNA, but also its corresponding su^- tRNA upon infection. In such cases two tRNA genes for each tRNA species are located within a transducing region of bacterial chromosome. It has been shown that two tRNA$_1^{Tyr}$ genes for su^+3 are in tandem separated by a short spacer; these are transcribed together as a single precursor RNA (Russel et al. 1970; Landy et al. 1974; Smith 1974).

We determined the number of tRNA$_2^{Gln}$ genes in λpsu^+2 in the following experiments (Kodaira et al. 1976; Inokuchi et al. 1979a). The su^+2 gene was first converted to an ocher suppressor by a second mutation in the anticodon (see above) and then an amber suppressor was selected from the remaining su^-2 genes. This can be achieved by selecting a better amber suppressing phage from λpsu^+2 (ocher), since the amber suppression by this phage is inefficient on the lac_{am} strain used. Thus, a transducing phage carrying su^+2 (ocher) and su^+2 (amber) was obtained. Upon infection, the stimulated production of su^-2-type tRNA$_2^{Gln}$ was no longer detected, whereas ocher- and amber-type tRNA were found in equal amounts. We concluded that the original transducing phage carried two copies of tRNA$_2^{Gln}$ genes and not more. As described above, λpsu^+2 carries a noninitiator tRNA^{Met} gene in addition to those tRNA$_2^{Gln}$ genes, and DNA heteroduplex analysis indicated that these three tRNA genes are located together forming a cluster (H. Yamagishi and H. Inokuchi, in prep.). λpsu^+2 (ocher) su^+2 (amber) segregates the transducing phages carrying one or the other suppressor, without the loss of noninitiator tRNA^{Met}, by unequal crossing over. All these results together suggested that the two tRNA$_2^{Gln}$ genes are in tandem and a noninitiator tRNA^{Met} gene is closely linked to them. Thus the order of these tRNA genes is: noninitiator tRNA^{Met}-tRNA$_2^{Gln}$-tRNA$_2^{Gln}$. DNA-tRNA hybridization experiments performed with separated l and r strands of the transducing phage DNA confirmed that all these tRNAs are transcribed from the l strand in the right-to-left direction on an ordinary phage map or in the counterclockwise direction on the bacterial chromosome (Kodaira et al. 1977). A transducing phage carrying su^- and su^+2^{U24} segregates a single-gene derivative carrying normal su^+2. Since the U24 mutation is located between the 5' terminal and the anticodon on the tRNA (see Fig. 3), the original su^+2 gene must be located in the transcriptional distal side of su^-2 (Fig. 6).

As mentioned above, λpsu^+B carries more tRNA genes, including three pairs of tRNA$_1^{Gln}$, tRNA$_2^{Gln}$, and noninitiator tRNA^{Met} genes, and a single tRNA$_{36}^X$ gene. The most likely order of these tRNA genes is: promoter 1-tRNA^{Met}-tRNA$_{36}^X$-tRNA$_1^{Gln}$-tRNA$_1^{Gln}$-(promoter 2?)-tRNA^{Met} - tRNA$_2^{Gln}$ - tRNA$_2^{Gln}$ (H. Inokuchi et al., in prep.). This is mainly based on the following observations. First, multimeric tRNA precursors are detected that consist of tRNA^{Met}-tRNA$_{36}^X$-tRNA$_1^{Gln}$-tRNA$_1^{Gln}$ or tRNA^{Met}-tRNA$_{36}^X$-tRNA$_1^{Gln}$ in that order with the 5'-triphosphate terminal at the noninitiator tRNA^{Met} side (Sakano and Shimura 1978; Shimura et al., this volume). Corresponding precursors are also detected in the λpsu^+B-infected cells with su^+B- and su^-B-type of tRNA$_1^{Gln}$ (H. Sakano, unpubl.). Second, a transducing phage, in which one of the tRNA$_1^{Gln}$ genes is marked as su^+B (ocher) and one of the tRNA$_2^{Gln}$ as su^+2 (amber), frequently segregates the

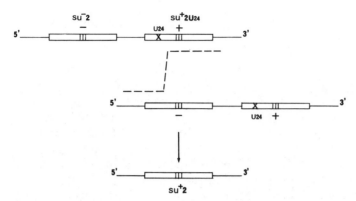

Figure 6 Segregation of su^+2 singlet from $su^-2 \cdot su^+2$U24 doublet by unequal crossing over (see text).

phages that had lost su^+B but not su^+2. Many of them had also lost the genes for tRNA$_{36}^X$ and su^-B tRNA$_1^{Gln}$, presumably due to the unequal crossing over at two noninitiator tRNA^{Met} genes. Analysis of other types of segregants, with their frequency of occurrence, also supported the above gene order. A second promoter may or may not be present between tRNA$_1^{Gln}$ and noninitiator tRNA^{Met} at the middle of the tRNA gene cluster. λpsu^+2 described above may be accounted as a segregant that was formed during isolation of the phage, since the original λdsu^+2 carried tRNA$_1^{Gln}$ and tRNA$_{36}^X$ genes in addition to noninitiator tRNA^{Met} and tRNA$_2^{Gln}$. We come to the conclusion that a cluster of seven tRNA genes in the above order, including four tRNA species, exists at the 15-minute position of E. coli chromosome.

In $su^+\beta$ transducing phage, we detected the genes for $su^+\beta$ and $su^-\beta$ tRNA^{Lys}, and for tRNA$_1^{Val}$. Although the detailed gene order has not yet been worked out, these three tRNA genes appear to form a cluster on the E. coli chromosome at 16 minutes. A dimeric precursor containing tRNA^{Lys} and tRNA$_1^{Val}$ has been identified in E. coli with a chain length of about 180 nucleotides (Shimura and Sakano 1977). If this is a precursor transcribed from the supL region, a tRNA$_1^{Val}$ gene is linked in tandem with the duplicate tRNA^{Lys} genes. Thus, a possible gene order is either tRNA^{Lys}-tRNA^{Lys}-tRNA$_1^{Val}$ or tRNA^{Lys}-tRNA$_1^{Val}$-tRNA^{Lys}.

Duplicate tRNA genes are not a specific situation for suppressor tRNAs, although the presence of a duplicate is one of the necessary conditions for a tRNA gene to become a suppressor. For instance, the genes for tRNA$_{2A}^{Val}$ and tRNA$_{2B}^{Val}$ are linked in tandem, since a dimeric precursor for these two tRNAs has been detected that has an approximate chain length of 190 nucleotides (T. Ikemura, unpubl.; Shimura and Sakano 1977). Similarly, precursors consisting of two or more tRNA copies of tRNA$_3^{Gly}$ or tRNA$_1^{Leu}$ have been identified in temperature-sensitive *E. coli* RNase P mutants. Multimeric transcripts consisting of two different tRNAs, such as tRNA$_2^{Gly}$ and tRNA$_3^{Thr}$ or tRNA$_3^{Ser}$ and tRNA$_2^{Arg}$ (Carbon et al. 1974; Sakano and Shimura 1978). Thus, not only genetic analysis, but also structural analysis of multimeric tRNA precursors provides pertinent information on the tandem linkage of tRNA genes (see Shimura et al., this volume). Analysis of rRNA operons has also revealed the tRNA genes associated with them (see Morgan et al., this volume). Among the rRNA operons, for instance, three have been shown to carry genes for both tRNA$_1^{Ile}$ and tRNA$_{1B}^{Ala}$ in their spacers between 16S and 23S rRNA (Morgan et al. 1977). Taking all these instances together, it appears that the tRNA genes in *E. coli* tend to exist in clusters of several genes for homologous or heterologous species of tRNAs. The tRNA genes in bacteriophages T4 (Wilson et al. 1972; McClain et al. 1972), T5 (Chen et al. 1976), and BF23 (Ikemura et al. 1978) also form clusters on their genomes.

It is not known how a single gene duplicates in tandem, but it must occur in many cases on the tRNA genes during the establishment of the *E. coli* chromosome. Tandem duplications may occur not only in a single gene, such as A to AA, but also in a pair of genes as a unit, resulting in an alternating arrangement of tRNA genes, such as AB to ABAB. After gene duplication takes place, one of the copies may be allowed to mutate to another type of tRNA, possibly via defective stages. For instance, the tRNA gene cluster of *supB-E* region described above might be formed in the following steps, although other pathways are also possible. As a unit, tRNA^{Met}-tRNA$_1^{Gln}$ (or tRNA^{Met}-tRNA$_{36}^X$-tRNA$_1^{Gln}$) was duplicated first, and then one of the tRNA$_1^{Gln}$ was changed to tRNA$_2^{Gln}$ by several steps of mutation. After this differentiation of tRNA$_1^{Gln}$ and tRNA$_2^{Gln}$ took place, they were duplicated in each unit; gene tRNA$_{36}^X$ was added to (or deleted from) one of the duplicated units during the process, resulting in a cluster of tRNA^{Met}-tRNA$_{36}^X$-tRNA$_1^{Gln}$-tRNA$_1^{Gln}$-tRNA^{Met}-tRNA$_2^{Gln}$-tRNA$_2^{Gln}$. Apparently the duplicate tRNA genes in tandem are stably maintained in the *E. coli* chromosome, although when they are transferred onto a transducing phage, the duplicate genes are appreciably unstable and tend to segregate one-gene derivatives by unequal crossing over. One-gene derivatives are usually fairly stable. In bacteria, the duplicate genes may be maintained to keep an appropriate level of gene dosage for each tRNA. Some tRNAs are relatively abundant, whereas some others are few in number and the relative contents of individual tRNAs are kept fairly constant in *E. coli* (Ikemura and Ozeki 1977). The content of a tRNA is approximately doubled by the introduction of an F' factor carrying the corresponding gene, due to the gene-dosage effect (Ikemura and Ozeki 1977). This conversely suggests that the bacteria must keep the ratio of various tRNA genes constant to maintain a balanced content of tRNAs.

Duplicate tRNA genes are also found at two entirely different positions on the bacterial chromosome, for instance, two tRNA$_3^{Gly}$ genes at 42 minutes and 94 minutes, two tRNA^{Tyr} genes at 27 minutes (tRNA$_1^{Tyr}$) and 88 minutes (tRNA$_2^{Tyr}$), or two tRNA^{fMet} genes at about 61 minutes (tRNA$_1^{fMet}$) and about 68 minutes (tRNA$_2^{fMet}$) (see Fig. 7). These gene locations may reflect earlier events in evolution, such as genome duplications proposed by Zipkas and Riley (1975). Translocation of a tRNA

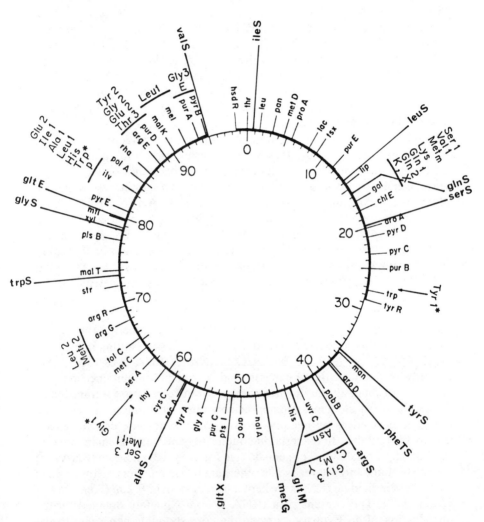

Figure 7 Distribution of tRNA and of aminoacyl-tRNA synthetase genes on the *E. coli* chromosome. The diagram is constructed from the maps of Ikemura and Ozeki (1977) and Bachmann et al. (1976). Synthetase genes are *ileS, leuS, glnS, serS, tyrS, pheTS, argS, gltM, metG, gltX, alaS, trpS, glyS, gltE*, and *valS*. tRNA genes are indicated at the outside of the marker genes, with bars showing chromosomal regions.

gene from tandem duplicates may also take place. In Figure 7, the locations of tRNA genes on the *E. coli* chromosome are indicated, including those for unidentified tRNAs (Ikemura and Ozeki 1977). For the sake of convenience, the genes for aminoacyl-tRNA synthetases are also added to the map, taking the positions from the contemporary chromosome map of *E. coli* (Bachmann et al. 1976). A detailed tRNA gene map can be completed by combining the gross and fine mapping. DNA sequence analysis of various tRNA genes, including promoters and spacers, is now in progress in many laboratories, and the detailed structure of tRNA operons will soon be elucidated. Since the variety of tRNA genes may stem from a few original genes, the genes for tRNAs may serve as good markers to follow the process of genome development in *E. coli* during its evolution. The genetic analysis of tRNAs is also interesting from this viewpoint.

ACKNOWLEDGMENTS

We are indebted to Dr. Dieter Söll for his critical reading of the manuscript. This work was supported by a Scientific Research Grant from the Ministry of Education of Japan.

REFERENCES

Abelson, J. N., M. L. Gefter, L. Barnett, A. Landy, R. L. Russell, and J. D. Smith. 1970. Mutant tyrosine transfer ribonucleic acids. *J. Mol. Biol.* **47**:15.

Andoh, T. and H. Ozeki. 1968. Suppressor gene su3$^+$ of *E. coli*, as a structural gene for tyrosine tRNA. *Proc. Natl. Acad. Sci.* **59**:792.

Aono, H., H. Inokuchi, and H. Ozeki. 1969. Genetic analysis of suppressor gene su3. II. On the mutants accepting amino acids other than tyrosine. *Jpn. J. Genet.* (Abstr.) **44**:382.

Bachmann, B. J., K. B. Low, and A. L. Taylor. 1976. Recalibrated linkage map of *Escherichia coli* K12. *Bacteriol. Rev.* **40**:116.

Carbon, J., S. Chang, and L. L. Kirk. 1974. Clustered tRNA genes in *Escherichia coli*: Transcription and processing. *Brookhaven Symp. Biol.* **26**:26.

Celis, J. E., M. L. Hooper, and J. D. Smith. 1973. Amino acid acceptor stem of *E. coli* suppressor tRNA^{Tyr} is a site of synthetase recognition. *Nat. New Biol.* **244**:261.

Chan, T-S., R. E. Webster, and N. D. Zinder. 1971. Suppression of UGA codon by a tryptophan tRNA. *J. Mol. Biol.* **56**:101.

Chakraburtty, K., A. Steinschneider, R. V. Case, and A. H. Mehler. 1975. Primary structure of tRNA^{Lys} of *E. coli* B. *Nucleic Acids Res.* **2**:2069.

Chen, M-J., J. Locker, and S. B. Weiss. 1976. The physical mapping of bacteriophage T5 transfer RNAs. *J. Biol. Chem.* **251**:536.

Crothers, D. M., T. Seno, and D. G. Söll. 1972. Is there a discriminator site in transfer RNA? *Proc. Natl. Acad. Sci.* **69**:3063.

Eggertson, G. and E. A. Adelberg. 1965. Map positions and specificity of suppressor mutations in *Escherichia coli* K-12. *Genetics* **52**:319.

Garen, A. 1968. Sense and nonsense in the genetic code. *Science* **160**: 149.

Ghysen, A. and J. E. Celis. 1974. Mischarging signal and double mutants of *Escherichia coli* sup3 tyrosine transfer RNA. *J. Mol. Biol.* **83**: 333.

Gorini, L. 1971. Informational suppression. *Annu. Rev. Genet.* **4**: 107.

Hirsh, D. 1971. Tryptophan transfer RNA as the UGA suppressor. *J. Mol. Biol.* **58**: 439.

Hirsh, D. and L. Gold. 1971. Translation of the UGA triplet *in vitro* by tryptophan transfer RNAs. *J. Mol. Biol.* **58**: 459.

Hopper, M. L., R. L. Russell, and J. D. Smith. 1972. Mischarging in mutant tyrosine transfer RNA. *FEBS Lett.* **22**: 149.

Ikemura, T. and M. Nomura. 1977. Expression of spacer tRNA genesin ribosomal RNA transcription units carried by hybrid colE1 plasmids in *E. coli*. *Cell* **11**: 779.

Ikemura, T. and H. Ozeki. 1977. Gross map location of *Escherichia coli* transfer RNA genes. *J. Mol. Biol.* **117**: 419.

Ikemura, T., K. Okada, and H. Ozeki. Clustering of transfer RNA genes in bacteriophage BF23. *Virology* **90**: 142.

Inokuchi, H. and H. Ozeki. 1976. Characterization of ochre suppressor su$^+$B in *Escherichia coli* K12. *Jpn. J. Genet.* (Abstr.) **51**: 412.

Inokuchi, H., J. E. Celis, and J. D. Smith. 1974. Mutant tyrosine transfer RNA of *E. coli*: Construction by recombination of a double mutant A1G82 chargeable with glutamine. *J. Mol. Biol.* **85**: 187.

Inokuchi, H., M. Kodaira, F. Yamao, and H. Ozeki. 1979a. Identification of transfer RNA suppressors in *Escherichia coli*. II. Duplicate genes for tRNA$_2^{Gln}$. *J. Mol. Biol.* **132**: 663.

Inokuchi, H., F. Yamao, H. Sakano, and H. Ozeki. 1975. Isolation and characterization of transducing phage λ carrying su$^+$2 in *Escherichia coli*. *Jpn. J. Genet.* (Abstr.) **50**: 466.

———. 1979b. Identification of transfer RNA suppressors in *Escherichia coli*. I. Amber suppressor su^+2, an anticodon mutant of tRNA$_2^{Gln}$. *J. Mol. Biol.* **132**: 649.

Ito, K. 1972. Regulatory mechanism of the tryptophan operon in *Escherichia coli*: Possible interaction between *trpR* and *trpS* gene products. *Mol. Gen. Genet.* **115**: 349.

Kano, Y., A. Matsushiro, and Y. Shimura. 1968. Isolation of the novel regulatory mutants of tryptophan biosynthetic system in *Escherichia coli*. *Mol. Gen. Genet.* **102**: 15.

Kaplan, S. 1971. Lysine suppressor in *Escherichia coli*. *J. Bacteriol.* **105**: 984.

Kim, S.-H., F. L. Suddath, G. J. Quigley, A. McPherson, J. L. Sussman, A. H. Wang, N. C. Seeman, and A. Rich. 1974. Three-dimensional tertiary structure of yeast phenylalanine transfer RNA. *Science* **185**: 435.

Kimura, M., H. Inokuchi, and H. Ozeki. 1976. Analysis of suppressor su$^+\beta$ in *Escherichia coli* K12. *Jpn. J. Genet.* (Abstr.) **51**: 417.

Kodaira, M., H. Inockuchi, and H. Ozeki. 1976. Tandem duplication of tRNA$_2^{Gln}$ genes in *Escherichia coli*. *Jpn. J. Genet.* (Abstr.) **51**: 419.

Kodaira, M., H. Inokuchi, H. Yamagishi, and H. Ozeki. 1977. Orientation of transcription for a tRNA gene cluster including *sup*E. *Jpn. J. Genet.* (Abstr.) **52**: 452.

Landy, A., C. Foeller, and W. Ross. 1974. DNA fragments carrying genes for tRNA$_1^{Tyr}$. *Nature* **249:**738.

McClain, W. H., C. Guthrie, and B. G. Barrell. 1972. Eight transfer RNAs induced by infection of *Escherichia coli* with bacteriophage T4. *Proc. Natl. Acad. Sci.* **69:**3703.

Morgan, E. A., T. Ikemura, and M. Nomura. 1977. Identification of spacer tRNA genes in individual ribosomal RNA transcription units of *Escherichia coli*. *Proc. Natl. Acad. Sci.* **74:**2710.

Ozeki, H., H. Sakano, S. Yamada, T. Ikemura, and Y. Shimura. 1974. Temperature-sensitive mutants of *Escherichia coli* defective in tRNA biosynthesis. *Brookhaven Symp. Biol.* **26:**89.

Rich, A. and P. R. Schimmel. 1977. Structural organization of complexes of transfer RNAs with aminoacyl transfer RNA synthetases. *Nucleic Acids Res.* **4:**1649.

Russell, R. L., J. N. Abelson, A. Landy, M. L. Gefter, S. Brenner, and J. D. Smith. 1970. Duplicate genes for tyrosine transfer RNA in *Escherichia coli*. *J. Mol. Biol.* **47:**1.

Sakano, H. and Y. Shimura. 1978. Characterization and *in vitro* processing of transfer RNA precursors accumulated in a temperature-sensitive mutant of *Escherichia coli*. *J. Mol. Biol.* **123:**287.

Sakano, H., S. Yamada, T. Ikemura, Y. Shimura, and H. Ozeki. 1974. Temperature-sensitive mutants of *Escherichia coli* for tRNA synthesis. *Nucleic Acids Res.* **1:**335.

Schedl, P. and P. Primakoff. 1973. Mutants of *Escherichia coli* thermosensitive for the synthesis of transfer RNA. *Proc. Natl. Acad. Sci.* **70:**2191.

Shimura, Y. and H. Ozeki. 1973. Genetic study on transfer RNA. *Adv. Biophys.* **4:**191.

Shimura, Y. and H. Sakano. 1977. Processing of tRNA precursors in *Escherichia coli*. In *Nucleic acid-protein recognition* (ed. H. J. Vogel), p. 293. Academic Press, New York.

Shimura, Y., H. Aono, H. Ozeki, A. Sarabhai, H. Lamfrom, and J. Abelson. 1972. Mutant tyrosine tRNA of altered amino acid specificity. *FEBS Lett.* **22:**144.

Smith, J. D. 1972. Genetics of transfer RNA. *Annu. Rev. Genet.* **6:**235.

———. 1974. Mutants which allow accumulation of tRNA^{Tyr} precursor molecules. *Brookhaven Symp. Biol.* **26:**1.

Smith, J. D., L. Barnett, S. Brenner, and R. L. Russell. 1970. More mutant tyrosine transfer ribonucleic acids. *J. Mol. Biol.* **54:**1.

Smith, J. D., J. N. Abelson, B. F. Clark, H. M. Goodman, and S. Brenner. 1967. Studies on *amber* suppressor tRNA. *Cold Spring Harbor Symp. Quant. Biol.* **31:**479.

Soll, L. 1974. Mutational alteration of tryptophan-specific transfer RNA that generate translation suppressors of the UAA, UAG and UGA nonsense codons. *J. Mol. Biol.* **86:**233.

Soll, L. and P. Berg. 1969. Recessive lethals: A new class of nonsense suppressors in *Escherichia coli*. *Proc. Natl. Acad. Sci.* **63:**392.

Steege, D. A. and B. Low. 1975. Isolation and characterization of lambda transducing bacteriophages for the $su1^+$ (supD$^-$) amber suppressor of *Escherichia coli*. *J. Bacteriol.* **122:**120.

Steege, D. A. and D. Söll. 1979. Suppression. In *Biological regulation and development* (ed. R. F. Goldberger), vol. I, p. 433. Plenum Press, New York.

Wilson, J. H., J. S. Kim, and J. N. Abelson. 1972. Bacteriophage T4 transfer RNA. III. Clustering of the genes for the T4 transfer RNA's. *J. Mol. Biol.* **71:**547.

Yamao, F., H. Inokuchi, and H. Ozeki. 1975. The specialized transducing phage lambda which carries su$^+$1 amber suppressor gene. *Jpn. J. Genet* (Abstr.) **50:**506.

―――. 1977. Defective mutants of suppressor su$^+$2 tRNA$_2^{Gln}$. *Jpn. J. Genet.* (Abstr.) **52:**488.

―――. 1978. Mischarging mutants of transfer RNA in *Escherichia coli*. In *Sixth International Biophysics Congress Abstract*, p. 83. International Union for Pure and Applied Biophysics, Kyoto, Japan.

Yaniv, M. and W. R. Folk. 1975. The nucleotide sequences of the two glutamine transfer ribonucleic acids from *Escherichia coli*. *J. Biol. Chem.* **250:**3243.

Yaniv, M., W. R. Folk, P. Berg, and L. Soll. 1974. A single mutational modification of a tryptophan-specific transfer RNA permits aminoacylation by glutamine and translation of the codon UAG. *J. Mol. Biol.* **86:**245.

Zipkas, D. and M. Riley. 1975. Proposal concerning mechanism of evolution of the genome of *Escherichia coli*. *Proc. Natl. Acad. Sci.* **72:**1354.

Applications of Temperature-sensitive Suppressors to the Study of Cellular Biochemistry and Physiology

Max P. Oeschger
Department of Microbiology
Schools of Medicine and Dentistry
Georgetown University, Washington, D.C. 20007

The value of suppressors as biochemical and physiological research tools was first demonstrated in the study of bacteriophage replication (Epstein et al. 1964; Wood et al. 1968; Kozak and Nathans 1972). Phages are ideally suited for investigation with this system because they can be freely moved from one bacterial strain to another. This feature permits ready identification of phage strains that carry an amber mutation in an essential gene because they can form plaques on suppressor-containing (Su^+) strains but not on Su^- strains. The amber gene product, which cannot be made in Su^- strains, can be identified by comparing the phage proteins produced in infections of Su^+ and Su^- cells. The role of the amber gene product in the phage replication cycle is established by determining the step at which development is arrested in Su^- cells.

The application of this approach to cellular processes was not possible until the isolation of temperature-sensitive suppressor strains, with which, depending upon the temperature, the same cell can be either Su^+ or Su^-. A bacterial strain carrying a temperature-sensitive suppressor and an amber mutation in an essential gene is viable, but only at permissive temperatures, and can be recognized by its temperature-sensitive phenotype. In such a strain, temperature affects synthesis, rather than specific activity, of the protein encoded by the amber gene. As a result, the number of molecules of a selected protein can be systematically varied within the intact cell. The ability to titrate protein levels in vivo with amber mutant temperature-sensitive suppressor strains makes possible a whole new approach to the study of cell biochemistry and physiology.

ISOLATION OF TEMPERATURE-SENSITIVE SUPPRESSORS

Initial Isolates

As an outgrowth of their work on suppression, Smith, Brenner, and their colleagues initiated a program to isolate strains that produced mutant

su^+3 tRNAs.[1] They had previously established the chromosomal location of the gene encoding the su^+3 tRNA (Stretton et al. 1967), shown that the suppressor activity resulted from a base change in the anticodon region of a tRNA^{Tyr} (Goodman et al. 1968), and constructed a transducing phage carrying the su^+3 gene (Smith et al. 1967). They isolated strains with additional mutations in the *tyrT* gene by mutagenesis of either su^+3 cells or su^+3 transducing phage particles (Abelson et al. 1970). To aid in the isolation and screening of mutants, they constructed a strain containing a number of experimentally useful amber mutations (Russell et al. 1970). In this strain, an amber mutation in the *galK* gene and a missense mutation in the *galE* gene provide positive selection for the loss of suppressor activity. (*galE* strains grown in the presence of galactose commit suicide; *galKgalE* strains survive. Hence, such a strain, when it is Su$^+$, is killed by galactose; when it is Su$^-$, it is insensitive to galactose.) An amber mutation in the *trp* operon provides positive selection for suppressor activity. With this combination they established systems for positive selection of both the presence and absence of suppressor activity in a single strain. In addition, they incorporated into the strain a *lacZ* amber mutation that, by biochemical assay of β-galactosidase activity, was used to quantitate the level of suppression.

Using these selection systems, Smith, Brenner, and their colleagues isolated mutant strains with temperature-sensitive suppression. They first confirmed that the temperature-sensitive suppressor activity of a number of their isolates was genetically linked to the suppressor tRNA locus. Subsequently they showed, by structural analysis of tRNA derived from such mutants, that base changes in the CCA stem rendered the tRNA molecule temperature-sensitive (Smith et al. 1970, 1971); this was the first demonstration of temperature-sensitive RNA. Temperature-sensitive suppressor derivatives of an su^+3 strain and an su^+4 strain were isolated independently by Ozeki et al. (1969) and by Gallucci et al. (1970).

su^+1 Temperature-sensitive Suppressors

Smith, Brenner, and Gallucci isolated temperature-sensitive suppressor mutants to study structure-function relationships in tRNA. Other investigators attempted to exploit the temperature-sensitive suppressor activity in these strains for the isolation and study of amber mutants. These early attempts met with limited success (Beckman and Cooper

---

[1] su^+3 is the name given to tyrosine-inserting amber suppressors encoded by the *tyrT* gene. (This suppressor is referred to as su$^+_{III}$ in the reports of Smith, Brenner, and colleagues [Smith et al. 1970].) su^+4 is the name given to the corresponding ocher suppressor tRNA encoded by the same gene. su^+1 is the name given to the serine-inserting amber suppressor encoded by the *supD* gene. For a review on the genetics of suppression, see Garen (1968). Gene symbols used in this paper are as listed in Bachmann et al. (1976).

1973; M. P. Oeschger, unpubl.). As a result, new temperature-sensitive suppressors were sought and isolated. Selecting for either the presence or absence of suppressor activity, Oeschger (Oeschger and Berlyn 1973; Oeschger and Woods 1976) and Nagata and Horiuchi (1973) isolated derivatives of su^+1 strains that were temperature sensitive for suppression. The mutations in these isolates map at the *supD* locus.

The *supD126* mutation of Nagata and Horiuchi and the *supD43,74* mutation of Oeschger, although affecting the same suppressor, produce temperature-sensitive suppressors with different properties. One difference is observed when suppressor activity is measured as a function of temperature. *supD126* maintains a suppressor efficiency of approximately 2% up to 37°C, where it begins to fall (Davidoff-Abelson and Mindich 1978). *supD43,74* exhibits approximately the same suppressor activity as *supD126* at 30°C (Davidoff-Abelson and Mindich 1978), but suppressor efficiency varies with temperature (Oeschger and Woods 1976). Below 37°C, the suppressor activity of *supD43,74* increases linearly with decreasing temperature, reaching 4% efficiency at 15°C (Oeschger and Wiprud 1980). This feature of *supD43,74* permits convenient and precise control over suppressor activity.

Another difference is found on temperature shift from 30°C to 42°C. When a *supD126* strain is transferred to 42°C there is progressive loss of suppressor activity with time. The activity decays with a half-life of approximately 22 minutes, resulting in a loss of 80% of the activity after 50 minutes—the equivalent of one cell division (Nagata and Horiuchi 1973). On the other hand, a *supD43,74* strain loses 80% of its suppressor activity within 2 minutes after transfer to 42°C and the residual suppressor activity decays with a 7-minute half-life (Oeschger and Woods 1976; Oeschger and Wiprud 1980). This feature of *supD43,74* makes it useful for studies where rapid cessation of synthesis of the protein under investigation is important. Such studies include functional bioassay of amber gene products to determine turnover rates, differential pulse-labeling of cellular proteins to identify amber gene products (in cell extracts), and, by in vivo dilution with cell growth, intracellular titration of amber gene products.

A third difference is found on temperature shift from 42°C to 30°C. At 42°C the suppressor activity of *supD126* is inactivated reversibly, for upon transfer to 30°C, strains carrying this suppressor rapidly regain activity (Wainwright and Beacham 1977). In contrast, the suppressor activity of *supD43,74* is irreversibly inactivated at 42°C (probably by degradation), and on transfer to 30°C suppressor activity is recovered only by de novo synthesis (Oeschger and Woods 1976). The rapid recovery of suppressor activity in strains containing *supD126* can be used to make the expression of normally constitutive genes inducible with temperature shift. This application may be useful in the investigation of assembly processes (such as incorporation of proteins into the membrane) and cascade reactions (such as initiation of chromosome replication). Thus, between them these two su^+1 temperature-sensitive derivatives possess virtually all the properties that one would desire for utilization of temperature-sensitive suppressors in performing molecular biological experiments. The one remaining requirement is a high level of suppressor efficiency.

ISOLATION AND STUDY OF AMBER MUTANTS

Mutant Isolation

Amber Mutants

The first report of the successful use of temperature-sensitive suppressor strains for the isolation of essential nonsense mutants was made by Beckman and Cooper (1973). That study outlined a basic strategy for the isolation and characterization of essential amber mutants using temperature-sensitive suppressor strains. With the isolation of more efficient temperature-sensitive su^+1 suppressor strains, programs to find amber mutations affecting specific functions were initiated. Nagata and Horiuchi (1974) used a strain containing their temperature-sensitive suppressor mutation (*supD126*) to isolate a strain with an amber mutation in the gene encoding DNA ligase. Subsequently, strains containing *supD126* were used to isolate derivatives with amber mutations: in a gene involved in the regulation of RNA polymerase synthesis (Nakamura and Yura 1975), the *uvrA* and *uvrB* genes (Morimyo et al. 1976), the *rho* gene (Inoko et al. 1977), and plasmid-encoded genes required for either plasmid replication (Hashimoto-Gotoh and Sekiguchi 1977) or regulation of plasmid number (Gustafson and Nordstrom 1978). Bassford et al. (1977) used a strain containing *supD43,74* to isolate derivatives with amber mutations in *bfe* and *tonB*, the genes encoding the vitamin B12 transport proteins of *Escherichia coli*.

Conditional-lethal Mutants

Temperature-sensitive suppressors have also proven useful as tools in the selection of strains carrying mutations that produce a lethal phenotype only in cells with a lesion in a second gene. Mutations in a number of genes that encode proteins involved in DNA metabolism are not inherently lethal. Examples of such mutations include *polA* (De Lucia and Cairns 1969; Gross and Gross 1969), *recA* (Willets et al. 1969), *recB* (Howard-Flanders and Boyce 1966), and *uvrB* (Van de Putte et al. 1965). Even though loss of the activity of any one of these genes is not lethal to the cell, simultaneous loss of one of these gene activities and *polA* is very often lethal (Gross et al. 1971; Monk and Kinross 1972). This phenomenon was exploited by Horiuchi and Nagata (1973), who, starting with a *supD126polA* amber mutant strain, isolated derivatives that were temperature sensitive for growth. Analysis of the temperature-sensitive isolates showed that many carried mutations that were lethal only in the absence of the *polA* gene product, DNA polymerase I. Genetic analysis of those isolates revealed several new genes affecting DNA metabolism (Horiuchi and Nagata 1973).

Although the same results could have been achieved with a strain carrying a temperature-sensitive *polA* mutation (Monk and Kinross 1972), genetic analysis of temperature-sensitive isolates derived in an amber *polA* temperature-sensitive suppressor strain is much simpler. The problem is to separate the two mutations (parental and induced) to determine whether the induced mutation alone causes the temperature-sensitive phenotype. The simplest solution to this problem is to remove or correct the parental lesion. This is most easily accomplished when it is an amber, for one need only introduce a non-temperature-sensitive suppressor via a plasmid; thus amber mutant temperature-sensitive suppressor strains have a marked advantage over temperature-sensitive mutant strains for such mutant searches.

Studies with Amber Mutations in Temperature-sensitive Suppressor Strains

DNA Metabolism

The amber mutation in *polA* derived by De Lucia and Cairns (1969) has been used in a temperature-sensitive suppressor strain to investigate the requirement for DNA polymerase I in a number of situations. First, as described above, there was the collection of mutant strains where the loss of a second gene function, in addition to *polA*, was lethal (Horiuchi and Nagata 1973). Horiuchi and Nagata (1974) subsequently observed that a mutation in one of their isolates markedly reduced the level of activity of DNA ligase. A low level of ligase activity combined with the loss of DNA polymerase I resulted in chromosomal degradation and loss of viability (Horiuchi et al. 1975). Similarly, Morimyo and Shimazu (1976) observed that cells deficient in polymerase I and the *uvrB* gene product are not viable. Tacon and Sherratt (1976) investigated the requirement for polymerase I in plasmid replication. They observed that ColE1 plasmid replication is independent of polymerase I.

Membrane Proteins

A number of experiments investigating membrane synthesis and membrane protein synthesis and stability have relied on either rapid loss or gain of suppressor activity following temperature shift. Using the *su3*-A81 temperature-sensitive suppressor of Smith et al. (1970, 1971) in a *tsx* amber mutant strain, Begg and Donachie (1973, 1977) investigated the topography of outer membrane growth. Temperature shift to 42°C was used to terminate synthesis of the *tsx* gene product, the receptor protein for bacteriophage T6. The organization and placement of bacteriophage T6 receptor proteins in the outer membrane was monitored in the period of growth following temperature shift by the addition of phage T6. Electron microscopy revealed that new growth, seen as regions free of attached phage T6 (therefore lacking T6 receptor protein), occurred at the cell poles.

Beacham et al. (1976) investigated the role of phospholipid synthesis in the excretion of periplasmic enzyme using a *supD126* mutant strain containing three additional mutations: *gpsA* (a mutation making glycerol essential for phospholipid synthesis), *phoR* (a mutation resulting in the constitutive synthesis of alkaline phosphatase), and a *phoA* amber (the structural gene for alkaline phosphatase). In these experiments, cells were grown under conditions where phospholipid synthesis was limited by omitting glycerol from the medium. The culture temperature was shifted from 40°C to 30°C to initiate synthesis of the *phoA* gene product, alkaline phosphatase. Their experiments showed that newly synthesized alkaline phosphatase entered the periplasmic space at the normal rate in the absence of new membrane synthesis.

Finally, Bassford et al. (1977), using *bfe* or *tonB* amber mutant *supD43,74* strains, investigated the functional stability of the *tonB* and *bfe* gene products—two vitamin B12 transport proteins found in the outer membrane. In temperature shift-up experiments where synthesis of amber gene products is terminated, they observed that although the *bfe* gene product was stable, it quickly lost some of its biological activities. The first activity to be lost was sensitivity to ColE2 and ColE3, followed by loss of the ability to absorb bacteriophage BF23, but the ability to transport vitamin B12 was unaffected. Similar experiments with the *tonB* amber mutant strain showed this gene product was unstable for all detectable properties.

RNA Polymerase

The regulation of RNA polymerase synthesis has been intensively investigated using temperature-sensitive suppressor *rpoB* amber mutant strains. Following transfer of such strains to 42°C, synthesis of the product of the *rpoB* gene, the β subunit of RNA polymerase, is terminated and the intracellular level of RNA polymerase progressively diluted by cell growth. Polymerase gene activity is measured by monitoring the rate of synthesis of the β' subunit as the level of polymerase falls. Glass et al. (1975) and Oeschger (1976) observed a dramatic stimulation in the rate of β'-subunit synthesis in cells growing with reduced levels of polymerase. These results provide strong evidence for autoregulation of RNA polymerase gene expression.

Nakamura and Yura (1975) used a *supD126* strain to isolate an amber mutant derivative in which synthesis of RNA polymerase is restricted at 42°C. Genetic analysis showed that the mutation is not linked to any known polymerase structural gene. The role of this gene in the regulation of polymerase gene expression remains to be established.

Glass (1977) used an *su3*-A81 temperature-sensitive suppressor strain to identify the amber peptide gene product of an *rpoB* amber mutation. His experiments prove that the amber mutation in the strain studied is located within the *rpoB* structural gene and that there is stimulation of synthesis of both the β amber fragment and the β' subunit in cells growing with reduced levels of polymerase.

Ribosomes

Delcuve et al. (1977), using specially constructed merodiploid strains, have generated amber mutations in ribosomal protein genes. They transduced their amber mutations into an *su3*-A81 temperature-sensitive suppressor strain for characterization. Although the suppressor activity of this strain is limited at 30°C, recombinants from a number of amber mutant isolates were successfully cultured.[2] They observed that ribosome assembly is defective in the temperature-sensitive suppressor recombinants at 42°C, confirming that the mutations affected ribosomal protein genes.

SUPPRESSOR-ENHANCING MUTATIONS

Isolation

Amber suppressors are the most efficient class of nonsense suppressors. Amber suppressor efficiencies range from 10% (*supE*) to 70% (*supU*). The temperature-sensitive suppressor strains so far isolated exhibit considerably less activity at permissive temperatures than their parental strains. The most efficient of all the temperature-sensitive suppressor strains appears to be the *supD43,74* isolate described by Oeschger and Woods (1976). This strain contains a suppressor-enhancing mutation that raises the suppressor efficiency 3.3-fold (see below) and results in 13% efficient suppression at 15°C (Oeschger and Woods 1976; Haggerty et al. 1978). Even this temperature-sensitive suppressor strain does not have sufficient activity to permit the study of general cellular biochemistry and regulation. The isolation of high-efficiency, temperature-sensitive suppressor strains poses a potentially difficult problem, for, as mentioned above, the parental suppressors themselves have limited efficiency.

To resolve this problem, a new approach to raising the efficiency of

[2]The temperature-sensitive suppressor strain used in this work contains an uncharacterized mutation that raises suppressor efficiency. This mutation is thought to be in the *tyrT* gene (G. Delcuve, unpubl., quoted in Delcuve et al. 1977).

temperature-sensitive suppressor strains was investigated. Mutations in genes that specifically enhance the efficiency of temperature-sensitive suppressors were sought. A number of such mutants have been isolated. Davidoff-Abelson and Mindich (1978) identified a mutation, *ups*, mapping between 25 and 27 minutes, which raises the efficiency of a number of suppressors, specifically *supD126* and *supD43,74*, approximately tenfold at 30°C. The effect of this mutation is strongly influenced by temperature, with maximal activity expressed at 30°C.

Two other strains containing suppressor-enhancing mutations (*sueA* and *sueB*) have been isolated and characterized (Haggerty et al. 1978; Oeschger et al. 1980). These mutations raise suppressor efficiency 3.3-fold (*sueA*) and 10-fold (*sueB*) (Table 1). Subsequently, it was discovered that the series of strains in which the *supD43,74* mutation was initially investigated (Oeschger and Woods 1976) contained an additional suppressor-enhancing mutation, *sueC*. The *sueC* mutation raises *supD43,74* suppressor efficiency 3.3-fold (Table 1). These mutations are distinct from one another, the *ups* mutation, and *supD*, with *sueA* mapping between 70 and 82 minutes, *sueB* between 6 and 14 minutes, *sueC* between 82 and 87 minutes, and *supD* at 43 minutes. As opposed to the *ups* mutation, the *sue* mutations exert the same relative enhancement at all temperatures (Fig. 1). Since the suppressor activity of *supD43,74* steadily increases with decreasing temperature, these mutations can make the suppressor quite efficient. Also, the effects of *sueA* and *sueB* are additive and when combined they give a 13-fold enhancement (Table 1). In contrast, the enhancement by *sueC* is multiplicative, and strains com-

Table 1 Suppressor enhancement by *sue* mutations

sue mutation	Relative suppressor efficiency
sue⁺	1
sueA	3.3
sueB	10
sueC	3.3
sueA sueB	13.3
sueA sueC	10
sueB sueC	27
sueA sueB sueC	40

The relative suppressor efficiencies of a set of *lacZ* (amber) *supD43,74* strains, carrying none to all the *sue* suppressor-enhancing mutations, were determined by comparing their abilities to suppress the *lacZ* amber mutation. The data used for the computation are the same as those presented in Figure 1.

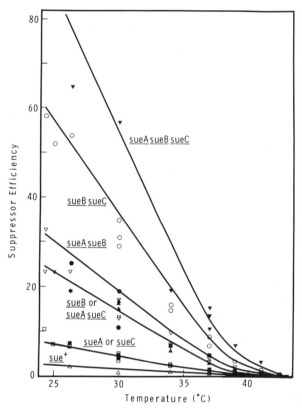

Figure 1 Suppressor efficiencies as a function of temperature. Suppressor activity was determined by measuring the rate of β-galactosidase synthesis in strains carrying a lacZ amber mutation. Log phase cultures of lacZ (amber) supD43,74 strains that were sue⁺, sueA, sueB, sueC, sueA sueB, sueA sueC, sueB sueC, or sueA sueB sueC were grown in parallel with a lacZ (amber) supD sueC strain, as previously described (Oeschger and Woods 1976). Isopropyl-β-D-thiogalactoside was added to induce lac operon expression. Portions of the culture were harvested at 20, 30, and 40 min after induction and β-galactosidase activity per cell determined (Oeschger and Woods 1976). All values are plotted relative to the suppressor efficiency of the supD sueC strain, which has been shown to be 31% (Haggerty et al. 1978). (△) sue⁺; (■) sueA; (▲) sueB; (□) sueC; (●) sueA sueB; (▽) sueA sueC; (○) sueB sueC; (▼) sueA sueB sueC.

bining supD43,74, sueC, and sueB provide up to 100% efficient suppression (Fig. 2).

It is clear that the sue mutations in combination with supD43,74 solve two of the major problems limiting use of temperature-sensitive suppressors. First, they provide a wide range of suppressor efficiency, from 100% in sueB sueC (and sueA sueB sueC) strains to 0.01% in sue⁺ strains

Figure 2 Suppressor efficiency of a *supD43,74* strain carrying *sueB* and *sueC* mutations. Suppression of a nonpolar amber mutation in the gene encoding the β subunit of RNA polymerase (*rpoB*) was used to measure suppressor efficiency. The β and β' subunits of RNA polymerase are synthesized from a single mRNA (Errington et al. 1974); therefore, the ratio of the β to β' subunits synthesized can be used as a direct measure of suppressor efficiency. The rates of β and β' subunit synthesis are determined by quantitation of the radioactivity incorporated into the subunits by pulse-labeling. This quantitation is performed by densitometry of autoradiographs of polyacrylamide-gel-fractionated cell extracts (Matzura et al. 1971; Oeschger and Berlyn 1975). Cells growing in log phase were pulse-labeled with [^{14}C]tyrosine as previously described (Oeschger 1978). Cell extracts were fractionated on 4% polyacrylamide gels. (*a*) Levels of β and β' subunits in cells growing at 15°C. The cells were labeled for 10 min and allowed to grow for an additional 3 hr in the presence of excess cold tyrosine. The ratio of the radioactivity in the β and β' proteins of the parental and *rpoB* amber mutant strains is the same

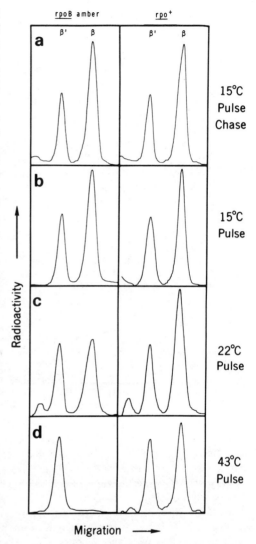

in both strains. This result demonstrates that the tyrosine content of the β and β' subunits is unchanged in the *rpoB* amber mutant. (*b*) Relative rates of synthesis of the two proteins at 15°C. Cultures were again labeled for 10 min but chased for only 5 min. Quantitation of the tracings shows the suppressor is 100% efficient at this temperature. (*c*) Relative rates of synthesis of the two proteins at 22°C. Cells were labeled for 6 min and chased for 3 min. Quantitation of the tracings shows the suppressor is 65% efficient. (*d*) The relative rates of synthesis of the two proteins at 43°C. Cells were transferred from 22°C to 43°C 120 min before labeling. The cells were labeled for 2 min followed by a 2-min chase. Quantitation of the tracings shows that the suppressor is less than 3% efficient. The amount of β' shown in *c* and *d* has been normalized to a constant value. The changes in ratio of β to β' subunit in the parental strain at 43°C is unexplained.

(Figs. 1 and 2). Second, a series of strains carrying different combinations of *sue* mutations can be used to obtain many levels of suppression at the same temperature. Hence, effects that could be attributed to temperature can be eliminated (Haggerty et al. 1978).

Application to the Isolation of Amber Mutant Strains

The high efficiency of temperature-sensitive suppressor *sue* strains permits the derivation of strains with amber mutations in any gene. Enhanced temperature-sensitive suppressor strains have been used in searches for amber mutations in the RNA polymerase structural genes (Ridley 1980), genes encoding the protein synthetic machinery (S. P. Ridley et al., unpubl.), and genes affecting stringent control (M. Cashel and M. P. Oeschger, unpubl.). The yield of essential amber mutant strains among temperature-sensitive derivatives is good, with approximately 20% of the isolates containing an amber mutation. As a result of this work, strains carrying amber mutations in the genes that encode the β' subunit of RNA polymerase, protein synthesis elongation factor G, and the *relA* gene product have been isolated.

Studies with Amber Mutant Strains

Two proteins, RNA polymerase and the *araC* protein, have already been studied in *sue supD* temperature-sensitive strains. Using strains containing amber mutations in *rpoB*, the gene encoding the β subunit of RNA polymerase, the level of polymerase has been titrated against growth rate and polymerase gene expression (M. P. Oeschger 1976 and unpubl.). The results of these studies show RNA polymerase is in twofold functional excess in optimally growing cells. Reduction in the level of polymerase results in specific stimulation of polymerase gene expression. The stimulation in the expression of the *rpoBC* operon, determined by the relative rate of β' subunit synthesis in a nonpolar *rpoB* amber mutant strain, reaches a maximum of 12-fold in cells growing with 20% of the normal level of polymerase. Marked stimulation in the rate of polymerase σ subunit synthesis is also observed. The degree of stimulation in the rates of β' subunit and *rpoC* mRNA synthesis was analyzed in these strains. The results indicate that the enhanced expression can be solely attributed to increased transcription of the *rpoBC* operon (Little and Dennis 1980).

The product of the *araC* gene regulates activity of the arabinose operon (*araBAD*). Haggerty et al. (1978) used a set of *sue supD43,74 araC* amber mutant strains to titrate the levels of *araC* protein against *araBAD* operon expression. Each of these strains provides a different level of suppressor activity at a given temperature. With this set of strains the level of *araC* protein was varied eightfold at each of a number of constant temperatures as well as titrated by growth of the strains at different temperatures. The combination of these two approaches ensured that any temperature effects would be recognized. The levels of *araC* protein ranged from 0.007 to 1.8 times the normal amount. The results of these experiments indicate that the normal level of *araC* protein is just sufficient to provide maximal expression of the *araBAD* operon.

CONCLUSIONS

Temperature-sensitive suppressor mutants with high efficiency, rapid loss of suppressor activity, rapid recovery of suppressor activity, multiple levels of suppression at any given temperature, and systematic variability of suppressor activity with temperature constitute a system that is ideal for a wide range of biochemical and physiological studies. High-efficiency temperature-sensitive suppressor strains permit the isolation and propagation of strains with amber mutations in any gene. The rapid inactivation of suppressor activity on temperature shift-up allows titration of the amber gene product by dilution with cell growth. Rapid inactivation of suppressor activity also allows the investigation of the functional stability of proteins. Rapid recovery of suppressor activity on temperature shift-down makes the synthesis of any amber gene protein product inducible. Most importantly, rapid alterations in suppressor activity permit differential labeling of cellular proteins to identify, quantitate, and study amber gene products in cell extracts, avoiding complications that may arise in enzymatic assays. Titration of amber gene products under steady-state conditions can be accomplished in two ways: First, by systematic variation of suppressor activity via culture temperature, and second, by the use of a set of *sue* temperature-sensitive suppressor strains that provide a wide range of suppressor activity at any temperature. These features permit the execution of systematic in vivo biochemical experiments with any cellular protein.

ACKNOWLEDGMENTS

I thank John Scaife for supplying a strain containing the *rpoB* amber mutation used in this work. I am especially indebted to Anne Morris Hooke, Patrice Cuccaro, Susan Porter Ridley, and Judith Toffenetti for their valuable criticism of the manuscript. The work from this laboratory was supported, in part, by National Science Foundation grant PCM-75-21305.

REFERENCES

Abelson, J. N., M. L. Gefter, L. Barnett, A. Landy, R. L. Russell, and J. D. Smith. 1970. Mutant tyrosine transfer ribonucleic acids. *J. Mol. Biol.* **47**:15.

Bachmann, B. J., K. B. Low, and A. L. Taylor. 1976. Recalibrated linkage map of *Escherichia coli* K-12. *Bacteriol. Rev.* **40**:116.

Bassford, P. J., C. A. Schnaitman, and R. J. Kadner. 1977. Functional stability of the *bfe* and *tonB* gene products in *Escherichia coli*. *J. Bacteriol.* **130**:750.

Beacham, I. R., N. S. Taylor, and M. Youell. 1976. Enzyme secretion in *Escherichia coli*: Synthesis of alkaline phosphatase and acid hexose phosphatase in the absence of phospholipid synthesis. *J. Bacteriol.* **128**:522.

Beckman, D. and S. Cooper. 1973. Temperature-sensitive nonsense mutations in essential genes of *E. coli*. *J. Bacteriol.* **116**:1336.

Begg, K. J. and W. D. Donachie. 1973. Topography of outer membrane growth in *E. coli*. *Nat. New Biol.* **245**:38.

———. 1977. Growth of the *Escherichia coli* cell surface. *J. Bacteriol.* **129**:1524.

Davidoff-Abelson, R. and L. Mindich. 1978. A mutation that increases the activity of nonsense suppressors in *Escherichia coli*. *Mol. Gen. Genet.* **159**:161.

Delcuve, G., T. Cabezón, A. Ghysen, A. Herzog, and A. Bollen. 1977. Amber mutations in *Escherichia coli* essential genes: Isolation of mutants affected in the ribosomes. *Mol. Gen. Genet.* **157**:149.

De Lucia, P. and J. Cairns. 1969. Isolation of an *E. coli* strain with a mutation affecting DNA polymerase. *Nature* **224**:1164.

Epstein, R. H., A. Bolle, C. M. Steinberg, E. Kellenberger, E. Boy de la Tour, R. Chevalley, R. S. Edgar, M. Susman, G. N. Denhardt, and A. Lielausis. 1964. Physiological studies of conditional lethal mutants of bacteriophage T4D. *Cold Spring Harbor Symp. Quant. Biol.* **28**:375.

Errington, L., R. E. Glass, R. S. Hayward, and J. G. Scaife. 1974. Structure and orientation of an RNA polymerase operon in *Escherichia coli*. *Nature* **249**:519.

Gallucci, E., G. Pacchetti, and S. Zangrossi. 1970. Genetic studies on temperature sensitive nonsense suppression. *Mol. Gen. Genet.* **106**:362.

Garen, A. 1968. Sense and nonsense in the genetic code. *Science* **160**:149.

Glass, R. E. 1977. Identification of an amber fragment of the β subunit of *Escherichia coli* RNA polymerase: A yardstick for measuring controls on RNA polymerase subunit synthesis. *Mol. Gen. Genet.* **151**:83.

Glass, R. E., M. Goman, L. Errington, and J. Scaife. 1975. Induction of RNA polymerase synthesis in *Escherichia coli*. *Mol. Gen. Genet.* **143**:79.

Goodman, H. M., J. Abelson, A. Landy, S. Brenner, and J. D. Smith. 1968. Amber suppression: A nucleotide change in the anticodon of a tyrosine transfer RNA. *Nature* **217**:1019.

Gross, J. and M. Gross. 1969. Genetic analysis of an *Escherichia coli* strain with a mutation affecting DNA polymerase. *Nature* **224**:1166.

Gross, J. D., J. Grunstein, and E. M. Witkin. 1971. Inviability of $recA^-$ derivatives of the DNA polymerase mutant of De Lucia and Cairns. *J. Mol. Biol.* **58**:631.

Gustafson, P. and K. Nordstrom. 1978. Temperature-dependent and amber copy mutants of plasmid R1*drd*-19 in *Escherichia coli*. *Plasmid* **1**:134.

Haggerty, D. M., M. P. Oeschger, and R. F. Schleif. 1978. In vivo titration of *araC* protein. *J. Bacteriol.* **135**:775.

Hashimoto-Gotoh, T. and M. Sekiguchi. 1977. Mutations to temperature sensitivity in R plasmid pSC101. *J. Bacteriol.* **131**:405.

Horiuchi, T. and T. Nagata. 1973. Mutations affecting growth of the *Escherichia coli* cell under a condition of DNA polymerase I-deficiency. *Mol. Gen. Genet.* **123**:89.

———. 1974. Lethality of the *Escherichia coli* K12 cell doubly deficient in DNA polymerase I and DNA strand-joining activity. *Mol. Gen. Genet.* **128**:105.
Horiuchi, T., T. Sato, and T. Nagata. 1975. DNA degradation in an amber mutant of *Escherichia coli* K12 affecting DNA ligase and viability. *J. Mol. Biol.* **95**:271.
Howard-Flanders, P. and R. P. Boyce. 1966. DNA repair and genetic recombination: Studies on mutants of *Escherichia coli* defective in these processes. *Radiat. Res.* (suppl.) **6**:156.
Inoko, H., K. Shigesada, and M. Imai. 1977. Isolation and characterization of conditional-lethal rho mutants of *Escherichia coli*. *Proc. Natl. Acad. Sci.* **74**:1162.
Kozak, M. and D. Nathans. 1972. Translation of the genome of a ribonucleic acid bacteriophage. *Bacteriol. Rev.* **36**:109.
Little, R. and P. Dennis. 1980. Regulation of RNA polymerase synthesis: Conditional lethal amber mutations in the β subunit gene. *J. Biol. Chem.* (in press).
Matzura, H., S. Molin, and O. Maaløe. 1971. Sequential biosynthesis of the β and β' subunits of the DNA-dependent RNA polymerase from *E. coli*. *J. Mol. Biol.* **59**:17.
Monk, M. and J. Kinross. 1972. Conditional lethality of *recA* and *recB* derivatives of a strain of *Escherichia coli* K-12 with a temperature-sensitive deoxyribonucleic acid polymerase I. *J. Bacteriol.* **109**:971.
Morimyo, M. and Y. Shimazu. 1976. Evidence that the gene *uvrB* is indispensable for a polymerase I deficient strain of *Escherichia coli* K12. *Mol. Gen. Genet.* **147**:243.
Morimyo, M., Y. Shimazu, and N. Ishii. 1976. Isolation and genetic analysis of amber *uvrA* and *uvrB* mutants. *J. Bacteriol.* **126**:529.
Nagata, T. and T. Horiuchi. 1973. Isolation and characterization of a temperature-sensitive amber suppressor mutant of *Escherichia coli* K12. *Mol. Gen. Genet.* **123**:77.
———. 1974. An amber *dna* mutant of *Escherichia coli* K12 affecting DNA ligase. *J. Mol. Biol.* **87**:369.
Nakamura, Y. and T. Yura. 1975. Evidence for a positive regulation of RNA polymerase synthesis in *Escherichia coli*. *J. Mol. Biol.* **97**:621.
Oeschger, M. P. 1976. Autogenous regulation of RNA polymerase synthesis in *E. coli*. *Fed. Proc.* **35**:1638.
———. 1978. Rich culture medium for the radiochemical labeling of proteins and nucleic acids. *J. Bacteriol.* **134**:913.
Oeschger, M. P. and M. K. B. Berlyn. 1973. Temperature sensitive Sul suppression mutants of *E. coli*. *Fed. Proc.* **32**:653.
———. 1975. Regulation of RNA polymerase synthesis in *Escherichia coli*: A mutant unable to synthesize the enzyme at 43°C. *Proc. Natl. Acad. Sci.* **72**:911.
Oeschger, M. P. and G. T. Wiprud. 1980. High-efficiency temperature-sensitive amber suppressor strains of *Escherichia coli* K12: Construction and characterization of recombinant strains with suppressor-enhancing mutations. *Mol. Gen. Genet.* (in press).
Oeschger, M. P. and S. L. Woods. 1976. A temperature-sensitive suppressor enabling the manipulation of the level of individual proteins in intact cells. *Cell* **7**:205.

Oeschger, M. P., N. S. Oeschger, G. T. Wiprud, and S. L. Woods. 1980. High-efficiency temperature-sensitive amber suppressor strains of *Escherichia coli* K12: Isolation of strains with suppressor-enhancing mutations. *Mol. Gen. Genet.* **177:** (in press).
Ozeki, H., H. Inokuchi, and H. Aono. 1969. Genetic analysis of suppressor gene Su3. I. Isolation of defective mutants. *Jpn. J. Genet.* **44:** 406.
Ridley, S. P. 1980. "Isolation and characterization of an *Escherichia coli* K12 strain carrying an amber mutation in the gene for β' subunit of RNA polymerase." Ph.D. thesis, Georgetown University, Washington, D.C.
Russell, R. L., J. N. Abelson, A. Landy, M. L. Gefter, S. Brenner, and J. D. Smith. 1970. Duplicate genes for tyrosine transfer RNA in *Escherichia coli*. *J. Mol. Biol.* **47:** 1.
Smith, J. D., L. Barnett, S. Brenner, and R. L. Russell. 1970. More mutant tyrosine transfer ribonucleic acids. *J. Mol. Biol.* **54:** 1.
Smith, J. D., J. N. Abelson, B. F. C. Clark, H. M. Goodman, and S. Brenner. 1967. Studies on *amber* suppressor tRNA. *Cold Spring Harbor Symp. Quant. Biol.* **31:** 479.
Smith, J. D., K. Anderson, A. Cashmore, M. L. Hooper, and R. L. Russell. 1971. Studies on the structure and synthesis of *Escherichia coli* tyrosine transfer RNA. *Cold Spring Harbor Symp. Quant. Biol.* **35:** 21.
Stretton, A. O. W., S. Kaplan, and S. Brenner. 1967. Nonsense codons. *Cold Spring Harbor Symp. Quant. Biol.* **31:** 173.
Tacon, W. and D. Sherratt. 1976. ColE plasmid replication in DNA polymerase I-deficient strains of *Escherichia coli*. *Mol. Gen. Genet.* **147:** 331.
Van de Putte, P., C. A. Van Sluis, J. Van Dillewijn, and A. Rorsch. 1965. The location of genes controlling radiation sensitivity in *Escherichia coli*. *Mutat. Res.* **2:** 97.
Wainright, M. and I. R. Beacham. 1977. The effect of translation and transcription inhibitors on the synthesis of periplasmic phosphatases in *E. coli*. *Mol. Gen. Genet.* **154:** 67.
Willetts, N. S., A. J. Clark, and B. Low. 1969. Genetic location of certain mutations conferring recombination deficiency in *Escherichia coli*. *J. Bacteriol.* **97:** 244.
Wood, W. B., R. S. Edgar, J. King, I. Lielausis, and M. Henninger. 1968. Bacteriophage assembly. *Fed. Proc.* **27:** 1160.

Characterization of Nonsense Suppressor tRNAs from *Saccharomyces cerevisiae*: Identification of the Mutational Alterations That Give Rise to the Suppressor Function

Peter W. Piper
Imperial Cancer Research Fund Laboratories
Lincoln's Inn Fields
London WC2A 2PX, England

Nonsense suppressor mutations of *Saccharomyces cerevisiae* have been identified from their ability to suppress auxotrophic markers that are known to correspond to either UAA, UAG, or UGA nonsense mutations. Each of these mutations, with only a few exceptions, causes suppression at only one of the nonsense codons. UAA (ocher) and UAG (amber) suppressors have also been characterized from analyses on the iso-1-cytochrome *c* of strains bearing ocher or amber mutations in the structural gene for this protein, and it has been found that these suppressors each insert one of just three amino acids during translational suppression of UAA or UAG codons (Gilmore et al. 1971; Sherman et al. 1973; Liebman et al. 1975, 1976; Brandriss et al. 1976; Sherman et al. 1979). These amino acids are tyrosine, serine, and leucine. The amino acid insertion specificities of *S. cerevisiae* UGA suppressors (Hawthorne 1976) are as yet unknown.

Evidence that nonsense suppression in yeast is mediated, as in prokaryotes, by altered tRNA molecules has come from two different lines of experimentation. On the one hand, it has been demonstrated that total tRNA isolated from suppressor-carrying yeast strains can permit the insertion of amino acids in response to nonsense codons in in vitro protein-synthesizing systems (Capecchi et al. 1975; Gesteland et al. 1976). On the other hand, it has been shown that certain suppressor mutations determine anticodon base changes in tRNAs from nucleotide sequencing studies of suppressor tRNAs and genes that encode them (Piper et al. 1976; Goodman et al. 1977; Piper 1978). Both approaches have been used in establishing the molecular identity of the tRNA products of individual suppressor genes, as well as the tRNA structural changes determined by the mutation that produces the nonsense suppressor function. It is also important to know the nature of a suppressor tRNA species if this molecule is to be purified for microinjection into mutant clones of cells from a higher eukaryote to observe whether or not it brings about a temporary phenotypic reversion in the cells of each clone in turn. Those clones having a nonsense mutation suppressed by the yeast suppressor that was used can then be selected (Kaltoft et al. 1976; Celis et al. 1979). The ability of yeast suppressor tRNAs to give in vivo suppression after introduction into mammalian cells has recently extended the scope of somatic cell genetics by providing a selection technique for nonsense mutant cell lines.

SEPARATION OF YEAST tRNAs ON TWO-DIMENSIONAL POLYACRYLAMIDE GELS

The ability of *S. cerevisiae* ocher and amber suppressors to insert tyrosine, serine, or leucine in response to the UAA or the UAG codon is an indication that these suppressors are probably derived by mutation of the tyrosine-, serine-, and leucine-accepting tRNAs of this organism. This is assuming that the suppressor mutations do not produce misacylation in vivo of the tRNAs whose structure they alter, a situation that does not hold for certain *Escherichia coli* tRNA mutations, such as the su^+7 tRNA^{Trp} (Yaniv et al. 1974; Soll 1974; Celis et al. 1976) and glutamine-mischarging mutants derived from su^+3 tRNA^{Tyr} (Celis et al. 1973). To investigate whether yeast ocher and amber suppressors might be derived from tRNA^{Tyr}, tRNA^{Ser}, and tRNA^{Leu}, we developed a routine method for purifying, in small amounts from a culture of a given yeast strain, those tRNAs^{Tyr}, tRNAs^{Ser}, and tRNAs^{Leu} that recognize a codon differing in not more than one base from the ocher or amber codon. Such a procedure has to be relatively rapid to enable ready comparisons among different strains.

All nonsense codons begin with U, so that nonsense-suppressing tRNAs will have A as the final anticodon nucleoside. Most yeast tRNAs with A as the 3'-terminal nucleoside of their anticodons have a hydrophobic moiety like i^{6}A to the 3' side of this A, a modification that causes the tRNA to be retained strongly by benzoylated DEAE-cellulose (BD-cellulose). Therefore, most yeast tRNAs that have a cognate codon differing from UAA or UAG in the second or third base positions are among the late-eluting species on BD-cellulose columns. The protocol we eventually adopted for purifying them involved taking the total tRNA of each small ^{32}P-labeled yeast culture and isolating from it those species that were not eluted from a BD-cellulose column by a buffer containing 0.85 M NaCl, 0.01 M MgCl$_2$, and 0.01 M sodium acetate (pH 5.0) (Piper et al. 1976; Piper and Wasserstein 1977; Piper 1978). This subfraction was then resolved into individual tRNAs by two-dimensional polyacrylamide gel electrophoresis; the relative migration of the various species giving rise to the characteristic gel pattern is illustrated in Figure 1. Among the tRNAs on these gels were identified the only tRNA^{Tyr} of yeast, as well as tRNA^{Ser} and tRNA^{Leu}, which between them can decode all the serine and leucine codons differing in a single base from either UAA or UAG (Piper and Wasserstein 1977).

Using this procedure for purifying these tRNAs in conjunction with RNA fingerprinting, we are in a position to investigate whether specific amber or ocher suppressor mutations determine structural changes in tRNA^{Tyr}, tRNA^{Ser}, and tRNA^{Leu}. It is also possible to purify small, non-radioactive amounts of these tRNAs on the two-dimensional gels either to test each individually for an ability to function as a suppressor in an in vitro protein-synthesizing system or to label these tRNAs in vitro for use as a specific hybridization probes in selecting *E. coli* clones containing the DNA of yeast tRNA genes inserted into plasmids (Beckmann et al. 1977). A suppressor tRNA derived by mutation of a single base in a fraction of one of these species might be expected to comigrate on these gels with the wild-type tRNA. At least in the instances of amber suppressors derived from the *SUP5* and *SUPRL1* loci, the gels do not resolve the wild-type tRNA and the suppressor derived from it.

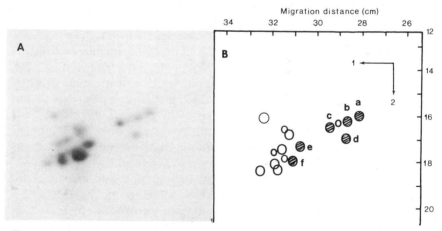

Figure 1 Separation by two-dimensional polyacrylamide gel electrophoresis of the yeast tRNAs retained by BD-cellulose in 0.85 M NaCl, 0.01 M MgCl$_2$, 0.01 M sodium acetate (pH 5.0). (*A*) Autoradiograph of the gel; (*B*) a map of the individual major tRNA species. (○) Unidentified species of tRNA; (⊘) identified tRNAs. The scale indicates the distances of migration in the first dimension (1) 10% gel and in the second dimension (2) 20% gel. The identified tRNAs are: a UUA- and/or a UUG-decoding tRNA^{Leu} (*a*); a UUG-decoding tRNA^{Leu} (tRNA$^{Leu}_3$) (*b*); a UCA-, UCC-, and UCU-decoding tRNA^{Ser} (tRNA$^{Ser}_2$) (*c*); tRNA$^{Ser}_{UCG}$ (*d*); tRNA^{Tyr} (*e*); and tRNA^{Phe} (*f*). Each of these tRNAs, except tRNA^{Phe}, has a cognate codon that is related to either UAA or UAG by a single-base substitution.

The yeast suppressors analyzed in this way fall into the tyrosine-inserting and serine-inserting classes listed in Table 1 and are derived from genetic loci that can be mutated to produce both the ocher- and amber-suppressing phenotypes. The loci that give rise to tyrosine-inserting suppressors can each give either an ocher or amber suppressor by a single-step mutation (Liebman et al. 1976), whereas those loci giving serine-inserting suppressors that are not recessive-lethal yield only ocher suppressors on single mutation, the amber suppressors from these loci being the result of double mutation (Sherman et al. 1979). The *SUPRL1* (or *SUP61*) locus, however, gives a recessive-lethal, serine-inserting amber suppressor by single mutation, and an ocher suppressor by double mutation.

CHARACTERIZATION OF TYROSINE-INSERTING SUPPRESSOR tRNAs

The first characterization of a yeast suppressor tRNA (Piper et al. 1976) was of the amber suppressor *SUP5-a*, one of the class of eight unlinked, highly efficient suppressors in *S. cerevisiae* (Table 1) that cause the insertion of tyrosine at UAG sites (Liebman et al. 1976). A parallel series of eight allelic suppressors cause tyrosine to be inserted at UAA sites (Gilmore et al. 1971). More recently, Olsen et al. (1977) have shown that yeast tRNA^{Tyr} hybridizes to eight endonuclease *Eco*RI fragments of yeast DNA and have been able to correlate these fragments with individual tyrosine-inserting suppressor loci. The tyrosine-inserting suppressors of *S. cerevisiae* constitute a minor fraction of the total tRNA^{Tyr} of strains bearing these suppressors. The mutation giving rise to the suppressor function can be investigated by sequencing the tRNA^{Tyr} of these strains, provided the fingerprinting systems resolve the additional ribonuclease-digestion products originating from the minor suppressor component from those derived from the wild-type sequence.

When the tRNA^{Tyr} of *SUP5-a* and *sup*$^+$ strains was examined by fingerprint analysis (Piper et al. 1976), a difference in the RNase T1 digests became apparent (Fig. 2a,b,c). A large oligonucleotide (marked C in Fig. 2b,c) was present in the digest of the *SUP5-a* tRNA^{Tyr} but was absent from the corresponding digest of the *sup*$^+$ tRNA^{Tyr}. This fragment has the sequence ACUCψAi^{6}AAψCUUGp and does not constitute one of the normal digestion products of yeast tRNA^{Tyr}. Its structure indicates that it was originally derived from a tRNA^{Tyr} with the anticodon CψA, instead of the normal GψA (Fig. 3a). Such a molecule would be expected to function as a UAG suppressor, the anticodon CUA having been identified in amber-suppressing tRNAs of *E. coli* (Goodman et al. 1968; Yaniv et al. 1974), phage T4 (Comer et al. 1975), and, as described below, in the recessive-lethal yeast *SUPRL1* suppressor.

Table 1 Classes of amber and ocher suppressors of *Saccharomyces cerevisiae* and the tRNAs from which they are derived

Class and amino acid insertion specificity	Act on	Genetic loci that mutate to these suppressors[a]	tRNA species mutated to give suppressor[b]	References
Tyrosine-inserting, highly efficient[c]	UAA or UAG	*SUP2, SUP3, SUP4, SUP5, SUP6, SUP7, SUP8,* and *SUP11*	tRNA^{Tyr} (*SUP4* and *SUP5*)	Piper et al. (1976) Goodman et al. (1977)
Serine-inserting, moderately efficient	UAA or UAG	*SUP16 (SUQ5), SUP17,* and *SUP18;* as in Sherman (1978)	a tRNA^{Ser} that copurifies with minor tRNA$^{Ser}_{UCG}$	C. Waldron et al. (unpubl.)
Serine-inserting, moderately efficient, and recessive-lethal	UAA or UAG	*SUPRL1 (SUP61)*	minor tRNA$^{Ser}_{UCG}$ (*SUPRL1*)	Piper (1978)
Leucine-inserting, very low efficiency	UAA or UAG	*SUP26,* etc., as in Sherman et al. (1979)	unknown	—

[a]Alternative designations in parentheses.
[b]The specific suppressor loci for which an identification of the tRNA giving rise to the suppressor has been performed are indicated in parentheses.
[c]An efficient suppressor is defined as one that can appreciably suppress nonsense mutations in the structural gene for iso-1-cytochrome *c*.

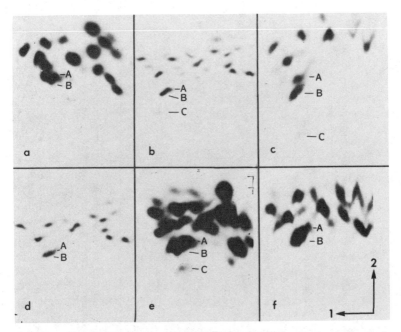

Figure 2 RNase T1 digests of tRNA^{Tyr} from different suppressing and nonsuppressing strains: sup^+ (SL210-3A) (a); two different cultures of SUP5-a (L-133) (b, c); SUP5-o (SL171-2C) (d); SUP5-o → a (L-58) (e); and SUP5-a → a' (L-133a) (f). Oligonucleotide separation was by electrophoresis on cellulose acetate in the first dimension (1) and ascending chromatography on DEAE-cellulose thin layers in the second dimension (2) using a 3% dialyzed homomixture (Piper et al. 1976). (A) pCUCUGp; (B) ψAi^{6}AAψCUUGp; (C) ACUCψAi^{6}AAψCUUGp, the anticodon fragment of the amber suppressor tRNA^{Tyr} present only in fingerprints b, c, and e. Other differences among the fingerprints are due to differences in the resolving power of homomixtures and fluctuations in the extent to which RNase T1 has cleaved the fragment Cm$_2^2$GCAAGp into Cm$_2^2$G > p and CAAGp.

The weak intensity of the mutant anticodon (spot C in Fig. 2b,c) indicated that only 5–8% of the total tRNA^{Tyr} of the SUP5-a strain existed in the form of the amber suppressor, a level consistent with this species being the product of one of the eight genes for tRNA^{Tyr}. As such it comprised no more than 0.15–0.35% of the total tRNA of the cell and was present at a level comparable to the amount of suppressor tRNA in a diploid strain heterozygous for the SUPRL1 amber suppressor (Piper 1978).

The tRNA^{Tyr} from other SUP5 suppressor alleles has also been investigated. The allele SUP5-o → a was obtained (Liebman et al. 1976) by mutation of the SUP5-o ocher suppressor in contrast to the SUP5-a allele, which was obtained directly from the sup^+ wild type. However, SUP5-o → a and SUP5-a are phenotypically indistinguishable and

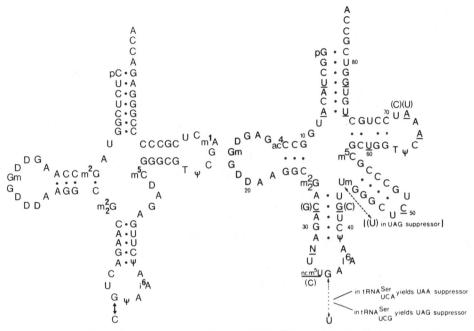

Figure 3 The primary structure of yeast tRNA^{Tyr} showing the base substitution present in the *sup5* amber suppressor (Piper et al. 1976). The *sup4* ocher suppressor has a G→U* substitution at the same position (Goodman et al. 1977). (*Right*) The nucleotide sequence of the major species of tRNA in the tRNA$^{Ser}_{UCG}$ gel spot. The sequence differences in the minor species are indicated by nucleotides in parentheses. Since the original RNA sequence analysis of these two molecules (Piper 1978) there has been some uncertainty about the true nature of the first anticodon or wobble bases present in each. The minor species has C in this position and mutates to a UAG suppressor by a CGA→CUA anticodon change (Piper 1978). The major species has a modified unstable U derivative as wobble base, this base existing substantially as ncm^{5}U in tRNA isolated by polyacrylamide gel electrophoresis (C. Waldron et al. unpubl.). Furthermore, this tRNA is almost certainly the one that mutates to give serine-inserting UAA suppressors, and studies are in progress to try to demonstrate a partial ncm^{5}UGA → U*UA anticodon change in this species from strains bearing such UAA suppressors.

sequencing of their tRNA^{Tyr} (Fig. 2b,c,e) indicated that they are identical (Piper et al. 1976). In the fingerprint of tRNA^{Tyr} from a strain bearing a less efficient amber suppressor *SUP5-a → a'*, which arose by intragenic mutation of *SUP5-a*, the anticodon RNase-T1-digestion fragment of the amber supressor tRNA^{Tyr} was undetectable (Fig. 2f) and it is possible that the second-site mutation at the *SUP5* locus caused reduced synthesis of suppressor tRNA in much the same way as certain mutations of the su^+3 tRNA^{Tyr} gene of *E. coli* impair tRNA maturation (Anderson and Smith 1972).

Tyrosine-inserting ocher suppressors of yeast are also, at least in certain cases, forms of tRNA^{Tyr} with an altered first anticodon base. Goodman et al. (1977) have cloned the suppressor gene from a yeast strain bearing the *SUP4-o* ocher suppressor and analyzed it by DNA sequencing. The sequence of this gene differs from that of the wild-type tRNA^{Tyr} genes by virtue of a GC → TA transversion at the base pair that codes for the wobble base of the tRNA^{Tyr} anticodon. The nature of the wobble base in yeast ocher suppressors has been the subject of considerable speculation, since these suppressors act only on UAA and not on UAG codons, unlike bacterial ocher suppressors, which recognize both UAA and UAG. There are not very many anticodon structures possible for *S. cerevisiae* tRNAs that have acquired the ability to read UAA, but not UAG, by anticodon mutation and that do not produce misreading of the genetic code (Piper and Wasserstein 1977). The available information would suggest that the wobble nucleoside of a UAA-specific suppressor tRNA may be mcm^{5}U, its 2-thiolated derivative mcm^{5}S, or, although only in a mutant tRNA^{Tyr}, I (inosine). The ocher suppressor gene sequenced by Goodman et al. (1977) must be transcribed into a tRNA^{Tyr} with a U derivative (U*) at the wobble position of the anticodon (Fig. 3a), although it is possible that other tyrosine-inserting ocher suppressors possess I and arise from a GC → AT transition in the DNA. In a fingerprint analysis of the tRNA^{Tyr} of an *SUP5-o* strain (Fig. 2d) we were unable to detect the fragment of sequence ACUU*ψAi^{6}AAψCUUGp that would be expected to migrate on the fingerprints in Figure 2 close to the corresponding fragment (C) from the amber suppressor. Here we were prevented by the limitations of the analysis systems from making a positive identification of inosine as the wobble nucleoside of this *SUP5-o* suppressor (Piper et al. 1976).

CHARACTERIZATION OF SERINE-INSERTING SUPPRESSOR tRNAs

Hawthorne and Leupold (1974) and Brandriss et al. (1975) have described yeast amber suppressors derived in diploid strains that could be maintained in the heterozygous state but not in haploid cells. One of these recessive-lethal suppressors, *SUPRL1-a* (alternatively designated *SUP61-a*), causes the insertion of serine at the UAG codon in a *cyc1* mutant (Brandriss et al. 1976). This led to the suggestion that the *SUPRL1* suppressor might be derived from a UCG-decoding tRNA^{Ser}, since a single anticodon base change (CGA → CUA) should give such a molecule the ability to read UAG codons, and UCG is the only serine codon related to UAG by a single-base change. Also purified total tRNA$^{Ser}_{UCG}$ extracted from a two-dimensional gel of *SUPRL1/sup$^+$* strain tRNAs can cause in vitro translational readthrough of UAG codons (Gesteland et al.

1976; Piper 1978). It was further proposed by Brandriss et al. (1976) that this mutant tRNA might be coded for by only a single gene in the haploid genome, and that its mutation to the suppressor causes lethality in haploid cells through a loss of the ability to translate the codon UCG. If this were the case, then *SUPRL1* would be analogous to certain recessive-lethal suppressors isolated in merodiploids of *E. coli* and *Salmonella typhimurium* (Soll and Berg 1969; Miller and Roth 1971; Soll 1974; Yaniv et al. 1974). We have investigated the structure of the tRNA species present in a gel spot that contains the UCG-reading tRNA component of *S. cerevisiae* (Fig. 1B, spot d) to test these proposals concerning the nature of the *SUPRL1* suppressor (Piper 1978).

The total tRNA in spot d on the two-dimensional gels was isolated from haploid and diploid yeast strains lacking *SUPRL1*. It is heterogeneous in structure and exists as major and minor nucleotide sequences, the former being present in approximately three times the amount of the latter tRNA. These forms will be referred to as the major $tRNA_{UCG}^{Ser}$ and the minor $tRNA_{UCG}^{Ser}$. Although their sequences differ from each other at only a few nucleotide positions, they are not base modification variants of the same sequence and must therefore be encoded as more than one gene in the haploid genome. The complete nucleotide sequence of the major $tRNA_{UCG}^{Ser}$ was derived (Fig. 3b) but it was only possible for us to unequivocally deduce the structure of the minor species in the vicinity of the anticodon, since we were unable to separate this form from the major $tRNA_{UCG}^{Ser}$ and other sequences. Differences shown in Figure 3b are tentative. The minor species, almost certainly from the sequence and genetic evidence described below, reads the UCG codon, whereas it is uncertain yet whether the major $tRNA_{UCG}^{Ser}$ species recognizes UCG, UCA, or both of these codons. The complete structure of a precursor to minor $tRNA_{UCG}^{Ser}$ that accumulates in the *ts*136 mutant has recently been determined by C. Guthrie (pers. comm.).

Figure 4 illustrates a fingerprint of an RNase T1 digest of wild-type $tRNA_{UCG}^{Ser}$. The structures of the numbered oligonucleotides and their derivation is documented elsewhere (Piper 1978). A small number of the oligonucleotides on this fingerprint are present in low yield, since they are derived only from the minor $tRNA_{UCG}^{Ser}$, whereas those fragments originating only from the major sequence or derived from both species are present in much higher yield. The corresponding fingerprint of $tRNA_{UCG}^{Ser}$ from a strain heterozygous for the recessive-lethal *SUPRL1-a* suppressor, MBD16 (Brandriss et al. 1976), is shown in Figure 4b. A distinct and reproducible difference between this fingerprint and that from the wild-type $tRNA_{UCG}^{Ser}$ was the presence, in fairly low yield, of an additional large oligonucleotide designated T24 in Figure 4b. The sequence of T24—ANUCUAi^{6}AAψCUCUUGp (Piper 1978)—shows that it cannot originate from a mutant form of the major $tRNA_{UCG}^{Ser}$ that differs in only a

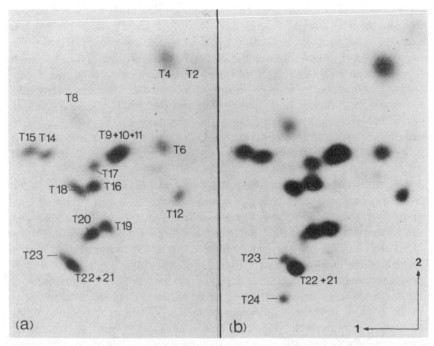

Figure 4 RNase T1 digests of wild-type tRNA$^{Ser}_{UCG}$ (a) and tRNA$^{Ser}_{UCG}$ from strain MBD16 (b). Fractionation was by electrophoresis at pH 3.5 on cellulose acetate in the first dimension (1) and homochromatography using a 3% dialyzed homomixture in the second dimension (2). The sequences of the numbered oligonucleotides are listed in Piper (1978).

single base from the wild-type species (Fig. 3b). Instead, this structure indicates a derivation from a tRNA having the same anticodon region structure as the minor tRNA$^{Ser}_{UCG}$ except for the substitution of U for G at position 35 and the lack of a ribose methylation at position 44. This conclusion and the absence of other changes on the fingerprints provide a strong indication that this altered tRNA is the *SUPRL1-a* suppressor, that it is derived from the minor tRNA$^{Ser}_{UCG}$, and that it has the ability to read the codon UAG through a G→U anticodon base change (Fig. 3b).

Differences were also found among the minor pancreatic RNase digestion products of the tRNA$^{Ser}_{UCG}$ of MBD16 that were consistent with this interpretation (Piper 1978), although a direct demonstration of the GC→TA transversion in the DNA postulated to give rise to the *SUPRL1-a* mutation must await the cloning and sequencing of the suppressor gene. So far we have been unable to separate the major and minor components of tRNA$^{Ser}_{UCG}$ from the presumed suppressor species in our *SUPRL1/sup$^+$* strain tRNA. This has prevented a direct demonstration that it is specifically the latter tRNA that is reading the UAG codon when total

$SUPRL1/sup^+$ tRNA$^{Ser}_{UCG}$ is added to in vitro protein-synthesizing systems. In the RNase T1 digests of tRNA$^{Ser}_{UCG}$ from an $SUPRL1/sup^+$ strain (Fig. 4b) the oligonucleotide T24 and the corresponding fragment from the wild-type minor tRNA$^{Ser}_{UCG}$, Ai^{6}AAψCUCUUmGp (T23), were present in equimolar yields, indicating that the suppressor species was present in an amount equal to that of the minor tRNA$^{Ser}_{UCG}$. In conjunction with the sequence data, this has led us to propose that $SUPRL1/sup^+$ strains carry one gene for the wild-type minor tRNA$^{Ser}_{UCG}$ as well as one copy of this gene that has mutated so as to code for the $SUPRL1$-a suppressor. Similarly, haploid cells will have only one gene for minor tRNA$^{Ser}_{UCG}$, this corresponding to the $SUPRL1$ locus. Other genes will in turn code for the major tRNA$^{Ser}_{UCG}$, and the abundance of this species relative to that of the minor form further suggests that it may be encoded as more than one gene in the haploid genome.

The suppressor mutation in $SUPRL1/sup^+$ cells affects the structure of a fraction of only one of the S. cerevisiae tRNA^{Ser} species and whether the lethality it causes in haploid cells is due to the total alteration of a unique coding tRNA presumably depends on whether there are other UCG-reading tRNAs in the cell. The tRNA with which this minor tRNA$^{Ser}_{UCG}$ copurifies, here called major tRNA$^{Ser}_{UCG}$, was previously thought to be another UCG-reading tRNA (Piper 1978), but evidence that it can mutate by single mutation to give ocher suppressors, described below, suggests that it might be a UCA-reading or a UCA- and UCG-reading tRNA. Recent studies (C. Waldron et al., unpubl.) have shown that it has a modified U in the first anticodon position. This nucleoside appears to exist largely as ncm^{5}U in the wild-type tRNA, although its nature in the wobble position of suppressor tRNAs is not known. Since one half of the minor tRNA$^{Ser}_{UCG}$ is altered in MBD16, haploid and homozygous $SUPRL1$ strains, if they were inviable, might be expected to totally lack the wild-type form of this molecule. The absence of wild-type copy of the minor tRNA$^{Ser}_{UCG}$ gene is presumably the condition that causes lethality in such cells. However, it might be suggested that these cells are inviable due to the amount of suppressor, or efficiency of suppression, being too high and that the suppressor may have been reduced to a level compatible with viability in $SUPRL1/sup^+$ strains by a gene-dosage effect. If the recessive lethality of $SUPRL1$ is due to oversuppression, revertants in which the suppressor has either been lost or its efficiency substantially reduced would not be expected to exhibit the haplo-lethality. One such revertant has been isolated by S. Liebman (pers. comm.) from the $SUPRL1/sup^+$ strain DBD339 (Brandriss et al. 1975). Although devoid of detectable suppressor activity, it still exhibits the lethal effect of $SUPRL1$, never yielding more than two live spores upon tetrad analysis, thereby showing that the lethality caused by $SUPRL1$ in haploid cells is not due to excessive suppression of amber codons. Instead, this lethality is probably a direct consequence of the minor tRNA$^{Ser}_{UCG}$ losing its ability to decode UCG codons.

Only one recessive-lethal, serine-inserting suppressor has been characterized in *S. cerevisiae*; it would appear from the searches of Hawthorne and Leupold (1974) and Brandriss et al. (1975) that there are not many genetic loci in this organism that can mutate to suppressors causing recessive-lethality. Certain other serine-inserting ocher and amber suppressors do not produce such lethality. The tRNA product of one of these suppressor genes has been extensively purified using column chromatography by C. Waldron (pers. comm.), using the in vitro assay systems for ocher and amber suppressors developed by R. Gesteland and his collaborators (Gesteland et al. 1976). Waldron found that purified *SUP16* (alternatively designated *SUQ5*) ocher or amber suppressing tRNAs copurify with one of the tRNAs^{Ser} of the yeast cell, but that this species is not the one that reads the UCU, UCC, and UCA codons (tRNA$_2^{Ser}$). We have electrophoresed two preparations of Waldron's purified *SUP16* ocher suppressor on two-dimensional polyacrylamide gels and have found that it is revealed by staining of the gels as a single spot almost exactly coinciding in position with purified ^{32}P-labeled tRNA$_{UCG}^{Ser}$ (C. Waldron and P. W. Piper, unpubl.).

We have fingerprinted tRNA$_2^{Ser}$ and tRNA$_{UCG}^{Ser}$ from strains carrying the *SUP16* ocher and amber suppressors and were unable to detect an altered species of tRNA^{Ser} (Piper 1978). However, the ability to detect mutational changes in a small fraction of a gel-purified tRNA, even when it is known from suppressor assays to contain the suppressor, is limited by the resolution of the fingerprints, and therefore not too much significance should be placed on these negative results. Furthermore, more recent work has indicated that the tRNA$_{UCG}^{Ser}$ analyzed in our sequence study was devoid of an additional tRNA component that comigrates with tRNA$_{UCG}^{Ser}$ when total yeast tRNA is separated on the two-dimensional gels. The samples in our original study were partially purified by column chromatography prior to gel electrophoresis and subsequent work has shown that the *SUP16* amber suppressor was lost at the column step. If the column stage is omitted, RNase-T1-digest fingerprints of tRNA$_{UCG}^{Ser}$ of strains possessing the *SUP16* amber and wild-type nonsuppressing phenotypes do differ (C. Waldron et al., unpubl.). Preliminary results indicate that the *SUP16* amber suppressor has the same nucleotide sequence as the major tRNA in the tRNA$_{UCG}^{Ser}$ gel spot, except at the first anticodon base, the nature of which is still under investigation. It is probably not a UCG-reading species, since it is difficult to envisage how such a molecule could mutate to a UAA-reading tRNA by what is apparently a single-step mutation (Sherman et al. 1979) giving rise to the *SUP16-o* and *SUP17-o* suppressors. It is possible that it is a tRNA^{Ser}, which selectively reads the UCA codon, since these suppressors are generated by anticodon mutation of a species that should possibly be designated tRNA$_{UCA}^{Ser}$, rather than, as previously, tRNA$_{UCG}^{Ser}$.

THE EFFECTS OF THE psi⁺ FACTOR AND OF ANTISUPPRESSOR MUTATIONS ON THE STRUCTURE AND FUNCTION OF YEAST SUPPRESSORS

The psi⁺ factor (Cox 1971) is a curious genetic element sometimes present in *S. cerevisiae*. It is inherited in non-Mendelian fashion and is known not to be associated with the mitochondrial DNA or the 2-μm circular DNA, although it is presumably a gene function of a cytoplasmic, self-replicating nucleic acid. It enhances severalfold the efficiency of tRNA ocher and some frameshift suppressors but is without effect on UAG or UGA suppressors, and it renders the tyrosine-inserting ocher suppressors so efficient that they become generally lethal to the cell (Liebman et al. 1975). Since the psi⁺ factor's mechanism of action was unknown, we investigated whether it exerted its effect by altering the molecular structure of certain of the tRNA molecules of the cell, and not just ocher suppressors, and we examined the modified nucleotide composition of tRNA fractions from isogenic psi⁺ and psi⁻ strains. Purified *SUP16-o* suppressor tRNAs from psi⁺ and psi⁻ strains are known to have comparable efficiencies as ocher suppressors in mammalian in vitro protein-synthesis systems (C. Waldron, unpubl.). In separating the minor components of RNase T2 digests on the two-dimensional thin-layer plates (Nishimura 1972), we could not detect any difference in the minor base compositions of tRNA^{Tyr}, tRNA^{Phe}, tRNA$^{Ser}_2$, tRNA$^{Ser}_{UCG}$, tRNA$^{Leu}_{UAA}$, and tRNA$^{Leu}_{UUG}$ from psi⁺ and psi⁻ strains (P. W. Piper, unpubl.). Unlinked nuclear mutations that modify suppressor function, antisuppressors, may affect levels and structures of suppressor tRNAs or they may operate by altering components of the ribosome. One antisuppressor mutation is known to cause gross undermodification of the A to i⁶A at the 3′ side of the anticodon of tRNA^{Tyr} (Laten et al. 1978).

CONCLUSIONS

The investigations described above have succeeded in establishing the molecular identity of some yeast tRNA nonsense suppressors; until recently, the nature of such suppressors was known only in prokaryotes. The tyrosine-inserting and serine-inserting ocher and amber suppressors of *S. cerevisiae* can now be purified to a reasonable degree of homogeneity for use in screening tests for nonsense mutants in other eukaryotic organisms. By introducing yeast suppressor tRNAs into a mammalian cell containing a nonsense mutation, it is possible to bring about a temporary correction of the genetic deficiency (Celis et al. 1979). So far, the molecular nature of UGA-suppressing tRNAs derived by single mutation in *S. cerevisiae* is unknown, although it is possible to obtain known UGA suppressors by further mutation of characterized ocher suppressors for use in in vitro tests for suppression in mammalian cells. Also unknown are the identities of the leucine-inserting ocher and amber suppressors of yeast. It should be possible to test whether these originate by mutation of UUG- and UUA-decoding tRNAs^{Leu} using suppressor assays in combination with RNA fingerprinting, since tRNA$^{Leu}_{UUG}$ and tRNA$^{Leu}_{UUA}$ are readily purified on two-dimensional gels (Piper and Wasserstein 1977).

ACKNOWLEDGMENTS

I thank K. Marcker, who kindly provided laboratory facilities for much of this work, during which period I was supported by a long-term European Molecular Biology Organization fellowship. I also thank S. Liebman and F. Sherman for supplying strains and C. Guthrie, S. Leibman, and C. Waldron for permission to cite their unpublished works.

REFERENCES

Anderson, K. W. and J. D. Smith. 1972. Still more mutant tyrosine transfer ribonucleic acids. *J. Mol. Biol.* **69:** 349.

Beckmann, J. S., P. F. Johnson, and J. Abelson. 1977. Cloning of yeast transfer RNA genes in *Escherichia coli*. *Science* **196:** 205.

Brandriss, M. C., L. Soll, and D. Botstein. 1975. Recessive lethal amber suppressors in yeast. *Genetics* **79:** 551.

Brandriss, M. C., J. W. Stewart, F. Sherman, and D. Botstein. 1976. Substitution of serine caused by a recessive lethal suppressor in yeast. *J. Mol. Biol.* **102:** 467.

Capecchi, M. R., S. H. Hughs, and G. M. Wahl. 1975. Yeast super-suppressors are altered tRNAs capable of translating a nonsense codon in vitro. *Cell* **6:** 269.

Celis, J. E., K. Kaltoft, A. Celis, R. G. Fenwick, and C. T. Caskey. 1979. Microinjection of tRNAs into somatic cells. In *Nonsense mutants and tRNA suppressors* (ed. J. E. Celis and J. D. Smith), p. 255. Academic Press, New York.

Celis, J. E., C. Coulondre, and J. H. Miller. 1976. Suppressor su+7 inserts tryptophan in addition to glutamine. *J. Mol. Biol.* **104:** 729.

Celis, J. E., M. L. Hooper, and J. D. Smith. 1973. Amino acid acceptor stem of *E. coli* suppressor tRNA^{tyr} is a site of synthetase recognition. *Nat. New Biol.* **244:** 261.

Comer, M. M., K. Foss, and W. H. McClain. 1975. A mutation of the wobble nucleotide of a bacteriophage T4 transfer RNA. *J. Mol. Biol.* **99:** 283.

Cox, B. S. 1971. A recessive lethal super-suppressor mutation in yeast and other Ψ phenomena. *Heredity* **26:** 211.

Gesteland, R. F., M. Wolfner, P. Grisafi, G. Fink, D. Botstein, and J. R. Roth. 1976. Yeast suppressors of UAA and UAG nonsense codons work efficiently in vitro via tRNA. *Cell* **7:** 381.

Goodman, H. M., M. V. Olson, and B. D. Hall. 1977. Nucleotide sequence of a mutant eukaryotic gene: The yeast tyrosine-inserting ocher suppressor SUP4-o. *Proc. Natl. Acad. Sci.* **74:** 5453.

Goodman, H. M., J. Abelson, A. Landy, S. Brenner, and J. D. Smith. 1968.

Amber suppression: A nucleotide change in the anticodon of a tyrosine transfer RNA. *Nature* **217**: 1019.

Gilmore, R. A., J. W. Stewart, and F. Sherman. 1971. Amino acid replacements resulting from super-suppression of nonsense mutants of iso-1-cytochrome *c* from yeast. *J. Mol. Biol.* **61**: 157.

Hawthorne, D. C. 1976. UGA mutations and UGA suppressors in yeast. *Biochimie* **58**: 179.

Hawthorne, D. C. and U. Leupold. 1974. Suppressors in yeast. *Curr. Top. Microbiol. Immunol.* **64**: 1.

Kaltoft, K., J. Zeuthen, F. Engbaek, P. W. Piper, and J. E. Celis. 1976. Transfer of tRNAs to somatic cells mediated by Sendai-virus-induced fusion. *Proc. Natl. Acad. Sci.* **73**: 2793.

Laten, H., J. Gorman, and R. M. Bock. 1978. N^6-isopentyladenosine deficiency in an antisuppressing strain of *Saccharomyces cerevisiae*. *Nucleic Acids Res.* **5**: 4329.

Liebman, S. W., F. Sherman, and J. W. Stewart. 1976. Isolation and characterization of amber suppressors in yeast. *Genetics* **82**: 251.

Liebman, S. W., J. W. Stewart, and F. Sherman. 1975. Serine substitutions caused by an ocher suppressor in yeast. *J. Mol. Biol.* **94**: 595.

Miller, C. G. and J. R. Roth. 1971. Recessive-lethal nonsense suppressors in *Salmonella typhimurium*. *J. Mol. Biol.* **59**: 63.

Nishimura, S. 1972. Minor components in transfer RNA: Their characterization, location, and function. *Prog. Nucleic Acid Res. Mol. Biol.* **12**: 50.

Olson, M. V., D. L. Montgomery, A. K. Hopper, G. S. Page, F. Horodyski, and B. D. Hall. 1977. Molecular characterisation of the tyrosine tRNA genes of yeast. *Nature* **267**: 639.

Piper, P. W. 1978. A correlation between a recessive lethal amber suppressor mutation in *Saccharomyces cerevisiae* and an anticodon change in a minor serine transfer RNA. *J. Mol. Biol.* **122**: 217.

Piper P. W. and M. Wasserstein. 1977. Separation of *Saccharomyces cerevisiae* tRNAs on two-dimensional polyacrylamide gels as applied to investigations on the mutational alterations of tRNA that produce nonsense suppressors. *Eur. J. Biochem.* **80**: 103.

Piper, P. W., M. Wasserstein, F. Engbaek, K. Kaltoft, J. E. Celis, J. Zeuthen, S. Liebman, and F. Sherman. 1976. Nonsense suppressors of *Saccharomyces cerevisiae* can be generated by mutation of the tyrosine tRNA anticodon. *Nature* **262**: 757.

Sherman, F., B. Ono, and J. W. Stewart. 1979. Use of the iso-1-cytochrome C system for investigating nonsense mutants and suppressors in yeast. In *Nonsense mutants and tRNA suppressors* (ed. J. E. Celis and J. D. Smith), p. 133. Academic Press, New York.

Sherman, F., S. W. Liebman, J. W. Stewart, and M. Jackson. 1973. Tyrosine substitutions resulting from suppression of amber mutants of iso-1-cytochrome *c* in yeast. *J. Mol. Biol.* **78**: 157.

Soll, L. 1974. Mutational alterations of tryptophan-specific transfer RNA that generate translation suppressors of the UAA, UAG, and UGA nonsense codons. *J. Mol. Biol.* **86**: 233.

Soll, L. and P. Berg. 1969. Recessive lethals: A new class of nonsense suppressors in *Escherichia coli*. *Proc. Natl. Acad. Sci.* **63**: 392.

Yaniv, M., W. R. Folk, P. Berg, and L. Soll. 1974. A single mutational modification of a tryptophan-specific transfer RNA permits aminoacylation by glutamine and translation of the codon UAG. *J. Mol. Biol.* **86:**245.

Zachau, H. G., D. Dutting, and H. Feldmann. 1966. Structures of two serine tRNAs from brewer's yeast. *Angew Chem. Int. Ed. Eng.* **5:**422.

i⁶A-deficient tRNA from an Antisuppressor Mutant of *Saccharomyces cerevisiae*

Howard Laten, John Gorman, and Robert M. Bock
Department of Biochemistry and Laboratory of Molecular Biology
University of Wisconsin
Madison, Wisconsin 53706

The determination of the functional roles of modified nucleotides in tRNA has been the goal of numerous investigations (most recently reviewed by McCloskey and Nishimura 1977; Agris and Söll 1977). Nearly all attempts to assess minor base function have suffered from the same shortcomings. First, mutations in tRNA modifying enzymes have been difficult to select. Second, the use of in vitro techniques to create specific alterations in tRNA modification and assess their effects on tRNA function may lead to ambiguous results (see Discussion).

Our goal was to screen for strains of the yeast *Saccharomyces cerevisiae* with mutations in genes for tRNA modifying enzymes. The genetic screening for these mutants was based on the probable correlation between loss of base modification and reduction in the efficiency of suppression of a UAA nonsense suppressing tRNA. The existence of such a correlation was suggested by early in vitro findings (Gefter and Russell 1969) and has since been supported by recent in vivo studies (Colby et al. 1976; Marinus et al. 1975).

We have isolated a mutant that contains 1.5% of the normal tRNA complement of i⁶A (isopentenyladenosine). This mutant affords the opportunity to compare in vitro results with in vivo observations. The mutation, which has been designated *mod5-1*, reduces the suppression by a dominant UAA suppressor, *SUP7-1* (Gilmore 1967), so that only the more easily suppressed of several UAA mutations are suppressed. *SUP7-1* is one of several efficient tyrosine-inserting UAA suppressors and most probably codes for an altered tRNA^{Tyr} (Gilmore et al. 1971; Capecchi et al. 1975; Gesteland et al. 1976; Piper et al. 1976; Olson et al. 1977; Goodman et al. 1977). Genes corresponding to these suppressors have been genetically mapped at eight different loci in the yeast genome (Gilmore 1967; Hawthorne and Mortimer 1968); yeast tRNA^{Tyr} hybridizes to eight distinct yeast DNA fragments generated by *Eco*RI digestion (Olson et al. 1977). This tRNA normally contains i⁶A adjacent to the 3' end of the anticodon, as do yeast tRNA$_1^{Ser}$, tRNA$_2^{Ser}$, and tRNA^{Cys}.

i⁶A is one of a few hypermodified derivatives of A that have been found in the tRNA of almost all organisms investigated (Hall 1971). Its location is restricted to the position adjacent to the 3' end of the anticodons of most tRNA species that recognize codons beginning with U (McCloskey and Nishimura 1977). A number of investigations have been undertaken to determine the role of this modified nucleoside in tRNA function. Fittler and Hall (1966) demonstrated that chemical modification of the i^{6}A moiety in yeast tRNA^{Ser} with iodine reduced the level of ribosome binding of the tRNA in response to a synthetic message. The level of in vitro aminoacylation of the modified tRNA was not altered. An analogous study involving bisulfite treatment of yeast tRNA^{Tyr} resulted in similar findings (Furuichi et al. 1970). Gefter and Russell (1969) isolated tRNA^{Tyr} deficient in ms^{2}i^{6}A from *Escherichia coli* infected with a transducing bacteriophage carrying the tRNA gene. They demonstrated that i^{6}A is required for the efficient binding of tRNA^{Tyr} to ribosomes in the presence of the appropriate codons. They also found that the kinetics of in vitro aminoacylation was not affected by the absence of i^{6}A. In a study involving the formation and dissociation of complexes between tRNAs with complementary anticodons, Grosjean et al. (1976) examined the relationship between modification of the purine adjacent to the 3' end of the anticodon of tRNA^{Phe} and the stability of the complex. They found that tRNA^{Phe} from *E. coli* (with ms^{2}i^{6}A) complexed with tRNA^{Glu} from *E. coli* as did tRNA^{Phe} from yeast (with base Y). Complexes between tRNA^{Phe} from *Mycoplasma* sp. (kid), which lacks a hypermodified constituent, and *E. coli* tRNA^{Glu}, and between yeast tRNA^{Phe} with base Y removed and *E. coli* tRNA^{Glu} were less stable. Thus, hypermodification of the base adjacent to the 3' end of the anticodon was correlated with an enhancement in the stability of anticodon-anticodon interaction. Litwack and Peterkofsky (1971) isolated tRNA with reduced levels (50%) of i^{6}A from a mevalonate-requiring mutant of *Lactobacillus acidophilus*. They were unable, however, to separate the i^{6}A-deficient tRNAs from the fully modified species. Contrary to the other reports, they found no difference in the level of polynucleotide-directed amino acid incorporation from specific tRNA species into protein when partially modified and fully modified tRNAs were compared. In addition, Kimball and Söll (1974) demonstrated that tRNA^{Phe} from *Mycoplasma* sp. (kid) was fully active in promoting phenylalanine incorporation in an amino-acid-incorporating system that is cell-free, tRNA-dependent, and polyuridylic-acid-directed, even though this tRNA species lacks i^{6}A or any related hypermodified base. These last two studies suggest that there is no requirement for i^{6}A or related modifications for in vitro polypeptide synthesis.

SCREENING FOR MUTANTS DEFECTIVE IN tRNA FUNCTION

The strategy for selection of mutants characterized by a reduction in efficiency of class-I ocher-specific suppressors in *S. cerevisiae* was analogous to that described by McCready and Cox (1973). The parent strain, 700:1, was canavanine-sensitive and independent of adenine, histidine, lysine, and tryptophan by virtue of the suppression by *SUP7-1* of the UAA nonsense mutations *can1-100, ade2-1, his5-2, lys1-1,* and *trp5-48.* Mutants with partial loss of suppression were selected either as canavanine-resistant or adenine-requiring revertants. Most antisuppressor mutants, including *mod5-1,* remained independent of histidine, lysine, and tryptophan.

A total of 150 mutants were isolated and placed into eight genetic complementation groups. The loci defined by the mutants were not linked to each other, nor were any linked to *SUP7-1.*

PHENOTYPIC CHARACTERISTICS OF *mod5-1* MUTANTS

The *mod5-1* mutation is recessive to the wild-type allele, segregates as a nuclear gene, and is unlinked to *SUP7-1*. Other than reducing suppressor efficiency, the mutation appeared to have little effect on its host, although *mod5-1* homozygotes failed to sporulate. Mutant strains grew as well as nonmutants in both complex (peptone, yeast extract, and dextrose) and simple (vitamins, salts, dextrose, and nutritional supplements) media at both moderate (28°C) and elevated (37°C) temperatures. Generation times and cell yields did not differ significantly.

BASE COMPOSITION FOR tRNA FROM MUTANT STRAINS

The nucleoside compositions of unfractionated tRNA extracted from six mutants were first determined by thin-layer chromatography of ^{3}H-labeled nucleoside derivatives (Randerath et al. 1974). There were no significant differences in nucleoside composition with respect to those nucleosides detectable by this procedure (Table 1).

Quantitation of i^6A is not amenable to the thin-layer procedure and was facilitated by Sephadex-LH20 chromatography. The elution profile in Figure 1A is representative of those recorded for tRNA digests from several nonmutant controls (with and without *SUP7-1*) and five of six mutants. The concentration of i^6A in unfractionated tRNA from these strains ranged from 60 mmoles/mole of tRNA to 120 mmoles/mole of tRNA. The elution profile in Figure 1B is representative of those recorded for tRNA digests from several different strains, each carrying the antisuppressor mutation *mod5-1*. Unfractionated tRNA from each of these strains yielded approximately 1.2 mmoles of i^6A per mole of tRNA, or about 1.5% of the above concentrations.

CORRELATION BETWEEN ANTISUPPRESSOR PHENOTYPE AND i^6A REDUCTION

To confirm the correlation of antisuppressor phenotype with i^6A deficiency, tRNA was isolated from 14 different yeast segregants

Table 1 Base compositions of unfractionated tRNAs from mutant strains

Nucleoside	tRNA source (%)					
	mod1	mod2	mod3	mod4	mod5	mod6
A	18.8	19.5	19.2	18.9	19.1	18.5
C	25.6	24.9	25.3	25.2	24.9	25.6
G	26.2	26.9	26.7	26.7	27.2	26.7
U	17.2	16.2	17.2	17.2	17.0	17.3
m^{1}A	0.73	0.72	0.60	0.68	0.52	0.59
t^{6}A	0.34	0.34	0.27	0.33	0.35	0.27
I	0.57	0.43	0.42	0.51	0.49	0.42
m^{3}C	0.09	0.06	0.07	0.09	0.06	0.07
m^{5}C	1.21	1.15	1.10	1.21	0.91	1.11
m^{1}G	0.68	0.79	0.69	0.67	0.63	0.66
m^{2}G	0.91	0.79	0.88	0.71	0.89	0.71
m$^2_2$G	0.61	0.61	0.52	0.64	0.57	0.59
m^{7}G	0.55	0.46	0.39	0.55	0.57	0.42
T	0.74	0.76	0.76	0.76	0.81	0.70
ψ	2.76	3.09	2.71	2.73	3.18	2.96
D	3.01	3.26	3.19	3.17	2.83	3.35

Nucleoside composition analysis by two-dimensional thin-layer chromatography of ^{3}H-labeled derivatives as described by Randerath et al. (1974). Data from two analyses for each tRNA sample. Values for the 4 major nucleosides are < ±4%; values for all others are < ±10%.

identified genetically as *mod5-1* and from five nonmutant segregants. All 19 segregants were derived from the same cross by dissection of individual yeast asci. Segregants resulted from the following diploid:

$$\frac{SUP7\text{-}1\,MOD5^+\,ade2\text{-}1\,his5\text{-}2\,lys1\text{-}1\,trp5\text{-}48}{sup^+\,mod5\text{-}1\,ade2\text{-}1\,his5\text{-}2\,lys1\text{-}1\,trp5\text{-}48} \qquad (1)$$

(Although a diploid homozygous for *SUP7-1* would have facilitated the identification of *mod5-1* segregants, such diploids sporulated poorly.) *SUP7-1 mod5-1* segregants required adenine and were canavanine resistant; *SUP7-1 MOD5$^+$* segregants were adenine independent and canavanine sensitive. In the absence of *SUP7-1*, the identification of *mod5-1* could not be directly made because mutant and nonmutant strains were indistinguishable in a genetic background lacking a class-I suppressor. Nonsuppressor segregants were therefore crossed to *SUP7-1 mod5-1* strains of opposite mating type and the phenotypes of the resulting diploids were analyzed. In the case of *sup$^+$ mod5-1* segregants, diploids heterozygous for *SUP7-1* and homozygous for *mod5-1* proved to require adenine and histidine; in the case of *sup$^+$ MOD5$^+$* segregants,

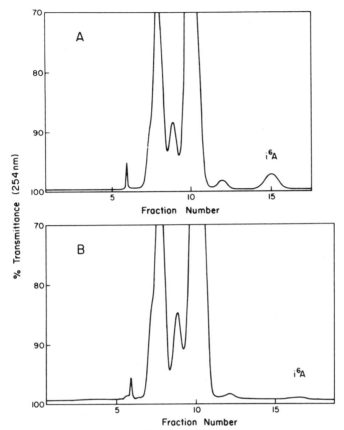

Figure 1 Separation of i⁶A from a mixture of tRNA nucleosides on Sephadex-LH20. (A) tRNA digest from nonmutant control; (B) tRNA digest from *mod5-1*. (Reprinted, with permission, from Laten et al. 1978.)

diploids heterozygous for both *SUP7-1* and *mod5-1* proved to require adenine but not histidine.

tRNA from 13 of the *mod5-1* segregants was combined into three pools; two pools each contained tRNA from four different segregants and the third pool contained tRNA from the remaining five segregants. If tRNA from any single mutant segregant contained the normal complement of i⁶A, the concentration of i⁶A in the pooled sample was expected to be at least 20% of that in the nonmutants. Nonmutant tRNA was not pooled. The concentrations of i⁶A in the pooled samples were approximately the same as that in the unpooled tRNA isolated from a single *mod5-1* segregant (Table 2), approximately 1.5% of nonmutant levels. These results indicated that the mutant segregants were uniformly deficient in i⁶A and that the loss of suppressor function was correlated with the reduction of i⁶A.

Table 2 Levels of i^{6}A in tRNA from yeast segregants genetically defined with respect to *mod5-1*

tRNA source	i^{6}A (mmoles/mole tRNA)
a (*mod5*)	1.1
b (*mod5*)	1.2
c (*mod5*)	1.4
1 (*mod5*)	1.2
2 (nonmutant)	82
3 (nonmutant)	118
4 (nonmutant)	81
5 (nonmutant)	99

For samples a and c, tRNA from each of four different *mod5-1* strains was combined; for sample b, tRNA from five different *mod5-1* strains was combined. In all other determinations, tRNA from different individual segregants derived from the same cross was used.

CHROMATOGRAPHIC BEHAVIOR OF i^{6}A-DEFICIENT tRNAs

Because of the strongly hydrophobic character of i^{6}A, the chromatographic behavior of those tRNA species normally containing this nucleoside should be greatly altered when the modification is absent. tRNA species containing i^{6}A are significantly retarded on BD-cellulose (benzoylated DEAE-cellulose) subjected to an NaCl gradient (Gillam et al. 1967). Figure 2 illustrates that tRNA^{Tyr} (Fig. 2A) and tRNA^{Ser} (Fig. 2B) species from the *mod5-1* mutant eluted from the BD-cellulose column much earlier than the corresponding fully modified tRNA species from a nonmutant strain. The behavior of arginine-accepting tRNA species, which do not contain i^{6}A, was not influenced by the presence of the *mod5-1* mutation (Fig. 2C). There was also no change in the chromatographic behavior of tRNA^{Phe}, which contains base Y adjacent to the 3' end of the anticodon (not shown). The major peak of tyrosine acceptor activity for the nonmutant in Figure 2A is considerably smaller than the peak for the mutant because tyrosine charging was inhibited by high salt concentrations present in assays of late column fractions (Kirkegaard 1969). The concentrations of the tyrosine acceptor activities in unfractionated tRNA from mutant and nonmutant strains were, in fact, comparable (see below). The trace of unmodified tRNA^{Tyr} in the nonmutant preparation most probably reflected incomplete isopentenylation in logarithmic phase cells. Although fully modified tRNA^{Ser} eluted from BD-cellulose in a single peak, this peak contains several unresolved

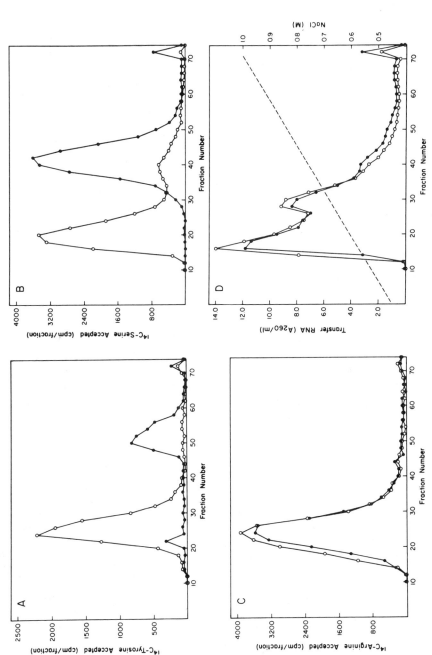

Figure 2 Comparison of tRNAs derived from nonmutant and *mod5-1* cells by BD-cellulose chromatography. (*A*) Tyrosine acceptance; (*B*) serine acceptance; (*C*) arginine acceptance; (*D*) absorbance profile. (●) tRNA derived from the nonmutant; (○) tRNA derived from the *mod5-1* mutant; (---) NaCl concentration. (Reprinted, with permission, from Laten et al. 1978.)

isoaccepting species. At least one of these, the one that would be expected to translate AGU and AGC codons, should not normally contain i^{6}A. In Figure 2B, most but not all of the serine-accepting species are shifted in the mutant preparation. The small amount of unshifted material was most likely due to this tRNA^{Ser}. The shift in the elution profiles of some species to earlier positions is also reflected in the absorbance profile (Fig. 2D). A portion (20%) of the tRNA from the mutant eluted from the column after fraction 34, whereas 27% of the tRNA from the nonmutant eluted after this fraction. This 7% difference is roughly equivalent to the sum of tyrosine, serine, and cysteine tRNAs found in unfractionated tRNA from yeast.

CYTOKININ ACTIVITY IN i^{6}A-DEFICIENT tRNA

i^{6}A and its related derivatives promote cell division (cytokinesis) in plants (Skoog and Armstrong 1970; Hall 1973). Although these cytokinins, as they are called, are present in virtually all organisms (Hall 1971), their hormonal activity has been conclusively demonstrated only in higher plants (Skoog and Armstrong 1970; Hall 1973). It has been suggested that tRNA is the primary, if not sole, source of naturally occurring cytokinins (Skoog et al. 1966). The only tRNA constituent in yeast known to have cytokinin activity is i^{6}A (Armstrong et al. 1969; Swaminathan 1975); a reduction in cytokinin activity should parallel the reduction in i^{6}A.

To further corroborate the reduction of i^{6}A and to determine if the i^{6}A-deficient tRNA exhibited significant cytokinin activity, mixtures of bases generated from unfractionated tRNA were assayed for their ability to stimulate plant tissue growth (Linsmaier and Skoog 1965; Skoog et al. 1966). The results of the tobacco-tissue-culture bioassay are presented in Table 3. tRNA from the nonmutant strain contained 0.63 kinetin

Table 3 Cytokinin activity in tRNA from nonmutant and *mod5-1* strains

Nonmutant strain						
tRNA conc. (mg/l)	350	70	14	2.8	0.56	0.11
Average fresh weight (gm/flask)	5.88	7.39	5.75	1.10	0.43	n.g.[a]
Cytokinin activity (KE/mg tRNA)	—	—	0.62	—	—	—
mod5-1 strain						
tRNA conc. (mg/l)	375	75	15	3	0.60	0.12
Average fresh weight (gm/flask)	8.00	2.45	0.41	n.g.	n.g.	n.g.
Cytokinin activity (KE/mg tRNA)	—	0.04	—	—	—	—

Kinetin standards								
Kinetin conc. (μg/ml)	0	1	2	3	5	7	10	15	20
Average fresh weight (gm/flask)	0.6	1.1	1.9	2.6	4.1	4.2	6.3	7.4	6.4

[a]No growth.

equivalents (KE; μg of kinetin required to give the same growth stimulation as the test sample) per milligram of tRNA. Because i^{6}A is ten times as potent as kinetin in this assay (Skoog et al. 1967), this represents 62 ng of i^{6}A/mg tRNA or 9.2 mmoles of i^{6}A/mole of tRNA. tRNA from the *mod5-1* strain contained 0.04 KE/mg tRNA or 0.6 mmoles of i^{6}A/mole of tRNA, 6.5% of that in the nonmutant. These concentrations are lower than those determined spectrophotometrically on Sephadex-LH20-purified i^{6}A, but the relationship between mutant and nonmutant i^{6}A levels is still maintained. The low values obtained from the bioassay are probably the result of chemical degradation during reactions required for sample preparation prior to the assay.

AMINO ACID ACCEPTANCE ACTIVITIES OF MUTANT tRNAs

Unfractionated tRNA from *mod5-1* and nonmutant cells were assayed for tyrosine- and serine-acceptor activities. tRNA from nonmutant cells accepted 26 and 53 pmoles/A$_{260}$ unit of tRNA for tyrosine and serine, respectively. From *mod5-1* cells, tRNA accepted 24 and 53 pmoles/A$_{260}$ unit of tRNA for tyrosine and serine, respectively. Thus, the loss of i^{6}A has no apparent effect on the level of amino acid charging of two tRNAs that normally contain this modification.

DISCUSSION

The application of in vitro findings to the problems of in vivo functions is limited by the difficulty of duplicating a cellular environment in a reaction mixture. Unlike previous efforts to determine the function of the i^{6}A modification in cell-free systems (Fittler and Hall 1966; Gefter and Russell 1969; Furuichi et al. 1970; Litwack and Peterkofsky 1971; Kimball and Söll 1974; Grosjean et al. 1976), we have generated a mutant that affords the opportunity to study the effect of i^{6}A deficiency in growing cells. That *mod5-1* cells apparently grow as well as nonmutant cells under conditions where suppression is not required for growth suggests that the i^{6}A-deficient tRNAs do not limit the rate of protein synthesis in the mutant. The efficiency of suppression by *SUP7-1*, however, may be a more sensitive indicator of the limitations imposed by i^{6}A-deficient tRNAs with respect to protein synthesis. The suppressor tRNA constitutes only a fraction of the total tRNA^{Tyr} and must compete with termination factors responding to the nonsense mutation. Thus, suppressor tRNA concentration could be growth limiting, or nearly so, under conditions requiring suppressor function; impairment of tRNA function would visibly affect growth under these conditions. The reduced suppressor efficiency in the mutant, as measured by growth on selective media, most probably results from the impaired functioning of the i^{6}A-deficient suppressor tRNA in protein synthesis. It does not result from decreased levels of i⁶A-deficient tRNA. Unfractionated tRNA from the mutant contains as much tRNA^Tyr and tRNA^Ser as nonmutant tRNA. The i⁶A deficiency, therefore, does not affect the relative rates of tRNA synthesis and degradation. In addition, *mod5-1* cells do not differ from nonmutant cells in accumulation of tRNA precursors (I. Park, pers. comm.). Our finding that the acceptor activities of the above two species are not reduced by the loss of i⁶A suggests that the translational deficiency in vivo is related to tRNA interactions subsequent to aminoacylation (ribosomal binding and codon-anticodon or elongation factor interactions). This conclusion is in agreement with the findings of Fittler and Hall (1966), Furuichi et al. (1970), Gefter and Russell (1969), and Grosjean et al. (1976). On the other hand, the negative findings of Litwack and Peterkofsky (1971) and Kimball and Söll (1974) are not unexpected in view of the subtle in vivo consequences of the reduction in i⁶A.

An i⁶A-deficient mutant has recently been found among a collection of antisuppressor mutants of *Schizosaccharomyces pombe* (G. Vögeli, pers. comm.). The mutation appears to be analogous to *mod5-1*. In addition, the *trpX* mutation in *E. coli* leads to the loss of ms²i⁶A in all tRNAs reading codons with U in the first position (Eisenberg et al., this volume). The observed derepression of *trp* operon enzymes in the *trpX* mutant (Yanofsky and Soll 1977) is apparently due to the inability of tRNA^Trp lacking ms²i⁶A to efficiently translate tandem tryptophan codons in the *trp* leader sequence RNA (Eisenberg et al., this volume).

SUMMARY

We have isolated a mutant of *S. cerevisiae* that contains 1.5% of the normal tRNA complement of i⁶A. The mutant was characterized by the reduction in efficiency of a tyrosine-inserting UAA nonsense suppressor. The chromatographic profiles of tRNA^Tyr and tRNA^Ser on BD-cellulose are consistent with the loss of i⁶A by these species. tRNA from the mutant exhibits 6.5% of the cytokinin biological activity expected for yeast tRNA. tRNAs from the mutant that normally contain i⁶A accept the same levels of amino acids in vitro as the fully modified species. With the exception of i⁶A, the levels of modified bases in unfractionated tRNA from the mutant appear to be normal. The loss of i⁶A apparently affects tRNA's role in protein synthesis at a step subsequent to aminoacylation.

ACKNOWLEDGMENTS

We thank F. H. Webb for excellent technical assistance and F. Skoog and N. Murai for their help with the cytokinin bioassay. The work was supported by National Institutes of Health grant GM-12395 and National Science Foundation grant GB-41275.

REFERENCES

Agris, P. F. and D. Söll. 1977. The modified nucleosides in transfer RNA. In *Nucleic acid-protein recognition* (ed. H. J. Vogel), *P&S Biomedical Sciences Symposia*, vol. 1, p. 321. Academic Press, New York.

Armstrong, D. J., F. Skoog, L. H. Kirkegaard, A. E. Hampel, R. M. Bock, I. Gillam, and G. M. Tener. 1969. Cytokinins: Distribution in species of yeast transfer RNA. *Proc. Natl. Acad. Sci.* **63**:504.

Capecchi, M. R., S. H. Hughes, and G. M. Wahl. 1975. Yeast supersuppressors are altered tRNAs capable of translating a nonsense codon *in vitro*. *Cell* **6**:269.

Colby, D. S., P. Schedl, and C. Guthrie. 1976. A functional requirement for modification of the wobble nucleotide in the anticodon of a T4 suppressor tRNA. *Cell* **9**:449.

Fittler, F. and R. H. Hall. 1966. Selective modification of yeast seryl tRNA and its effect on the acceptance and binding functions. *Biochem. Biophys. Res. Commun.* **25**:441.

Furuichi, Y., Y. Wataya, H. Hayatsu, and T. Yukita. 1970. Chemical modification of yeast tyrosyl tRNA with bisulfite. A new method to modify isopentenyladenosine residue. *Biochem. Biophys. Res. Commun.* **41**:1185.

Gefter, M. and R. L. Russell. 1969. Role of modifications in tyrosine transfer RNA: A modified base affecting ribosome binding. *J. Mol. Biol.* **39**:145.

Gesteland, R. F., M. Wolfner, M. Grisafi, G. Fink, D. Botstein, and J. R. Roth. 1976. Yeast suppressors of UAA and UAG nonsense codons work efficiently *in vitro* via tRNA. *Cell* **7**:381.

Gillam, I. C., S. Millward, D. Blew, M. von Tigerstrom, E. Wimmer, and G. M. Tener. 1967. The separation of soluble ribonucleic acids on benzoylated diethylaminoethylcellulose. *Biochemistry* **6**:3043.

Gilmore, R. A. 1967. Super-suppressors in *S. cerevisiae*. *Genetics* **56**:641.

Gilmore, R. A., J. W. Stewart, and F. Sherman. 1971. Amino acid replacements resulting from super-suppression of nonsense mutants of iso-l-cytochrome c from yeast. *J. Mol. Biol.* **61**:157.

Goodman, H. M., M. V. Olson, and B. D. Hall. 1977. Nucleotide sequence of a mutant eukaryotic gene: The yeast tyrosine-inserting ochre suppressor *SUP4-o*. *Proc. Natl. Acad. Sci.* **74**:5453.

Grosjean, H., D. G. Söll, and D. M. Crothers. 1976. Studies of the complex between transfer RNAs with complementary anticodons. I. Origins of enhanced affinity between complementary triplets. *J. Mol. Biol.* **103**:499.

Hall, R. H. 1971. *The modified nucleosides in nucleic acids*. Columbia University Press, New York.

———. 1973. Cytokinins as probe of developmental processes. *Annu. Rev. Plant Physiol.* **24**:415.

Hawthorne, D. C. and R. K. Mortimer. 1968. Genetic mapping of nonsense suppressors in yeast. *Genetics* **60**:735.

Kimball, M. E. and D. Söll. 1974. The phenylalanine tRNA from *Mycoplasma* sp. (kid): A tRNA lacking hypermodified nucleosides functional in protein synthesis. *Nucleic Acids Res.* **1**:1713.

Kirkegaard, L. H. 1969. "I. Large scale purification of yeast tRNA. II. Procedures and problems in the amino acid acceptor assay." Ph.D. thesis, University of Wisconsin, Madison.

Laten, H., J. Gorman, and R. M. Bock. 1978. Isopentenyladenosine-deficient tRNA from an antisuppressor mutant of *S. cerevisiae*. *Nucleic Acids Res.* **5**: 4329.

Linsmaier, E. M. and F. Skoog. 1965. Organic growth factor requirements of tobacco tissue cultures. *Physiol. Plant.* **18**: 100.

Litwack, M. and A. Peterkofsky. 1971. Transfer ribonucleic acid deficient in N^6-(Δ^2-isopentenyl) adenosine due to mevalonic acid limitation. *Biochemistry* **10**: 994.

Marinus, M. G., N. R. Morris, D. Söll, and T. C. Kwong. 1975. Isolation and partial characterization of three *E. coli* mutants with altered transfer ribonucleic acid methylases. *J. Bacteriol.* **122**: 257.

McCloskey, J. A. and S. Nishimura. 1977. Modified nucleosides in transfer RNA. *Accts. Chem. Res.* **10**: 403.

McCready, S. J. and B. S. Cox. 1973. Antisuppressors in yeast. *Mol. Gen. Genet.* **124**: 305.

Olson, M. V., D. L. Montgomery, A. K. Hopper, G. S. Page, F. Horodyski, and B. D. Hall. 1977. Molecular characterization of the tyrosine tRNA genes of yeast. *Nature* **267**: 639.

Piper, P., M. Wasserstein, F. Engbaek, K. Kaltoft, J. Celis, J. Zeuthen, S. Liebman, and F. Sherman. 1976. Nonsense suppressors of *S. cerevisiae* can be generated by mutation of the tyrosine tRNA anticodon. *Nature* **262**: 757.

Randerath, K., E. Randerath, L. S. Y. Chia, and B. Nowak. 1974. Base analysis of ribonucleotides by chemical tritium labelling: An improved mapping procedure for nucleoside trialcohols. *Anal. Biochem.* **59**: 263.

Skoog, F. and D. J. Armstrong. 1970. Cytokinins. *Annu. Rev. Plant Physiol.* **21**: 359.

Skoog, F., D. J. Armstrong, J. D. Cherayil, A. E. Hampel, and R. M. Bock. 1966. Cytokinin activity: Localization in transfer RNA preparations. *Science* **154**: 1354.

Skoog, F., H. Hamzi, A. Szweykowska, N. Leonard, K. Carraway, T. Fujii, J. Helgeson, and R. Loeppky. 1967. Cytokinins: Structure/activity relationships. *Phytochemistry* **6**: 1169.

Swaminathan, S. 1975. "Studies on the structure, subcellular localization and biochemical effects of cytokinins in tRNAs." Ph.D. thesis, University of Wisconsin, Madison.

Yanofsky, C. and L. Soll. 1977. Mutations affecting tRNA^{Trp} and its charging and their effect on regulation of transcription termination at the attenuator of the tryptophan operon. *J. Mol. Biol.* **113**: 663.

Nonsense Suppressor tRNA in *Schizosaccharomyces pombe*

Jürg Kohli, Fiorella Altruda, Tai Kwong, Antoni Rafalski, Ronald Wetzel, and Dieter Söll
Department of Molecular Biophysics and Biochemistry
Yale University
New Haven, Connecticut 06520

Geoffrey Wahl
Department of Biology
University of Utah
Salt Lake City, Utah 84112

Urs Leupold
Institute of General Microbiology
University of Bern
Bern, Switzerland

The study of informational suppression in prokaryotes played an important role in the elucidation of the genetic code and of the detailed mechanism of protein biosynthesis (for a review, see Steege and Söll 1978). Some areas in the molecular biology of bacteria have benefited greatly from studies of suppression: the unraveling of the steps of tRNA biosynthesis, the formation and function of modified nucleotides in tRNA, and structure-function relations in tRNA (for reviews, see Smith 1976; McClain 1977; Steege and Söll 1978). Our understanding of the modes of tRNA recognition by aminoacyl-tRNA synthetases (Söll and Schimmel 1974; Ozeki et al., this volume) or by tRNA-modifying enzymes (McClain 1977) has been refined by the genetic opportunities provided by nonsense suppression. Our knowledge of informational suppression in eukaryotes is most advanced in yeasts. An excellent summary of the earlier genetic data has been provided by Hawthorne and Leupold (1974). In the past few years, biochemical studies have begun to characterize the suppression mechanism, mostly in *Saccharomyces cerevisiae*. In this organism, nonsense suppressor tRNAs of the amber (UAG) and ocher (UAA) types have been characterized in vitro (Capecchi et al. 1975; Gesteland et al. 1976). Both tyrosine- and serine-inserting amber suppressor tRNAs have been sequenced (Piper, this volume), and the nucleotide sequence of the gene for an ocher suppressor tRNA^{Tyr} is known (Goodman et al. 1977). This report reviews the state of suppressor genetics in *Schizosaccharomyces pombe* and describes the identification and the nucleotide sequences of two opal suppressor tRNAs from this organism.

GENETIC CHARACTERIZATION OF NONSENSE SUPPRESSORS

A simplified scheme describing the status of suppressor genetics of *S. pombe* is presented in Figure 1. The nonsense suppressors were obtained by reversion of putative nonsense mutations in a number of genes (Barben 1966). In this report we restrict the discussion of nonsense mutations to those in genes *ade6* and *ade7*, which confer not only adenine auxotrophy but also a red color (by accumulation of a red pigment) to *S. pombe* cells. Efficient suppression of such nonsense mutations by an additional mutation in a suppressor gene, which is not closely linked to *ade6* or *ade7*, restores wild-type-like, white colony color as well as adenine prototrophy.

Independent suppressor isolates were then checked for their suppression patterns against all available nonsense mutations. The resulting classification of suppressors is outlined in Table 1. The rather weak suppressors *sup1* and *sup2* suppress both ocher and opal mutations. When considered with other genetic data, this fact indicates that *sup1* and *sup2* do not code for tRNAs. Possibly they define structural genes for ribosomal proteins like the *ram* suppressors in *Escherichia coli* (Rosset and Gorini 1969).

The efficient suppressor isolates specific for opal mutations were allocated to four individual genes. The locations of three of these genes on the chromosome maps of *S. pombe* have been described (Kohli et al. 1977). These four opal suppressors could be classified into two groups (*sup3-e* and *sup9-e* vs *sup8-e* and *sup10-e*) due to slight differences in their suppression patterns, differential sensitivity to antisuppressors (Table

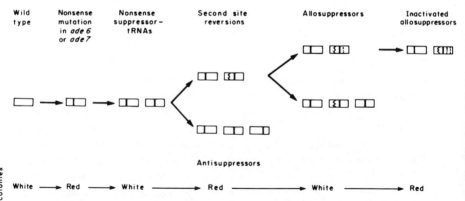

Figure 1 Suppressor genetics in *S. pombe*. (□) Individual genes; (→) single mutation steps; (▱) resulting mutations. At the bottom, the phenotypes of the strains are indicated. Red (auxotroph) colonies occur due to expression of the nonsense mutations in *ade6* or *ade7*, whereas white stands for the (prototroph) wild-type or the suppressed state of the *ade* mutation.

Table 1 Phenotypes of nonsense suppressors and antisuppressors

Suppressors	Suppression of nonsense mutants		Sensitivity to antisuppressors				
	opal	ocher	sin1	sin7	sin8	cyh1	sin10
sup1[a]	(+)[b]	(+)	(+)	−[c]	−	+[d]	+
sup2[a]	(+)	(+)	(+)	−	−	−	+
sup3-e	+	−	+	−	(+)	−	−
sup9-e	+	−	+	−	+	−	−
sup8-e	+	−	−	−	+	−	−
sup10-e	+	−	−	−	+	−	−
sup3-i	−	(+)	+	+	+	−	−
sup8-i	−	(+)	−	−	+	−	−

[a] Not tRNA; all others are tRNA suppressors.
[b] (+) Denotes weak.
[c] − Denotes absence of suppression or antisuppression.
[d] + Denotes strong.

1), and unique properties of their fine-structure maps (Hawthorne and Leupold 1974).

To obtain specific ocher suppressors, both opal suppressors *sup3-e* and *sup8-e* were crossed into strains carrying ocher mutations and subjected to selection for ocher suppression. In this way *sup3-i* and *sup8-i* were obtained; both are inefficient but specific ocher suppressors, derived from their respective parent opal suppressors by single-base substitutions (Hawthorne and Leupold 1974). This result, as well as other genetic data, strongly suggested a tRNA nature for *sup3*, *sup8*, *sup9*, and *sup10*.

FINE-STRUCTURE MAPS OF SUPPRESSOR tRNA GENES

As the next step in the genetic analysis of suppressor tRNA genes, suppressor strains were mutagenized and clones that lacked suppressor function were isolated. Two interesting classes of new mutations were further characterized: antisuppressors and second-site reversions within the suppressor tRNA genes (Fig. 1). These additional mutations within or close to the tRNA gene abolish suppressor function, possibly by interfering with tRNA biosynthesis or tRNA function. Such mutants constitute valuable tools for the detailed study of these processes. These second-site mutations made the construction of fine-structure maps of tRNA genes possible (Hawthorne and Leupold 1974; F. Hofer et al.; P. Munz et al.; both unpubl.). As an example, the maps of *sup3-e* are given in Figure 2. They were obtained by measuring frequencies of recombination in meiosis as well as by determining the rates of methylmethanesulfonate-

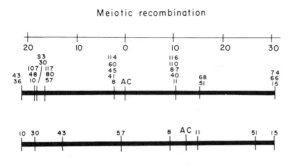

Figure 2 Genetic fine-structure maps of *sup3-e*. For details on tRNA fine-structure mapping, see Hawthorne and Leupold (1974). The scale for the meiotic recombination map refers to the frequency of prototrophic recombinants per 10^6 ascospores. The map obtained from methylmethanesulfonate-induced mitotic recombination was given the same total length (actual recombination data are not shown). This facilitates the comparison of the relative order and distances of sites in both maps.

induced mitotic recombination. The order of the mutation sites agrees fairly well between the two maps, whereas the relative distances between sites vary considerably. Mutations at a few specific sites showed pronounced marker effects, that is, abnormally high recombination frequencies as compared to the majority of sites (Munz and Leupold 1979; P. Thuriaux et al., unpubl.). These phenomena are a reflection of some intrinsic properties of the recombination mechanisms involved, which may be revealed by comparing the DNA sequences of the mutant suppressor genes in question.

CHARACTERIZATION OF ANTISUPPRESSORS AND ALLOSUPPRESSORS

Parallel with the second-site reversions, strains were obtained that had lost suppressor function due to mutations in genes unlinked to the suppressor tRNA gene (Fig. 1). These antisuppressor mutations were found to map in a surprisingly large number of different genes. They were checked against all available suppressors (Thuriaux et al. 1975). Some representative results are shown in Table 1. Mutations in the gene *sin1* specifically affect *sup3* and *sup9*, but not *sup8* and *sup10*, among the tRNA suppressors. Janner et al. (1978) have demonstrated that *sin1⁻* strains are unable to form i⁶A in their tRNAs. The lack of this modified nucleoside, which is found adjacent to the 3′ end of the anticodon in tRNAs, shows no serious effect on the phenotypes of suppressor-free strains. *sin7⁻* mutations affect only the weak ocher suppressor *sup3-i* but not the strong opal suppressor *sup3-e* or any other suppressor. Obviously, antisuppressors of this type allow assessment of the impact of slight structural changes in macromolecules on their function in vivo; these effects may be hard to detect otherwise, due to their more modulating than disrupting action on the particular function. In contrast to the antisuppressors discussed above, the *sin8⁻* alleles inactivate all tRNA suppressors, but not *sup1* and *sup2*. The reversed pattern is demonstrated by mutations in the genes *sin10* and *cyh1*. *cyh1* mutants were isolated as cycloheximide-resistant strains and were subsequently shown to have antisuppressor activity. An alteration in a ribosomal protein of the large subunit is caused by this mutation (Coddington and Fluri 1977). Besides the demonstrated changes in a ribosomal protein (in *cyh1*) and presumably in a tRNA-modifying enzyme (in *sin1*), a number of other components of the protein-synthesizing machinery may be affected in antisuppressor mutants, e.g., aminoacyl-tRNA synthetases or release factors.

Recently, an attempt was made to restore suppressor activity to strains carrying a suppressor inactivated through a second-site reversion by induction of a third mutation (Fig. 1). Mutations of this type, which by themselves show no nonsense suppressor activity, are called allosuppressors. As expected, two kinds of allosuppressors were isolated. Some strains carry a third mutation in the suppressor tRNA gene that obviously counteracts the effect of the second-site reversion. In other cases, a mutated third gene is involved. In some of these strains, temperature-sensitive allosuppressors were obtained. Some of these mutants were specifically blocked in the cell-division cycle at the restrictive temperature (Nurse et al. 1980). This new set of mutations has not yet been characterized in full detail, but they promise some insight into the regulation mechanisms that connect protein synthesis and cell division.

By an additional mutation step (selection for red colonies, Fig. 1), inactivated allosuppressor strains were obtained that now carry tRNA genes with four mutations (P. Munz, unpubl.).

CHARACTERIZATION OF SUPPRESSOR tRNAs IN VITRO

At the beginning of the biochemical studies on *S. pombe* suppressors, the following questions had to be investigated: (1) What is the nature of the different classes of nonsense mutations (amber, ocher, or opal)? (2) Are the efficient suppressors tRNA species? (3) Which amino acids are inserted by the suppressors?

The tRNA nature of *S. cerevisiae* nonsense suppressors was proven through the development of in vitro assays for amber and ocher suppression (Capecchi et al. 1975; Gesteland et al. 1976). tRNA isolated from *S. pombe* strains *sup3-e* and *sup8-e* was inactive in both systems. Subsequently we developed a new in vitro system for suppression by eukaryotic tRNA. It relies on the formation of a readthrough protein when rabbit globin mRNA is translated in a wheat germ system in the presence of suppressor tRNAs (Kohli et al. 1979). This should provide an assay for ocher and opal suppression, since rabbit α-globin mRNA terminates in UAA (Proudfoot et al. 1977) and β-globin mRNA terminates in UGA (Proudfoot 1977; Efstratiadis et al. 1977). We tested this prediction with already characterized suppressor tRNAs. Amber suppressor tRNAs showed no effect, whereas ocher suppressor tRNAs yielded a readthrough polypeptide that was easily separated from globin by SDS-polyacrylamide gel electrophoresis. The results of the in vitro translation of separated α- and β-globin mRNAs in the presence of highly purified *S. pombe sup3-e* suppressor tRNA or *S. cerevisiae* ocher suppressor tRNA are presented in Figure 3. Ocher suppressor tRNA yields a readthrough product with α-mRNA while *sup3-e* clearly induces a readthrough product only with β-mRNA. Thus, *sup3-e* strains code for an opal suppressor tRNA and the nonsense mutations suppressed in vivo by *sup3-e* are interpreted to be of UGA nature.

In further experiments we demonstrated that *sup8-e* strains code for a different opal suppressor tRNA (Wetzel et al. 1979). In addition tRNA was tested from the weak suppressors *sup3-i* and *sup8-i*. No activity was found in an amber in vitro system (M. Capecchi, pers. comm.), whereas weak suppression was found in the wheat germ globin system (J. Chan and J. Kohli, unpubl.). The tentative conclusion is that *sup3-i* and *sup8-i* code for ocher suppressor tRNAs, but this needs to be corroborated by experiments with purified tRNAs. Apparently the weakness of these suppressors in vivo is paralleled by inefficient suppression in vitro.

Independent confirmation of our results concerning the nature of *sup3-e* was obtained by other investigators. The opal suppressor tRNA isolated from *sup3-e* has been used by Cremer et al. (1979) to identify an opal nonsense mutation in the thymidylate kinase gene of the herpes simplex virus. In vitro translation studies by Atkins (this volume) showed this tRNA to be able to suppress a known opal mutation in the bacteriophage f2.

AMINO ACID SPECIFICITY OF THE SUPPRESSOR tRNA SPECIES

To identify the amino acids inserted by *sup3-e* and *sup8-e*, we decided to purify the suppressor tRNA species and then determine their amino acid acceptor activity. Since suppressor tRNAs often originate from minor isoacceptors and since redundancy of the particular tRNA genes was suggested by the genetic data, we anticipated a very low suppressor tRNA concentration in unfractionated tRNA. Therefore we started with a large amount of crude tRNA from which the suppressor tRNA was purified by standard chromatographic techniques on benzoylated DEAE-cellulose,

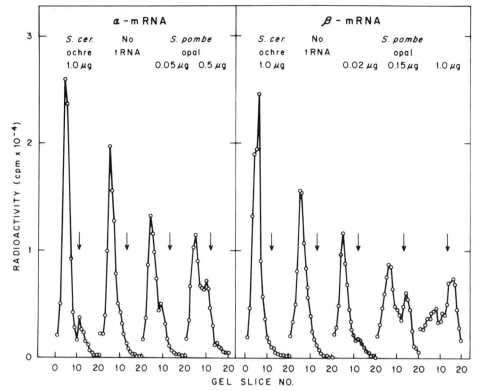

Figure 3 Readthrough of the termination signals of rabbit α- and β-globin mRNAs by suppressor tRNAs. A wheat germ extract (25 μl) was programmed with either mRNA in the presence of fractionated S. cerevisiae ocher suppressor tRNA or highly purified S. pombe sup3-e opal suppressor tRNA or in the absence of additional tRNA. The radioactive polypeptides formed were then electrophoresed in 15% SDS-polyacrylamide gels in the presence of a fluoresceinated myoglobin marker (arrows). The gels were sliced and the radioactivity in the slices determined. For details and for a discussion of the short, unexpected readthrough protein of α-mRNA with sup3-e tRNA, see Kohli et al. (1979).

Sepharose 4B, and RPC-5. The globin mRNA readthrough test (described above) was used as assay. As a final purification step, two-dimensional polyacrylamide gel electrophoresis was used. Enzymatic aminoacylation showed the pure sup3-e tRNA to be a tRNA^{Ser} (Kohli et al. 1979) and the sup8-e tRNA to be a tRNA^{Leu} (Wetzel et al. 1979). The high purification factor of more than 500-fold over unfractionated tRNA for both suppressors confirmed our assumption that these suppressor tRNAs constitute less than 0.5% of total tRNA.

Preliminary purification results indicate that sup9-e corresponds to sup3-e, whereas sup10-e is like sup8-e (J. Chan and J. Kohli, unpubl.).

Thus, the four *S. pombe* opal suppressor tRNAs appear to originate from two sets of redundant tRNA genes.

NUCLEOTIDE SEQUENCE OF THE SUPPRESSOR tRNAs

The availability of pure, unlabeled suppressor tRNA species permitted their nucleotide sequence determination with the modern microscale postlabeling methods (Wetzel et al. 1979; Rafalski et al. 1979). The primary structure of the suppressor tRNA^{Leu} and tRNA^{Ser} are shown in Figure 4. Like other known *S. pombe* tRNAs (McCutchan et al. 1978), the two suppressor tRNA species differ significantly (in about 15% of the nucleotide positions) from the known *S. cerevisiae* isoacceptor. The *sup3-e* sequence resembles the overall structure of other tRNAs^{Ser} with the exception of the first base pair of the D stem (UA), which in all other known tRNAs^{Ser} is a GC pair. Similarly, *sup8-e* tRNA has the general features expected from a tRNA^{Leu}. Rather surprising, though, is the occurrence of an AC mispairing situation at the end of the anticodon stem, giving rise to a 9-nucleotide anticodon loop instead of the standard 7-nucleotide loop. A similar case has been described for an active suppressor tRNA carrying multiple mutations resulting in mischarging of the tRNA (Ozeki, this volume). Both suppressor tRNAs possess the anticodon U*CA expected for an opal suppressor that arises through a

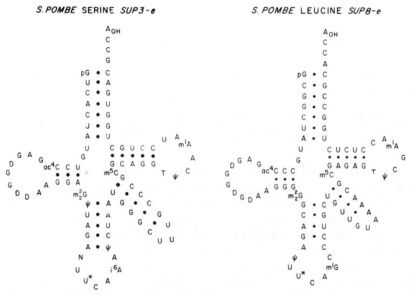

Figure 4 The nucleotide sequences of the two purified *S. pombe* opal suppressor tRNAs. N and U* are modified uridines of undetermined structure.

single-base change in the second position of the anticodon of a tRNA^{Ser} or a tRNA^{Leu}, respectively. The base U* could be mcm^{5}U or mcm^{5}s^{2}U. These nucleosides are thought to restrict codon–anticodon interactions to codons ending in A, whereas an unmodified U could read codons ending in A and G. Examples of such a coding restriction are found in yeast tRNA^{Arg} (Weissbach and Dirheimer 1978), and yeast glutamate tRNA (Yoshida et al. 1971) where mcm^{5}U or its 2-thio derivative are found in the first position of the anticodon. Further work is necessary to establish the structure of this modified nucleotide and also of the unknown U-derivative N at position 32 of the *sup3-e* tRNA.

SYNOPSIS OF THE BIOCHEMICAL AND GENETIC DATA

Although our biochemical results on the weak suppressors *sup3-i* and *sup8-i* need confirmation, complementary genetic data permit a description of the nature and origin of the *S. pombe* tRNA nonsense suppressors. Figure 5 describes the single-base substitutions in the anticodons of either tRNA^{Ser} or a tRNA^{Leu} in the wild-type strain leading to the creation of opal suppressor tRNAs. Final support of this scheme will come from the sequence determination of the unmutated wild-type tRNAs. The situation contrasts to the one found with another opal suppressor tRNA; the *E. coli* suppressor tRNA^{Trp} differs from wild type in a base substitution only in the D stem (Hirsh 1971) and still decodes UGA and the tryptophan codon, UGG. The two *S. pombe* suppressors may represent cases of specific UGA decoding or alternatively the reading of the UGG codon

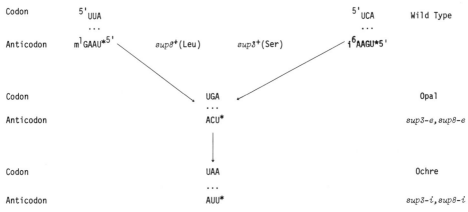

Figure 5 Description of the proposed base changes in the anticodon leading to suppressor tRNAs in *S. pombe*. For further explanations, see text.

could be minimized in vivo by the excess of tRNA^{Trp}. Structural analysis of the hypermodified base in the first position of the anticodon and functional tests of the tRNA should reveal the underlying mechanism for restricting decoding to triplets ending in A. A case similar to the *S. pombe* suppressors was described for the opal suppressor tRNA^{Arg} from bacteriophage T4 (Kao and McClain 1977). A further base substitution in the opal suppressors creates the inefficient ocher suppressors (Fig. 5). We cannot exclude the possibility that these weak suppressors also read amber codons, because a distinct class of UAG nonsense mutations has not been described in *S. pombe* so far. On the other hand, *S. cerevisiae* ocher suppressors are known to be inactive in amber suppression, unlike the bacterial ocher suppressors (Hawthorne and Leupold 1974).

It is well known that the modification of the base on the 3' side of the anticodon influences the stability of dimeric tRNA complexes with complementary anticodons (Grosjean et al. 1978) and the decoding efficiency of a tRNA (Gefter and Russell 1969). In vivo results in *S. pombe* confirm these observations: The lack of i^{6}A in tRNAs caused by mutation of the *sin1* gene (Janner et al. 1978) inactivates only *sup3* coding for a tRNA^{Ser} carrying this modification, but not *sup8*, which carries m^{1}G in the respective position (Fig. 5). Studies of additional antisuppressor strains may identify the biological function of other nucleotide modifications in these suppressor tRNAs.

DISCUSSION AND OUTLOOK

It is evident from the nucleotide sequences of genes and mRNAs (for a compilation, see Steege and Söll 1978), from the isolation of nonsense mutants (Hawthorne and Leupold 1974; Capecchi et al. 1977; Gesteland et al. 1977; Cremer et al. 1979), and from the study of nonsense suppression (Hawthorne and Leupold 1974) that the three codons UAA (ocher), UAG (amber), and UGA (opal) are chain-termination signals for protein synthesis in eukaryotes. *S. cerevisiae* suppressor tRNAs recognizing amber and ocher terminators have been characterized before (Capecchi et al. 1975; Gesteland et al. 1976; Piper, this volume). Based on a new assay for in vitro suppression, we have now characterized serine- and leucine-inserting opal suppressor tRNAs in *S. pombe*. In contrast to the *E. coli* opal suppressor tRNA^{Trp} (Hirsh 1971), they appear to have acquired their suppressor function due to a base change in the second position of their anticodon.

For a detailed understanding of the mechanisms of decoding and of chain termination in protein synthesis, it would be desirable to study the kinetics of suppression in vitro. Unfortunately, none of the eukaryotic in vitro suppression systems developed so far meets all the requirements for such studies. All of them consist of heterologous components, a fact that greatly increases the number of possible artifacts. To date, eukaryotic suppressor tRNAs have only been characterized in yeasts. It is hoped that the recently established cell-free protein-synthesizing system from *S. cerevisiae* (C. McLaughlin and K. Moldave, pers. comm.) can be adapted to the assay of yeast suppressor tRNAs. This may permit comparisons of in vitro and in vivo suppression characteristics and may also provide a tool to study the changes in tRNA function due to structural changes in the tRNA sequence (second-site reversions of suppressors) or undermodification of tRNA (antisuppressor mutants defective in tRNA modifying enzymes). In addition, such a homologous protein-synthesizing system could be the basis for a biochemical analysis of mutant proteins affecting translation (e.g., ribosomal nonsense suppressors, antisuppressor mutations affecting ribosomal proteins, and release factors).

The knowledge of the nucleotide sequence of the two *S. pombe* suppressor tRNAs provides the basis for meaningful studies on the structure elucidation of the many second-site revertants of these tRNAs and on the biochemical characterization of the antisuppressors. The results of such studies may provide much deeper insights into the mechanisms of eukaryotic protein synthesis, as was the case with similar studies in *E. coli* (Smith et al. 1970; Schedl and Primakoff 1973).

Another promising line of research is the analysis of the structure of suppressor tRNA genes by cloning and DNA sequencing techniques. The nucleotide sequences of suppressor tRNA genes carrying further mutations (second-site reversions, allosuppressors, inactivated allosuppressors) should allow some insight into the characteristics of the recombination mechanisms in *S. pombe*, since a wealth of genetic recombination data are available from fine-structure maps (F. Hofer et al., unpubl.) and from studies on conversion at specific sites within suppressor tRNA genes (Munz and Leupold 1979). Cloned suppressor tRNA genes can also be mutated artificially and the result of the changes assessed in vivo after transformation into suitably marked *S. pombe* strains.

ACKNOWLEDGMENTS

This work was supported by grants from the National Institutes of Health and the National Science Foundation. J. K. was a postdoctoral fellow of the Schweizerischer Nationalfond.

REFERENCES

Barben, H. 1966. Allelspezifische Suppressormutationen von *Schizosaccharomyces pombe*. *Genetica* **37:** 109.

Capecchi, M. R., S. H. Hughes, and G. M. Wahl. 1975. Yeast supersuppressors

are altered tRNAs capable of translating a nonsense codon *in vitro*. *Cell* **6:**269.
Capecchi, M. R., R. A. von der Haar, N. E. Capecchi, and M. M. Sveda. 1977. The isolation of a suppressible nonsense mutant in mammalian cells. *Cell* **12:**371.
Coddington, A. and R. Fluri. 1977. Characterization of the ribosomal proteins from *Schizosaccharomyces pombe* by two-dimensional polyacrylamide gel electrophoresis. Demonstration that a cycloheximide resistant strain, *cyh1*, has an altered 60S ribosomal protein. *Mol. Gen. Genet.* **158:**93.
Cremer, K. J., M. Bodemer, W. P. Summers, W. C. Summers, and R. F. Gesteland. 1979. *In vitro* suppression of UAG and UGA mutants in the thymidine kinase gene of herpes simplex virus. *Proc. Natl. Acad. Sci.* **76:**430.
Efstratiadis, A., F. C. Kafatos, and T. Maniatis. 1977. The primary structure of rabbit β-globin mRNA as determined from cloned DNA. *Cell* **10:**571.
Gefter, M. L. and R. L. Russell. 1969. Role of modifications in tyrosine transfer RNA: A modified base affecting ribosome binding. *J. Mol. Biol.* **39:**145.
Gesteland, R. F., N. Wills, J. B. Lewis, and T. Grodzicker. 1977. Identification of amber and ocher mutants of the human virus Ad2$^+$ND1. *Proc. Natl. Acad. Sci.* **74:**4567.
Gesteland, R. F., M. Wolfner, P. Grisafi, G. Fink, D. Botstein, and J. R. Roth. 1976. Yeast suppressors of UAA and UAG nonsense codons work efficiently *in vitro* via tRNA. *Cell* **7:**381.
Goodman, H. M., M. V. Olson, and B. D. Hall. 1977. Nucleotide sequence of a mutant eukaryotic gene: The yeast tyrosine-inserting ocher suppressor SUP4-O. *Proc. Natl. Acad. Sci.* **74:**5453.
Grosjean, J. H., S. DeHenau, and D. M. Crothers. 1978. On the physical basis for ambiguity in genetic coding interactions. *Proc. Natl. Acad. Sci.* **75:**610.
Hawthorne, D. C. and U. Leupold. 1974. Suppressors in yeast. *Curr. Top. Microbiol. Immunol.* **64:**1.
Hirsh, D. 1971. Tryptophan transfer RNA as the UGA suppressor. *J. Mol. Biol.* **58:**439.
Janner, F., G. Vögeli, and R. Fluri. 1978. The effect of an antisuppressor on tRNA in the yeast *Schizosaccharomyces pombe*. *Experientia* **34:**943 (Abstr.).
Kao, S. and W. H. McClain. 1977. UGA suppressor of bacteriophage T4 associated with arginine transfer RNA. *J. Biol. Chem.* **252:**8254.
Kohli, J., H. Hottinger, P. Munz, A. Strauss, and P. Thuriaux. 1977. Genetic mapping in *Schizosaccharomyces pombe* by mitotic and meiotic analysis and induced haploidization. *Genetics* **87:**471.
Kohli, J., T. Kwong, F. Altruda, D. Söll, and G. Wahl. 1979. Characterization of a UGA suppressing serine tRNA from *Schizosaccharomyces pombe* with the help of a new *in vitro* assay system for eukaryotic suppressor tRNAs. *J. Biol. Chem.* **254:**1546.
McClain, W. H. 1977. Seven terminal steps in a biosynthetic pathway leading from DNA to transfer RNA. *Accts. Chem. Res.* **10:**418.
McCutchan, T., S. Silverman, J. Kohli, and D. Söll. 1978. Nucleotide sequence of phenylalanine transfer RNA from *Schizosaccharomyces pombe*: Implications for transfer RNA recognition by yeast phenylalanyl-tRNA synthetase. *Biochemistry* **17:**1622.
Munz, P. and U. Leupold. 1979. Gene conversion in nonsense suppressors of

Schizosaccharomyces pombe. I. The influence of the genetic background and of three mutant genes (*rad2, mut1* and *mut2*) on the frequency of post-meiotic segregation. *Mol. Gen. Genet.* **170:** 145.

Nurse, P., P. Fantes, P. Munz, and P. Thuriaux. 1980. Control integrating growth rate and the initiation of mitosis in fission yeast. *Heredity* (Abstr.) (in press).

Proudfoot, N. J. 1977. Complete 3' noncoding region sequences of rabbit and human β-globin messenger RNAs. *Cell* **10:** 559.

Proudfoot, N. J., S. Gillam, M. Smith, and J. Longley. 1977. Nucleotide sequence of the 3'-terminal third of rabbit α-globin messenger RNA: Comparison with human α-globin messenger RNA. *Cell* **11:** 807.

Rafalski, A., J. Kohli, and D. Söll. 1979. The nucleotide sequence of a UGA suppressor serine tRNA from *Schizosaccharomyces pombe*. *Nucleic Acids Res.* **6:** 2683.

Rosset, R. and L. Gorini. 1969. A ribosomal ambiguity mutation. *J. Mol. Biol.* **39:** 95.

Schedl, P. and P. Primakoff. 1973. Mutants of *Escherichia coli* thermosensitive for the synthesis of transfer RNA. *Proc. Natl. Acad. Sci.* **70:** 2091.

Smith, J. D. 1976. Transcription and processing of transfer RNA precursors. *Prog. Nucleic Acid Res. Mol. Biol.* **16:** 25.

Smith, J. D., L. Barnett, S. Brenner, and R. L. Russell. 1970. More mutant tyrosine transfer ribonucleic acids. *J. Mol. Biol.* **54:** 1.

Söll, D. and P. R. Schimmel. 1974. Aminoacyl-tRNA synthetases. In *The enzymes* (ed. P. Boyer), vol. 10, p. 489. Academic Press, New York.

Steege, D. A. and D. G. Söll. 1978. Suppression. In *Biological regulation and control* (ed. R. F. Goldberger), p. 433. Plenum Press, New York.

Thuriaux, P., M. Minet, F. Hofer, and U. Leupold. 1975. Genetic analysis of antisuppressor mutants in the fission yeast *Schizosaccharomyces pombe*. *Mol. Gen. Genet.* **142:** 251.

Weissenbach, J. and G. Dirheimer. 1978. Pairing properties of the methyl ester of 5-carboxymethyl uridine in the wobble position of yeast $tRNA_3^{Arg}$. *Biochim. Biophys. Acta* **518:** 530.

Wetzel, R., J. Kohli, F. Altruda, and D. Söll. 1979. Identification and nucleotide sequence of the *sup8-e* UGA suppressor leucine tRNA from *Schizosaccharomyces pombe*. *Mol. Gen. Genet.* **172:** 221.

Yoshida, M., K. Takeishi, and T. Ukita. 1971. Structural studies on a yeast glutamic acid tRNA specific to GAA codon. *Biochim. Biophys. Acta* **228:** 153.

Interactions between UGA-suppressor tRNA^{Trp} and the Ribosome: Mechanisms of tRNA Selection

Richard H. Buckingham
Institut de Biologie Physico-Chimique
13 rue Pierre et Marie Curie
75005 Paris, France

Charles G. Kurland
Molecular Biology Institute
Wallenberg Laboratory
University of Uppsala
Uppsala, Sweden

The notion that codon–anticodon interactions are the sole determinants for the selection of aminoacyl-tRNAs on the ribosome has long been central to our ideas about the mechanism of protein synthesis. There is, however, at least one nonsense suppressor tRNA that seems to contradict this generalization. The anomalous tRNA species is the UGA suppressor form of tRNA^{Trp}. This tRNA was shown by Hirsh (1971) to contain a normal tryptophan anticodon (CCA) and a single-base change in the D stem, where G24 is replaced by an A. Thus, we are obliged to explain how a base change far from the anticodon can transform this tRNA from one that reads the codon UGG to one that can read both the UGG codon and the nonsense codon UGA.

One simple way to explain the behavior of the UGA suppressor is to assume that the D-stem alteration changes the structure of the tRNA so that its affinity for the UGA codon is enhanced. Two experiments have been performed to detect such an effect. First, Högenauer (1974) measured the binding of the oligonucleotide UpGpA to wild-type and mutant tRNAs. No significant differences between the affinities of these two tRNA species for the nonsense codon analog were detected by equilibrium dialysis. Similarly, Buckingham (1976) compared the affinities of the wild-type and suppressor forms of tRNA^{Trp} for tRNA^{Pro}, which possesses an anticodon complementary to that of the tRNA^{Trp} species. As shown in Figure 1, both wild-type and suppressor tRNA^{Trp} are retarded to the same extent by tRNA^{Pro} fixed to a column bed. These experiments suggest that the codon specificities of the two tRNA^{Trp} species are not detectable in the absence of ribosomes.

Given such a conclusion it is natural to ask how the ribosome might contribute to the tRNA selection without interfering with the primary

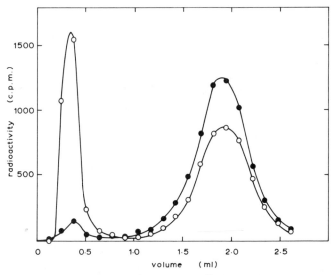

Figure 1 Chromatography of tRNA^{Trp} on a column containing immobilized tRNA^{Pro}. tRNA^{Pro} (major, E. coli, 10 A$_{260}$ units) was periodate-oxidized and linked covalently to hydrazine-activated P-200 polyacrylamide beads (200 μl wet volume). To a column (4.5-mm diameter) of this material, a sample containing ^{3}H-labeled wild-type tRNA^{Trp} (CC*) and ^{14}C-labeled suppressor tRNA^{Trp} (CC*), i.e., lacking the terminal A, was applied and eluted with 20 mM Na cacodylate (pH 7.4), 10 mM MgCl$_2$, 0.7 M NaCl at 20°C. ^{3}H (○) and ^{14}C (●) were measured in the fractions (130 μl). For further details, see Buckingham (1976).

role of the codon in this process. Relevant here is a model for tRNA selection that is based on the idea that the codon–anticodon interaction can influence the conformation of the tRNA. In particular, it is suggested that proper matching of codon and anticodon normally facilitates the accommodation of the whole tRNA to a ribosomal binding site (Kurland et al. 1975; Kurland 1978). It may be inferred from this model that the D-stem substitution in the UGA suppressor has altered the conformational coupling between the anticodon loop and sites more distant from the anticodon so that this tRNA can translate any codon that begins with the two nucleotides U and G. Accordingly, one prediction of the model is that UGA suppressor should be able to translate the cysteine codons UGU and UGC.

Verification of this prediction was obtained by studying the translation of poly(UUUUUG), which contains the cysteine codon UGU (Buckingham and Kurland 1977). Data from experiments, such as those summarized in Table 1, show that when cysteine is withdrawn from the in vitro translating system, cysteine codons can be mistranslated as tryptophan and that such misreading is severalfold more probable under these conditions with suppressor tRNA^{Trp} than with wild-type tRNA^{Trp}.

Table 1 Misincorporation of tryptophan in response to UGU codons: The effect of omitting different amino acids on phenylalanine or tryptophan incorporation directed by poly(UUUUUG)

Amino acids	Phenylalanine incorporated (pmoles)	Tryptophan incorporated (pmoles)	
		wild type	suppressor
All	467	13.1	12.3
−Valine	397	11.1	13.4
−Leucine	419	11.6	11.8
−Cysteine	336	16.8	23.8
−Tryptophan	409	—	—

Aliquots (50 µl) containing 0.11 A_{260} unit of poly(UUUUUG); 50 nmoles each of phenylalanine (unlabeled or ^{14}C-labeled, 19.9 mCi/mmole), valine, leucine, cysteine, and glycine (except as noted); 10 nmoles tryptophan (unlabeled or ^{14}C-labeled, 52 mCi/mmole); 60 mM Tris-HCl (pH 7.8); 100 mM NH$_4$Cl; 11 mM MgCl; 1 mM ATP; 0.4 mM GTP; 4 mM phosphoenolpyruvate; 1 mM dithiothreitol; 0.1 mM leucovorin; 0.1 mM methionine; 50 µg MRE600 supernatant protein; 1.2 A_{260} units ribosomes; total tRNA from wild-type *E. coli* (0.75 A_{260} unit, including 12 pmoles tRNA^{Trp}); and purified wild-type or suppressor tRNA^{Trp} (42 pmoles) were incubated for 30 min at 37°C. Incorporation of radioactivity was determined into hot-trichloroacetic-acid-insoluble material. The background values observed in the absence of mRNA (3.2 pmoles tryptophan or 6.5 pmoles phenylalanine) have been subtracted. Phenylalanine incorporation was measured in the presence of wild-type tRNA. The stimulation of tryptophan is calculated as the difference between incorporation in the absence of cysteine and that with all amino acids. Data are from Buckingham and Kurland (1977).

Our account of the different codon specificities of the wild-type and UGA-suppressor forms of tRNA^{Trp} is based on the idea that the conformational states of the tRNAs are distributed differently in response to the relevant codon–anticodon interactions. Therefore, we might expect there to be characteristic differences between the conformational states of the two tRNAs, even in the absence of ribosomes. A number of indirect experiments confirm this expectation.

Favre et al. (1975) have compared the kinetics of photochemical cross-linking between s^{4}U8 and C13 with wild-type and suppressor tRNA^{Trp}. Data such as those in Table 2 indicate that the rate of such cross-linking is approximately 30% slower with the mutant tRNA than with the wild-type tRNA under the same conditions. The implication of this result is that there is a conformational difference between the two tRNAs in the neighborhood of the D stem, where the substitution at position 24 has occurred in the mutant.

A more striking difference between the two tRNA species in the absence of ribosomes is seen when they are degraded with polynucleotide phosphorylase. It has been shown previously that there are multiple tRNA states in solution differing in their susceptibility to attack by this enzyme (Thang et al. 1967, 1971; Danchin and Thang 1972). Therefore, the data showing that the extent of phosphorolysis of mutant tRNA^{Trp} at 37°C is one third that of wild-type tRNA^{Trp} (R. H. Buckingham et al., unpubl.) strongly suggest that the two tRNAs differ markedly in their distribution among the conformational states that are sensitive or resistant to the enzymatic attack of polynucleotide phosphorylase (see Fig. 2).

Although these results indicate that significant differences exist between the two tRNA^{Trp} species with respect to their occupancies of conformational states, these data do not by themselves show that such differences are relevant to the codon specificities of the two tRNAs. Fortunately, more recent experiments by Vacher and Buckingham (1978) provide the connection between the two sets of phenomena. They have studied the codon specificities of wild-type and suppressor tRNA^{Trp} after the photochemical cross-linking of the tRNA, as mentioned above. The functional consequence of the introduction of the cross-link is to restrict

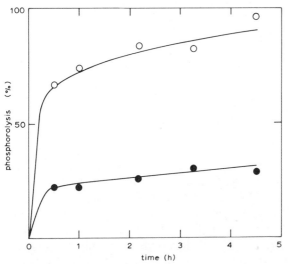

Figure 2 Phosphorolysis of wild-type and suppressor tRNA^{Trp}. A mixture of ^{3}H-labeled wild-type tRNA^{Trp} (CCA*) and ^{14}C-labeled suppressor tRNA^{Trp} (CCA*) containing 0.34 nmole of each tRNA in a total volume of 212 μl of 50 mM Tris-HCl (pH 8.1), 1 mM MgCl$_2$, 10 mM phosphate, was digested with 6 units of polynucleotide phosphorylase (1 unit releases 1 μmole ADP from poly[A] per hour at 37°C). Aliquots were removed at the times shown, precipitated with 10% trichloroacetic acid and the insoluble ^{3}H and ^{14}C radioactivity determined by filtration on glass-fiber filters and liquid-scintillation counting. The percentage of phosphorolysis for ^{3}H-labeled wild-type tRNA (○) and ^{14}C-labeled suppressor tRNA (●) is measured as the decrease in acid-insoluble radioactivity.

Table 2 Kinetics of photochemical cross-linking between s⁴U8 and C13 in wild-type and suppressor tRNA^Trp

	tRNA^Trp		Unfractionated tRNA	
	wild-type	suppressor	wild-type	suppressor
Experiment 1	0.65 ± 0.05	0.85 ± 0.05	1.00 ± 0.05	—
Experiment 2	0.47 ± 0.05	0.67 ± 0.05	1.00 ± 0.05	1.00 ± 0.05

The data are presented as the time in min. needed to obtain 50% of the final level of cross-linking and are normalized with respect to unfractionated tRNA. The experiments were conducted as described by Favre et al. (1975) under the following ionic conditions. Experiment 1: 50 mM Na cacodylate buffer (pH 7.0), 50 mM NaCl, 5 mM MgCl$_2$; experiment 2: 25 mM Na cacodylate buffer (pH 7.0), 25 mM NaCl, 10 mM MgCl$_2$.

the codon specificity of the UGA suppressor to that of wild-type tRNA^Trp, while having little or no effect on that of the normal tRNA^Trp species. In other words, the introduction of an artificial conformational constraint on the mutant tRNA restricts the range of codons that are translatable. Such results provide a strong argument in favor of the conformational selection theory.

ACKNOWLEDGMENTS

We thank M. Grunberg-Managó for her constant advice and encouragement. This work was supported by grants to M. Grunberg-Managó from the Centre National de la Recherche Scientifique (Groupe de Recherche no. 18) and the Délégation Générale à la Recherche Scientifique et Technique (convention no. 74.7.0356), as well as grants from The Swedish Cancer Society and Natural Sciences Research Council.

REFERENCES

Buckingham, R. H. 1976. Anticodon conformation and accessibility in wild-type and suppressor tryptophan tRNA from *E. coli*. *Nucleic Acids Res.* **3:** 965.

Buckingham, R. H. and C. G. Kurland. 1977. Codon specificity of UGA suppressor tRNA^Trp from *E. coli*. *Proc. Natl. Acad. Sci.* **74:** 5496.

Danchin, A. and M. N. Thang. 1972. Multiple states in macromolecules. II. Entropic behaviour of tRNA degraded by polynucleotide phosphorylase. *FEBS Lett.* **19:** 297.

Favre, A., R. H. Buckingham, and G. Thomas. 1975. tRNA tertiary structure in solution as probed by the photochemically induced 8-13 cross-link. *Nucleic Acids Res.* **2:** 1421.

Hirsh, D. 1971. Tryptophan transfer RNA as the UGA suppressor. *J. Mol. Biol.* **58:** 439.

Högenauer, G. 1974. Binding of UGA to wild-type and suppressor tryptophan tRNA from *E. coli*. *FEBS Lett.* **39**:310.

Kurland, C. G. 1978. The role of guanine nucleotides in protein biosynthesis. *Biophys. J.* **22**:373.

Kurland, C. G., R. Rigler, M. Ehrenberg, and C. Blomberg. 1975. Allosteric mechanism for codon-dependent tRNA selection on ribosomes. *Proc. Natl. Acad. Sci.* **72**:4248.

Thang, M. N., B. Belchev, and M. Grunberg-Manago. 1971. Phosphorolysis of tRNA. Multiple conformational states of tRNA in solution. *Eur. J. Biochem.* **19**:184.

Thang, M. N., W. Guschlbauer, H. G. Zachau, and M. Grunberg-Manago. 1967. Degradation of transfer tRNA by polynucleotide phosphorylase. I. Mechanism of phosphorolysis and structure of tRNA. *J. Mol. Biol.* **26**:403.

Vacher, J. and R. H. Buckingham. 1979. Effect of photochemical cross-link $S^4U(8)$–$C(13)$ on suppressor activity of su^+-tRNA^{Trp} from *E. coli*. *J. Mol. Biol.* **129**:287.

Use of Protein Synthesis In Vitro to Study Codon Recognition by *Escherichia coli* tRNA^{Leu} Isoaccepting Species

Emanuel Goldman* and G. Wesley Hatfield
Department of Medical Microbiology, College of Medicine
University of California, Irvine
Irvine, California 92717

In the past, studies of codon recognition by tRNA isoaccepting species have employed the techniques of ribosome binding (Nirenberg et al. 1967), protein synthesis in vitro directed by synthetic polyribonucleotides (Khorana et al. 1967), and direct nucleotide interactions between tRNAs containing complementary anticodons (Grosjean et al. 1976). The availability of completely sequenced, natural mRNAs has facilitated the study of codon recognition in vitro under conditions that more closely resemble the situation in vivo than were previously possible. One can now compare the relative efficiencies of different acylated tRNA isoaccepting species to transfer their amino acids into isolated peptides of known sequence that contain residues encoded by known codons. The efficiency of codon recognition is thus deduced from a functional test involving synthesis of a bona fide natural polypeptide chain directed by a natural mRNA.

The validity of this approach depends upon the presumptions that the amino acid isolated in a single peptide is in fact transferred only from the individual aminoacylated tRNA isoaccepting species added to the reaction and that no exchange of amino acid occurs between different isoaccepting species present in the extract. This constraint has been satisfied by two different approaches. The first is the use of extracts of *Escherichia coli* mutants that are temperature-sensitive in one of the aminoacyl-tRNA synthetase enzymes (Mitra et al. 1977). These extracts are dependent on added aminoacylated tRNA for protein synthesis to occur, and no exchange of the homologous amino acid should be possible. The second approach is the addition of a large excess of the non-radioactive amino acid homologous to the radioactive aminoacylated tRNA (Goldman and Hatfield 1979). Although exchange between

*Present address: Department of Microbiology, New Jersey Medical School, Newark, New Jersey 07103.

different tRNA isoaccepting species is possible in this approach, the specific activity of the free amino acid intermediate is vastly diluted by the excess of nonradioactive amino acid; hence radioactivity isolated in peptides is transferred effectively from only the radioactive aminoacylated tRNA species added to the reaction. Thus, the latter approach employs aminoacylated tRNA species as tracers for protein synthesis, whereas the former requires aminoacylated tRNA dependence.

tRNA-DEPENDENT CONDITIONS

Experiments using extracts of cells that possess a temperature-sensitive valyl-tRNA synthetase led to the suggestion that valine codons were read operationally as a two-letter code (Mitra et al. 1977), and to the hypothesis that codon recognition overall was governed by a two-out-of-three mechanism (Lagerkvist 1978).

When we examined protein synthesis in an analogous extract from an *E. coli* temperature-sensitive in the leucyl-tRNA synthetase, we obtained evidence that suggested that misreading of the genetic code occurs during such tRNA-dependent protein synthesis (Holmes et al. 1978). Substantial incorporation of other amino acids into protein was observed even though no leucine incorporation occurred in the leucyl-tRNA-synthetase-deficient extracts. Tai et al. (1978) also observed substantial incorporation of other amino acids into protein in extracts of temperature-sensitive cells in either the glutamyl- or valyl-tRNA synthetase enzymes.

To determine the nature of the protein synthesized in these extracts, we performed the following experiment (Holmes et al. 1978). [^{14}C]Tyrosine-labeled protein directed by bacteriophage MS2 RNA was synthesized in vitro in leucyl-tRNA-synthetase-deficient extracts, in a system reconstituted by adding individual leucyl-tRNA^{Leu} isoaccepting species. The products of the reaction were digested with trypsin, subjected to electrophoresis at pH 3.5, and autoradiographed. Previous experiments (Goldman and Lodish 1971; Mitra et al. 1977) have defined the electrophoretic migration under these conditions of three tyrosine-containing tryptic peptides of the MS2 coat protein—T8 (tyrosine at position 42), T9 (tyrosine at position 58), and T1 (tyrosine at the COOH-terminus, position 129). The remaining coat tyrosine (position 85) is in the T10 peptide, which is insoluble and expected to remain at the origin. (Peptide nomenclature is as in Weber and Konigsberg 1967). The T9 peptide has been observed to split into two bands under these conditions; one migrates just ahead of the T8 peptide, and the other is more mobile (Mitra et al. 1977).

In the absence of added leucyl-tRNA^{Leu}, the T8 and T9 peptides were synthesized but not the COOH-terminal T1 peptide (Fig. 1). Addition of only leucyl-tRNA$_3^{Leu}$ permitted the T1 peptide to be synthesized. The spots from this electrophoretograph were cut out and quantitated in the

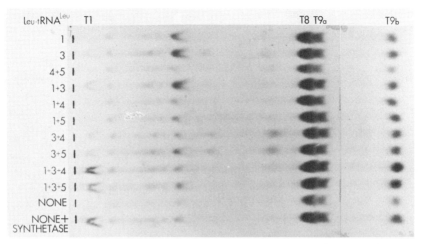

Figure 1 Autoradiograph of digests of phage-MS2-RNA-directed protein synthesized in vitro and labeled with [^{14}C]tyrosine. Individual, purified tRNA^{Leu} isoaccepting species were prepared as described (Holmes et al. 1975), aminoacylated with nonradioactive leucine, and added as indicated to MS2-RNA-directed protein-synthesis reactions containing [^{14}C]tyrosine in extracts of cells that possess a temperature-sensitive leucyl-tRNA synthetase. The products of the reaction were digested with trypsin, subjected to electrophoresis at pH 3.5, and autoradiographed. (Reprinted, with permission, from Holmes et al. 1978).

codon:			CUC		UUA	
residue:	fMet	Leu	Tyr	Tyr	Leu	Tyr
position:	9		42	58	86	129
peptide:			T8	T9		T1

Schematic diagram of certain residues in the MS2 coat protein gene taken from data of Lin et al. (1967) and Min Jou et al. (1972).

scintillation counter. In the absence of added leucyl-tRNA^{Leu}, the T8 and T9 peptides were synthesized at a level of about 30% of the amounts made in the maximum sample reconstituted with several leucyl-tRNA^{Leu} isoaccepting species, even though these peptides are located downstream of a leucine residue at position 9 of the coat protein. The simplest explanation for this result is that misreading of the position-9 CUC leucine codon is occurring in vitro under the forced conditions of tRNA-dependence. With added leucyl-tRNA$_3^{Leu}$, the COOH-terminal T1 peptide was synthesized at a level of 17% of the maximum reconstituted sample, and the combination of leucyl-tRNA$_1^{Leu}$ and leucyl-tRNA$_3^{Leu}$ permitted synthesis of T1 at 28% of the maximum reconstituted sample, even though there is a UUA-encoded leucine residue at position 86. However, the sequenced anticodons of tRNA$_1^{Leu}$ and tRNA$_3^{Leu}$ correspond to CUG

(Dube et al. 1970) and CUC (Blank and Söll 1971) codons, respectively. The completion of the coat protein with only added leucyl-tRNA$_3^{Leu}$ suggests that either this isoaccepting species is capable of reading UUA as well as CUA, CUU, and CUC codons (all present in the coat protein gene), or that misreading of the position-86 UUA codon is occurring in vitro under tRNA-dependent conditions. Thus, the leucyl-tRNA-synthetase-deficient extracts were able to synthesize phage MS2 peptides downstream of leucine codons that should not have been read in the absence of the aminoacylated tRNA^{Leu} upon which the system was dependent. A possible explanation is that if the favored aminoacylated tRNA for a given codon is not available, ribosomes might accept a different tRNA with a lower equilibrium constant as the next best available choice. Therefore, under conditions in vitro where one forces the system, mistakes that do not ordinarily occur in vivo may be made.

USE OF tRNAs AS TRACERS

To examine codon recognition in vitro for the leucine family of tRNA isoaccepting species of *E. coli*, we have added [^{14}C]leucyl-tRNA^{Leu} isoaccepting species as tracers and have shown that, in the presence of an excess of nonradioactive leucine, there was no measurable transfer of radioactive leucine from one isoaccepting species to another (Goldman and Hatfield 1979), thereby ensuring that the [^{14}C]leucine in the isolated peptide was transferred only from the added [^{14}C]leucyl-tRNA^{Leu} species. Since both the nucleotide sequence (Min Jou et al. 1972) and the amino acid sequence (Lin et al. 1967) of the phage MS2 coat-protein gene are known and the amino acid sequence of the MS2 replicase protein has been deduced from the nucleotide sequence (Fiers et al. 1976), we were able to identify specific peptides by isolating them through peptide mapping procedures and then determining the presence or absence of radioactive marker amino acids run in parallel tracks—in effect, by a kind of partial amino acid composition (Goldman and Hatfield 1979).

Two peptides from the phage MS2 coat protein were isolated in the following experiment (Goldman et al. 1979). MS2 RNA-directed protein synthesis reactions containing different [^{14}C]leucyl-tRNA^{Leu} isoaccepting species or various radioactive marker amino acids were digested with trypsin, subjected to electrophoresis at pH 1.9, and autoradiographed (Fig. 2). The region of paper corresponding to band A was excised and subjected to electrophoresis at pH 3.5 (Fig. 3, I). This second dimension of electrophoresis contained a peptide with the appropriate pattern of radioactive marker amino acids for the T2 peptide. This band was excised and subjected to electrophoresis at pH 4.7 (Fig. 3, II). Although the T2 peptide appeared to be pure at this stage, to be certain, the band was again excised and subjected to descending chroma-

Figure 2 Autoradiograph of digests of phage-MS2-RNA-directed protein synthesized in vitro and labeled with various radioactive amino acids or [^{14}C]leucyl-tRNA^{Leu} isoaccepting species. Reactions contained extracts of wild-type *E. coli* and either radioactive amino acids or equal amounts of each purified tRNA^{Leu} isoaccepting species aminoacylated with [^{14}C]leucine, as indicated below. The products of the reaction were digested with trypsin, subjected to electrophoresis at pH 1.9, and autoradiographed. Details of sample preparation and processing have been described (Goldman and Hatfield 1979). (*1*) [^{14}C]Leucine blank (no phage MS2 RNA added); (*2*) [^{14}C]leucyl-tRNA$_1^{Leu}$; (*3*) [^{14}C]leucyl-tRNA$_3^{Leu}$; (*4*) [^{14}C]leucyl-tRNA$_4^{Leu}$; (*5*) [^{14}C]leucyl-tRNA$_5^{Leu}$; (*6*) [^{14}C]leucine; (*7*) [^{14}C]arginine; (*8*) [^{14}C]tryptophan; (*9*) [^{14}C]lysine; (*10*) [^{14}C]glutamic acid; (*11*) [^{14}C]tyrosine; (*12*) [^{14}C]phenylalanine; (*13*) [^{14}C]histidine; (*14*) [^{35}S]methionine. (Reprinted, with permission, from Goldman et al. 1979.)

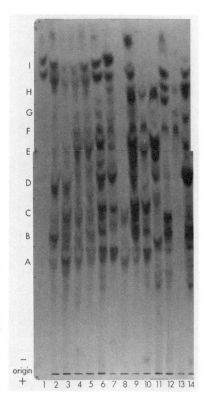

tography (Fig. 3, III). The peptide spots were cut out and counted, and the observed incorporation of marker amino acids confirms this to be the pure T2 peptide, since no other MS2 peptide contains leucine, arginine, tryptophan, and glutamic acid, and is missing all the other markers tested (Goldman and Hatfield 1979).

The T2 peptide contains one leucine residue encoded by CUU. [^{14}C]Leucyl-tRNA$_3^{Leu}$ was at least four times more efficient than [^{14}C]leucyl-tRNA$_1^{Leu}$ in inserting [^{14}C]leucine into this peptide, whereas [^{14}C]leucyl-tRNA$_4^{Leu}$ and [^{14}C]leucyl-tRNA$_5^{Leu}$ did not significantly transfer leucine into this peptide. According to the wobble hypothesis (Crick 1966), tRNA$_1^{Leu}$ (anticodon CAG) ought not to recognize the CUU codon, since this would involve a forbidden UC base pair in the wobble position; but tRNA$_3^{Leu}$ (anticodon GAG) ought to be the preferred isoacceptor, since UG pairing is permissible. The experimental finding for this codon at this location in the message is that the UG base pair in the wobble position is indeed highly preferred over the UC base pair; the latter is, however, capable of occurring, but with much lower frequency.

The region of paper corresponding to band D of Figure 2 was excised

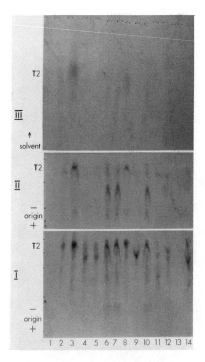

Figure 3 Peptide mapping the T2 peptide. (*I*) Second dimension. Region A in Fig. 2 was excised, sewn onto a new sheet of paper, subjected to electrophoresis at pH 3.5, and autoradiographed. (*II*) Third dimension. Region T2 in *I* was excised and subjected to electrophoresis at pH 4.7. (*III*) Fourth dimension. Region T2 in *II* was subjected to descending chromatography in BPAW (butanol, pyridine, acetic acid, and water). Migration of the visible dye marker (Brilliant Cresyl Blue) to just behind the solvent front is indicated by the series of dots at the top of *III*. (Reprinted, with permission, from Goldman et al. 1979.)

and subjected to electrophoresis at pH 3.5 (Fig. 4, I). This second-dimension electrophoresis contained a peptide with the appropriate pattern of marker radioactive amino acids for the T6 peptide. This peptide was excised and subjected to electrophoresis at pH 4.7 (Fig. 4, II), followed by descending chromatography (Fig. 4, III). The peptide spots were cut out and counted, and the observed incorporation of marker amino acids confirms this to be the pure T6 peptide since no other MS2 peptide contains leucine, lysine, and methionine and is missing all the other markers tested.

The T6 peptide contains two leucines in tandem encoded by CUC and CUA. [^{14}C]Leucyl-tRNA$_1^{Leu}$ and [^{14}C]leucyl-tRNA$_3^{Leu}$ were both about equally efficient in inserting [^{14}C]leucine into these residues, whereas [^{14}C]leucyl-tRNA$_4^{Leu}$ and [^{14}C]leucyl-tRNA$_5^{Leu}$ failed to insert leucine. Since the sequenced anticodon of tRNA$_1^{Leu}$ is CAG, corresponding to a CUG codon, the fact that this species inserts leucine into a peptide containing residues encoded by CUC and CUA codons means that unorthodox base pairing is occurring, since neither CC nor AC pairing was deemed likely by the original wobble hypothesis. Furthermore, though the sequenced anticodon of tRNA$_3^{Leu}$ is GAG, corresponding to a CUC codon, this species is not favored over tRNA$_1^{Leu}$, despite the fact that a CUC-encoded leucine appears in this peptide.

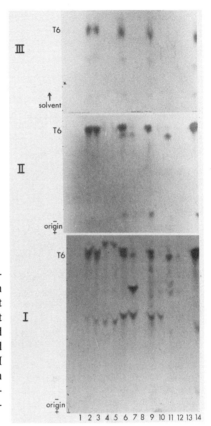

Figure 4 Peptide mapping the T6 peptide. (*I*) Second dimension. Region *D* in Fig. 2 was excised, sewn onto a new sheet of paper, subjected to electrophoresis at pH 3.5, and autoradiographed. (*II*) Third dimension. Region T6 in *I* was excised and subjected to electrophoresis at pH 4.7. (*III*) Fourth dimension. Region T6 in *II* was subjected to descending chromatography in BPAW. (Reprinted, with permission, from Goldman et al. 1979.)

Several leucine-containing peptides from the replicase protein have also been isolated, indicating that $tRNA_4^{Leu}$ and $tRNA_5^{Leu}$ both recognize UUA and UUG codons, with $tRNA_4^{Leu}$ apparently slightly preferred for the UUA codon, whereas $tRNA_1^{Leu}$ and $tRNA_3^{Leu}$ are about equal for the CUC codon. We have not detected mistranslation in the phenylalanine (UUU, UUC) and leucine (UUA, UUG) series of codons. A summary of the relative transfer of leucine from leucyl-$tRNA^{Leu}$ isoaccepting species into specific coat- and replicase-protein peptides that we have isolated, identified, and quantitated is shown in Table 1. Table 2 summarizes and idealizes the codon responses of the $tRNA^{Leu}$ isoaccepting species to leucine codons on the basis of these data (Goldman et al. 1979).

CONCLUDING REMARKS

We have described studies of codon recognition by purified isoacceptor tRNAs determined by protein synthesis in vitro directed by sequenced,

Table 1 Summary of relative transfer of leucine from leucyl-tRNA^{Leu} isoaccepting species into specific peptides

Peptide	Codons	Maximum incorporation from leucyl-tRNA^{Leu} (%)			
		1	3	4	5
T2[a]	CUU	20	100	5	4
T6[b]	CUC, CUA	100	82	4	3
Replicase 2[c]	UUG, CUC	38	55	91	100
Replicase 5[d]	UUA	3	5	100	61
Replicase 9[e]	UUG	3	3	100	94
Replicase 11[f]	CUC	100	76	8	9
R17 coat chymotryptic[g]	UUA	0	10	100	34

[a]The 17-amino-acid tryptic peptide beginning at position 67 of the phage MS2 coat protein (Lin et al. 1967; Min Jou et al. 1972).
[b]The 7-amino-acid tryptic peptide beginning at position 107 of the phage MS2 coat protein.
[c]The 6-amino-acid tryptic peptide beginning at position 49 of the phage MS2 replicase protein (Fiers et al. 1976).
[d]The 7-amino-acid tryptic peptide beginning at position 88 of the phage MS2 replicase protein.
[e]The 8-amino-acid tryptic peptide beginning at position 138 of the phage MS2 replicase protein.
[f]The 12-amino-acid tryptic peptide beginning at position 186 of the phage MS2 replicase protein.
[g]The 10-amino-acid chymotryptic peptide beginning at position 86 of the R17 coat protein (Weber 1967; Nichols and Robertson 1971).
Reprinted, with permission, from Goldman et al. (1979).

natural mRNA. In tRNA-dependent systems, we have obtained evidence for mistranslation, since leucyl-tRNA-synthetase-deficient extracts were able to synthesize phage MS2 peptides downstream of leucine codons that should not have been read in the absence of the aminoacylated tRNA^{Leu}, upon which the system was dependent. We propose that if the favored aminoacylated tRNA for a given codon is not available, ribosomes might accept a different tRNA with a lower equilibrium constant as the next best available choice.

To determine codon recognition, rather than use a tRNA-dependent system, we have added [^{14}C]leucyl-tRNA^{Leu} isoaccepting species as tracers, and have shown that in the presence of an excess of non-radioactive leucine, there is no transfer of radioactive leucine from one isoaccepting species to another. Phage-MS2-specific peptides containing leucine residues encoded by known codons have been isolated and identified, and the relative abilities of the leucyl-tRNA^{Leu} isoaccepting

Table 2 Summary of idealized codon responses by tRNA^{Leu} isoaccepting species

	tRNA^{Leu}			
Codon^a	1	3	4	5
CUC	++	++	−	−
CUA^b	(++)	(++)	−	−
CUU	±	++	−	−
UUA	−	−	++	+
UUG	−	−	++	++

^{a}No CUG-encoded leucine residues were obtained.
^{b}No CUA-encoded leucine residue was obtained in a peptide by itself, but in combination with the CUC codon; hence, the codon responses to CUA are deduced rather than determined directly and, therefore, are placed in parentheses. (Reprinted, with permission, from Goldman et al. 1979.)

species to transfer leucine into these peptides were compared. Sequenced tRNA$_1^{Leu}$ and sequenced tRNA$_3^{Leu}$ are of about equal efficiency in their ability to recognize CUC and CUA codons, whereas tRNA$_3^{Leu}$ is highly preferred for the CUU codon. tRNA$_4^{Leu}$ and tRNA$_5^{Leu}$ both recognize UUA and UUG codons, with tRNA$_4^{Leu}$ slightly preferred for the UUA codon. We conclude that:

1. Wobble is greater than permitted by the wobble hypothesis.
2. There is still some discrimination in the third code letter, and that the CUX (CUC, CUA, CUU, or CUG) portion of the leucine family of six codons is not read by a simple two-out-of-three mechanism.
3. A Watson-Crick pair (CG) between codon and anticodon does not appear to be preferred over an unorthodox pair (CC) in the wobble position.
4. A standard wobble pair (UG) between codon and anticodon is preferred over an unorthodox pair (UC).
5. The extensive wobble observed in the CUX leucine codon series is not paralleled in the UUX leucine (UUG, UUA) and phenylalanine (UUU, UUC) codon series, where mistranslation would be the consequence of such wobble.

The fact that tRNA$_1^{Leu}$ and tRNA$_3^{Leu}$ can respond to the same codons, albeit with varying efficiencies, is consistent with our observations of tRNA^{Leu} utilization in vivo (Holmes et al. 1977), indicating that tRNA$_1^{Leu}$

could substitute for (preferred) $tRNA_3^{Leu}$ under different growth conditions.

ACKNOWLEDGMENTS

This work was supported by grants from the National Science Foundation (PCM 75-23482) and the National Institutes of Health (GM-24330). E. G. was a Lievre Senior Fellow (D303) of the California Division–American Cancer Society. G. W. H. was the recipient of a U.S. Public Health Service Career Development Award (GM-70530). Special acknowledgment is also made to W. Michael Holmes, who collaborated and contributed in this work.

REFERENCES

Blank, H. U. and D. Söll. 1971. The nucleotide sequence of two leucine tRNA species from *Escherichia coli* K12. *Biochem. Biophys. Res. Commun.* **43:**1192.
Crick, F. H. C. 1966. Codon-anticodon pairing: The wobble hypothesis. *J. Mol. Biol.* **19:**548.
Dube, S. K., K. A. Marcker, and A. Yudelevich. 1970. The nucleotide sequence of a leucine transfer RNA from *E. coli. FEBS Lett.* **9:**168.
Fiers, W., R. Contreras, F. Duerinck, G. Haegeman, D. Iserentant, J. Merregaert, W. Min Jou, F. Molemans, A. Raeymaekers, A. Van den Berghe, G. Volckaert, and M. Ysebaert. 1976. Complete nucleotide sequence of bacteriophage MS2 RNA: Primary and secondary structure of the replicase gene. *Nature* **260:**500.
Goldman, E. and G. W. Hatfield. 1979. Use of purified isoacceptor tRNAs for the study of codon-anticodon recognition *in vitro* with sequenced natural messenger RNA. *Methods Enzymol.* **59:**292.
Goldman, E., W. M. Holmes and G. W. Hatfield. 1979. Specificity of codon recognition by *Escherichia coli* $tRNA^{Leu}$ isoaccepting species determined by protein synthesis in vitro directed by phage RNA. *J. Mol. Biol.* **129:**567.
Goldman, E. and H. F. Lodish. 1971. Inhibition of replication of ribonucleic acid bacteriophage f2 by superinfection with bacteriophage T4. *J. Virol.* **8:**417.
Grosjean, H., D. G. Söll, and D. M. Crothers. 1976. Studies on the complex between transfer RNAs with complementary anticodons. I. Origins of enhanced affinity between complementary triplets. *J. Mol. Biol.* **103:**499.
Holmes, W. M., G. W. Hatfield, and E. Goldman. 1978. Evidence for misreading during tRNA-dependent protein synthesis *in vitro. J. Biol. Chem.* **253:**3482.
Holmes, W. M., E. Goldman, T. A. Miner, and G. W. Hatfield. 1977. Differential utilization of leucyl-tRNAs by *Escherichia coli. Proc. Natl. Acad. Sci.* **74:**1393.
Holmes, W. M., R. E. Hurd, B. R. Reid, R. A. Rimerman, and G. W. Hatfield. 1975. Separation of transfer ribonucleic acid by sepharose chromatography using reverse salt gradients. *Proc. Natl. Acad. Sci.* **72:**1068.
Khorana, H. G., H. Büchi, H. Ghosh, N. Gupta, T. M. Jacob, H. Kössel, R. Morgan, S. A. Narang, E. Ohtsuka, and R. D. Wells. 1967. Polynucleotide synthesis and the genetic code. *Cold Spring Harbor Symp. Quant. Biol.* **31:**39.

Lagerkvist, U. 1978. "Two out of three": An alternative method for codon reading. *Proc. Natl. Acad. Sci.* **75**:1759.

Lin, J., C. M. Tsung, and H. Fraenkel-Conrat. 1967. The coat protein of the RNA bacteriphage MS2. *J. Mol. Biol.* **24**:1.

Min Jou, W., G. Haegeman, M. Ysebaert, and W. Fiers. 1972. Nucleotide sequence of the gene coding for the bacteriophage MS2 coat protein. *Nature* **237**:82.

Mitra, S. K., F. Lustig, B. Åkesson, U. Lagerkvist, and L. Strid. 1977. Codon-anticodon recognition in the valine codon family. *J. Biol. Chem.* **252**:471.

Nichols, J. L. and H. D. Robertson. 1971. Sequences of RNA fragments from the bacteriophage f2 coat protein cistron which differ from their R17 counterparts. *Biochim. Biophys. Acta* **228**:676.

Nirenberg, M., T. Caskey, R. Marshall, R. Brimacombe, D. Kellogg, B. Doctor, D. Hatfield, J. Levin, F. Rottman, S. Pestka, M. Wilcox, and F. Anderson. 1967. The RNA code and protein synthesis. *Cold Spring Harbor Symp. Quant. Biol.* **31**:11.

Tai, P. C., B. J. Wallace, and B. D. Davis. 1978. Streptomycin causes misreading of natural messenger by interacting with ribosomes after initiation. *Proc. Natl. Acad. Sci.* **75**:275.

Weber, K. 1967. Amino acid sequence studies on the tryptic peptides of the coat protein of the bacteriophage R17. *Biochemistry* **6**:3144.

Weber, K. and W. Konigsberg. 1967. Amino acid sequence of the f2 coat protein. *J. Biol. Chem.* **242**:3563.

Nontriplet tRNA–mRNA Interactions

John F. Atkins*
Cold Spring Harbor Laboratory
Cold Spring Harbor, New York 11724

Genetic decoding proceeds overwhelmingly in a triplet manner. However, studying the exceptions to this rule may help in the elucidation of the rules governing mRNA–tRNA interaction of the ribosome, the function of some modified bases in tRNA, and whether any proteins derived from nontriplet reading are utilized by cells or viruses. In addition, since it is possible that the primitive genetic code used predominately nontriplet codon–anticodon interactions (Woese 1972; Crick et al. 1976), studies on the extent of the alterations required to allow the present-day code to be read in a nontriplet manner at certain codons may provide some insight into the evolution of the code. This chapter summarizes the current status of the investigation on these topics. The first part considers mutationally produced nontriplet reading, and the second part considers the low-level nontriplet reading normally encountered at certain codons.

EXTRAGENIC SUPPRESSORS FOR FRAMESHIFT MUTANTS

Mutationally produced nontriplet reading has been sought by selecting for extragenic suppressors of frameshift mutations. The pioneering studies of Crick et al. (1961) showed that insertion or deletion of one or two bases in a gene causes the ribosomes to shift reading frame; they termed these alterations frameshift mutations. These studies revealed the general nature of the code and were confirmed by the amino acid sequencing studies of altered proteins by Streisinger et al. (1967) and also by a variety of other approaches. However, these early studies on frameshift mutations used mutant phages; it was not until Ames and Whitfield (1967) described a method for isolating frameshift mutants in bacteria (for a review, see Roth 1974) that extragenic suppressors were isolated (Riyasaty and Atkins 1968; Yourno et al. 1969; Riddle and Roth 1970).

Recently, L. Gold and colleagues (pers. comm.) have determined the DNA sequence in the region of phage T4 where the original frameshift mutants were isolated. The phage mutants have not yet been tested for suppression by the characterized bacterial suppressors. All the bacterial suppressors have been isolated in *Salmonella*, but it now seems worthwhile,

*Present address: Department of Biology, University of Utah, Salt Lake City, Utah 84112.

in view of the wealth of mutants and the current sequencing studies, to transfer selected suppressors to *Escherichia coli* for suppression tests with the phage T4 *rII* and lysozyme frameshift mutants. To date, one of the suppressors, *sufG*, has been transferred to *E. coli*, and certain of the lysozyme mutants have been found to be suppressible by it (Kohno and Roth 1978).

Most of our knowledge of the genetics of the bacterial frameshift suppressors comes from the work of J. R. Roth and his colleagues and is summarized in Table 1. The majority of the suppressors described in bacteria are frameshift-specific, act in runs of repeat bases, and suppress +1 frameshift mutations. However, the generality of these characteristics is doubtful, since many of the frameshift mutants for which suppressors were selected were isolated after mutagenesis with mutagens that act predominately on a limited range of sequences. The majority of the suppressors isolated for the first recognized, suppressible frameshift mutant were frameshift-specific and map in the *sufS* gene (see Table 1), but a small minority also suppressed UGA and mapped in the *supK* gene (Atkins and Ryce 1974; Uomini and Roth 1974). Previously it was found that the UGA suppressor alleles of *supK* were recessive (Reeves and Roth 1971). Pope and Reeves (1978) have shown that in *supK* strains several tRNAs are undermethylated, probably lacking a methyl ester (see Dudock et al. 1978). Thus, it is possible that the tRNAs mediating the nontriplet and UGA reading are different species.

J. Yourno and his colleagues (see Table 1) have determined the amino acid inserted by one frameshift suppressor and have deduced from amino acid sequencing studies a partial nucleotide sequence around several suppressible and nonsuppressible frameshift mutants. These studies were performed on histidinol dehydrogenase, the product of the *hisD* gene where most of the characterized suppressible frameshift mutants are located. T. Kohno et al. (pers. comm.) have now determined the nucleotide sequence of the *hisD* gene and the amino acid sequence of its product, and thus a better understanding of the characteristics of the sequences where frameshift suppression can occur should soon be available. Two alternatives for the mechanism of suppression have been proposed. Riddle and Carbon (1973) have shown that the glycine-inserting suppressor, *sufD*, has an extra C in the region of the CCC anticodon and one of the possible mechanisms they considered is that there is quadruplet base-pairing with the mRNA. The second alternative, which has been discussed by Yourno and Tanemura (1970), Riddle and Carbon (1973), and Kurland (1979), is that the extra nucleotide acts as a spacer so that the tRNA occupies an unusually large space on the message, leading to one base in the mRNA not forming part of a codon. Either alternative suggests that the tRNA directly controls the extent of mRNA movement (see Watson 1975). Another possibility that is very unlikely to be applicable to most frameshift suppressors but may apply to

Table 1 Bacterial frameshift extragenic suppressors

Locus	Map position[a]	Cotransducible marker	Dominant (D) or recessive (R)	Altered tRNA	Type of alteration	Sequence of suppressible mutant	Mutagens used for isolation of suppressible frameshift mutants	Other features
sufA	77	xyl[b]	D[b]	tRNA^{Pro} maj[c]		CCC,[d,e]	ICR-191	
sufB	46	hisW[b]	D[b]	tRNA^{Pro} min[c]			ICR-191[f]	sufA sufB doubles inviable
sufC	15	nag[g]	R[g]				ICR-191	
sufD	63	lys[b]	D[b]	tRNA$_1^{Gly}$	anticodon loop + c[h]	GGG,[s]	ICR-191	hisT required for suppression of one frameshift mutant[h]
sufE	89	rif[b]	D[b]				ICR-191	
sufF	12	pur E[b]	R[b]				ICR-191	
sufG	15	nag[i]	D[i]	tRNA$_{II}^{Gly}$			proflavin	
sufH	52	cys A[i]	D[q]	tRNA^{Lys} (AAAA/G?)[j]		AAAA + A[j]	proflavin	
sufI	12						proflavin	
sufJ	88	rif	D[r]				ICR-191 mitomycin C[r]	hisT required for suppression of one frameshift mutant
supK	62	lys[k,l]	R[k,l]		uridine-5-oxyacetic acid methylester?[m]		X ray[n]	also suppresses UGA weak fs suppression
sufM	77	xyl[j]					ICR-191	sufA and sufM may be allelic
sufS	88	rif	R[t]				9 aminoacridine	trmA (m^{5}U) maps in the same area and allelic tests are required[u]
sufT	59	recA[t]	D[t]				X ray[n,t]	
							spontaneous[t]	suppresses a frameshift mutant that is of opposite site sign to a frameshift mutant suppressed by sufS[t]

References: [a]Sanderson and Hartman 1978; [b]Riddle and Roth 1972a,b; [c]Riddle and Roth 1970; [d]Yourno and Tanemura 1970; [e]Yourno and Kohno 1972; [f]Oeschger and Hartman 1970; [g]G. Roberts and J. R. Roth, unpubl.; [h]Riddle and Carbon 1973; [i]Kohno and Roth 1978; [j]T. Kohno and J. R. Roth, in prep.; [k]Reeves and Roth 1971; [l]Atkins and Ryce 1974; [m]Pope and Reeves 1978; [n]Riyasaty and Atkins 1968; [o]Uomini and Roth 1974; [p]L. Bossi and J. R. Roth, pers. comm.; [q]T. Kohno, unpubl.; [r]L. Bossi, unpubl.; [s]Yourno 1972; [t]D. Hughes and S. Thompson, unpubl.; [u]Björk 1975.

some and may well be responsible for at least part of the low-level frameshifting mediated by nonmutant tRNAs (see below) is overlapping (stuttering) reading.

O. Uhlenbenck and G. Bruce (pers. comm.) have developed a technique that allows the replacement of the anticodon region of certain tRNAs by short oligmers of predetermined length and sequence. Also, the alteration of the critical regions of particular tRNA genes may now be feasible. The addition of the resulting tRNAs to an in vitro protein-synthesizing system that allows for an assay of frameshift suppression (see below) has not yet been attempted but may be instructive.

Other aspects requiring further investigation are suppressors of -1 frameshift mutations (see Atkins and Ryce 1974) and the tRNA modification defects in the recessive frameshift-specific suppressors (Riddle and Roth 1972b). However, recent genetic studies by L. Bossi et al. (unpubl.) have revealed an interesting means of assessing the in vivo functional importance of one tRNA modification. The bacterial mutation *hisT*, which abolishes the cell's ability to form ψ in the anticodon arm of many tRNAs, has striking effects on the activity of some nonsense and frameshift suppressors. In some cases a suppressor tRNA is made more efficient, in other cases less efficient, due to lack of anticodon arm ψs. The effects appear to be context- or site-specific. Lack of ψ residues impairs the suppressor tRNA's ability to read some sites but not others. The tRNA involved seems to recognize aspects of the mutant site outside of the codon being read. In one case the base adjacent to the 3' end of the codon is critical to reading efficiency (L. Bossi, unpubl.). It seems possible that the modified tRNA base is an important aspect of how some adjacent tRNAs "fit" together on the ribosome during protein synthesis. The findings of Bossi with *hisT* mutants are reminiscent of the ribosome binding studies of Taniguichi and Weissmann (1978) with mutants of phage Qβ. These studies show that formylmethionyl-tRNA plays an essential role in the formation of the 70S initiation complex; in addition, these studies suggest that if the base 3' of the AUG initiation codon is A, the interaction of formylmethionyl-tRNA with the initiation codon is strengthened. The 5' adjacent base of the tRNA anticodon is U, but whether it is directly involved in pairing with the A-base 3' of the codon in a quadruplet interaction remains to be determined. To assess the in vivo significance of the findings of Taniguichi and Weissmann (1978) the 92 prokaryotic translation initiation sequences currently known have been examined. No obvious correlation of any other feature with the presence or absence of A in the 3' position adjacent to the AUG initiators was detected.

Extragenic suppressors mapping in 14 genes have recently been described for frameshift mutants in the *his4* gene of yeast. Based on their cross suppression patterns, these suppressors fall into three classes; double-label chromatography experiments show that different glycine tRNAs are altered in some of the suppressors in one of these classes. The wild-type *his4* gene sequence has now been determined (Culbertson et al. 1980; Cummins et al. 1980) and sequencing studies of the suppressible frameshift mutants and further characterization of their suppressors are in progress. Despite some initial similarities to the bacterial frameshift suppressors, the yeast suppressors should prove interesting in view of the rather different types of tRNA modifications and ribosome structures in eukaryotes.

LOW-LEVEL NONTRIPLET READING

Leakiness of Frameshift Mutants

The low level of nontriplet reading normally encountered (as distinct from the nontriplet reading that is mutationally produced) at certain codons has been documented in several very different ways. The first is an investigation of the leakiness of frameshift mutants. In the detailed account of their earlier pioneering studies on phage frameshift mutants, Barnett et al. (1967) noted that certain frameshift mutants, or combinations of frameshift mutants, produced minute plaques due to a low level of the active gene product. This activity is not due to activity of the N-terminal fragment, since these frameshift mutants were not restricted to the distal part of the gene. Barnett et al. tentatively explained that the origin of the active gene product was due to a low level of quadruplet reading at certain, but not all, UGA codons. More recently, Hirsh and Gold (1971) showed that tRNA^{Trp} reads UGA at a low level; but whether tRNA^{Trp} can read UGA in a quadruplet manner and, if so, which base in the 3' adjacent position is involved, have not been investigated. Barnett et al. (1967) also considered, and did not rule out, the possibility of termination and reinitiation in the correct frame (cf. Sarabhai and Brenner 1967). Shortly afterwards some bacterial frameshift mutants were also observed to be leaky (Riyasaty and Atkins 1968; Newton 1970). Translation termination and reinitiation in the correct frame was shown not to be the explanation for the leakiness of many of these frameshift mutants, but it was not determined whether the nontriplet reading was occurring downstream at out-of-phase nonsense codons or at certain special sequences (Atkins et al. 1972). The degree of nontriplet reading was influenced by ribosome mutants *strA* and *ram* previously known to influence the fidelity of triplet reading (see Gorini 1974).

A similar situation with regard to *relA* mistranslation has recently been discovered. Mutants defective in the *relA* gene show amino acid substitutions when the acylation of a tRNA species is limited (Edelmann and Gallant, 1977; O'Farrell, 1978; Parker et al. 1978). This presumably occurs by triplet misreading at codons calling for the aminoacyl-tRNA in short supply. However, increased leakiness of certain frameshift mutations (including two of opposite sign and at two different nonsense mutations) has been observed during partial inhibition of the synthesis or activation of specific amino acids (Gallant and Foley 1980). The simplest interpretation is that frameshifting can occur in conjunction with triplet misreading, at least in *relA* mutants, as predicted on theoretical grounds by Kurland (1979).

Two-base Recognition with Triplet Decoding

It has been proposed that, during triplet reading, recognition of two of the three bases of a codon is often sufficient (Lagerkvist 1978; Mitra et al. 1979). Goldman and Hatfield (this volume) review this topic and discuss their recent data, which is also relevant to this point.

The Relative Concentration of Specific, Normal tRNAs in Cell-free Protein-synthesis Reactions May Affect the Amount of Nontriplet Reading

One explanation for the leakiness of frameshift mutants was that some normal tRNAs can read certain sequences in a nontriplet manner. By increasing the relative amount of such a tRNA, the amount of nontriplet reading may be increased. Fortuitously, in vitro protein synthesis experiments initiated to study a different problem (Atkins and Gesteland 1975) have given results that are consistent with a frameshift model. If a proportion of the ribosomes translating a gene-shift reading frame are near the distal end of a gene, then they will bypass the normal 0-frame termination codon and may produce a protein considerably longer than the 0-frame product. For the in vitro translation studies, RNA from the phage MS2 (or R17 or f2) was used as template in an *E. coli* S30 system containing the endogenous tRNAs. About 5% of the synthetase made in this system migrates on electrophoresis in SDS-polyacrylamide gels as if it were 4000 daltons larger than the rest of the synthetase. This product must result from initiation at the beginning of the synthetase gene, as an equivalent, truncated polypeptide from this gene is not seen when RNA from a synthetase amber mutant is used as template and prior initiation is not compatible with the nucleotide sequence. Heterogeneity of the template RNA is not the explanation for the larger form of the synthetase, as it is not produced in a mammalian translation system that yields the main synthetase (Atkins et al. 1975). It follows from an examination of the nucleotide sequencing studies of Fiers et al. (1976) and in vitro translation studies with nonsense suppressors, that the larger form of the synthetase cannot be derived from inphase reading past a leaky termination codon. In recent studies utilizing purified normal *E. coli* tRNAs (Atkins et al.

1979a), it was shown that the proportion of the larger (66K) form of the synthetase is increased by adding tRNA^{Thr} ACU/ACC and decreased by adding minor tRNA^{Pro}. The nucleotide sequence of MS2 RNA (Fiers et al. 1976) is consistent with a fraction of the ribosomes shifting into the -1 reading frame near the end of the synthetase gene and continuing in that frame to generate the larger form of the synthetase (Fig. 1).

Three minor proteins larger than the coat, but that are derived in part from the translation of the coat gene, are synthesized in extracts of *E. coli* with the endogenous tRNA pools. The smallest of these is labeled 20K in Figure 1. Addition of tRNA^{Ser} (AGU/AGC) enhanced synthesis of a protein that migrated with the 20K protein in electrophoresis in SDS-polyacrylamide gels. At higher concentrations of tRNA$^{Ser}_3$ other proteins are detected (Fig. 1). Synthesis of the tRNA$^{Ser}_3$-stimulated proteins is competed by tRNA$^{Ala}_{1B}$ (GC$^{U}_{G}$A). Amino-acid-sequencing studies (Atkins 1979a) have shown that the 20K tRNA^{Ser}-stimulated protein has the same amino terminal region as coat protein and that it is encoded in part by the -1 frame near the end of the coat gene. Synthesis of the largest protein derived from reading the coat gene is not stimulated by the addition of tRNA$^{Ser}_3$ but is enhanced by the addition of large amounts of tRNA$^{Pro}_{minor}$ or tRNA^{Gln}. Its synthesis also initiates at the start of the coat gene but is shifted into the $+1$ reading frame near the end of the coat gene. The lysis gene overlaps in the $+1$ frame of the 3' end of the coat gene (Model et al. 1979; Atkins et al. 1979b; Beremand and Blumenthal 1979). Thus, the largest protein derived from reading the coat gene has sequences in its carboxy-terminal region encoded by the lysis gene. However, despite the fact that these in vitro translation data strongly support a frameshift model for the origin of the proteins, it is not clear, especially for the tRNA$^{Ser}_3$-stimulated polypeptides, that nontriplet tRNA–mRNA interactions are involved. An alternative is overlapping reading. The frameshift site has not yet been identified, but near the end of the coat gene there are adjacent alanine codons GCA and GCA. The competitor tRNA for the AGU/AGC decoding tRNA$^{Ser}_3$-mediated effects is tRNA$^{Ala}_{1B}$ and it is possible that the tRNA$^{Ser}_3$ is reading in a triplet manner the AGC that is part of the two alanine codons. Further sequencing studies are required to delimit the coding possibilities.

The tRNA-mediated effects were detected in an in vitro translation system. However, Beremand and Blumenthal (1979) have detected a protein in phage-infected cells that is apparently the same size as the largest coat-related protein detected in the in vitro translation system. Studies using coat and lysis mutants suggest that both proteins are the same. It is unknown whether the hybrid protein has any role in infected cells or whether it is merely an unavoidable byproduct of selection pressure for high translation speed. Also unknown is the degree of relationship between this phenomenon and the leakiness of frameshift mutants discussed above.

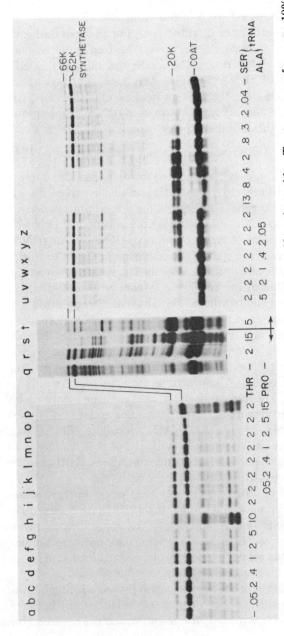

Figure 1 tRNA competition for the enhancement of synthesis of specific polypeptides. Tracts *a–p* are from one 10% gel with *h* and *p* being from a longer exposed autoradiograph. The other tracts are from a 15% gel. The amounts of tRNA in µg added to the 12.5-µl protein synthesis reaction mixtures are shown below the tracts. The tRNAs added were tRNA$_3^{Ser}$, tRNA$_{IB}^{Ala}$, major tRNA^{Thr}, and minor tRNA^{Pro}.

Another RNA phage, PP7, which is similar in size to MS2, has an additional protein in its virus particle (Dhaese and Van Montagu 1978). Sequencing studies are required to determine if this protein arises by a frameshifting mechanism. Selection for small size in these viruses presumably forced the maximum utilization of sequence information. In eukaryotes, the splicing of a transcript in different reading frames could provide an alternative way of bypassing termination codons. However, it is also possible that ribosomes shifting frame may sometimes be involved. Avian sarcoma virus polymerase is produced from a precursor that, though derived from translating the *gag* and *pol* genes, is not produced by in-phase readthrough of a leaky termination codon (Weiss et al. 1978). Whether *pol* is produced by a splicing mechanism or ribosomes shifting frame has not yet been determined.

ACKNOWLEDGMENTS

I thank R. Gesteland for making this work possible in several ways and for the pleasure of working in his laboratory; S. Thompson for the initial collaboration that started this investigation; B. Reid for the numerous tRNAs, and J. D. Watson for support. Unpublished information was kindly provided by T. Kohno, L. Bossi, J. Roth, S. Thompson, C. Kurland, J. Gallant, M. Goldman, R. Reeves, O. Uhlenbeck, M. Beremand, and T. Blumenthal.

REFERENCES

Ames, B. N. and H. J. Whitfield, Jr. 1967. Frameshift mutagenesis in *Salmonella*. *Cold Spring Harbor Symp. Quant. Biol.* **31**:221.

Atkins, J. F. and S. Ryce. 1974. UGA and non-triplet suppressor reading of the genetic code. *Nature* **249**:527.

Atkins, J. F. and R. F. Gesteland. 1975. The synthetase gene of the RNA phages R17, MS2 and f2 has a single UAG terminator codon. *Mol. Gen. Genet.* **139**:19.

Atkins, J. F., D. Elseviers, and L. Gorini. 1972. Low activity of β-galactosidase in frameshift mutants of *Escherichia coli*. *Proc. Natl. Acad. Sci.* **69**:1192.

Atkins, J. F., R. F. Gesteland, B. R. Reid, and C. W. Anderson. 1979a. Normal tRNAs promote ribosomal frame shifting. *Cell* **18**:119.

Atkins, J. F., J. B. Lewis, C. W. Anderson, and R. F. Gesteland. 1975. Enhanced differential synthesis of proteins in a mammalian cell-free system by addition of polyamines. *J. Biol. Chem.* **250**:5688.

Atkins, J. F., J. A. Steitz, C. W. Anderson, and P. Model. 1979b. Binding of mammalian ribosomes to MS2 phage RNA reveals an overlapping gene encoding a lysis function. *Cell* **18**:247.

Barnett, L., S. Brenner, F. H. C. Crick, R. G. Shulman, and R. J. Watts-Tobin. 1967. Phase-shift and other mutants in the first part of the r11B cistron of bacteriophage T4. *Philos. Trans. R. Soc. Lond. B* **252**:487.

Beremand, M. N. and T. Blumenthal. 1979. Overlapping genes in RNA phage: A new protein implicated in lysis. *Cell* **18**:257.

Björk, G. R. 1975. Transductional mapping of gene *trmA* responsible for the production of 5-methyluridine in transfer ribonucleic acid of *Escherichia coli*. *J. Bacteriol.* **124**:92.

Crick, F. H. C., L. Barnett, S. Brenner, and R. J. Watts-Tobin. 1961. General nature of the genetic code for proteins. *Nature* **192**:1227.

Crick, F. H. C., S. Brenner, A. Klug, and G. Pieczenik. 1976. A speculation on the origin of protein synthesis. *Origins Life* **7**:389.

Culbertson, M. R., K. M. Underbrink, and G. R. Fink. 1980. Frameshift suppression in *Saccharomyces cerevisiae*, II. Genetic properties of group II suppressors. *Genetics* (in press).

Cummins, C. M., R. F. Gaber, M. R. Culbertson, R. Mann, and G. R. Fink. 1980. Frameshift suppression in *Saccharomyces cerevisiae*, III. Isolation and genetic properties of group III suppressors. *Genetics* (in press).

Dhaese, P. and M. Van Montagu. 1978. The gene products of *Pseudomonas aeruginosa* RNA bacteriophage PP7. *Abst. IV Int. Congr. Virol.*, p. 593.

Dudock, B., J. Lesiewicz, and M. Y. Wang. 1978. A new class of tRNA modification reactions in *E. coli*: tRNA methyl esters. *Fed. Proc.* **37**:1732.

Edelmann, P. and J. Gallant. 1977. Mistranslation in *E. coli*. *Cell* **10**:131.

Fiers, W., R. Contreras, F. Duerinck, G. Haegeman, D. Iserentant, J. Merregaert, W. Min Jou, F. Molemans, A. Raeymaekers, A. Van Den Berghe, G. Volckaert, and M. Ysebaert. 1976. Complete nucleotide sequence of bacteriophage MS2 RNA: Primary and secondary structure of the replicase gene. *Nature* **260**:500.

Gallant, J. and D. Foley. 1980. On the causes and prevention of mistranslation. In *Ribosomes: Structure, function, and genetics* (ed. G. Chambliss et al.), p. 615. University Press, Baltimore.

Gorini, L. 1974. Streptomycin and misreading of the genetic code. In *Ribosomes* (ed. M. Nomura et al.), p. 791. Cold Spring Harbor Laboratory, Cold Spring Harbor, New York.

Hirsh, D. and L. Gold. 1971. Translation of a T_4 lysozyme UGA triplet *in vitro* by tryptophan transfer RNAs. *J. Mol. Biol.* **58**:459.

Kohno, T. and J. R. Roth. 1978. A frameshift suppressor which acts at runs of A residues in the message. *J. Mol. Biol.* **126**:37.

Kurland, C. G. 1979. Reading frame errors on ribosomes. In *Nonsense mutations and tRNA suppressors* (ed. J. E. Celis and J. D. Smith), p. 97. Academic Press, New York.

Lagerkvist, U. 1978. "Two out of three": An alternative method for codon reading. *Proc. Natl. Acad. Sci.* **75**:1759.

Mitra, S. K., F. Lustig, B. Åkesson, T. Axberg, P. Elias, and U. Lagerkvist. 1979. Relative efficiency of anticodons in reading the valine codons during protein synthesis *in vitro*. *J. Biol. Chem.* **254**:6397.

Model, P., R. E. Webster, and N. D. Zinder. 1979. Characterization of op3, a lysis defective mutant of bacteriophage f2. *Cell* **18**:235.

Newton, A. 1970. Isolation and characterization of frameshift mutations in the *lac* operon. *J. Mol. Biol.* **49**:589.

Oeschger, N. and P. E. Hartman. 1970. ICR induced frameshift mutations in the histidine operon of *Salmonella*. *J. Bacteriol.* **101**:490.

O'Farrell, P. H. 1978. The suppression of defective translation by ppGpp and its role in the stringent response. *Cell* **14**:545.

Parker, J., J. W. Pollard, J. D. Friesen, and C. P. Stanners. 1978. Stuttering: High-level mistranslation in animal and bacterial cells. *Proc. Natl. Acad. Sci.* **75**:1091.

Pope, W. T. and R. H. Reeves. 1978. Purification and characterization of a tRNA methylase from *Salmonella typhimurium*. *J. Bacteriol.* **136**:191.

Reeves, R. H. and J. R. Roth. 1971. A recessive UGA suppressor. *J. Mol. Biol.* **56**:523.

Riddle, E. L. and J. Carbon. 1973. Frameshift suppression: A nucleotide addition in the anticodon of a glycine transfer RNA. *Nat. New Biol.* **242**:230.

Riddle, D. L. and J. R. Roth. 1970. Suppressors of frameshift mutations in *Salmonella typhimurium*. *J. Mol. Biol.* **54**:131.

———. 1972a. Frameshift suppressors. II. Genetic mapping and dominance studies. *J. Mol. Biol.* **66**:483.

———. 1972b. Frameshift suppressors. III. Effects of suppressor mutations on transfer RNA. *J. Mol. Biol.* **66**:495.

Riyasaty, S. and J. F. Atkins. 1968. External suppression of a frameshift mutant in *Salmonella*. *J. Mol. Biol.* **34**:541.

Roth, J. R. 1974. Frameshift mutations. *Annu. Rev. Genet.* **8**:319.

Sanderson, K. E. and P. E. Hartman. 1978. Linkage map of *Salmonella typhimurium*, Edition V. *Microbiol. Rev.* **42**:471.

Sarabhai, A. and S. Brenner. 1967. A mutant which reinitiates the polypeptide chain after chain termination. *J. Mol. Biol.* **27**:145.

Streisinger, G., Y. Okada, J. Emrich, J. Newton, A. Tsugita, E. Terzaghi, and M. Inouye. 1967. Frameshift mutations and the genetic code. *Cold Spring Harbor Symp. Quant. Biol.* **31**:77.

Taniguichi, T. and C. Weissmann. 1978. Site directed mutations in the initiator region of the bacteriophage Qβ coat cistron and their effect on ribosome binding. *J. Mol. Biol.* **118**:533.

Uomini, J. R. and J. R. Roth. 1974. Suppressor-dependent frameshift mutants of bacteriophage P22. *Mol. Gen. Genet.* **134**:237.

Watson, J. D. 1975. *Molecular biology of the gene*, 3rd Ed., p. 335. W. A. Benjamin, Menlo Park, California.

Weiss, S. R., P. B. Hackett, H. Oppermann, A. Ullrich, L. Levintow, and J. M. Bishop. 1978. Cell-free translation of avian sarcoma virus RNA: Suppression of the *gag* termination codon does not augment synthesis of the joint *gag/pol* product. *Cell* **15**:607.

Woese, C. R. 1972. The emergence of genetic organization. In *Exobiology* (ed. C. Ponnamperuma), p. 301. North-Holland, Amsterdam.

Yourno, J. 1972. Externally suppressible +1 "glycine" frameshift: Possible quadruplet isomers for glycine and proline. *Nat. New Biol.* **239**:219.

Yourno, J. and T. Kohno. 1972. Externally suppressible proline quadruplet CCC^U. *Science* **175**:650.

Yourno, J. and S. Tanemura. 1970. Restoration of in-phase translation by an unlinked suppressor of a frameshift mutation in *Salmonella typhimurium*. *Nature* **225**:422.

Yourno, J., D. Barr, and S. Tanemura. 1969. Externally suppressible frameshift mutant of *Salmonella typhimurium*. *J. Bacteriol.* **100**:453.

Other Roles of tRNA

Comments on the Role of Aminoacyl-tRNA in the Regulation of Amino Acid Biosynthesis

H. Edwin Umbarger
Department of Biological Sciences
Purdue University
West Lafayette, Indiana 47907

The report by Eidlic and Neidhardt (1965) that reduced charging of acceptor $tRNA^{Val}$ in *Escherichia coli* was accompanied by derepression of the isoleucine and valine biosynthetic enzymes is as important to bacterial physiology as the earlier finding by Cohen and Jacob (1959) of mutations (*trpR*) unlinked to the *trp* gene cluster that lead to derepression of a tryptophan biosynthetic enzyme. The discovery of *trpR*, along with analogous mutations affecting arginine biosynthesis (Gorini et al. 1962; Maas 1962), was certainly instrumental in justifying Jacob and Monod's extension (1962) of the operator-repressor model of gene control to biosynthetic systems. Although the Eidlic-Neidhardt observation was not immediately followed by the formulation of an equally useful model, it was followed by a flurry of activity, primarily with bacteria, in many laboratories reporting the existence or nonexistence of a similar effect of limited charging of other tRNAs on the respective pathways.

Neidhardt and his coworkers were not guilty of the most common interpretation—that when an inverse correlation was found between the level of charging by a given amino acid and the amount of biosynthetic enzyme activity, it was not the free amino acid that bound to the repressor but an aminoacyl-tRNA. Some workers suggested that certain aminoacyl-tRNAs had distinct regulatory roles but, at most, only minor roles in protein synthesis. A more operational interpretation was that the aminoacyl-tRNA synthetase was involved in the level of expression of the cognate amino acid biosynthetic pathway and that the data did not allow the decision between the direct involvement of the synthetase itself, an involvement of the product, aminoacyl-tRNA, or the involvement of some unknown product of the synthetase. Neither did it permit the decision whether control was transcriptional or translational.

Very seldom was it suggested that previously unsuspected control parameters not involving operator or repressor might be defective when the charging level was low. This situation illustrated a point made before;

the repressor-operator model was remarkable for the number of observations that could be pulled together under a single concept (Umbarger 1969). At the same time, this conceptual umbrella became an intellectual shroud stifling the development of either reasonable alternative or supplementary concepts of control of gene expression.

There were, it is true, a limited number of observations that were, at the time, probably justifiably taken as evidence of some direct role of tRNA in regulation. Since regulation was assumed to occur via repression, it followed that tRNA was involved in the generation of a repression signal. The first of these was the discovery that the mutation in the *hisT* locus of *Salmonella typhimurium* resulted in a modification of the single isoacceptor species of tRNA^{His} as well as derepression of the *his* operon (Singer et al. 1972). Although an alteration in tRNA^{His} might have been accompanied by limited charging or limited capacity for peptidyl transfer, it was thought that *hisT* tRNA^{His} was normal in both charging and amino acid incorporation (Lewis and Ames 1972). Therefore, it did seem that there was a specific regulatory role for presumably charged tRNA^{His} that was independent of its role in the translation process.

Actually, consideration of the possible roles for histidyl-tRNA in regulation of the *his* operon did provide one of the exceptions to my claim that the Jacob-Monod repressor also repressed new ideas. This exception is an unpublished model proposed by Martin et al. (1967)—the "CAU-CAU" model. It proposed that the first structural gene in the *his* operon was preceded by a sequence rich in histidine codons followed by two stop codons, e.g., UGA-UGA. Thus, when histidine was in excess, the polyhistidine peptide would be completed and translation would stop at the UGA. When histidine was limiting, the stop codon would not be reached, but rather the sequence reading "stop-stop" (UGA-UGA) would be read out of phase as "start with formylmethionine" (UG-AUG-A), and translation could proceed into the *hisG* message. This speculative model is interesting in light of the recent finding by Barnes (1978) and by F. Blasi (pers. comm.) of a transcribed region of DNA preceding the *his* operon that would specify a histidine-rich polypeptide. Even though this model was a much less sophisticated one than that of Lee and Yanofsky (1977), to which I shall refer later, it is interesting that it presaged the discovery of a leader region by several years. It also predicted translation of the leader.

AMINO ACID BIOSYNTHETIC PATHWAYS APPARENTLY REGULATED BY tRNA CHARGING

It might be appropriate to list the amino acid biosynthetic pathways for which there are reports of derepression, again primarily in bacteria, resulting from limited charging. These are listed in the top portion of

Table 1. The lower portion of Table 1 lists amino acids for which limited charging of a given tRNA by its cognate amino acid was not accompanied by derepression of the biosynthetic pathway for the amino acid. Several amino acids appear in both parts of the table, since opposite results were obtained under different conditions, in different organisms, or for different enzymes in the same pathway. These discrepancies will be considered later. Finally, the table notes amino acids, analogs of which were shown to be less inhibitory to *hisT* mutants of *S. typhimurium* than to wild type. This pleiotropic effect could be explained by derepression of the amino acid biosynthetic pathway; but as the findings of Quay and Oxender (this volume) indicate, there are other possibilities.

Space does not permit a critical appraisal of the evidence underlying the claims cited in Table 1, but commentary on some of the examples is appropriate. It might be of interest to contrast the results obtained with

Table 1 Reported effects of limited charging of tRNAs on amino acid biosynthetic enzymes

Amino acid	References
Amino acid biosynthetic enzymes elevated	
His[a]	Roth et al. (1966)
Trp	Morse and Morse (1976)
Val	Eidlic and Neidhardt (1965)
Leu	Alexander et al. (1971); Low et al. (1971)
Ile	Szentirmai et al. (1968); Iaccarino and Berg (1971)
Thr	Nass et al. (1969); Johnson et al. (1977)
Asn	Arfin et al. (1977)
Glu	Lapointe et al. (1975)
Lys[a]	Boy et al. (1976)
Arg	Williams (1973)
Tyr[a]	Heinonen et al. (1972)
Met	Cherest et al. (1975)
Amino acid biosynthetic enzymes not elevated	
Ser	Low et al. (1971)
Gly	Folk and Berg (1970)
Gln	Körner et al. (1974)
Phe	Neidhardt (1966)
Lys[a]	Boy et al. (1976)
Arg	Hirshfield et al. (1968)
Tyr[a]	Schlesinger and Nester (1969); Ravel et al. (1965)
Met	Gross and Rowbury (1971); Ahmed (1973)

[a]*hisT* mutations cause resistance to certain analogs of these amino acids (Singer et al. 1972).

two systems, histidine biosynthesis in *S. typhimurium* and tryptophan biosynthesis in *E. coli*. It has been strikingly clear since Ames and Hartman and their collaborators (Roth et al. 1966; Roth and Ames 1966) studied the *hisR* and *hisS* mutants that tRNA^{His} or histidyl-tRNA synthetase or both were involved somehow in regulation of the *his* operon. In contrast, the evidence for an analogous involvement in regulation of the *trp* operon was weaker and often contradictory. Indeed, unequivocal evidence for involvement of tRNA^{Trp} or tryptophanyl-tRNA synthetase was possible only in strains in which the repressor-operator interaction was abolished by a *trpR* mutation (Morse and Morse 1976). In other words, the tRNA-mediated control was masked by the *trpR*-mediated control, which had been shown to function in vitro in the complete absence of either tryptophanyl-tRNA synthetase or tRNA^{Trp} (Squires et al. 1973).

Perhaps there is a lesson to be learned here when it is recalled that regulation of the *his* operon appears to be analogous to that of the *trp* operon in the absence of the *trpR* gene. Have the failures to demonstrate tRNA-mediated control in at least some other systems been due to a masking of the effect by either a repressor-mediated control or to some other amino-acid-specific control? The discrepant results with regulation of the tyrosine biosynthetic enzymes are in accord with this possibility. Examination of strains defective in charging tRNA^{Tyr} (Schlesinger and Nester 1969), as well as a comparison between tyrosine analogs that can be transferred to tRNA^{Tyr} and those that appear to repress (Ravel et al. 1965), pointed to a lack of a tRNA-mediated control. However, there remains one report of an elevation of a tyrosine biosynthetic enzyme upon limited charging of tRNA^{Tyr} in a *tyrR* (repressor-negative) strain (Heinonen et al. 1972). Perhaps the *tyrR*-dependent control masks the tRNA-dependent control.

Evidence both for and against a tRNA-mediated control also exists in the case of the arginine pathway. It is interesting that the mutants in which limited arginyl-tRNA synthetase activity had no effect on regulation were selected from among cells in a medium in which the cells were already physiologically derepressed (Hirshfield et al. 1968). In contrast, mutants in which reduced charging was accompanied by increased enzyme levels were selected in cells that had to acquire high enzyme levels before growth was possible (Williams 1973). Perhaps examination of the effect of limited charging should be undertaken in *argR* strains. All of the enzymes should be examined in the case of arginine biosynthesis, since only three enzymes are specified by a multicistronic operon.

There are numerous precedents for different enzymes in a single pathway responding differently to regulatory signals. One example is the behavior of the lysine biosynthetic enzymes. Only a single lysine biosynthetic enzyme, dihydropicolinate reductase, has been found to be derepressed upon limited charging in one strain of *E. coli* but not in another (Boy et al. 1976). Regulation of the other lysine- (and diaminopimelate-) forming enzymes appeared not to be affected by synthetase function. However, the *hisT* mutation that leads to an altered tRNA^{Lys} appears to prevent the derepression of diaminopimelate decarboxylase (but not that of the other enzymes) that normally occurs upon lysine limitation (Boy et al. 1978).

Some comment should probably be made concerning the inclusion of glutamate in the top portion of Table 1. The experiments of LaPointe et al. (1975) did, indeed, show an increase in the level of both glutamate synthetase and glutamine synthetase upon shifting an *E. coli* strain containing a temperature-sensitive glutamyl-tRNA synthetase to the restrictive temperature. However, these experiments were all performed with cells grown in a richly supplemented medium in which it is to be expected that amino acid biosynthetic enzymes would generally be under a severe metabolic depression. Thus, a more proper interpretation would be that the experiments showed that a partial relief of metabolic depression of those enzymes occurred when tRNA^{Glu} charging was restricted.

The inclusion of glutamine in the lower portion of Table 1 implies no relationship between the regulation of glutamine synthetase formation and either glutaminyl-tRNA synthetase or tRNA^{Gln}. There is a report, however, that an excess of tRNA^{Gln} caused by a mutation in *glnT* was accompanied by an increased formation of glutamine synthetase (Morgan et al. 1977). Actually, this mutant was selected for suppression of the effect of a temperature-sensitive glutaminyl-tRNA synthetase. Since the *glnT* mutation (a duplication of the tRNA^{Gln} structural gene) was not transferred out of the revertant in which it arose, one cannot yet be certain that the glutamine synthetase elevation and the *glnT* duplication were causally related. There could have been two independent mutations that, in concert, allowed effective suppression.

The methionine biosynthetic enzymes have been shown both in *E. coli* and *S. typhimurium* not to be affected by the charging level of tRNA^{Met} nor by methionyl-tRNA synthetase function (Ahmed 1973; Gross and Rowbury 1971). In view of the finding that in yeast there is both a methionine-dependent control (involving methionyl-tRNA) and a *S*-adenosylmethionine-dependent control (Cherest et al. 1975), it would be worthwhile reexamining the effect of methionyl-tRNA synthetase function in *met* strains presumably lacking a repressor that may also be *S*-adenosylmethionine-dependent.

THE MECHANISM OF "tRNA-MEDIATED" CONTROL

Given the now well-established fact that there is an effect of synthetase-tRNA function in regulation of at least some amino acid biosynthetic enzymes, the question arises concerning the mechanism of the effect. Until direct evidence to the contrary is uncovered for some system, the simple-minded notion that it is the aminoacyl-tRNA, rather than the amino acid, that serves as corepressor can probably be eliminated. With the precedence of one system, tryptophan biosynthesis, and perhaps a second, histidine biosynthesis, a mechanism acting quite independently of operator-repressor control seems more attractive. We should also be prepared to recognize more than one mechanism underlying the synthetase-tRNA involvement. Perhaps this state of mental preparedness will be hard to sustain in light of the simplicity of the model proposed by Lee and Yanofsky (1977) to account for the effect of limiting charging of tRNA^{Trp} on *trp* operon expression. Indeed, it may be quite tempting to extend this model to other systems, whether the evidence justifies it or not.

The Lee and Yanofsky model takes into account the fact that in both *E. coli* and *S. typhimurium* two kinds of transcripts are formed. One, the leader, is a transcript of about 140 nucleotides. The second contains the leader and is extended into the *trpE* transcript, which begins about 20 nucleotides beyond the point at which the leader ends when transcription is terminated early. Whether transcription is terminated at the end of the leader sequence or continues into the structural genes depends in part on whether tRNA^{Trp} is sufficiently charged. The nucleotide sequence of the region immediately preceding the site of transcription termination is markedly similar to sequences preceding transcription termination sites in other systems, such as λ. The sequence at the 3' terminal of such transcripts can be postulated to form stable stem-and-loop structures. It seems likely that this structure, which precedes a poly(U) sequence, has a strong tendency to cause a pausing of the polymerase that had formed it and finally a termination of transcription. In vitro, this transcription termination occurs without additional factors. In vivo, it appears that the rho termination factor is important. Platt (1978) has suggested that rho may be needed for release of the transcript from the termination complex. The mechanism of termination is probably not a necessary part of the model, since transcription termination at such stem-and-loop regions appears to be a general property of RNA polymerase and is not tryptophan-specific.

The leader transcript is interesting because it specifies a 14-amino-acid peptide, beginning with an AUG, about 30 nucleotides beyond the site of transcription initiation and ending at an in-phase stop codon about 40 nucleotides before the stem-and-loop region. Of particular interest is the fact that the putative leader peptide would contain two tandem tryptophan residues at positions 10 and 11. The Lee and Yanofsky model (1977) further assumes, with good reason, that the leader is indeed translated. The distance between the stop codon and the potential stem-and-loop secondary structure is such that its formation would not be hindered by polymerases translating the leader to the termination codon and releasing a 14-amino-acid peptide. Complete translation might be expected to occur only if tryptophanyl-tRNA would be sufficiently plentiful to allow the relatively high demand for tryptophan (2 of 14 codons) to be met. If tryptophanyl-tRNA were limiting, ribosomes would pause at the tryptophan codon. If this were to happen, it is postulated that a different stem-and-loop structure (or some alternative secondary structure) could occur. This different structure would interfere with the one thought to be responsible for transcription termination. Since the early structure is not a terminating structure, the polymerase is not halted, but rather it proceeds into the *trpE* gene. In other words, the frequency with which transcription terminates early would be a function of the frequency with which the ribosomes can proceed to the UGA codon, disrupt the early stem and loop, and thus allow formation of the terminating stem and loop.

In essence, the model gives no special role to either the synthetase or to the tRNA. Rather, it allows for transcription to be enhanced when, for whatever reason, the rate of incorporation of tryptophan becomes rate-limiting in protein synthesis. This model is one that can be extended to any amino acid biosynthetic transcript in which the leader could specify a peptide that would tip the scale, so to speak, in the discharging of the cognate tRNA. As we now know from the work of Barnes (1978) and of DiNocera et al. (1978), this is the case for the *his* operon leader.

tRNA PLAYS NO SPECIAL ROLE IN REGULATION

As noted above, the applicability of the model also depends on there being no special role for either synthetase or tRNA. To what extent is this true for either tryptophan or histidine biosynthesis?

Study of regulation of the *his* operon in *S. typhimurium* has revealed quite clearly that strains containing *hisT* lesions have derepressed levels of the histidine biosynthetic enzymes (Chang et al. 1971). tRNA^{His} from such strains has been shown to lack the two ψ residues normally present in the anticodon loop (Singer et al. 1972). That the presence of this kind of tRNA does, in fact, enhance expression, almost certainly at the level of transcription termination, of a normal *his* operon was shown by the in vitro studies of Artz and Broach (1975). It did not enhance expression of *his* DNA if the template lacked the attenuator, the postulated site of leader termination. Thus, there is some direct or indirect relationship between the *kind* of tRNA^{His} present and the attenuator. Until now it has been thought that *hisT* tRNA^{His} functions normally in charging, ribosome binding, and chain elongation (Singer et al., 1972). If this is indeed so, there must be a special role for tRNA^{His} in regulation. However, if it has any tendency to retard ribosome travel along a transcript rich in histidine codons, the Lee and Yanofsky (1977) model could readily accommodate the *hisT* mutant behavior.

There are two additional findings that must be explained if the Lee and Yanofsky model is to account for *his* operon expression. The work of Wyche et al. (1974) showed that a strain diploid for the structural gene for histidyl-tRNA synthetase (*hisS*) has an increased level of *his* operon expression. The question here is whether the elevated histidyl-tRNA synthetase serves to trap some histidyl-tRNA and thus actually retards ribosome travel when there is high demand for histidine or whether the synthetase plays a specific role in overcoming transcription termination at the attenuator. Some considerations favor the former, but the question should be studied further.

It should also be recalled that Artz and Broach (1975) were unable to obtain an uncoupled in vitro synthesis of *his* mRNA with $hisO^+$ DNA as template. In contrast, when *his* DNA lacking the attenuator was used as template, an uncoupled synthesis was possible. On the surface, this observation in terms of the Lee and Yanofsky model (1977) implies that the attenuator not only terminates transcription if the entire leader is translated, but it also somehow prevents transcription into the structural genes if the leader cannot be translated at all. It may be that the sequence of the leader region of the *his* operon will allow us to postulate a variety of secondary structures. If there were a secondary structure in the leader sequence that was prevented upon the initiation of translation and that prevented formation of the chain-terminating structure in the vicinity of the attenuator, the dependence of transcription upon translation would be understandable. Indeed, it is not hard to imagine a leader region in which the stem and loop that stops transcription would persist if translation initiation did not occur, would be masked if translation were halted at a midpoint in the leader, and would be restored if the putative leader peptide were translated to completion. Under such conditions it might well appear that tRNA charging played a positive and a negative control role. In reality, it would only be that early termination and continuation into the structural genes was a function of the extent to which translation of the leader occurred.

The presence of the transcribed leader regions preceding the structural genes for the histidine and tryptophan biosynthetic enzymes, as we have seen, can be correlated well with the regulatory behavior of the two systems. Quite intriguing therefore is the analogous leader region preceding the structural gene (*pheA*) for a phenylalanine biosynthetic enzyme that is also rich in phenylalanine codons (Zurawski et al. 1978), since as Table 1 implies, restricted $tRNA^{Phe}$ charging has been shown not to lead to derepression of phenylalanine biosynthetic enzymes (Neidhardt 1966). However, that study was apparently based on the behavior of the phenylalanine-sensitive DAHP (3-deoxy-D-*arabino*-heptulosonate-7-

phosphate) synthetase, the gene for which (*aroG*) is unlinked to the *pheA* gene. The behavior of chorismate mutase *P*-prephenate dehydratase (the *PheA* product) upon limiting charging of tRNA^{Phe} has apparently not been reported.

For the present, it seems most likely that the effects of aminoacyl-tRNA synthetase or tRNA or both that have been observed on the levels of amino acid biosynthetic enzymes are independent of either a repressor-operator interaction or a promoter-positive control element interaction. The evidence is strong, however, for only two such cases. One goal will be to ascertain whether some effects of synthetase function have been missed because of a masking by a control at the level of transcription initiation. Another goal, of course, will be to ascertain whether those regulatory systems that have been detected can in fact be explained by a model similar to that proposed by Lee and Yanofsky (1977) for tryptophan biosynthesis or whether some other mechanism (even the unlikely repressor-corepressor interaction) is responsible.

THE EFFECT OF LIMITED CHARGING ON OTHER SYSTEMS

The recent findings by Oxender's group (Quay et al. 1977) that at least one amino acid transport system is subject to some of the responses found for the biosynthetic enzymes, including derepression upon limiting charging of a tRNA, make it imperative that other transport systems should be so analyzed. Since altered repressibility of biosynthetic enzymes is generally tested by adding amino acids, the inability to show normal repression may really be due to the altered transport of the amino acids. It is also obvious that the existence of genes for the transport system provides another whole array of control sites to be analyzed.

Although there are good reasons for reservations concerning a specific role for either tRNAs or the synthetases in the regulation of amino acid biosynthetic enzymes, there is one area in which it is too early to dismiss the idea of a regulatory role for either the synthetases or tRNAs, and that is in the formation of the synthetases themselves. Since the review of a few years ago by Neidhardt et al. (1975) covered both the specific amino acid control and the general metabolic control of synthetase formation and is still timely, this topic will be considered only briefly here.

AMINO-ACID-SPECIFIC REGULATION OF tRNA SYNTHETASE FORMATION

It is well established that, for at least some of the aminoacyl-tRNA synthetases, either a transient or a permanent derepression of the synthetase occurs upon restricting the supply of the cognate amino acid. The derepression is sometimes compared with that of the amino acid biosynthetic enzymes. In no case, however, has it appeared that a repressor gene affecting the biosynthetic enzymes also affected the synthetase. In contrast, there have been several examples in which mutations affecting tRNA or the synthetase affected the derepression. Table 2 lists the aminoacyl-tRNA synthetases that have been shown to be subject to such a control.

In some cases, initial experiments failed to demonstrate a derepression upon amino acid restriction because of the synthetase's limited stability in the absence of the amino acid. Thus, derepression occurring as a result of amino acid deficiency was masked by a decay that was also due to the amino acid deficiency. Arginyl-tRNA synthetase is such an enzyme, and it required the use of density-shift experiments to distinguish between enzyme made before and after arginine deprivation (Williams and Neidhardt 1969).

Neidhardt et al. (1975) have discussed the question of possible regulatory effectors for regulation of aminoacyl-tRNA synthetase formation. The tRNA substrates are good candidates, and the possibility that the synthetase itself serves to regulate further synthetase formation is attractive. Mutations have been described that led to elevated levels of seryl- (Clarke et al. 1973), threonyl- (Paetz and Nass 1973), methionyl- (Cassio 1975), and leucyl- (LaRossa et al. 1977) tRNA synthetases. In some

Table 2 Reported effects of amino acid restriction on the level of the corresponding aminoacyl-tRNA synthetase activities

Amino acid	References
Derepression of the synthetase	
His	McGinnis and Williams (1972)
Leu	McGinnis and Williams (1971)
Met	Archibold and Williams (1972)
Pro	Archibold and Williams (1972)
Ser	Clarke et al. (1973)
Thr	Archibold and Williams (1972)
Val	McGinnis and Williams (1971)
Arg	Williams and Neidhardt (1969)
Ile	Nass and Neidhardt (1967); McGinnis and Williams (1971)
Phe	Nass and Neidhardt (1967)
Tyr	Dale and Nester (1971)
No derepression of the synthetase	
Gln	Körner et al. (1974)
Ser	Pizer et al. (1972)

cases these elevated levels could have been due to promoter mutations or to gene amplification, as has been reported for several synthetases, rather than to mutations in regulatory genes. One of two kinds of mutation-elevating leucyl-tRNA synthetases was due to an unlinked lesion in a locus (*leuY*) that may be a regulatory gene.

A lesion affecting isoleucyl-tRNA synthetase has recently been studied by Fayerman (1978). Mutations in *ilvU* prevent the derepression of isoleucyl-tRNA synthetase usually accompanying isoleucine limitation. They do not affect the biosynthetic enzymes, however. Although the interaction between *ilvU* and the synthetase remains unclear, dominance and deletion studies indicated that the *ilvU* product retarded the conversion of the minor species (as resolved by RPC-5 chromatography) of both tRNA^{Val} and tRNA^{Ile} to the major species. The mutant form of *ilvU* caused a greater retardation. Deletion of *ilvU* resulted in a rapid conversion of the minor to the major species so that only the major species of both acceptors appeared. Since the derepression of isoleucyl-tRNA synthetase still occurred when the *ilvU* gene was deleted, it is probably the altered *ilvU* product itself and not the amount of the minor species that prevents derepression.

METABOLIC CONTROL OF SYNTHETASE FORMATION

The specific-amino-acid-mediated control over synthetase formation may be masked not only by stability problems but also by the phenomenon of metabolic control studied primarily by Neidhardt et al. (1975). Thus, it might be that the reduced growth rate encountered during restricted amino acid supply would evoke the growth-rate-mediated depression of synthetase formation to the same (or greater) extent as the elevation of synthesis evoked by amino acid limitation itself. Certainly the question requires more study with perhaps newer techniques. For example, the recent application of the O'Farrell two-dimensional gel system to actual demonstration of synthetase proteins by Neidhardt et al. (1977) should prove valuable in answering questions that could not be answered otherwise. Another approach would be to link the promoter region of an aminoacyl-tRNA synthetase structural gene to the *lac* operon and to study its expression in vitro.

The attempt has been made here to help sound the death knell for the concept that tRNAs play specific roles in repression of amino acid biosynthetic enzymes. This view seems justified even if it is biased. At the same time, no attempt has been made either to sustain or eliminate the possibility that tRNAs or even the synthetases themselves may have regulatory roles in aminoacyl-tRNA synthetase formation. However, biology might prove to be a great deal more interesting if I am wrong on both counts.

ACKNOWLEDGMENTS

I wish to thank my colleagues, Jean Brenchley and Luther Williams who, although not necessarily agreeing with all my views, called many of the points discussed here to my attention during the course of numerous lively discussions.

REFERENCES

Ahmed, A. 1973. Mechanism of repression of methionine biosynthesis in *Escherichia coli*. I. The role of methionine, S-adenosyl methionine, and methionyl-transfer ribonucleic acid in repression. *Mol. Gen. Genet.* **123**:299.

Alexander, R. R., J. M. Calvo, and M. Freundlich. 1971. Mutants of *Salmonella typhimurium* with an altered leucyl-transfer ribonucleic acid synthetase. *J. Bacteriol.* **106**:213.

Archibold, E. R. and L. S. Williams. 1972. Regulation of synthesis of methionyl-prolyl- and threonyl-transfer ribonucleic acid synthetases of *Escherichia coli*. *J. Bacteriol.* **109**:1020.

Arfin, S., D. R. Simpson, C. S. Chiang, I. L. Audrulis, and G. W. Hatfield. 1977. A role for asparaginyl-tRNA in the regulation of asparagine synthetase in a mammalian cell line. *Proc. Natl. Acad. Sci.* **74**:2367.

Artz, S. W. and J. R. Broach. 1975. Histidine regulation in *Salmonella typhimurium*: An activator-attenuator model of gene regulation. *Proc. Natl. Acad. Sci.* **72**:3453.

Barnes, W. N. 1978. DNA sequence of the histidine operon control region: Seven histidine codons in a row. *Proc. Natl. Acad. Sci.* **75**:4281.

Boy, E., F. Borne, and J.-C. Patte. 1978. Effect of mutations affecting lysyl-tRNA[Lys] on the regulation of lysine biosynthesis in *Escherichia coli*. *Mol. Gen. Genet.* **159**:33.

Boy, E., F. Reinisch, C. Richaud, and J.-C. Patte. 1976. Role of lysyl-tRNA in the regulation of lysine biosynthesis in *Escherichia coli* K-12. *Biochimie* **58**:213.

Cassio, D., 1975. Role of methionyl-transfer ribonucleic acid in the regulation of methionyl-transfer ribonucleic acid synthetase of *Escherichia coli* K-12. *J. Bacteriol.* **123**:589.

Chang, G. W., J. R. Roth, and B. N. Ames. 1971. Histidine regulation in *Salmonella typhimurium*. VIII. Mutations in the *hisT* gene. *J. Bacteriol.* **108**:410.

Cherest, H., Y. Surdin-Kerjan, and H. de Robichon-Szulmajster. 1975. Methionine- and S-adenosyl methionine-mediated repression in a methionyl-transfer ribonucleic acid synthetase mutant of *Saccharomyces cerevisiae*. *J. Bacteriol.* **123**:428.

Clarke, S. J., B. Low, and W. Konigsberg. 1973. Isolation and characterization of a regulatory mutant of an aminoacyl-transfer ribonucleic acid synthetase in *Escherichia coli* K-12. *J. Bacteriol.* **113**:1096.

Cohen, G. and F. Jacob. 1959. Sur la repression de la synthese des enzymes intervenant dans la formation du tryptophane chez *E. coli*. *C. R. Acad. Sci.* **248**:3490.

Dale, B. A. and E. W. Nester. 1971. Regulation of tyrosyl-transfer ribonucleic acid synthetase in *Bacillus subtilis. J. Bacteriol.* **108:**586.

DiNocera, P. P., F. Blasi, R. DiLauro, R. Frunzio, and C. B. Bruni. 1978. Nucleotide sequence of the attenuator region of the histidine operon of *Escherichia coli* K-12. *Proc. Natl. Acad. Sci.* **53:**4276.

Eidlic, L. and F. C. Neidhardt. 1965. Role of valyl-sRNA synthetase in enzyme repression. *Proc. Natl. Acad. Sci.* **53:**539.

Fayerman, J. 1978. "Role of the *ilvU* locus in *Escherichia coli*." Ph.D thesis, Purdue University, West Lafayette, Indiana.

Folk, W. R. and P. Berg. 1970. Isolation and characterization of *Escherichia coli* mutants with altered glycyl transfer ribonucleic acid synthetases. *J. Bacteriol.* **102:**193.

Gorini, L., W. Gunderson, and M. Burger. 1962. Genetics of regulation of enzyme synthesis in the arginine biosynthetic pathway of *Escherichia coli*. *Cold Spring Harbor Symp. Quant. Biol.* **26:**173.

Gross, T. S. and R. J. Rowbury. 1971. Biochemical and physiological properties of methionyl-sRNA synthetase mutants of *Salmonella typhimurium*. *J. Gen. Microbiol.* **65:**5.

Heinonen, J., S. W. Artz, and H. Zalkin. 1972. Regulation of the tyrosine biosynthetic enzymes in *Salmonella typhimurium*: Analysis of the involvement of tyrosyl-transfer ribonucleic acid and tyrosyl-transfer ribonucleic acid synthetase. *J. Bacteriol.* **112:**1254.

Hirshfield, I. N., R. DeDeken, P. C. Horn, D. A. Hopwood, and W. K. Maas. 1968. Studies on the mechanism of repression of arginine biosynthesis in *Escherichia coli*. III. Repression of enzymes of arginine biosynthesis in arginyl-tRNA synthetase mutants. *J. Mol. Biol.* **35:**83.

Iaccarino, M. and P. Berg. 1971. Isoleucine auxotrophy as a consequence of a mutationally altered isoleucyl tRNA synthetase. *J. Bacteriol.* **105:**527.

Jacob, F. and J. Monod. 1962. On the regulation of gene activity. *Cold Spring Harbor Symp. Quant. Biol.* **26:**193.

Johnson, E. J., G. N. Cohen, and I. Saint-Girons. 1977. Threonyl-transfer ribonucleic acid synthetase and the regulation of the threonine operon in *Escherichia coli*. *J. Bacteriol.* **129:**66.

Körner, A., B. B. Magee, B. Liska, K. B. Low, E. A. Adelberg, and D. Söll. 1974. Isolation and characterization of a temperature-sensitive *Escherichia coli* mutant with altered glutaminyl-transfer ribonucleic acid synthetase. *J. Bacteriol.* **120:**154.

Lapointe, J., G. Delcuve, and L. Duplain. 1975. Derepressed levels of glutamate synthase and glutamine synthetase in *Escherichia coli* mutants altered in glutamyl-transfer ribonucleic acid synthetase. *J. Bacteriol.* **123:**843.

LaRossa, R., G. Vögeli, K. B. Low, and D. Söll. 1977. Regulation of biosynthesis of aminoacyl-tRNA synthetase and of tRNA in *Escherichia coli*. II. Isolation of regulatory mutants affecting leucyl-tRNA synthetase levels. *J. Mol. Biol.* **117:**1033.

Lee, F. and C. Yanofsky. 1977. Transcription termination at the *trp* operon attenuators of *Escherichia coli* and *Salmonella typhimurium*: RNA secondary structure and regulation of termination. *Proc. Natl. Acad. Sci.* **74:**4365.

Lewis, J. A. and B. N. Ames. 1972. Histidine regulation in *Salmonella typhi-*

murium. XI. The percentage of transfer RNA^{His} charged *in vivo* and its relation to the repression of the histidine operon. *J. Mol. Biol.* **66:**131.

Low, B., F. Gates, T. Goldstein, and D. Söll. 1971. Isolation and partial characterization of temperature-sensitive *Escherichia coli* mutants with altered leucyl- and seryl-transfer ribonucleic acid synthetases. *J. Bacteriol.* **108:**742.

Maas, W. 1962. Studies on repression of arginine biosynthesis in *Escherichia coli*. *Cold Spring Harbor Symp. Quant. Biol.* **26:**183.

Martin, R. G., B. N. Ames, and P. E. Hartman. 1967. Regulation of the histidine operon. *Int. Congr. Biochem. Abstr.* **2:**259.

McGinnis, E. and L. S. Williams. 1971. Regulation of synthesis of the aminoacyl-transfer ribonucleic acid synthetases for the branched chain amino acids of *Escherichia coli*. *J. Bacteriol.* **108:**254.

———. 1972. Regulation of histidyl-transfer ribonucleic acid synthetase formation in a histidyl transfer ribonucleic acid synthetase mutant of *Salmonella typhimurium*. *J. Bacteriol.* **111:**739.

Morgan, S., A. Körner, B. K. Low, and D. Söll. 1977. Regulation of biosynthesis of aminoacyl-tRNA synthetases and of tRNA in *Escherichia coli*. I. Isolation and characterization of a mutant with elevated levels of tRNA^{Gln}. *J. Mol. Biol.* **117:**1013.

Morse, D. and A. Morse. 1976. Dual control of the tryptophan operon by the repressor and tryptophanyl-tRNA synthetase. *J. Mol. Biol.* **103:**209.

Nass, G. and F. C. Neidhardt. 1967. Regulation of formation of amino-acyl-ribonucleic acid synthetases in *Escherichia coli*. *Biochim. Biophys. Acta* **134:**347.

Nass, G., K. Poralla, and H. Zähner. 1969. Effect of the antibiotic borrelidin on the regulation of threonine biosynthetic enzymes in *E. coli*. *Biochem. Biophys. Res. Commun.* **34:**84.

Neidhardt, F. C. 1966. Roles of amino acid activating enzymes in cellular physiology. *Bacteriol. Rev.* **30:**701.

Neidhardt, F. C., J. Parker, and W. G. McKeever. 1975. Function and regulation of aminoacyl-tRNA synthetases in prokaryotic and eukaryotic cells. *Annu. Rev. Microbiol.* **29:**215.

Neidhardt, F. C., P. L. Bloch, S. Pedersen, and S. Reeh. 1977. Chemical measurement of steady state levels of ten aminoacyl-transfer ribonucleic acid synthetases in *Escherichia coli*. *J. Bacteriol.* **129:**378.

Paetz, W. and G. Nass. 1973. Biochemical and immunological characterization of threonyl-tRNA synthetase of two borrelidin-resistant mutants of *Escherichia coli* K-12. *Eur. J. Biochem.* **35:**331.

Pizer, L. I., J. McKitrick, and T. Tosa. 1972. Characterization of a mutant of *E. coli* with elevated levels of seryl-tRNA synthetase. *Biochem. Biophys. Res. Commun.* **49:**1351.

Platt, T. 1978. Regulation of gene expression in the tryptophan operon of *Escherichia coli*. In *The operon* (ed. J. H. Miller and W. S. Reznikoff), p. 263. Cold Spring Harbor Laboratory, Cold Spring Harbor, New York.

Quay, S. C., E. L. Kline, and D. L. Oxender. 1977. Role of the leucyl-tRNA synthetase in regulation of transport. *Proc. Natl. Acad. Sci.* **72:**3921.

Ravel, J. M., M. N. White, and W. Shive. 1965. Activation of tyrosine analogs in relation to enzyme repression. *Biochem. Biophys. Res. Commun.* **20:**352.

Roth, J. R. and B. N. Ames. 1966. Histidine regulatory mutants in *Salmonella typhimurium*. II. Histidine regulatory mutants having altered histidyl-tRNA synthetases. *J. Mol. Biol.* **22**:325.

Roth, J. R., D. N. Anton, and P. E. Hartman. 1966. Histidine regulatory mutants in *Salmonella typhimurium*. I. Isolation and general properties. *J. Mol. Biol.* **22**:305.

Schlesinger, S. and E. W. Nester. 1969. Mutants of *Escherichia coli* with an altered tyrosyl-transfer ribonucleic acid synthetase. *J. Bacteriol.* **100**:167.

Singer, C. E., G. R. Smith, R. Cortese, and B. N. Ames. 1972. Mutant tRNA^{His} ineffective in repression and lacking two pseudouridine modifications. *Nat. New Biol.* **238**:72.

Squires, C. L., J. K. Rose, C. Yanofsky, H.-L. Yang, and G. Zubay. 1973. Tryptophanyl-tRNA and tryptophanyl-tRNA synthetase are not required for *in vitro* repression of the tryptophan operon. *Nat. New Biol.* **245**:131.

Szentirmai, A., M. Szentirmai, and H. E. Umbarger. 1968. Isoleucine and valine metabolism of *Escherichia coli*. XV. Biochemical properties of mutants resistant to thiaisoleucine. *J. Bacteriol.* **95**:1672.

Umbarger, H. E. 1969. Regulation of amino acid metabolism. *Annu. Rev. Biochem.* **38**:323.

Williams, L. S. 1973. Control of arginine biosynthesis in *Escherichia coli*: Role of arginyl-transfer ribonucleic acid synthetase in repression. *J. Bacteriol.* **113**:1419.

Williams, L. S. and F. C. Neidhardt. 1969. Synthesis and inactivation of aminoacyl-tRNA synthetases during growth of *Escherichia coli*. *J. Mol. Biol.* **43**:529.

Wyche, J. H., B. Ely, T. A. Cebula, M. C. Sneed, and P. E. Hartman. 1974. Histidyl-transfer ribonucleic acid synthetase in positive control of the histidine operon in *Salmonella typhimurium*. *J. Bacteriol.* **117**:708.

Zurawski, G., K. Brown, D. Killingly, and C. Yanofsky. 1978. Nucleotide sequence of the leader region of the phenylalanine operon of *Escherichia coli*. *Proc. Natl. Acad. Sci.* **75**:4271.

Role of tRNA^{Trp} and Leader RNA: Secondary Structure in Attenuation of the *trp* Operon

Stephen P. Eisenberg, Larry Soll, and Michael Yarus
Department of Molecular, Cellular, and Developmental Biology
University of Colorado
Boulder, Colorado 80309

Expression of several bacterial amino acid biosynthesis operons is controlled at least in part by attenuation, or transcription termination, which occurs at a site in the operon's leader region between the promoter and the 5' end of the first structural gene (Brenchley and Williams 1975; Bertrand and Yanofsky 1976; Zurawski et al. 1978a; Barnes 1978). Attenuation in the tryptophan (*trp*) biosynthetic operon of *Escherichia coli* has been studied mainly by Yanofsky and his colleagues. In this chapter, we review data characterizing *trp* operon attenuation, with emphasis on the involvement of tryptophan tRNA (tRNA^{Trp}) in the attenuation process.

THE ATTENUATOR IS A REGULATORY LOCUS

Yanofsky and his colleagues established the existence of the attenuator site by showing that in unstarved cells, only 10% of the RNA polymerase molecules that initiate transcription actually transcribe the structural genes. The other 90% terminate transcription about 140 bp from the promoter site. They also found that when cells are starved of tryptophan, termination of transcription at this site is relieved (Yanofsky 1976). Thus, the attenuator is a regulatory site allowing the cell to respond to the level of tryptophan (or some metabolically related compound) in its environment by allowing transcription of the entire operon or terminating transcription before the structural genes.

INVOLVEMENT OF CHARGED tRNA^{Trp}

Morse and Morse (1976) used *trpS* mutants to establish that the attenuation process is dependent not on the presence of free tryptophan, but rather on the proper functioning of tryptophanyl-tRNA synthetase in aminoacylating tRNA^{Trp} and, more recently, Yanofsky and Soll (1977) showed that the extent of charging of tRNA^{Trp} in the cell determines the degree of regulation at the attenuator. Several mutations that alter the structure of tRNA^{Trp} also affect regulation at the attenuator site. These will be discussed in detail later.

SEQUENCE OF THE LEADER REGION

Knowledge of the sequence of nucleotides in the leader region of the *trp* operon has been helpful in establishing a mechanism for attenuation. The sequence, determined for both the DNA (Lee et al. 1978) and its RNA transcript (Squires et al. 1976), contains an AUG (initiator) codon at nucleotides 27–29 (which can function as a ribosome binding site in vitro [Platt et al. 1976]), two tandem UGG (tryptophan) codons (nucleotides 54–56 and 57–59), and a UGA (termination) codon (nucleotides 69–71) all in the same phase. If the RNA transcribed from this region were to be translated normally, a polypeptide 14 amino acid residues in length would result and the two tryptophan residues would be at positions 10 and 11 (Lee and Yanofsky 1977). Lee and Yanofsky (1977) also discovered two alternative secondary-structure hairpin loops in the leader region RNA by analyzing RNase T1 partial digestion products of the full-length transcript. One possible structure goes from nucleotide 74 to nucleotide 119 and contains a 12-bp stem and a non-base-paired 22-nucleotide loop (see Fig. 1). The other structure includes nucleotides from position 114 to position 134 with a GC-rich, 8-bp stem and a 4-base loop (Fig. 1). This second hairpin loop is the structure (preceding the short stretch of uridylates) that acts to retard the elongating RNA polymerase, allowing it to dissociate from the DNA (Stauffer et al. 1978; Pribnow 1978). The stabilities of the two hairpin structures were calculated based on the combined rules of Tinoco et al. (1973) and Borer et al. (1974). For the first structure (i.e., nucleotides 74–119), $\Delta G = -12.2$ kcal/mole, and for the terminator (or second) hairpin loop (nucleotides 114–134), $\Delta G = -20.1$ kcal/mole (Lee and Yanofsky 1977; Zurawski et al. 1978a).

A MODEL FOR ATTENUATION

These findings regarding the sequence and secondary structure of the leader region RNA along with the involvement of charged tRNA^{Trp} in the attenuation process led Lee and Yanofsky (1977) to their model for attenuation. Translation of the RNA leader occurs during synthesis with the ribosome following closely behind the RNA polymerase molecule. If charged tRNA^{Trp} is in plentiful supply, translation can occur through the *trp* codons to the UGA nonsense codon (nucleotides 69–71). The presence of the ribosome at that position should prevent the formation of the hairpin loop in the region from nucleotide 74 to nucleotide 119, since the ribosome probably covers about 25 bases or at least 10 bases on either side of the aminoacyl and peptidyl sites (Steitz 1978). This allows the

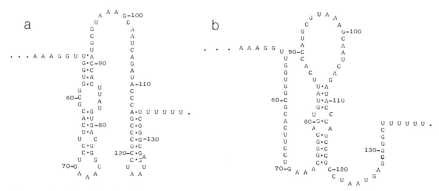

Figure 1 Possible hairpin loops in leader region RNA. (a) Two compatible hairpin structures. On the left is the 54–91 loop; on the right, the terminator loop. (b) A hairpin loop, which would preclude the existence of either of the two structures in a (see text).

terminator loop to form (114–134) and attenuation to occur. If charged tRNA^{Trp} is in short supply due to tryptophan starvation, the translating ribosome should become stalled at the UGG codons, the 74–119 hairpin would be allowed to form, and formation of the terminator loop would be prevented. Transcription of the remainder of the operon occurs in this situation.

ANOTHER HAIRPIN LOOP

D. Pribnow (pers. comm.) has pointed out that there is the possibility of forming yet another hairpin loop in the region from nucleotide 54 through nucleotide 91, which includes the two tryptophan codons (Fig. 1). The calculated ΔG for this structure is −13 kcal/mole; its presence is compatible with the existence of the terminator hairpin loop but is not compatible with the hairpin structure in the 74–119 region.

Presuming the existence of this loop from residue 54 through 91 allows one to understand several unexplained results. First, the fact that the two tryptophan codons (UGG and UGG) in the leader RNA are somewhat resistant to RNase T1 digestion (Lee and Yanofsky 1977) is explainable, since those nucleotides would be protected from digestion by secondary structure base-pairing in this newly proposed hairpin structure. Second, the very frequent occurrence of termination in vitro (Lee and Yanofsky 1977) makes one suspect that formation of the 74–119 nucleotide hairpin is prevented to allow the termination hairpin loop to form. Since the presence of the 54–91 hairpin loop is incompatible with the 74–119 structure, the formation of the former structure would prevent the latter hairpin from forming. Finally, in vivo, when translation of the leader RNA is prevented upstream from the *trp* codons, i.e., by mutating the leader region so that translational initiation is prevented, attenuation occurs that cannot be relieved through tryptophan starvation (Zurawski et al. 1978b). This observation can be explained in a way analogous to the in vitro situation mentioned above. Ribosomes are prevented from reaching the *trp* codons, thus the 54–91 hairpin loop can form, preventing the 74–119 hairpin from forming and allowing the existence of the terminator hairpin loop.

MUTATIONS ALTERING tRNA^{Trp} STRUCTURE AFFECT ATTENUATION

Several mutations alter the structure of tRNA^{Trp} and at the same time affect regulation of *trp* operon expression at the attenuator.

trpX

trpX maps outside the structural gene for tRNA^{Trp} but affects the mobility of tRNA^{Trp} on benzoylated DEAE-cellulose (Yanofsky and Soll 1977). tRNA^{Trp} from the *trpX* strain was sequenced along with tRNA^{Trp} from a normal (wild-type) strain and a single difference was discovered. The *trpX*-derived tRNA^{Trp} lacks modifications on the normally hypermodified base next to the anticodon (Eisenberg et al. 1979a) (Fig. 2).

Replacing the tRNA^{Trp} gene in the *trpX* strain with the gene for the UGA suppressor tRNA^{Trp} (a mutant of tRNA^{Trp} that can translate UGG and UGA codons [Hirsh 1971; Hirsh and Gold 1971]) restores normal attenuation (Yanofsky and Soll 1977) and this suppressor in the *trpX* strain also lacks modifications on the normally hypermodified base.

Figure 2 shows tRNA^{Trp} in the cloverleaf pattern and indicates the change at position 37 due to the *trpX* mutation and the base change that makes tRNA^{Trp} a UGA suppressor.

Normally, the hypermodified base ms^{2}i^{6}A is found in all *E. coli* tRNAs that respond to codons beginning with a U (Nishimura et al. 1969), presumably due to the relevant modifying enzymes recognizing the A in the third position of the tRNA's anticodon (Roberts and Carbon 1974). We found that tryptophanyl-, phenylalanyl-, and tyrosyl-tRNAs (which all respond to codons starting with U) are chromatographically affected by the *trpX* mutation. Others tested that normally do not have the ms^{2}i^{6}A were not affected (Eisenberg et al. 1979a). This is consistent with the proposal that *trpX* is the gene for the enzyme activity that modifies A to become ms^{2}i^{6}A.

The function of the hypermodified base ms^{2}i^{6}A has been studied by several groups. Gefter and Russell (1969) showed that a lack or even a partial lack of modifications on the normally hypermodified base reduced the efficiency of tRNA^{Tyr} binding to ribosomes in the presence of the appropriate trinucleotide codon. Recently, Grosjean et al. (1976) have

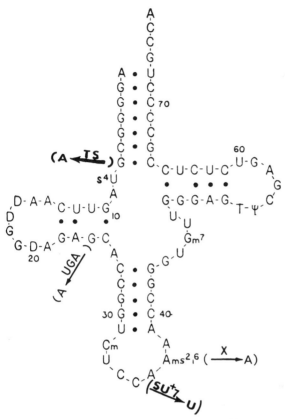

Figure 2 The nucleotide sequence of tRNA^{Trp}. Wild-type tRNA^{Trp} (Hirsh 1971) is shown with changes due to the following mutations: *trpX* ($\xrightarrow{X}$A) (Eisenberg et al. 1979a); UGA suppressor (A$\xleftarrow{UGA}$) (Hirsh 1971); temperature sensitive (A$\xleftarrow{TS}$) (Eisenberg et al. 1979b); and su^+7 ($\xrightarrow{Su+7}$U) (Yaniv et al. 1974).

provided a possible physical explanation for more efficient binding of fully modified tRNAs to ribosomes. They studied the binding interaction between tRNAs with complementary anticodons and found that tRNA^{Phe} (anticodon GAA) from yeast or *E. coli*, each having a hypermodified base next to the anticodon, bind much more strongly to tRNA^{Glu} (anticodon UUC) than does tRNA^{Phe} from *Mycoplasma*, which does not have a hypermodified base. Finally, Laten et al. (1978) found that an antisuppressor mutant of *Saccharomyces cerevisiae* contains tRNA that completely lacks the modification of the normally hypermodified base i^{6}A. Since the suppressor tRNA normally has an i^{6}A, the loss of that hypermodification apparently causes a reduced efficiency of suppression, presumably at the step involved with ribosomal interaction.

We can now explain how the *trpX* mutation affects attenuation. As described above, tRNA^{Trp} isolated from *trpX* bacteria differs from wild type only at position 37, the base adjacent to the anticodon on its 3' side. In wild-type tRNA^{Trp}, this position is occupied by the hypermodified base ms^{2}i^{6}A, whereas in the mutant, this A is completely unmodified. Since this modification has been strongly implicated as a requirement for efficient tRNA-ribosome interactions, its absence, due to the *trpX* mutation, should lead to a loss of regulation at the attenuator, presumably because of inefficient translation at the tandem tryptophan codons in the leader-sequence RNA. This is consistent, therefore, with the translation model of Lee and Yanofsky (1977).

The UGA suppressor tRNA^{Trp} must have an unusual ability to interact with the ribosome in response to codons of the group UGX (Buckingham and Kurland 1977), since the tRNA can efficiently read the codons UGG, UGA, and UGU, one of the cysteine codons. We suggest that because of this unusual property, the UGA suppressor tRNA^{Trp} can efficiently translate the tandem UGG codons despite the lack of hypermodifications.

The analysis of the *trpX* mutation presented here is apparently the first result to relate directly attenuation to translational efficiency, due to the well-documented function for the hypermodified base ms^{2}i^{6}A and the previously known effect of the UGA suppressor mutation on the ribosomal interaction.

Temperature-sensitive tRNA^{Trp}

Another mutation affecting tRNA^{Trp} structure and *trp* operon attenuation maps in the gene for tRNA^{Trp} and confers temperature sensitivity on the cell (Yanofsky and Soll 1977). This temperature-sensitive tRNA^{Trp} has been sequenced and the base change due to the temperature-sensitive mutation is indicated in Figure 2 (Eisenberg et al. 1979b).

As mentioned above, temperature-sensitive tRNA^{Trp} affects attenuation, and Yanofsky and Soll (1977) concluded that the effect is due to defective charging of temperature-sensitive tRNA^{Trp} in vivo based on the results of two experiments.

First, the presence of this tRNA in the cell caused deattenuation in unstarved cells. This indicates that the temperature-sensitive tRNA^{Trp} is deficient at one (or several) step in the attenuation process. Second, the strain carrying temperature-sensitive tRNA^{Trp}, when starved for isoleucine, exhibited normal attenuation. This starvation procedure presumably causes the pool of charged tRNAs to increase (except, of course, for isoleucyl-tRNA), since this prevention of protein synthesis keeps charged tRNAs from being depleted. Thus, the assumption was that in the temperature-sensitive cells unstarved for isoleucine, tRNA^{Trp} is uncharged.

However, there are several indications that this tRNA is in fact charged in vivo. First, the rate of enzymatic charging in vitro of the temperature-sensitive tRNA^{Trp} is equal to that of the wild type (Eisenberg and Yarus 1980). Second, total RNA synthesis in stringent cells containing temperature-sensitive tRNA^{Trp} is not inhibited, indicating the absence of uncharged tRNA^{Trp}; starving these cells for tryptophan or other amino acids does, in fact, severely reduce the synthesis of RNA due to the stringent response (Yanofsky and Soll 1977; L. Soll and M. Wilson, pers. comm.). Thus, it appears that temperature-sensitive tRNA^{Trp} is charged in vivo.

Why is it then that cells that contain the temperature-sensitive tRNA^{Trp} attenuate when starved for isoleucine. We mentioned earlier that preventing ribosomes from translating the leader region RNA (of the *trp* operon) prevents expression of the *trp* operon. Starving for any of the amino acids coded for upstream from the *trp* codons could cause attenuation in the same way, i.e., if ribosomes are prevented from reaching the *trp* codons, the loop at residue 54 through residue 91 remains intact and the terminator loop can form (see above). Isoleucine is one of those upstream amino acids. Thus, it could be that attenuation in the temperature-sensitive strain occurs in response to isoleucine starvation because translation of the leader sequence RNA upstream from the UGG codons is prevented.

The intracellular level of tRNA^{Trp} in the temperature-sensitive strain is about 25% of the level in wild-type strains (Eisenberg et al. 1979b) and this low level of tryptophanyl-tRNA^{Trp} could prevent attenuation (i.e., a minimum level of tRNA^{Trp} in the cell, greater than that of the temperature-sensitive tRNA^{Trp}, might be necessary for normal attenuation). This hypothesis is improbable, however, since the partial diploid strain, carrying genes for both wild-type and temperature-sensitive tRNA^{Trp} does not show normal attenuator activity (Yanofsky and Soll 1977), even though the total tRNA^{Trp} level in these cells (wild type plus temperature sensitive) is probably greater than that in normal $trpT^+$ (wild-type tRNA^{Trp})-carrying strains.

Having eliminated inefficient charging and low levels of temperature-sensitive tRNA^{Trp} as possible causes for deattenuation, we must assume that this tRNA is inefficient in one of the steps in the translation process (efficient translation being required for attenuation) following the charging reaction. Furthermore, there is an implication that the step affected occurs after the tRNA binds to the ribosome, since decreased attenuator activity in the diploid strain (see above) suggests that the temperature-sensitive tRNA^{Trp} competes with the wild type, presumably in the tRNA-ribosome binding reaction.

su^+7 tRNA^{Trp}

su^+7, an amber suppressor derived from tRNA^{Trp}, has a single-base change in the anticodon (Fig. 2) that allows it to read UAG (nonsense) codons instead of UGG, the codon for tryptophan (Yaniv et al. 1974). The amino acid specificity of the tRNA is also altered so that it can be charged with glutamine as well as tryptophan in vitro and in vivo (R. G. Knowlton and M. Yarus 1976; and in prep.). If a cell culture of an su^+7-carrying strain is starved for tryptophan, UGG codons cannot be translated, but surprisingly in this situation apparent termination of transcription at the attenuator occurs (Yanofsky and Soll 1977). This seems to contradict the prediction of the model of Lee and Yanofsky (1977), which states that efficient translation of the tandem UGG codons in the leader RNA is required for attenuation.

Fortunately, an explanation exists that resolves the apparent contradiction. Morse and Morse (1976) demonstrated that full expression of the *trp* operon requires an $relA^+$ (stringent) strain, or put another way, in an $relA^-$ (relaxed) strain starved for tryptophan, the expression of the *trp* operon (determined by presence of *trp* operon mRNA) is greatly reduced relative to the isogenic stringent strain. Yanofsky and Soll (1977) reported that total RNA synthesis proceeds at a high rate in tryptophan-starved cells carrying the su^+7 gene on F'14, whereas the rate is severely reduced in starved cells that lack su^+7. Thus, the su^+7-carrying strains appear relaxed, i.e., RNA synthesis is not severely reduced when the cells (carrying su^+7) are starved for an amino acid.

M. Yarus et al. (in prep.; Yarus 1979) have recently shown that the presence of an active su^+7 gene in the cell mimics the effect of an *relA* mutation in at least two ways. First, the rates of synthesis and subsequent levels of magic spot compounds MS I and MS II in the su^+7 strain are depressed twofold compared to magic spot levels in the strain lacking su^+7. Second, in su^-7 strains, the rate of stable RNA accumulation in cells starved for various amino acids is severely reduced by the stringent response (compared to unstarved controls), but the rate is not strongly affected in su^+7 strains. Furthermore, relaxation in these strains has been correlated with the level of su^+7 tRNA found in the cell; i.e., relaxed strains have a high level of su^+7 tRNA, whereas stringent strains carrying the su^+7 gene have (biochemically measured) (Yarus et al. 1977) lower levels of the su^+7 tRNA. Thus, su^+7 tRNA apparently relaxes the cell and this results in a (relative) decrease in expression of the *trp* operon, which was mistaken for attenuation.

SUMMARY

In this discussion we have reviewed the facts known concerning attenuation in the *trp* operon. This is clearly an important subject, since there is now evidence that many biosynthetic operons in bacteria are regulated by an analogous mechanism (Brenchley and Williams 1975; Barnes 1978; Zurawski et al. 1978).

The model proposed by Lee and Yanofsky (1977) is now well-supported by all that is known from the tRNA^{Trp} structural variants and the apparent contradictory result from the su^+7 tRNA^{Trp} experiment (described earlier) has been reconciled with the model.

The additional hairpin loop (between nucleotides 54–91) may have an important function in stabilizing the terminator hairpin loop (114–134) in situations where the ribosome is kept from translating the codons in the leader RNA upstream from the tandem *trp* codons.

ACKNOWLEDGMENTS

The authors wish to thank David Pribnow and Charles Yanofsky for many useful discussions.

REFERENCES

Barnes, W. 1978. DNA sequence from the histidine operon control region. *Proc. Natl. Acad. Sci.* **75**:4281.

Bertrand, K. and C. Yanofsky. 1976. Regulation of transcription termination in the leader region of the tryptophan operon of *Escherichia coli* involves tryptophan or its metabolic product. *J. Mol. Biol.* **103**:339.

Borer, P. N., B. Dengler, I. Tinoco, Jr., and O. C. Uhlenbeck. 1974. Stability of ribonucleic acid double-stranded helices. *J. Mol. Biol.* **86**:843.

Brenchley, J. E. and L. S. Williams. 1975. Transfer RNA involvement in the regulation of enzyme synthesis. *Annu. Rev. Microbiol.* **29**:251.

Buckingham, R. H. and C. G. Kurland. 1977. Codon specificity of UGA suppressor tRNA^{Trp} from *Escherichia coli*. *Proc. Natl. Acad. Sci.* **74**:5496.

Eisenberg, S. P. and M. Yarus. 1980. The structure and aminoacylation of a temperature-sensitive tRNA^{Trp} (*E. coli*). *J. Biol. Chem.* **255**:1128.

Eisenberg, S. P., L. Soll, and M. Yarus. 1979a. The effect of an *E. coli* regulatory mutation on tRNA structure. *J. Mol. Biol.* **135**:111.

Eisenberg, S. P., L. Soll, and M. Yarus. 1979b. The purification and sequence of a temperature-sensitive tryptophan tRNA. *J. Biol. Chem.* **254**:5562.

Gefter, M. L. and R. L. Russell. 1969. Role of modifications in tyrosine transfer RNA: A modified base affecting ribosome binding. *J. Mol. Biol.* **39**:145.

Grosjean, H., D. G. Söll, and D. M. Crothers. 1976. Studies of the complex between transfer RNAs with complementary anticodons. I. Origins of enhanced affinity between complementary triplets. *J. Mol. Biol.* **103**:499.

Hirsh, D. 1971. Tryptophan transfer RNA as the UGA suppressor. *J. Mol. Biol.* **58**:439.

Hirsh, D. and L. Gold. 1971. Translation of the UGA triplet *in vitro* by tryptophan transfer RNA's. *J. Mol. Biol.* **58**:459.

Knowlton, R. G. and M. Yarus. 1976. Amino acid accepting specificity of Su$^+7$ tRNA. *Fed. Proc.* **35**:1735.

Laten, H., J. Gorman, and R. M. Bock. 1978. Isopentenyladenosine-deficient tRNA from an antisuppressor mutant of *Saccharomyces cerevisiae*. *Nucleic Acids Res.* **5**:4329.

Lee, F. and C. Yanofsky. 1977. Transcription termination at the *trp* operon attenuators of *Escherichia coli* and *Salmonella typhimurium*: RNA secondary structure and regulation of termination. *Proc. Natl. Acad. Sci.* **74**:4365.

Lee, F., K. Bertrand, G. Bennett, and C. Yanofsky. 1978. Comparison of the nucleotide sequences of the initial transcribed regions of the tryptophan operons of *Escherichia coli* and *Salmonella typhimurium*. *J. Mol. Biol.* **121**:193.

Morse, D. E. and A. N. C. Morse. 1976. Dual-control of the tryptophan operon is mediated by both tryptophanyl-tRNA synthetase and the repressor. *J. Mol. Biol.* **103**:209.

Nishimura, S., Y. Yamada, and H. Ishikura. 1969. The presence of 2-methylthio-N^6-(Δ^2-isopentenyl) adenosine in serine and phenylalanine transfer RNA's from *Escherichia coli*. *Biochim. Biophys. Acta* **179**:517.

Platt, T., C. Squires, and C. Yanofsky. 1976. Ribosome-protected regions in the leader-*trpE* sequence of *Escherichia coli* tryptophan operon messenger RNA. *J. Mol. Biol.* **103**:411.

Pribnow, D. 1978. Genetic control signals in DNA. In *Biological regulation and development* (ed. R. Goldberger), vol. I, p. 219. Plenum Press, New York.

Roberts, J. W. and J. Carbon. 1974. Molecular mechanism for missense suppression in *E. coli*. *Nature* **250**:412.

Squires, C., F. Lee, K. Bertrand, C. L. Squires, M. J. Bronson, and C. Yanofsky. 1976. Nucleotide sequence of the 5′ end of tryptophan messenger RNA of *Escherichia coli*. *J. Mol. Biol.* **103**:351.

Stauffer, G. V., G. Zurawski, and C. Yanofsky. 1978. Single base-pair alterations in the *Escherichia coli trp* operon leader region that relieve transcription termination at the *trp* attenuator. *Proc. Natl. Acad. Sci.* **75**:4833.

Steitz, J. A. 1978. Genetic signals and nucleotide sequences in messenger RNA. In *Biological regulation and development* (ed. R. Goldberger), vol. I, p. 349. Plenum Press, New York.

Tinoco, I., Jr., P. N. Borer, B. Dengler, M. D. Levine, O. C. Uhlenbeck, D. M. Crothers, and J. Gralla. 1973. Improved estimation of secondary structure in ribonucleic acids. *Nat. New Biol.* **246**:40.

Yaniv, M., W. R. Folk, P. Berg, and L. Soll. 1974. A single mutational modification of a tryptophan-specific transfer RNA permits aminoacylation by glutamine and translation of the codon UAG. *J. Mol. Biol.* **86**:245.

Yanofsky, C. 1976. Regulation of transcription initiation and termination in the control of expression of the tryptophan operon in *E. coli*. In *Molecular mechanisms in the control of gene expression* (ed. D. P. Nierlich et al.), vol. 5, p. 75. Academic Press, New York.

Yanofsky, C. and L. Soll. 1977. Mutations affecting tRNA^{Trp} and its charging and their effect on regulation of transcription termination at the attenuator of the tryptophan operon. *J. Mol. Biol.* **113**:663.

Yarus, M. 1979. Relaxation of stable RNA synthesis by a plasmid-borne locus. *Mol. Gen. Genet.* **170**:309.

Yarus, M., R. Knowlton, and L. Soll. 1977. Aminoacylation of the ambivalent Su$^+$7 amber suppressor tRNA. In *Nucleic acid-protein recognition* (ed. H. J. Vogel), p. 391. Academic Press, New York.

Zurawski, G., K. Brown, D. Killingly, and C. Yanofsky. 1978a. Nucleotide sequence of the leader region of the phenylalanine operon of *Escherichia coli*. *Proc. Natl. Acad. Sci.* **75**:4271.

Zurawski, G., D. Elseviers, G. V. Stauffer, and C. Yanofsky. 1978b. Translational control of transcription termination at the attenuator of the *Escherichia coli* tryptophan operon. *Proc. Natl. Acad. Sci.* **75**:5988.

Role of tRNA^{Leu} in Branched-chain Amino Acid Transport

Steven C. Quay
Department of Pathology
Massachusetts General Hospital, Harvard Medical School
Boston, Massachusetts 02114

Dale L. Oxender
Department of Biological Chemistry
The University of Michigan Medical School
Ann Arbor, Michigan 48109

The intracellular levels of most nutrients are carefully controlled to meet the varying demands for these during the normal growth of the cell. To control the intracellular level of an amino acid, the cell must balance the processes that lead to increases in the nutrient concentration, such as transport and biosynthesis, with those that tend to decrease it, such as metabolism and macromolecular synthesis.

The regulation of the levels of the branched-chain amino acids, leucine, isoleucine, and valine, in bacteria and mammalian cells has been the subject of numerous studies (for reviews, see Umbarger 1973; Quay and Oxender 1979, 1980). These studies indicate a complex system for regulation of biosynthesis and transport involving amino acids, tRNA molecules and their synthetases, transcriptional termination factors, and the products of structural genes for regulatory loci. This review discusses the experimental evidence that leads to the hypothesis that tRNA^{Leu} is involved in branched-chain amino acid transport in both bacteria and mammalian cells.

ROLE OF AMINOACYL-tRNA SYNTHETASES

Aminoacyl-tRNA synthetases and their cognate tRNA species have been implicated in the regulation of amino acid biosynthetic enzymes for histidine (Brenner and Ames 1971), leucine (Low et al. 1971), isoleucine and valine (Umbarger 1973), arginine (Williams 1973), tryptophan (Bertrand et al. 1975), and methionine (Brenchley and Williams 1975). In a comprehensive study, the regulation of transport was tested in strains with mutations in the valyl-, isoleucyl-, and leucyl-tRNA synthetases to test whether these proteins were effectors in this regulation mechanism.

Strain NP29 *valS*^{ts} exhibits a temperature-sensitive phenotype that is due to the production of a temperature-sensitive valyl-tRNA synthetase

(Eidlic and Neidhardt 1965). At 31°C the strain has repressed isoleucine and valine biosynthetic enzymes when grown in excess branched-chain amino acids. When the culture is shifted to 37°C with no change in the composition of the medium, protein synthesis ceases after a small increase in cell mass, valyl-tRNA is partially deacylated, and the expression of operons for the biosynthesis of isoleucine and valine is derepressed (Eidlic and Neidhardt 1965). These results were confirmed by measuring the activity of the *ilvA* gene product, threonine deaminase. During these same experiments, branched-chain amino acid transport was monitored and was found not to derepress as growth became limited for valyl-tRNA. This is consistent with the finding that limitation for valine alone did not alter transport activity (Quay and Oxender 1976).

A strain with an altered form of the isoleucyl-tRNA synthetase (CU1018) has been isolated and contains at least three mutations that lead to derepression of enzymes for the biosynthesis of isoleucine and valine. This strain grows slowly in minimal medium; this slow growth can be increased by adding isoleucine to the media. The *ilv* operon in this mutant is refractory to repression by branched-chain amino acids (Szentirmai et al. 1968). The threonine deaminase activity was very high in cells grown in minimal medium (MM) and was only slightly lowered by growth with the branched-chain amino acids. On the other hand, the regulation of transport showed a normal repression response. This again confirmed our earlier experiments, which indicated that the regulation of transport was independent of cellular isoleucine levels and was probably also independent of the level of tRNA^{Ile} aminoacylation (Quay and Oxender 1976).

Strain KL231 contains a mutation that results in the production of a temperature-sensitive leucyl-tRNA synthetase (Low et al. 1971). When this strain was grown in medium supplemented with the branched-chain amino acids at a permissive temperature (31°C), transport for leucine was repressed. Shifting the culture to 38°C led to restricted growth, which ceased after a 50% increase in mass. These cells showed a large derepression of transport activity, which was prevented by protein synthesis inhibitors. A derepression in the enzymes for biosynthesis of leucine (Low et al. 1971) accompanied the transport increase, confirming that a deficit for leucyl-tRNA^{Leu} had occurred.

However, a potential source of error in measuring transport in a temperature-sensitive tRNA synthetase mutant of a prototrophic strain is the large increase in the internal amino acid levels upon derepression of the biosynthetic operons. For this reason, a strain was also studied with a deletion of the leucine biosynthetic operon.

Cultures of EB144 (*ara-leu* Δ *1101*) and EB143 (*ara-leu* Δ *1101, leuS1*) were grown at 36°C in a glucose minimal medium containing 0.2 mM L-leucine. The first-order growth-rate constant for both strains was 0.91/hour under these conditions. The temperature of the cultures was shifted to 41°C, increasing the growth-rate constant of EB144 to 1.06/hour and changing the growth-rate constant of EB143 to essentially 0 after an increase of about 55% in protein. The activities of several transport systems were measured after both cultures had incubated at 41°C for a period of time that allowed about 55% increase in cell mass in both cultures (24 min for strain EB144; 60 min for strain EB143). The regulation of histidine and proline transport is leucine independent and the uptake by these shock-sensitive and membrane-bound systems, respectively, served as a control during the temperature change. From the activities of these transport systems, the differential rate of synthesis (Δ uptake units/Δ mg of protein) during balanced growth at 37°C (which is identical to the specific activity) and the differential rate of synthesis during the growth increment at 41°C were calculated.

The results in Table 1 are typical of several experiments. They show that the change in temperature has a large effect on leucine, isoleucine, and valine transport activity in strain EB143 with an altered leucyl-tRNA synthetase, resulting in a five- to tenfold increase in transport activity compared to the isogenic parental strain. The temperature change had no effect on either proline or histidine uptake.

The possibility that the increase in branched-chain amino acid transport mediated by the leucyl-tRNA synthetase was due to activation of pre-existing transport components was tested. Strain EB143 was grown at 36°C in the glucose–basal salts medium containing 0.2 mM leucine, and when the culture was shifted to 41°C, either chloramphenicol (200 mg/liter) or rifampin (200 mg/liter) was added. Although the control cells (without antibiotics) derepressed transport, the presence of an inhibitor of protein synthesis (chloramphenicol) or an inhibitor of RNA polymerase initiation (refampin) prevented the increase in branched-chain amino acid transport.

Since these strains contain no leucine biosynthetic enzymes, we could not monitor the expression of this operon. However, the genes coding for the enzymes of valine and isoleucine biosynthesis are intact. Since the synthesis of *ilvA* gene product threonine deaminase derepresses in response to a limitation for leucyl-tRNA (Freundlich et al. 1962), we measured this enzyme in strains EB143 and EB144 at both permissive and nonpermissive temperatures (Table 1). The derepression of threonine deaminase in strain EB143 at 41°C, even in the presence of excess leucine, isoleucine, and valine, provides further evidence that, in fact, the activation of leucine has become the growth-rate-limiting step when the temperature is raised from 36°C to 41°C.

Since the periplasmic branched-chain amino acid binding proteins are required for branched-chain amino acid transport as shown by kinetic, genetic, and biochemical studies (Rahmanian and Oxender 1972; Rahmanian et al. 1973), the quantity of these proteins in the shock fluid from

Table 1 Expression of transport and a biosynthetic enzyme activity in strains EB143 and EB144

Strain	Growth conditions[a]	Growth-rate constant/hr	Transport activities[b]					Threonine deaminase[d]
			Leu	Ile	Val	His	Pro	
EB143, ara-leu Δ1101leuS1	36°C	0.91	0.23	0.15	0.20	3.68	0.19	26
EB144, ara-leu Δ1101	36°C	0.91	0.20	0.14	0.16	2.86	0.14	16
EB143, ara-leu Δ1101leuS1	41°C	—[c]	1.08	1.78	2.08	3.03	0.20	450
EB144, ara-leu Δ1101	41°C	1.06	0.22	0.16	0.17	3.64	0.12	14

[a]Growth in MOPS-G (morpholinopropanesulfonic acid-G) plus 0.2 mM L-leucine.
[b]Transport was assayed at 1 μM leucine or isoleucine and 3 μM valine, histidine, or proline. Specific activity represents mmoles taken up per min/kg of cells dry weight.
[c]A growth constant could not be accurately determined due to restricted growth that ceased after a 55% increase in cellular mass.
[d]Sp. act. represents μmole of α-ketobutyrate formed per min/g of cellular protein. The growth media included 0.4 mM L-leucine and L-isoleucine and 1 mM L-valine.

EB143 and EB144 was determined. The cells for these studies were grown in the same manner as those for the transport assays except on a larger scale. Although the parental strain remains repressed for the synthesis of the binding protein when grown in 0.2 mM L-leucine at either 36°C or 41°C (Table 2), the production of the protein is derepressed in the mutant at 41°C. These results are entirely consistent with a role of this protein in the rate-limiting step for branched-chain amino acid transport and the regulation of the expression of this protein by the intracellular level of leucyl-tRNA^{Leu} or the leucyl-tRNA synthetase itself.

ROLE OF LEUCYL-tRNA IN TRANSPORT REGULATION

To distinguish between these two alternatives, mutants (*hisT*) with a defect in the maturation of leucyl-tRNA were used. If these strains showed defects in transport regulation, one could conclude that the synthetase alone was not responsible for regulation. The *hisT* locus codes for an enzyme that catalyzes the modification of specific U residues to ψ residues in the anticodon region of several tRNA species, including tRNA^{Tyr}, tRNA^{His}, and tRNA^{Leu} (Allaudeen et al. 1972; Cortese et al. 1974). However the TψC-Pu loop in the tRNA of these mutants is entirely normal (Singer et al. 1972). The regulation of histidine, isoleucine-valine, and leucine biosynthetic enzymes in these *Salmonella typhimurium* mutants is altered in such a way that they are no longer sensitive to a limitation for their cognate amino acids (Cortese et al. 1974). Experiments to impose a limitation for leucine did not cause a derepression of the *leuA* gene product, α-isopropyl malate synthetase, in a *leu⁻ hisT* strain like it did in the isogenic *leu⁻* parent. Under the same growth conditions the transport activity of the *leu⁻ hisT* strain also showed little sensitivity to the leucine level in the growth medium. These

Table 2 Branched-chain amino acid binding proteins in strains EB143 and EB144

Strain	Growth conditions	Leucine binding activity
EB143, *ara-leu* Δ *1101 leuS1*	36°C	0.36
EB144, *ara-leu* Δ *1101*	36°C	0.46
EB143, *ara-leu* Δ *1101 leuS1*	41°C	1.80
EB144, *ara-leu* Δ *1101*	41°C	0.39

Proteins with leucine-binding activity were obtained by osmotic shock and were assayed at 10 μM L-leucine and 4°C. The sp. act. is expressed as μmole of leucine bound per g of protein.

results can be taken as evidence that the repression of transport requires fully maturated tRNA that is aminoacylated with leucine.

To extend these observations in *S. typhimurium*, transport regulation in *Escherichia coli* strains containing the *hisT* locus was studied in the four major states of growth important in terms of transport regulation, i.e., growth in rich medium, MM with excess branched-chain amino acids, MM alone, and leucine limitation.

The three transitions between steady states of growth, occurring when bacteria are transferred between rich medium and MM supplemented with leucine, isoleucine, and valine (MM + LIV), between MM + LIV and MM, and between MM and valine limitation, resulted in an increase in leucine, isoleucine, and valine transport in the parental strain, T31-4. Valine limitation was used here to produce a leucine deficit, since the precursor of the leucine biosynthetic system, α-ketoisovaleric acid, arises from valine by transamination. The largest derepression came about during the transition from MM + LIV to MM, as had been observed before (Quay and Oxender 1976). This derepression was specific for the leucine, isoleucine, and valine transport systems, since proline uptake was not affected by this transition. On the other hand, the derepression accompanying the transition from rich medium to MM + LIV seemed less specific with respect to both substrate (all four amino acids: leucine, isoleucine, valine, and proline, derepressed) and regulatory mutation (both strains had virtually identical derepression patterns).

The derepression observed in the parental strain undergoing the transition from MM + LIV to MM ranged from fourfold (isoleucine) to almost sevenfold (valine). The same transition in the *hisT* strain, strain T31-H, led to a much smaller derepression of transport, from 1.5-fold to slightly more than twofold. The decreased capacity for derepression of transport in this *hisT* strain is similar to the findings concerning transport of branched-chain amino acid biosynthesis in the *hisT* mutants of *S. typhimurium* LT2 (Bresalier et al. 1975).

The same defects in regulation observed for transport were found upon examining the leucine-binding activity of these strains: low capacity for derepression in the *hisT* strain.

The initial work (Quay et al. 1975; Quay and Oxender 1976) had indicated that the *leuS* locus, which codes for the leucyl-tRNA synthetase, is required for maintenance of transport repression. One could imagine that repression results from leucyl-tRNA^{Leu} or the leucyl-tRNA synthetase acting as a negative effector, or alternatively, that uncharged tRNA^{Leu} is a positive effector in transport regulation. The results presented here concerning the *hisT* locus indicate that the leucyl-tRNA synthetase is not directly involved in transport regulation and that tRNA^{Leu} with ψ residues is required for full transport derepression, i.e., deacylated tRNA$^{Leu}_{\psi}$ is a positive effector in transport. This is consistent with similar conclusions concerning leucine and isoleucine-valine biosynthesis in *S.*

typhimurium (Bresalier et al. 1975). Alternatively, leucyl-tRNA^{Leu} could act as a negative effector in regulation.

Significant advances in the understanding of the regulation of histidine and tryptophan biosynthetic operons have recently been made (Artz and Broach 1975; Bertrand et al. 1975). In addition to regulation by the classical mechanism of inhibition of RNA polymerase initiation by a corepressor-aporepressor complex (Jacob and Monod 1961), there appear to be methods for changing the frequency of RNA transcriptional termination, apparently at an attenuator site in the DNA proximal to the structural genes (the leader region) (Bertrand et al. 1975). The protein factor rho, originally identified as important in the relief of mutational polarity (Beckwith 1963), seems to be involved in transcriptional termination at the attenuator site in the *trp* leader region (Korn and Yanofsky 1976).

TRANSPORT REGULATION BY RHO FACTORS

We have shown previously (Quay and Oxender 1977) that rho-dependent termination of mRNA transcription is important for leucine-specific repression of branched-chain amino acid transport, although rho-independent regulation, presumably by a corepressor-aporepressor-type mechanism, must also occur. The use of strains containing mutations in both *rho* and *hisT* allowed us to test whether these regulatory elements interact during transport repression.

A strain, T31-H-480, containing both *hisT76* and *rho221* mutations, was used to examine transport regulation during growth-media transitions, as described earlier. The transition from MM + LIV to MM did not lead to any change in transport capacity, although this strain is partially derepressed in MM + LIV, as is the *rho* mutant alone.

These data indicate that the *rho221* genotype leads to derepression under conditions of excess leucine, isoleucine, and valine, whereas in MM the *hisT* genotype predominates. These findings are most easily interpreted if one assumes that two different mechanisms of maintaining repression predominate under the two different growth conditions: wild-type *rho* maintains repression with excess leucine by transcriptional termination, and wild-type tRNA^{Leu} assists derepression under MM growth conditions. In addition, the finding that *rho* does not elevate expression in a *hisT* mutant growing in MM suggests that *rho* interacts with tRNA^{Leu}.

TRANSPORT REGULATION IN MAMMALIAN CELLS

Since early work has shown that in animal cells protein synthesis was required for the regulation of transport activity and since we had found that leucyl-tRNA was important for leucine transport regulation in *E. coli*, we examined the possible role of aminoacyl-tRNA synthetases in mammalian cell transport regulation. The availability of several temperature-sensitive aminoacyl-tRNA synthetase mutants in Chinese hamster ovary cell lines (Thompson et al. 1973, 1975, 1977; Wasmuth and Caskey 1976) provided us with the opportunity to test this possibility. When a temperature-sensitive leucyl-tRNA synthetase mutant, tsH1 (Thompson et al. 1973, 1975, 1977), was shifted from a normal growth temperature of 35°C to a marginally permissive temperature for growth (38°C), a significant enhancement in the initial rate of uptake of the L-system prototype amino acid, leucine, and other L-system amino acids was observed (Moore et al. 1977). In contrast, the uptake of alanine, a prototype for the A-system amino acids, showed no significant difference relative to the parental cell line. In a similar manner, a temperature-sensitive asparaginyl-tRNA synthetase mutant strain RJK-4 (Wasmuth and Caskey 1976) exhibited increased transport activity of A-system amino acids when the growth temperature was shifted from 33°C to 39.5°C.

Preliminary kinetic studies suggest that the V_{max} for transport is increased when the mutants are grown under amino-acid-limiting conditions. It appears from these studies that when the Chinese hamster cells are starved for an A-system amino acid, they respond by increasing the transport activity of the A-system; when an L-system amino acid is growth limiting, the L-system transport activity is increased. These results suggest that the aminoacyl-tRNA synthetases or their related products may play a role in mammalian cell amino acid transport regulation in a manner similar to that found for prokaryotic organisms.

SUMMARY

Early in our work on transport regulation we observed that a temperature-sensitive leucyl-tRNA synthetase mutation led to a derepression of branched-chain amino acid transport under conditions in which the aminoacylation of tRNA^{Leu} was rate limiting for growth. At that time, we speculated that those results were consistent with a role for aminoacyl-tRNA^{Leu} or the synthetase acting as a negative effector in regulation or for the uncharged tRNA^{Leu} to act as a positive modulator. Since then, the finding that mutations in the tRNA^{Leu} maturation process, such as $hisT$ or $relA$ (Quay and Oxender 1979), prevent transport derepression has led to the conclusion that mature, fully modified, tRNA^{Leu} is an effector in transport derepression. In addition, the prospect that transport regulation involves the termination of RNA transcription at a site proximal to the structural genes, called the attenuator site, seems likely, since a mutation in rho greatly enhances the derepression of transport (Quay and Oxender 1977).

These aspects of transport regulation are summarized in Figure 1. We

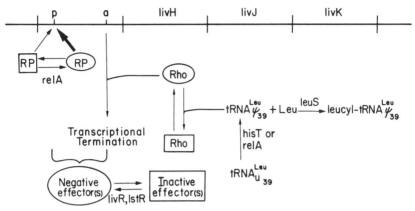

Figure 1 Model of regulation of branched-chain amino acid transport in *E. coli*. (RP) DNA-dependent RNA polymerase; (p) promoter site; (a) attenuator site; (livH) *livH* component of the LIV-I transport system; (livJ) *livJ* structural gene, LIV binding protein; (livK) *livK* structural gene, leucine-specific binding protein; (leuS) *leuS* gene product, leucyl-tRNA synthetase; (relA) *relA* gene product, stringent factor; (hisT) *hisT* gene product, ψ synthetase; (livR) *livR trans*-dominant regulator of transport; (lstR) *lstR* regulator of leucine-specific transport. The relative ordering of *livH*, *livJ*, and *livK* was arbitrary.

do not know whether the *relA* locus controls transport by alteration of the DNA-dependent RNA polymerase specificity (Travers 1976a,b) or by preventing pseudouridination at position 39 in tRNA^{Leu} (Kitchingman et al. 1976; Kitchingman and Fournier 1977).

Four growth conditions must be considered in determining the relative importance of these various regulatory effectors to the overall rate of transport biosynthesis. The growth conditions are rich medium, MM containing excess leucine, MM, and growth-rate-limiting levels of leucine. The work presented here suggests that the *relA* function is most important in the shift from rich medium to MM. On the other hand, *rho* and *hisT* mutations have their greatest effect in shifts from MM to limitation for leucine. This is in contrast with the normal physiological regulation of transport in wild-type *E. coli* that indicated maximum derepression for transport is present in MM. Since leucine biosynthesis undergoes the greatest derepression only on shifting from MM to leucine limitation (Quay and Oxender 1976), a hierarchy of regulation is suggested, with transport levels increasing before biosynthesis is derepressed.

ACKNOWLEDGMENT

This work was supported, in part, by U.S. Public Health Service grant GM-11024.

REFERENCES

Allaudeen, H. S., S. K. Yang, and D. Söll. 1972. Leucine tRNA$_1$ from *hisT* mutant of *Salmonella typhimurium* lacks two pseudouridines. *FEBS. Lett.* **28**:205.

Artz, S. W. and J. R. Broach. 1975. Histidine regulation in *Salmonella typhimurium*: An activator-attenuator model of gene regulation. *Proc. Natl. Acad. Sci.* **72**:3453.

Beckwith, J. 1963. Restoration of operon activity by suppressors. *Biochim. Biophys. Acta* **76**:162.

Bertrand, K., L. Korn, F. Lee, T. Platt, C. L. Squire, C. Squires, and C. Yanofsky. 1975. New features of the regulation of the tryptophan operon. *Science* **189**:22.

Brenchley, J. E. and L. S. Williams. 1975. Transfer RNA involvement in the regulation of enzyme synthesis. *Annu. Rev. Microbiol.* **29**:251.

Brenner, M. and B. N. Ames. 1971. The histidine operon and its regulation. In *Metabolic regulation* (ed. H. J. Vogel), vol. 6, p. 349. Academic Press, New York.

Bresalier, R. S., A. A. Rizzino, and M. Freundlich. 1975. Reduced maximal levels of depression of the isoleucine, valine, and leucine enzymes in *hisT* mutants of *Salmonella typhimurium*. *Nature* **253**:279.

Cortese, R., R. A. Landsberg, R. A. Von der Haar, H. E. Umbarger, and B. N. Ames. 1974. Pleiotrophy of *hisT* mutants blocked in pseudouridine synthesis in tRNA: Leucine and isoleucine-valine operons. *Proc. Natl. Acad. Sci.* **71**:1857.

Eidlic, L. and F. C. Neidhardt. 1965. Role of valyl-sRNA synthetase in enzyme repression. *Proc. Natl. Acad. Sci.* **53**:539.

Freundlich, M., R. O. Burns, and H. E. Umbarger. 1962. Control of isoleucine, valine, and leucine biosynthesis. I. Multivalent repression. *Proc. Natl. Acad. Sci.* **48**:1804.

Jacob, F. and J. Monod. 1961. Genetic regulatory mechanisms in the synthesis of proteins. *J. Mol. Biol.* **3**:318.

Kitchingman, G. R. and M. J. Fournier. 1977. Modification-deficient transfer ribonucleic acids from relaxed control *Escherichia coli*: Structures of the major undermodified phenylalanine and leucine transfer RNAs produced during leucine starvation. *Biochemistry* **16**:2013.

Kitchingman, G. R., E. Webb, and M. J. Fournier. 1976. Unique phenylalanine transfer of ribonucleic acids in relaxed control *Escherichia coli*: Genetic origin and some functional properties. *Biochemistry* **15**:1848.

Korn, L. J. and C. Yanofsky. 1976. Polarity suppressors defective in transcription termination at the attenuator of the tryptophan operon of *Escherichia coli* have altered Rho factor. *J. Mol. Biol.* **106**:231.

Low, B., F. Gates, T. Goldstein, and D. Söll. 1971. Isolation and partial characterization of temperature-sensitive *Escherichia coli* mutants with altered leucyl- and seryl-transfer ribonucleic acid synthetases. *J. Bacteriol.* **108**:742.

Moore, P. A., D. W. Jayme, and D. L. Oxender. 1977. A role for aminoacyl-tRNA synthetases in the regulation of amino acid transport in mammalian cell lines. *J. Biol. Chem.* **252**:7427.

Quay, S. C. and D. L. Oxender. 1976. Regulation of branched-chain amino acid transport in *Escherichia coli*. *J. Bacteriol.* **127**:1225.

―――. 1977. Regulation of amino acid transport in *Escherichia coli* by transcriptional terminator factor rho. *J. Bacteriol.* **130:**1024.

―――. 1979. The *relA* locus specifies a positive effector in branched-chain amino acid transport regulation. *J. Bacteriol.* **137:**1059.

―――. 1980. Regulation of membrane transport. In *Biological regulation and development* (ed. R. F. Goldberger), vol. 2. Plenum Press, New York. (In press.)

Quay, S. C., E. L. Kline, and O. L. Oxender. 1975. Role of the leucyl-tRNA synthetase in regulation of transport. *Proc. Natl. Acad. Sci.* **72:**3921.

Rahmanian, M. and D. L. Oxender. 1972. Derepressed leucine transport activity in *Escherichia coli*. *J. Supramol. Struct.* **1:**55.

Rahmanian, M., D. R. Claus, and D. L. Oxender. 1973. Multiplicity of leucine transport by systems in *Escherichia coli* K-12. *J. Bacteriol.* **116:**1258.

Singer, C. E., G. R. Smith, R. Cortese, and B. N. Ames. 1972. Mutant tRNA ineffective in repression and lacking two pseudouridine modifications. *Nat. New Biol.* **238:**72.

Szentirmai, A., M. Szentirmai, and H. E. Umbarger. 1968. Isoleucine and valine metabolism of *Escherichia coli*. XV. Biochemical properties of mutants resistant to thiaisoleucine. *J. Bacteriol.* **95:**1672.

Thompson, L. H., J. L. Harkins, and C. P. Stanners. 1973. A mammalian cell mutant with a temperature-sensitive leucyl-transfer RNA synthetase. *Proc. Natl. Acad. Sci.* **70:**3094.

Thompson, L. H., D. J. Lofgen, and B. M. Adair. 1977. CHO cell mutants for arginyl-, asparagyl-, glutaminyl- and methionyl-transfer RNA synthetases: Identification and initial characterization. *Cell* **11:**157.

Thompson, L. H., C. P. Stanners, and L. Siminovitch. 1975. Selection by [^{3}H] amino acids of CHO-cell mutants with altered leucyl- and asparagyl-transfer RNA synthetases. *Som. Cell Genet.* **1:**187.

Travers, A. 1976a. RNA polymerase specificity and the control of growth. *Nature* **263:**641.

―――. 1976b. Template selection by *E. coli* RNA polymerase holoenzyme. *FEBS Lett.* **69:**195.

Umbarger, H. E. 1973. Genetic and physiological regulation of isoleucine, valine, and leucine formation in the *Enterobacteriaceae*. In *Genetics of industrial organisms* (ed. A. Vanek et al.), vol. 1, p. 195. Academic Publishing, Prague.

Wasmuth, J. J. and C. T. Caskey. 1976. Selection of temperature-sensitive CHL asparagyl-tRNA synthetase mutants using the toxic lysine analog, S-2-aminoethyl-L-cysteine. *Cell* **9:**655.

Williams, L. S. 1973. Control of arginine biosynthesis in *Escherichia coli*: Role of arginyl-transfer ribonucleic acid synthetase in repression. *J. Bacteriol.* **113:**1419.

Biochemistry and Biology of Aminoacyl-tRNA-Protein Transferases

Richard L. Soffer
Departments of Biochemistry and Medicine
Cornell University Medical College
New York, New York 10021

The critical role played by aminoacyl-tRNA in protein biosynthesis has tended to obscure its participation as an aminoacyl donor in a number of other biological processes. The enzymes that catalyze these reactions have been termed aminoacyl-tRNA transferases (Soffer 1974). The reactions differ fundamentally from protein synthesis in that the reactive moiety of the aminoacyl residue is its carboxyl group and there is no requirement for ribosomes, GTP, or template nucleic acids. These transfer reactions can be classified on the basis of the acceptor molecule to which the aminoacyl residue is transferred from tRNA (Soffer 1974). Lennarz (1972) is largely responsible for the discovery that certain bacteria contain particulate activities that transfer lysine and/or alanine from tRNA into an ester linkage with the 3′-OH group of phosphatidylglycerol to yield aminoacyl phosphatidylglycerol. Strominger (1970) provided most of the evidence that similar enzymes account for the biosynthesis of interpeptide bridges in various bacterial mureins. These latter reactions are initiated by aminoacylation of the ϵ-amino group in a lysine residue of one peptide chain and can be extended by sequential enzymatic aminoacylation of primary α-amino groups until ultimate peptide bond formation with a D-alanine residue of a different peptide chain in the complex peptidoglycan molecule. Details concerning donor and substrate specificity in these reactions and appropriate original author citations have been presented previously (Soffer 1974). The third category of aminoacyl transfer reactions is those catalyzed by aminoacyl-tRNA-protein transferases. The discovery of these enzymes emerged as a consequence of the unexpected observations by H. Kaji et al. (1963) and A. Kaji et al. (1963, 1965a,b) that ribosome-free extracts of rat liver and *Escherichia coli* could incorporate arginine and leucine or phenylalanine, respectively, into protein by reactions that were tRNA-dependent. During the past 12 years our laboratory has established the enzymological basis for these reactions by isolating and characterizing aminoacyl-tRNA-protein transferases that catalyze the transfer of certain aminoacyl residues from tRNA into peptide linkage with specific NH_2-terminal residues of protein or peptide acceptors. We have also employed genetic technology for examining the cellular function of these enzymes. The historical elucidation of the reactions and the problems encountered, as well as the resultant strategies devised for investigating their biochemical and physiological properties, have recently been reviewed extensively elsewhere (Deutch et al. 1978). This chapter briefly summarizes most of the established facts and elaborates on newer developments that are particularly relevant to examining the biological role of aminoacyl-tRNA-protein transferases.

ARGINYL-tRNA-PROTEIN TRANSFERASE

Arginyl-tRNA-protein transferase (arginyl transferase, EC 2.3.2.8) is a soluble enzyme found in the cytoplasm of all nucleated eukaryotic cells, including protists such as *Saccharomyces cerevisiae* and *Blastocladiella emersonii*, plants, and numerous mammalian organs and cell lines from various species (see Deutch et al. [1978] for comprehensive citations). It is present at lower concentrations in nuclear and mitochondrial sap (R. L. Soffer, unpubl.). Most studies have been performed with preparations from rabbit liver (Soffer 1970a), which, though enriched 7000-fold with respect to the crude soluble fraction, are still probably quite impure. The purified enzyme displays a molecular weight of 45,000–50,000 as estimated by gel filtration or glycerol gradient centrifugation (Soffer 1970b), although it exists as a part of a much larger molecular weight complex in unfractionated extracts (R. L. Soffer, unpubl.). It specifically catalyzes the transfer of arginine from tRNA into peptide linkage with unblocked NH_2-terminal aspartyl, glutamyl, and cystinyl residues of protein (Soffer 1971a) or peptide (Soffer 1973a; Deutch and Soffer 1975) acceptors provided these moieties are not sterically shielded as they are, for example, in intact IgG molecules (Soffer and Capra 1971). Reaction requirements include the presence of a monovalent cation and a thiol compound as well as a suitable acceptor substrate and the pH range for activity is broad (~7.4–9.8) with an optimum of 9.0 (Soffer 1970a). Puromycin, cycloheximide, and EDTA are not inhibitory. Because of spontaneous deacylation of arginyl-tRNA in the alkaline pH range, stoichiometric acylation of acceptors is conveniently achieved by using a coupled system that includes arginine, L-arginyl-tRNA synthetase, the transferase, tRNA, and requirements for both the activation and transfer of arginine (Soffer 1970a). Arginyl-tRNA from *E. coli* or rat liver are equally satisfactory donor substrates and it seems likely that all isoaccepting species can participate in the reaction, since virtually complete transfer of arginine directly from tRNA to protein can be obtained (Soffer 1970a). A free α-amino group in the arginyl moiety of arginyl-tRNA is necessary for reaction of the donor substrate (R. L. Soffer and S. Pestka, unpubl.). The requirements for acceptor substrates can be studied with proteins at limiting concentrations by their ability to stimulate near stoichiometric incorporation of arginine in the coupled system. This method will not detect polypeptides that have already been extensively arginylated in vivo. It is established that the NH_2-terminal aspartyl residues in bovine serum albumin (Soffer and Horinishi 1969; Soffer 1970a) and bovine thyroglobulin (Soffer 1971b) can be nearly stoichiometrically arginylated in vitro and there is evidence that they are arginylated at the 5–10% level in vivo (M. J. Leibowitz and R. L. Soffer 1971a and unpubl.). Porcine β-melanocyte-stimulating hormone and synthetic human-type angiotensin II contain NH_2-terminal aspartic acid and this residue can be stoichiometrically arginylated in vitro (Soffer 1975). The biological activity of the hormone on frog epithelial cells is not altered by this enzymatic addition, whereas the pressor and contractive effects of the angiotensin II derivative are diminished. It is not known whether the modified form of either of these peptides occurs in vivo.

Acceptor specificity is best studied using defined small peptides and measuring their ability to inhibit competitively the direct transfer of radioactive arginine from tRNA to a bona fide protein acceptor (Soffer 1973a). Acceptance by the peptide can easily be confirmed by paper electrophoresis. Among 17 dipeptides containing different residues linked to alanine, only Glu-Ala, Asp-Ala, and cystinyl-bis-Ala are acceptors by these criteria, indicating that an NH_2-terminal dicarboxylic amino acid residue is an absolute determinant of acceptor specificity. Stereoconfiguration is also an absolute determinant, since dipeptides containing D-amino acids are not acceptors, even with a dicarboxylic amino acid NH_2 terminal. An advantage of the inhibition assay (Soffer 1973a) for acceptance is that it reflects relative, as well as absolute, determinants, since the inhibition constant (K_i) value determined for the low-molecular-weight substrate is an inverse measure of its affinity for the enzyme. For example, among dipeptides containing acceptor NH_2 terminals, those linked to an acidic residue (e.g., Glu-Glu, Glu-Asp) display markedly higher K_i values than when linked to a basic or neutral amino acid. Thus, the penultimate residue is a relative determinant of acceptor specificity that, if acidic, diminishes affinity between the enzyme and substrate. Glutamic acid itself is an acceptor of very low affinity. Glutamine and the γ-methyl ester of glutamic acid are not acceptors suggesting an absolute requirement for an unblocked γ-carboxylic acid group in the NH_2-terminal residue. The α-methyl ester of glutamic acid and particularly isoglutamine exhibit higher affinities for the enzyme than does glutamic acid, indicating that blockage of the carboxylic acid group in the α position is a favorable relative determinant. It is also noteworthy that the K_i value for Glu-Val is 600 times that of Glu-Val-Phe, implying that the farther the carboxyl terminal from the acceptor residue, the better the substrate, at least with small peptides. Glu-Lys and Glu-ϵ-Lys exhibit comparable acceptor properties, as do Glu-Ala and γ-Glu-Ala (Soffer 1973a). The latter result requires confirmation, since it is inconsistent with the adverse effect on acceptor properties of free glutamic acid caused by methylation or amidation of its γ-carboxyl group and with the observation (R. L. Soffer, unpubl.) that glutathione is not an acceptor.

Despite its ubiquitous distribution, nothing of substance is known concerning the function of arginyl-tRNA-protein transferase in cellular metabolism. Kaji (1976) reported some acceptance in vitro by a nonhistone fraction of chromatin and speculated that the enzyme might play a regulatory role by modification of chromosomal proteins. Although this may or may not be the case, acceptance of arginine by crude protein fractions is the rule rather than the exception and simply indicates that such fractions contain polypeptides with acidic NH_2 terminals that were not arginylated in vivo. A new approach for studying the cellular biology of this enzyme has very recently been developed in our laboratory (R. L. Soffer and M. Savage, unpubl.). We have isolated, after mutagenesis with ethyl methane sulfonate, a mutant of *S. cerevisiae* that completely lacks arginyl-tRNA-protein transferase activity when assayed at 37°C but contains about 8% of the wild-type activity at 25°C (Table 1). The strain is called Ate-1, since the structural gene has been designated *ate* for arginine transfer enzyme. This strain will grow at 25°C but not at 37°C. However, derivative strains selected for their ability to grow in enriched and minimal medium at 37°C still exhibit virtually no enzyme activity

Table 1 Arginyl-tRNA-protein transferase in yeast

Strain	Growth temperature (°C)	Enzyme activity (pmoles/min/mg)	
		25°C	37°C
S288C	25	3.8	14.1
S288C	37	4.1	17.8
Ate-1	25	0.3	<0.1
Ate-1	37	—	—
PR-1	25	1.6	0.3
PR-1	37	0.2	<0.1
PR-2	25	1.1	0.2
PR-2	37	<0.1	<0.1

Activities were determined as described by Soffer and Deutch (1975) using bovine serum albumin (75 μM) as the protein acceptor.

when assayed at the higher temperature or when grown at the higher temperature and assayed at 25°C. These results indicate that the *ate-1* mutation, although resulting in a temperature-sensitive enzyme, is not responsible for the conditionally lethal phenotype; that the activity is temperature sensitive in vivo, as well as in vitro; and that arginyl-tRNA-protein transferase is not required for cell viability. It is interesting and potentially useful, however, that these derivative strains possess approximately 30% of the wild-type activity when grown and assayed at 25°C. Separation of the *ate-1* and conditionally lethal mutations has also been achieved by genetic crosses and preliminary results suggest that the *ate* gene is centromere linked (R. L. Soffer et al., unpubl.). We are currently transferring the *ate-1* mutation into an established wild-type genome to define precisely the phenotype specifically resulting from a mutation in the *ate* gene.

LEUCYL-PHENYLALANYL-tRNA-PROTEIN TRANSFERASE

Leucyl-phenylalanyl-tRNA-protein transferase (leucyl transferase, EC 2.3.2.6) is a soluble enzyme found in a number of gram-negative bacteria (Deutch et al. 1978). It has been purified as much as 20,000-fold with respect to a crude soluble fraction of *E. coli* (Soffer 1973b) but its absolute purity at that stage has not been assessed and it is too unstable to be of much practical use. Therefore, most of our work has been done with a preparation enriched approximately 1200-fold (Soffer 1973b). The molecular weight of the purified enzyme, estimated by glycerol gradient centrifugation and gel filtration, is 25,000 and 14,000, respectively, implying unusual physical characteristics (Soffer 1974). As is the case with the eukaryotic enzyme, the bacterial activity in crude soluble extracts elutes as a much larger molecule during gel filtration than does the purified enzyme. It has been established biochemically and genetically (Scarpulla et al. 1976) that the same enzyme protein catalyzes the transfer of leucine, phenylalanine, or methionine from tRNA into peptide linkage with NH_2-terminal arginyl, lysyl, or histidyl residues of protein (Leibowitz and Soffer 1971b) or peptide (Soffer 1973b) acceptors. The possibility (Kaji et al. 1965a) has not been excluded that tryptophanyl-tRNA may also be a donor substrate. The enzyme is partially stabilized by 0.12 M $(NH_4)_2SO_4$ and reaction requirements include a monovalent cation and a thiol compound, as well as a suitable acceptor. There is a broad, slightly alkaline pH dependence. Puromycin and various divalent cations are inhibitory (Leibowitz and Soffer 1970).

Both the aminoacyl residue and the polyribonucleotide chain contribute to donor specificity in the reaction catalyzed by this enzyme. Phenylalanine but not valine can be transferred from *E. coli* tRNA^{Val} and a free α-amino group is required, since acetylation in phenylalanyl-tRNA abolishes substrate activity. The enzyme can utilize p-fluorophenylalanyl-tRNA though the reaction rate is slower than with phenylalanyl-tRNA. The 3'-pentanucleotide fragment of *E. coli* tRNA^{Phe} acylated with phenylalanine is not a substrate (Leibowitz and Soffer 1971b). K_m values differ for the transfer of leucine from various isoaccepting species of *E. coli* tRNA^{Leu} (Rao and Kaji 1974). Methionine can be transferred from noninitiator *E. coli* tRNA^{Met}, but not from tRNA^{fMet} (Scarpulla et al. 1976).

Acceptor specificity has been delineated by examining the ability of defined small peptides to inhibit competitively the transfer of radioactive aminoacyl residues from tRNA to α-casein (Soffer 1973b; Scarpulla et al. 1976). The donor aminoacyl-tRNA does not influence this specificity and the results obtained in the inhibitory assay have been confirmed by the presence or absence of peptide-dependent radioactive products separated during paper electrophoresis. Among 19 dipeptides containing different NH$_2$-terminal residues linked to alanine only Arg-Ala, Lys-Ala, and His-Ala are acceptors. The affinity of His-Ala for the enzyme is much lower than that of Arg-Ala or Lys-Ala. It has been chemically established that phenylalanine is linked by a peptide bond to the NH$_2$-terminal arginine residues of arginylated albumin as well as α- and β-casein (Leibowitz and Soffer 1971b) and to the NH$_2$-terminal lysyl residue of Lys-Ala-Ala (Soffer 1973b). Ala-Lys is not an acceptor suggesting that linkage is through the α-amino group of the lysyl residue. Dipeptides with D-aminoacyl residues are not acceptors even with a basic NH$_2$-terminal amino acid establishing that stereoconfiguration as well as a basic NH$_2$-terminal residue are absolute acceptor determinants. Dipeptides containing NH$_2$-terminal arginine attached to arginine or lysine exhibit lower K_i values than dipeptides with arginine attached to acidic or neutral amino acids, indicating that a penultimate basic residue represents a favorable relative determinant enhancing the affinity of the substrate for the transferase. Free arginine is a substrate with low affinity for the enzyme. Its methyl ester displays a higher affinity. Free lysine does not appear to function as an acceptor.

Acceptance in vitro by *E. coli* ribosomal preparations has been repeatedly observed (Momose and Kaji 1966; Leibowitz and Soffer 1971c; Soffer and Savage 1974). With salt-washed ribosomes the specific acceptance is higher for 30S particles than for 50S particles but still does not approach molar equivalence even under conditions of stoichiometric acylation (Leibowitz and Soffer 1971c). The only 30S ribosomal protein whose NH$_2$-terminal structure (Met-Lys-Pro) bears any relationship to the acceptor determinants of the transferase is S6; this protein from a transferaseless mutant (Soffer and Savage 1974) contains NH$_2$-terminal methionine (Deutch et al. 1978), indicating that the methionyl residue is derived from the conventional initiation process in protein synthesis rather than by posttranslational modification. The most likely explanations for acceptance by ribosomes in vitro are that they are contaminated with submolar equivalents of nonribosomal acceptors or that limited proteolysis during their preparation generates structurally appropriate NH$_2$ terminals. Acceptance by soluble extracts from the wild-type and a transferaseless mutant has also been studied in vitro using radioactive phenylalanyl-tRNA as a donor and subjecting the reaction mixtures to slab gel electrophoresis and autoradiography (Soffer and Savage 1974). At least 20 radioactive bands were observed and 6–8 of these appeared to be more heavily labeled in extracts from the transferaseless strain, suggesting that in the wild type they may have been partially acylated in vivo.

The cellular function of the *E. coli* transferase has been investigated by examining the properties of a mutant, MS845, that lacks the activity. This mutant was isolated by assaying survivors after treatment with nitrosoguanidine (Soffer and Savage 1974). (It is now deposited with the *E. coli* Genetic Stock Center at Yale University). The structural gene has been designated *aat* (aminoacyl transferase). This gene is probably located between 50 and 60 minutes on the *E. coli* map, since enzyme activity can be restored in merodiploids constructed with an episome (F142) bearing this portion of the *E. coli* genome (Soffer and Savage 1974; Deutch et al. 1978). However, this map assignment must be considered tentative, since the episome also contains an ocher suppressor. The transferaseless mutant, like its parental K12 strain W4977, is a proline auxotroph. The most striking physiologic phenotype of the mutant occurs when it is grown in minimal salts medium containing 0.5% glycerol and 2 mM L-proline. Under these conditions it exhibits a normal generation time but ceases growth at about 40% of the level achieved by the parental strain and undergoes a lag of about 12 hours before resuming growth in fresh medium (Soffer and Savage 1974). This growth phenotype is due to an increased rate of proline catabolism in the transferaseless mutant. It can be abolished by the presence of 10–20 mM L-proline in the medium and mimicked by W4977 when grown in 0.1–0.2 instead of 2 mM L-proline (Deutch and Soffer 1975). The increased rate of proline catabolism is due to a five- to tenfold elevated level of proline oxidase activity in the transferaseless mutant (Deutch and Soffer 1975). This biochemical difference is unrelated to proline auxotrophy, since it persists in protorophic derivatives of the parental strain and the transferaseless mutant obtained by introduction of an episome bearing the genes for proline biosynthesis. The initial kinetics of proline oxidase activity induced by proline in both prototrophic derivatives is identical but differential synthesis persists for much longer in the transferaseless strain (Deutch et al. 1977, 1978). The growth phenotype was originally used to select spontaneous revertants by recycling the transferaseless mutant in medium containing 2 mM L-proline. After eight cycles, strains that grew normally were isolated from four separate mutant clones and each was found to have regained transferase activity (Soffer and Savage 1974). In addition, since the transferaseless mutant can use proline much more efficiently as a sole source of nitrogen than can the wild type, it has been possible to obtain true transferase revertants among survivors of the penicillin selection technique carried out under these conditions (Deutch et al. 1978).

The observation that loss of the transferase is associated with a marked increase in proline oxidase activity led us to speculate that the transferase might regulate proline oxidase activity by catalyzing posttranslational modification of the proline-specific enzyme in the oxidative reaction (Deutch and Soffer 1975). Proline is metabolized to glutamic acid by bacteria. Δ^1-Pyrroline-5-carboxylate is an intermediate in this process and the particulate activity catalyzing its formation is coupled to the respiratory chain and commonly refered to as proline oxidase (Frank and Ranhand 1964). To test our hypothesis, it was necessary to isolate the responsible enzyme from the transferaseless mutant and to determine whether (1) it was an acceptor in the transfer reaction and (2) its higher activity in the mutant was due to an increase in catalytic activity of the proline oxidase molecule itself or to an increased number of proline oxidase molecules with the same catalytic activity as those in the parental and revertant strains.

Recently, we developed a procedure for solubilizing the enzyme from membranes of the transferaseless mutant and purifying it to a nearly homogeneous state (Scarpulla and Soffer 1978). It is a flavoprotein containing flavin-adenine dinucleotide and consists of a single type of polypeptide chain with a molecular weight of about 124,000. In the isolated state it will not use oxygen as proximate electron acceptor; therefore, we have designated it as proline dehydrogenase rather than proline oxidase (Scarpulla and Soffer 1978). In our earlier work with this enzyme, acceptance by a protein in quite highly purified preparations, which comigrated with the dehydrogenase activity during polyacrylamide gel electrophoresis under nondenaturing conditions, led us to conclude, tentatively, that the dehydrogenase protein was probably a bona fide acceptor (Deutch et al. 1978). However, we have now established that this is not the case with the nearly homogenous preparation, the purity of which is approximately 95%, as judged by densitometer scanning of stained polyacrylamide gels after electrophoresis in the reduced, denatured state (Soffer and Scarpulla 1979). The NH_2-terminal residue of proline dehydrogenase is glycine, which is not an acceptor determinant. Under conditions of stoichiometric acylation with radioactive phenylalanine our preparation accepts only 0.06 molar equivalents, based on the dehydrogenase m.w. of 124,000. This acceptance is due to a minor contaminating protein ($M_r \sim 100,000$) easily distinguished from the proline dehydrogenase protein by autoradiography after gel electrophoresis in the reduced, denatured state. Furthermore, we have demonstrated by two independent criteria that proline oxidase molecules in the parental and transferaseless strains have the same catalytic potential and that the higher level of activity in the transferaseless mutant is a consequence of an increased number of enzyme molecules.

First, the level of competing antigen in a competition radioimmunoassay for proline dehydrogenase is higher in crude solubilized extracts of the mutant as compared with the parental and revertant strains in direct proportion to the differences in their catalytic activities (Fig. 1). The ratio of catalytic activity to competing antigen is essentially identical and equal to the specific activity of the pure enzyme (7–8 units/mg; Scarpulla and Soffer 1978). In addition to demonstrating that individual dehydrogenase molecules from the different strains possess similar catalytic activity, these data also confirm those obtained by conventional mixing experiments in ruling out the possibility that the transferaseless mutant exhibits high dehydrogenase activity because it is lacking an inhibitor present in wild-type cells.

Second, when subjected to the identical purification procedure, proline dehydrogenase activity from the parental strain was enriched 765-fold but was only 26% pure, as judged by densitometry, whereas the corresponding figures for the mutant were 350-fold and 95% pure. From these results it can be calculated that the specific activity of the pure dehydrogenase protein in both strains is about 7.7 units/mg and that the proline dehydrogenase protein accounts for approximately 0.27% of crude homoge-

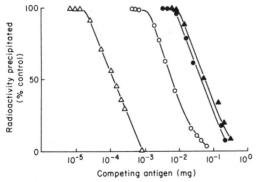

Figure 1 Competition radioimmunoassay of proline dehydrogenase protein. The procedure was that described by Scarpulla and Soffer (1978) using antibodies developed in rabbits against the nearly homogenous enzyme isolated from the transferaseless mutant. Competing antigens were the purified enzyme (△; 7.3 catalytic units/mg) and crude, solubilized fractions from the mutant (○; 0.112 units/mg), parental (▲; 0.0114 units/mg), and revertant (●; 0.0136 units/mg) strains.

nate protein in the mutant as compared with 0.034% of that in its parent. These results establish that, contrary to our speculation, the proline dehydrogenase protein is not a substrate in the transfer reaction and that regulation of proline dehydrogenase activity by the transferase is indirect and mediated by a mechanism involving the number of enzyme molecules in the cell.

Several other phenotypic characteristics appear to be associated with loss of transferase activity, as judged by their presence in the transferase-less mutant and absence in the parental and revertant strains (Deutch et al. 1977). These include (1) an abnormal morphology that can be reversed by addition of D-alanine or D-glutamic acid to the medium, (2) a twofold increase in the growth rate with aspartic acid as a sole source of nitrogen, (3) a slight accumulation of enterochelin and related derivatives in the culture medium, and (4) approximately 50% decreases in the specific activities of L-phenylalanyl-tRNA synthetase and tryptophanase. Homogenous preparations of the latter two enzymes from the mutant do not function as acceptors in the transfer reaction.

DISCUSSION

From a biochemical point of view the aminoacyl-tRNA-protein transferases are of special interest because they recognize determinants on the polyribonucleotide chain of aminoacyl-tRNA, the identity of the donor aminoacyl residue, its unblocked α-amino group and identity, stereoconfiguration, and subtle structural modifications of the acceptor molecule. These cognitive properties are especially impressive in the case of the *E. coli* transferase, since it appears to be an unusually small enzyme protein. The mechanism of interaction among these enzymes and their donor and acceptor substrates should therefore be a rewarding area of investigation, were it not for the lability of highly enriched enzyme preparations and the very extensive purification presumably required to achieve homogeneity. From a biological point of view the enzymes are interesting because they are widely distributed yet little is known about their cellular function. Arginyl-tRNA-protein transferase appears to be ubiquitous in nucleated eukaryotic cells and leucyl-phenylalanyl-tRNA-protein transferase is present in many gram-negative bacteria. Although it appears to be absent from gram-positive bacteria (Deutch et al. 1978), the possibility has not been investigated that these organisms contain similar transferases specific for other donor aminoacyl residues.

When we first realized (Soffer 1968) that there was an enzyme that catalyzed the transfer of arginine from tRNA to specific protein acceptors, we speculated that the function of such a transferase might be to regulate other enzyme activities by catalyzing direct posttranslational aminoacylation of the responsible enzyme proteins. Our recent work on the *E. coli* transferaseless mutant suggests that this putative regulatory role of the transferases may not be quite so mechanistically simple. Particularly in the case of proline dehydrogenase activity, it is clear that control by the transferase is not a consequence of direct aminoacylation but is mediated by a process involving the number of dehydrogenase molecules. The markedly increased number of such molecules in the transferaseless mutant is unlikely to be due to a decreased rate of their degradation, since we have repeatedly failed to detect protease activity in highly purified transferase preparations. In addition, the transferaseless mutant contains somewhat decreased levels of L-phenylalanyl-tRNA synthetase and tryptophanase activities and the responsible enzyme proteins isolated from it are not acceptors in the transfer reaction. These results suggest that the transferase may be involved in controlling the biosynthesis of other enzymes. Thus, either the transferase itself, and/or its physiological acceptor substrates may conceivably function as regulatory molecules that affect the biosynthesis of specific enzymes at the transcriptional or translational level. In this context it is noteworthy that the various phenotypic characteristics that we have thus far identified in the transferaseless mutant all seem to relate to the general area of amino acid metabolism. It will be especially interesting to determine whether the eukaryotic enzyme, despite its difference in donor and acceptor specificities, plays a similar role in cellular metabolism. Our recent isolation of a yeast mutant lacking the enzyme activity provides the opportunity for experimental examination of this question.

ACKNOWLEDGMENTS

I am grateful to my talented collaborators who, over the years, have shared the curiosity and frustration inherent in our investigation of these novel enzymes. These colleagues include Hiroo Horinishi, Michael J. Leibowitz, Margaret Savage, Richard Jagger, Charles E. Deutch, Emily Sonnenblick, and Richard C. Scarpulla. Recent research on this project has been supported by grant BMS-74-23970 from the National Science Foundation.

REFERENCES

Deutch, C. E. and R. L. Soffer. 1975. Regulation of proline catabolism by leucyl, phenylalanyl-tRNA-protein transferase. *Proc. Natl. Acad. Sci.* **72**:405.

Deutch, C. E., R. C. Scarpulla, and R. L. Soffer. 1978. Posttranslational NH_2-terminal aminoacylation. *Curr. Top. Cell Regul.* **13**:1.

Deutch, C. E., R. C. Scarpulla, E. B. Sonnenblick, and R. L. Soffer. 1977. Pleiotropic phenotype of an *Escherichia coli* mutant lacking leucyl, phenylalanyl-transfer ribonucleic acid-protein transferase. *J. Bacteriol.* **129**:544.

Frank, L. and B. Ranhand. 1964. Proline metabolism in *Escherichia coli*: The proline catabolic pathway. *Arch. Biochem. Biophys.* **107**:325.

Kaji, H. 1976. Amino-terminal arginylation of chromosomal proteins by arginyl-tRNA. *Biochemistry* **15**:5121.

Kaji, A., H. Kaji, and G. D. Novelli. 1963. A soluble amino acid-incorporating system. *Biochem. Biophys. Res. Commun.* **10**:406.

―――. 1965a. Soluble amino acid-incorporating system. Preparation of the system and nature of the reaction. *J. Biol. Chem.* **240**:1185.

―――. 1965b. Soluble amino acid incorporating system. Soluble nature of the system and the characterization of the radioactive product. *J. Biol. Chem.* **240**:1192.

Kaji, H., G. D. Novelli, and A. Kaji. 1963. A soluble amino acid-incorporating system from rat liver. *Biochim. Biophys. Acta* **76**:474.

Leibowitz, M. J. and R. L. Soffer. 1970. Purification and properties of a leucyl, phenylalanyl transfer ribonucleic acid-protein transferase from *Escherichia coli*. *J. Biol. Chem.* **245**:2066.

―――. 1971a. Site of acylation of bovine serum albumin in the leucine, phenylalanine-transfer reaction. *J. Biol. Chem.* **246**:4431.

―――. 1971b. Substrate specificity of leucyl, phenylalanyl-transfer ribonucleic acid-protein transferase. *J. Biol. Chem.* **246**:5207.

―――. 1971c. Modification of a specific ribosomal protein catalyzed by leucyl, phenylalanyl-tRNA: Protein transferase. *Proc. Natl. Acad. Sci.* **68**:1866.

Lennarz, W. J. 1972. Studies on the biosynthesis and function of lipids in bacterial membranes. *Acc. Chem. Res.* **5**:361.

Momose, K. and A. Kaji. 1966. Soluble amino acid-incorporating system. Further studies on the product and its relationship to the ribosomal system for incorporation. *J. Biol. Chem.* **241**:3294.

Rao, P. M. and H. Kaji. 1974. Utilization of isoaccepting leucyl-tRNA in the soluble incorporation system and protein synthesizing systems from *E. coli*. *FEBS Lett.* **43**:199.

Scarpulla, R. C. and R. L. Soffer. 1978. Membrane-bound proline dehydrogenase from *Escherichia coli*. Solubilization, purification and characterization. *J. Biol. Chem.* **253**:5997.

―――. 1979. Regulation of proline dehydrogenase activity in *E. coli* by leucyl, phenylalanyl-tRNA-protein transferase. *J. Biol. Chem.* **254**:1724.

Scarpulla, R. C., C. E. Deutch, and R. L. Soffer. 1976. Transfer of methionyl residues by leucyl, phenylalanyl-tRNA-protein transferase. *Biochem. Biophys. Res. Commun.* **71**:584.

Soffer, R. L. 1968. The arginine transfer reaction. *Biochim. Biophys. Acta* **155**:228.

―――. 1970a. Purification and properties of the arginyl transfer ribonucleic acid-protein transferase from rabbit liver cytoplasm. *J. Biol. Chem.* **245**:731.

―――. 1970b. Aminoacyl-tRNA-protein transferases. A novel class of enzymes catalyzing peptide bond formation. *Trans. N.Y. Acad. Sci.* (2) **32**:974.

―――. 1971a. Protein acceptor specificity in the arginine transfer reaction. *J. Biol. Chem.* **246**:1602.

―――. 1971b. Arginylation of bovine thyroglobulin. *J. Biol. Chem.* **246**:1481.

―――. 1973a. Peptide acceptors in the arginine transfer reaction. *J. Biol. Chem.* **248**:2918.

———. 1973b. Peptide acceptors in the leucine, phenylalanine transfer reaction. *J. Biol. Chem.* **248:**8424.

———. 1974. Aminoacyl-tRNA transferases. *Adv. Enzymol.* **40:**91.

———. 1975. Enzymatic arginylation of β-melanocyte-stimulating hormone and of angiotensin II. *J. Biol. Chem.* **250:**2626.

Soffer, R. L. and J. D. Capra. 1971. Enzymatic probe for accessibility of NH_2-terminal residues in immunoglobulins. *Nat. New Biol.* **223:**44.

Soffer, R. L. and C. E. Deutch. 1975. Arginyl-tRNA-protein transferase in eukaryotic protists. *Biochem. Biophys. Res. Commun.* **64:**926.

Soffer, R. L. and H. Horinishi. 1969. General characteristics of the arginine-transfer reaction in rabbit liver cytoplasm. *J. Mol. Biol.* **43:**163.

Soffer, R. L. and M. Savage. 1974. A mutant of *Escherichia coli* defective in leucyl, phenylalanyl-tRNA-protein transferase. *Proc. Natl. Acad. Sci.* **71:**1004.

Strominger, J. L. 1970. Penicillin-sensitive enzymatic reactions in bacterial cell wall synthesis. *Harvey Lect.* **64:**179.

tRNAs as Primers for Reverse Transcriptases

James E. Dahlberg
Department of Physiological Chemistry
University of Wisconsin-Madison
Madison, Wisconsin 53706

RNA-directed DNA polymerases (reverse transcriptases) are the enzymes responsible for synthesis of the DNA copy of RNA tumor virus genomic RNA. The reverse transcriptases associated with avian sarcoma/leukosis viruses (ASV/ALV) differ from the enzymes of the murine viruses, such as murine leukemia virus (MuLV) (for reviews, see Temin 1971; Temin and Baltimore 1972; Verma 1977; Taylor 1977; Bishop 1978). Nevertheless, the enzymes from these two sources share many important features. They are both coded for by their respective RNA genomes and are present in mature virus particles. After virus infection of susceptible cells, the enzymes initiate synthesis of a cDNA (complementary DNA) copy of the virion genomic RNA. Like other DNA polymerases, reverse transcriptases are unable to initiate DNA synthesis de novo; rather, they add deoxyribonucleoside triphosphates to preexisting primer RNAs. The primer RNAs are present in virions, tightly associated with the template, genomic RNA (Verma et al. 1971; Duesberg et al. 1971; Canaani and Duesberg 1972; Faras et al. 1973b).

IDENTIFICATION OF PRIMER tRNAs

Identification and analysis of the RNA primers is greatly facilitated by the fact that the virions of RNA tumor viruses contain all of the macromolecular components necessary for DNA synthesis. These include the primer RNAs, the viral coded enzymes, and the RNA template, which is the viral genome. This genome is a 70S RNA dimer consisting of two identical subunits of 35S RNA, each of which is 8000–10,000 nucleotides long (Fan and Paskind 1974; Billeter et al. 1974; Beemon et al. 1976). After disruption of virions in vitro with mild detergent and addition of deoxyribonucleoside triphosphates, the enzyme is able to synthesize a cDNA copy of the genomic RNA. Thus, isolation of intact virus particles constitutes a rapid method of purification of the DNA synthesizing system.

The primer RNAs for avian and murine RNA tumor viruses are identified by several criteria. First of all, the ability of 70S genomic RNA to serve as template-primer complex can be destroyed by heating the RNA above 70–75°C; however, this activity can be restored by reannealing the 35S subunit RNA to a small RNA that is released at this temperature (Dahlberg et al. 1974; Faras et al. 1974b; Taylor et al. 1974). In avian viruses, the released RNA is tRNA^{Trp} and in murine viruses it is tRNA^{Pro} (Waters et al. 1975; Waters 1975; Peters et al. 1977; Waters and Mullin 1977; Harada et al. 1975, 1979). Second, when DNA synthesis is restricted to the addition of a single radioactive nucleotide to the presumptive primer, the only radiolabeled molecules observed have two-dimensional gel electrophoretic mobilities identical to tRNA^{Trp} or tRNA^{Pro}, respectively. Finally, the 3' ends of these tRNAs are extended by participation in DNA synthesis (Faras et al. 1974a; Peters et al. 1977).

The nucleotide sequences of the two primer tRNAs were determined for the virion RNAs and for comparable cellular molecules. These results showed that the virion-associated tRNAs were identical to the corresponding cell RNAs, indicating that cellular molecules were taken into the virion during budding of the particle from the cell membrane (Sawyer et al. 1974; Peters et al. 1977; Harada et al. 1975, 1979).

STRUCTURES OF PRIMER tRNAs

The nucleotide sequences of the primer tRNA for the murine and avian viruses are presented in Figure 1, a and b, respectively. The anticodons of the two tRNAs show that these tRNAs are specific for the amino acids proline and tryptophan, respectively; these conclusions are confirmed by aminoacylation of the purified molecules. The only unusual feature of the two primer tRNAs is the existence of the sequence GψψC, instead of the GTψC sequence usually found in loop 4 of tRNA molecules. Whether or not the ψψ residues themselves have a direct role in the priming activity of these two tRNAs is unknown. Priming activity of tRNA^{Trp} does, however, require the 3'-terminal 29 nucleotides of the molecule, including the entire ψψ loop and stem (Cordell et al. 1979).

PRIMER-TEMPLATE INTERACTION

Experiments directed toward investigating the mechanism of priming by these tRNAs have centered upon interaction between the primers and their respective templates and reverse transcriptases. In addition, chemical modification or specific cleavages of the tRNAs have been used to probe how these tRNAs work as primers.

The regions of the primer tRNAs responsible for binding to the template RNAs were identified by RNase digestion of the primer-template complexes under conditions that allow for isolation of RNA-RNA duplexes (Cordell et al. 1976; Eiden et al. 1976; Peters and Dahlberg

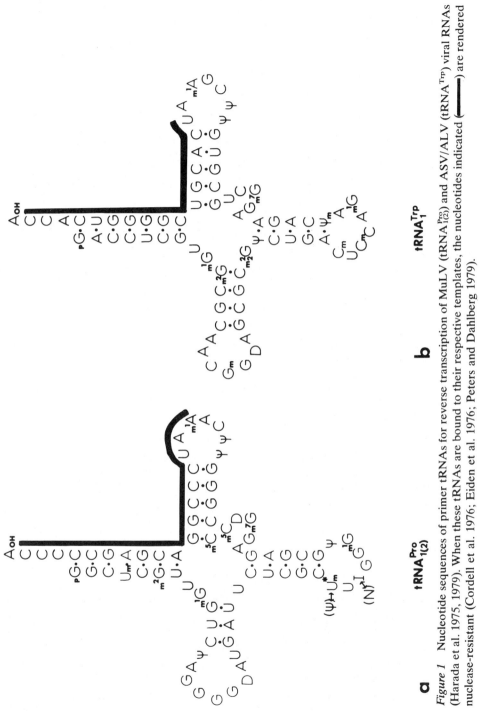

Figure 1 Nucleotide sequences of primer tRNAs for reverse transcription of MuLV (tRNA$_{1(2)}^{Pro}$) and ASV/ALV (tRNA$_1^{Trp}$) viral RNAs (Harada et al. 1975, 1979). When these tRNAs are bound to their respective templates, the nucleotides indicated (———) are rendered nuclease-resistant (Cordell et al. 1976; Eiden et al. 1976; Peters and Dahlberg 1979).

1979). Fingerprint analysis of such RNase-resistant fragments, labeled in the tRNA strand, revealed that the 3' 16-18 nucleotides of the primers were bound to the template. Interestingly, the 3'-terminal C-A bond was not protected against RNase in either the avian-primer or murine-template-primer complexes, although studies on the structure of the template RNAs indicated that U was present in the template immediately adjacent to the first nucleotide copied into the DNA (Coffin and Haseltine 1977b; Coffin et al. 1978; Cordell et al. 1978). The regions of tRNA^{Trp} and tRNA^{Pro} that bind tightly to the avian and murine RNA virus templates are indicated in Figure 1. Although nucleotides to the 5' side of m^{1}A are not protected against RNase digestion, they do appear to be required for priming activity (Cordell et al. 1979). The reason for this requirement is not well understood; it is possible that these nucleotides may interact in a weaker manner with other sequences in the template or that they may interact directly with the reverse transcriptase.

The primer tRNA binding sites in the templates were located relative to the 3' poly(A) of ASV/ALV RNA and MuLV RNA (Taylor and Illmensee 1975; Staskus et al. 1976; Peters and Dahlberg 1979). In these experiments, poly(A)-containing fragments of the template were assayed for their ability to bind radioactive primer tRNA or to support tRNA-primed DNA synthesis. Only essentially full-length template fragments were able to do these things, indicating that the primer tRNA bound and functioned far from the poly(A), i.e., near the 5' end of the template.

The initiation of DNA synthesis at a site near the 5' end of the template was confirmed by comparison of template oligonucleotide sequences and the initial DNA sequence (Staskus et al. 1976; Coffin and Haseltine 1977a; Haseltine et al. 1977; Shine et al. 1977; Coffin et al. 1978). Reverse transcription of ASV/ALV template-primer complexes leads to the production of a 101-nucleotide-long DNA molecule attached to the primer tRNA^{Trp}; in MuLVs the DNA attached to tRNA^{Pro} is 134 nucleotides long. These DNA fragments, called strong-stop cDNA, accumulate as a result of the enzyme coming to the 5' end of the template. The lengths of the strong-stop cDNAs are usually taken as the distances between the end of the primer binding site and the end of the template (Haseltine et al. 1976; Coffin and Haseltine 1977a). Strictly speaking, however, this is only the minimum distance between the primer and the end of the template, since the entire primer binding site has not been shown to be a contiguous sequence in the template RNA. For example, it is possible that the sequence complementary to the 3' end of the primer may be adjacent to the initial template region, with the remainder of the primer binding site being several (hundreds?) nucleotides removed (Fig. 2). Although there is no direct evidence for such a model, it would explain the fact that the template RNA does not protect the 3'-terminal C—A bond in the primer against RNase digestion.

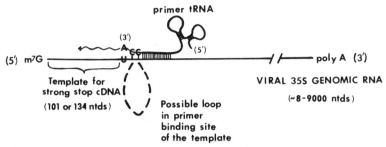

Figure 2 Schematic representation of the interaction between primer tRNAs and template RNAs. Synthesis of DNA starts 134 (MuLV) or 101 (ASV/ALV) nucleotides from the 5' end of the template. The primer binding site is located within a few hundred nucleotides of the 5' end of the template. The possible loop in the template is included to emphasize that the structure of the primer tRNA binding site is unknown and may be more complex than simply a contiguous complementary sequence. After synthesis of cDNA reaches the 5' end of the template, the nascent chain is elongated using the 3' end of the template RNA. Presumably a sequence repeated at both the 5' and 3' ends of the template RNA functions to facilitate this switch.

PRIMER–REVERSE TRANSCRIPTASE INTERACTION

The association between the avian virus reverse transcriptase and its primer, tRNA^{Trp}, has been shown to be strong and specific. In fact, the avian myeloblastosis virus reverse transcriptase can select tRNA^{Trp} plus tRNA$_4^{Met}$ (and lesser amounts of tRNA^{Pro}) from a mixture of chicken cell total tRNA (Panet et al. 1975b). This selectivity depends on the assay used, since methods that do not distinguish between weak and strong binding show that the enzyme actually binds many other tRNAs (Cavalieri and Yamaura 1975; Hizi et al. 1977; Panet and Berliner 1978). A functional significance of tRNA^{Trp} binding to reverse transcriptase (and by inference the in vivo role of the tRNA) is indicated by the finding that tRNA^{Trp} binding and DNA polymerase activity are comparably diminished for the reverse transcriptases of several mutants of ASVs that are temperature-sensitive for replication (Panet et al. 1978).

Preparations of the smaller subunit of avian reverse transcriptase (m.w. about 65,000) do not bind tRNA^{Trp} in vitro (Grandgenett et al. 1976; Haseltine et al. 1977). This observation has led to the conclusion that binding is a function of the larger subunit (m.w. about 95,000). Although that is likely, one cannot rule out a possible role of the larger subunit in activating the binding capacity of the smaller one.

The binding of the tRNA is sensitive to inhibitors of reverse transcriptase, such as N-ethyl maleimide and antibody to the enzyme (Haseltine and Baltimore 1976; Haseltine et al. 1977). Aminoacylation, oxidation, or removal of the 3' end of tRNA^{Trp} does not appreciably affect the binding of the tRNA to avian virus reverse transcriptase, nor does cleavage of the tRNA in the anticodon loop (Haseltine et al. 1977; J. Hu and J. Dahlberg, unpubl.). In contrast, tRNA fragments consisting of the sequence from the 3' end to the m^{7}G residue or to the anticodon are inactive (Haseltine et al. 1977). Not unexpectedly, the binding appears to be sensitive to more than primary sequence in the tRNA. There is evidence that binding is altered when the tRNA is annealed to the template (Cordell et al. 1979).

In contrast to the situation with avian RNA tumor viruses, no specific interaction between murine virus reverse transcriptase and tRNA has been observed. It is unclear whether this lack of binding specificity is an inherent difference between the enzymes or whether it results from alteration (or loss of specificity) during purification of the enzyme (Panet and Berliner 1978).

OTHER VIRION tRNAs

Virions of RNA tumor viruses contain, in addition to the tRNA primer, a variety of other small RNAs that are presumably taken up during budding of the virus through the cell membrane. The tRNAs found in virions are a specific subset of the cellular tRNAs (Rosenthal and Zamecnik 1973; Faras et al. 1973a; Sawyer and Dahlberg 1973; Wang et al. 1973; Sawyer et al. 1974; Waters 1975; Peters et al. 1977). Although the patterns of tRNAs differ between viruses, the genomic RNA probably is not entirely responsible for this selection (Levin and Seidman 1978; Sawyer and Hanafusa 1979; G. G. Peters et al., in prep.). It seems more likely that the virion-associated reverse transcriptase binds a subset of cellular tRNAs and transports them into the budding virus. Consistent with this view is the fact that the number of reverse transcriptase molecules per virion (80–100) approximates the total number of tRNA molecules (Sawyer and Dahlberg 1973; Panet et al. 1975a). The strongest evidence in support of the idea that the enzyme is responsible for the selection of virion tRNAs is the fact that strains of avian RNA tumor viruses that lack reverse transcriptase also lack the normal set of virion tRNAs. Instead, these defective virions contain complex collections of tRNAs comparable to those of the host cell (Sawyer and Hanafusa 1979; G. G. Peters et al., in prep.). In fact, under appropriate conditions purified avian virus enzyme can be made to select the virion pattern of tRNAs from cellular tRNAs in vitro (J. Hu and J. Dahlberg, unpubl.).

CONCLUDING REMARKS

The two tRNAs discussed here were the first ones for which a primer function was clearly defined. Recently, G. G. Peters (pers. comm.) showed that tRNA^{Pro} also functions as a primer for avian spleen necrosis virus. Very interestingly, Peters has also found that the primer RNA for mouse mammary tumor virus DNA synthesis is a third tRNA, tRNA$_3^{Lys}$.

Many questions remain unanswered in this field. We still do not understand why tRNAs are used as primers in initiation of DNA synthesis or whether there is any significance as to which tRNAs are used to carry out this role. In addition, we do not know whether these primers act in vivo to initiate synthesis of the second, positive (virion sense) strand of DNA and we can only speculate about the possible roles of primer tRNAs in initiation of cellular DNA synthesis.

ACKNOWLEDGMENTS

I thank E. Lund for help with the manuscript. My research was supported by National Institutes of Health grant CA-15166.

REFERENCES

Beemon, K. L., A. J. Faras, A. T. Haase, P. Duesberg, and J. E. Maisel. 1976. Genomic complexities of murine leukemia and sarcoma, reticuloendotheliosis and visna viruses. *J. Virol.* **17:**525.

Billeter, M. A., J. T. Parsons, and J. M. Coffin. 1974. The nucleotide sequence complexity of avian tumor virus RNA. *Proc. Natl. Acad. Sci.* **71:**3560.

Bishop, J. M. 1978. Retroviruses. *Annu. Rev. Biochem.* **47:**35.

Canaani, E. and P. Duesberg. 1972. Role of subunits of 60 to 70S avian tumor virus ribonucleic acid in its template activity for the viral deoxyribonucleic acid polymerase. *J. Virol.* **10:**23.

Cavalieri, L. F. and I. Yamaura. 1975. *E. coli* tRNAs as inhibitors of viral reverse transcription in vitro. *Nucleic Acids Res.* **2:**2315.

Coffin, J. M. and W. A. Haseltine. 1977a. Terminal redundancy and origin of replication of Rous sarcoma virus RNA. *Proc. Natl. Acad. Sci.* **74:**1908.

———. 1977b. Nucleotide sequence of Rous sarcoma virus RNA at the initiation site of DNA sequence. The 102nd nucleotide is U. *J. Mol. Biol.* **117:**805.

Coffin, J. M., T. G. Hageman, A. M. Maxam, and W. A. Haseltine. 1978. Structure of the genome of Moloney murine leukemia virus: A terminally redundant sequence. *Cell* **13:**761.

Cordell, B., R. Swanstrom, H. M. Goodman, and J. M. Bishop. 1979. tRNA^{Trp} as primer for RNA-directed DNA polymerase: Structural determinants of function. *J. Biol. Chem.* **254:**1866.

Cordell, B., S. R. Weiss, H. E. Varmus, and J. M. Bishop. 1978. At least 104 nucleotides are transposed from the 5' terminus of the avian sarcoma virus genome to the 5' terminai of smaller viral mRNAs. *Cell* **15:**79.

Cordell, B., E. Stavnezer, R. Friedrich, J. M. Bishop, and H. M. Goodman. 1976. Nucleotide sequence that binds primer for DNA synthesis to the avian sarcoma virus genome. *J. Virol.* **19:**548.

Dahlberg, J. E., R. C. Sawyer, J. M. Taylor, A. J. Faras, W. E. Levinson, H. M.

Goodman, and J. M. Bishop. 1974. Transcription of DNA from the 70S RNA of Rous sarcoma virus. I. Identification of a specific 4S RNA which serves as primer. *J. Virol.* **13:**1126.

Duesberg, P. H., K. V. D. Helm, and E. Canaani. 1971. Comparative properties of RNA and DNA templates for the DNA polymerase of Rous sarcoma virus. *Proc. Natl. Acad. Sci.* **68:**2505.

Eiden, J. J., K. Quade, and J. L. Nichols. 1976. Interaction of tryptophan transfer RNA with Rous sarcoma virus 35S RNA. *Nature* **259:**245.

Fan, H. and M. Paskind. 1974. Measurement of the complexity of cloned Moloney murine leukemia virus 60 to 70S RNA: Evidence for a haploid genome. *J. Virol.* **14:**421.

Faras, A. J., A. C. Garapin, W. E. Levinson, J. M. Bishop, and H. M. Goodman. 1973a. Characterization of the low-molecular-weight RNAs associated with the 70S RNA of Rous sarcoma virus. *J. Virol.* **12:**334.

Faras, A. J., J. M. Taylor, W. E. Levinson, H. M. Goodman, and J. M. Bishop. 1973b. RNA-directed DNA polymerase of Rous sarcomas virus: Initiation of synthesis of 70S viral RNA as template. *J. Mol. Biol.* **79:**163.

Faras, A. J., J. E. Dahlberg, R. C. Sawyer, F. Harada, J. M. Taylor, W. E. Levinson, J. M. Bishop, and H. M. Goodman. 1974. Transcription of DNA from the 70S RNA of Rous sarcoma virus. II. Structure of a 4S RNA primer. *J. Virol.* **13:**1134.

Grandgenett, D. P., A. C. Vora, and A. J. Faras. 1976. Different states of avian myeloblastosis virus DNA polymerase and their binding capacity to primer tRNA^{Trp}. *Virology* **75:**26.

Harada, F., G. Peters, and J. E. Dahlberg. 1980. The primer RNA for Moloney murine leukemia virus DNA synthesis: Nucleotide sequence and aminoacylation of tRNA^{Pro}. *J. Biol. Chem.* (in press).

Harada, F., R. C. Sawyer, and J. E. Dahlberg. 1975. A primer ribonucleic acid for initiation of *in vitro* Rous sarcoma virus deoxyribonucleic acid synthesis. *J. Biol. Chem.* **250:**3487.

Haseltine, W. A. and D. Baltimore. 1976. In vitro replication of RNA tumor viruses. In *Animal virology. ICN-UCLA Symposium on Molecular and Cell Biology* (ed. D. Baltimore et al.), vol. 4, p. 175. Academic Press, New York.

Haseltine, W. A., A. M. Maxam, and W. Gilbert. 1977. Rous sarcoma virus genome is terminally redundant: The 5' sequence. *Proc. Natl. Acad. Sci.* **74:**989.

Haseltine, W. A., D. G. Kleid, A. Panet, E. Rothenberg, and D. Baltimore. 1976. Ordered transcription of RNA tumor virus genomes. *J. Mol. Biol.* **106:**109.

Haseltine, W. A., A. Panet, D. Smoler, D. Baltimore, G. Peters, F. Harada, and J. E. Dahlberg. 1977. Further studies on the interaction between tRNA^{Trp} and avian myeloblastosis virus DNA polymerase. *Biochemistry* **16:**3625.

Hizi, A., J. P. Leis, and W. K. Joklik. 1977. The RNA-dependent DNA polymerase of avian sarcoma virus B77; binding of viral and nonviral ribonucleic acids to the α, β_2, and $\alpha\beta$ forms of the enzyme. *J. Biol. Chem.* **252:**6878.

Levin, J. G. and J. G. Seidman. 1979. Selective packaging of host tRNA's by murine leukemia virus particles does not require genomic RNA. *J. Virol.* **29:**328.

Panet, A. and H. Berliner. 1978. Binding of tRNA to reverse transcriptase of RNA tumor viruses. *J. Virol.* **26:**214.

Panet, A., D. Baltimore, and T. Hanafusa. 1975a. Quantitation of avian RNA tumor virus reverse transcriptase by radioimmunoassay. *J. Virol.* **16**:146.

Panet, A., G. Weil, and R. R. Friis. 1978. Binding of tryptophanyl-tRNA to the reverse transcriptase of replication-defective avian sarcoma viruses. *J. Virol.* **28**:434.

Panet, A., W. A. Haseltine, D. Baltimore, G. Peters, F. Harada, and J. E. Dahlberg. 1975b. Specific binding of tryptophan transfer RNA to avian myelobastosis virus RNA-dependent DNA polymerase (reverse transcriptase). *Proc. Natl. Acad. Sci.* **72**:2535.

Peters, G. G. and J. E. Dahlberg. 1979. RNA-directed DNA synthesis in Moloney murine leukemia virus: Interaction between the primer tRNA and the genome RNA. *J. Virol.* **31**:398.

Peters, G. G., F. Harada, J. E. Dahlberg, A. Panet, W. A. Haseltine, and D. Baltimore. 1977. Low-molecular-weight RNAs of Moloney murine leukemia virus: Identification of the primer for RNA-directed DNA synthesis. *J. Virol.* **21**:1031.

Rosenthal, L. J. and P. C. Zamecnik. 1973. Amino acid acceptor activity of the "70S-associated" 4S RNA from avian myeloblastosis virus. *Proc. Natl. Acad. Sci.* **70**:1184.

Sawyer, R. C. and J. E. Dahlberg. 1973. Small RNAs of Rous sarcoma virus: Characterization by two-dimensional polyacrylamide gel electrophoresis and fingerprint analysis. *J. Virol.* **12**:1226.

Sawyer, R. C. and H. Hanafusa. 1979. Comparison of the small RNAs of polymerase-deficient and polymerase-positive Rous sarcoma virus and another species of avian retrovirus. *J. Virol.* **29**:863.

Sawyer, R. C., F. Harada, and J. E. Dahlberg. 1974. Virion-associated RNA primer for Rous sarcoma virus DNA synthesis: Isolation from uninfected cells. *J. Virol.* **13**:1302.

Shine, J., A. P. Czernilofsky, R. Friedrich, J. M. Bishop, and H. M. Goodman. 1977. Nucleotide sequence at the 5' terminus of the avian sarcoma virus genome. *Proc. Natl. Acad. Sci.* **74**:1473.

Staskus, K. A., M. S. Collett, and A. J. Faras. 1976. Initiation of DNA synthesis by the avian oncornavirus RNA-directed DNA polymerase: Structural and functional localization of the major species of primer RNA on the oncornavirus genome. *Virology* **71**:162.

Taylor, J. M. 1977. An analysis of the role of tRNA species as primers for the transcription into DNA of RNA tumor virus genomes. *Biochim. Biophys. Acta* **473**:57.

Taylor, J. and R. Illmensee. 1975. Site on the RNA of an avian sarcoma virus at which primer is bound. *J. Virol.* **16**:553.

Taylor, J. M., D. E. Garfin, W. E. Levinson, J. M. Bishop, and H. M. Goodman. 1974. Tumor virus ribonucleic acid directed deoxyribonucleic acid synthesis: Nucleotide sequence at the 5' terminus of nascent deoxyribonucleic acid. *Biochemistry* **13**:3159.

Temin, H. M. 1971. Mechanism of cell transformation by RNA tumor viruses. *Annu. Rev. Microbiol.* **25**:609.

Temin, H. M. and D. Baltimore. 1972. RNA-directed DNA synthesis and RNA tumor viruses. *Adv. Virus Res.* **17**:129.

Verma, I. M. 1977. The reverse transcriptases. *Biochim. Biophys. Acta* **473**:1.

Verma, I. M., N. L. Meuth, E. Bromfeld, K. F. Manly, and D. Baltimore. 1971. Covalently linked RNA-DNA molecule as initial product of RNA tumour virus DNA polymerase. *Nat. New Biol.* **233**:131.

Wang, S., R. M. Kothari, M. Taylor, and P. Hung. 1973. Transfer RNA activities of Rous sarcoma and Rous associated viruses. *Nat. New Biol.* **242**:133.

Waters, L. C. 1975. Transfer RNAs associated with the 70S RNA of AKR murine leukemia virus. *Biochem. Biophys. Res. Commun.* **65**:1130.

Waters, L. C. and B. C. Mullin. 1977. Transfer RNA in RNA tumor viruses. *Prog. Nucleic Acid Res. Mol. Biol.* **20**:131.

Waters, L. C., B. C. Mullin, T. Ho, and W. K. Yang. 1975. Ability of tryptophan tRNA to hybridize with 35S RNA of avian myeloblastosis virus and prime reverse transcription in vitro. *Proc. Natl. Acad. Sci.* **72**:2155.

In Vitro Selective Binding of tRNAs to rRNAs of Vertebrates

Wen K. Yang
Biology Division
Oak Ridge National Laboratory
Oak Ridge, Tennessee 37830

David L. R. Hwang
Graduate School of Biomedical Sciences
University of Tennessee-Oak Ridge
Oak Ridge, Tennessee 37830

In this paper, we review briefly some previous studies, including those from our laboratories, concerning primer tRNAs of neoplasia-related retroviruses and then describe the results of a series of experiments demonstrating that in an in vitro hybridization reaction, selective tRNA species are capable of binding to rRNAs of various vertebrates. The experiments were carried out as a result of an initial observation made in a control study on the binding of primer tRNAs to retroviral genomic RNAs. Although the biological significance of selective tRNA binding to rRNAs in vitro is still in question, there seems to be a striking similarity between this phenomenon and the interaction of primer tRNAs with genomic RNAs of endogenous retroviruses. Thus, a novel experimental approach to the study of cellular expression and genetic origin of retroviruses has become apparent to us. The applicability of this experimental approach, especially to cancer research, is discussed here.

PRIMER tRNA BINDING TO GENOMIC RNAs OF RETROVIRUSES

Virions of retroviruses have been shown to contain a large quantity of tRNA (Beaudreau et al. 1964; Bonar et al. 1967; Trávnicek 1968; Carnegie et al. 1969; Erikson and Erikson 1970; Bishop et al. 1970; Rosenthal and Zamecnik 1973a,b; Elder and Smith 1973; Gallagher and Gallo 1973; Larsen et al. 1973; Sompayrac and Maaløe 1973; Sawyer and Dahlberg 1973; Waters et al. 1975). Aminoacylation and absorbance determinations indicate that at least 200 tRNA molecules, in addition to one 60S–70S RNA molecule, are included in a virion particle. The virion-included tRNA includes tRNA species selected from the host cell; some of them are concentrated ten times or more in the virion than in the cell. As extracted from the virions, these tRNA molecules are mostly free, but some are in complexes with viral genomic RNAs (Erikson and Erikson 1971; Waters and Mullin 1977).

Early enzymological studies of retroviral polymerases demonstrated that these enzymes cannot function without a primer molecule (Spiegelman et al. 1970; Baltimore and Smoler 1971; Flügel and Wells 1972). In a study of reverse transcription, Verma and colleagues (1971) observed that the first deoxyribonucleotide incorporated is covalently linked to a priming RNA of small size. Canaani and Duesberg (1972) reported that the primer constitutes approximately 20% of the total 70S-associated 4S RNA and can be differentially dissociated from the viral genomic RNA template at 63°C, which is higher than the melting temperature of nonpriming 4S RNA. On the basis of an earlier finding of Erikson and Erikson (1971) that 4S RNA molecules, with sequences characteristic of tRNA, are constituents of the 70S RNA complex within the virion particle, possible involvement of a tRNA in this genetic function of retroviruses was suspected. Definite experimental evidence for this was obtained by Dahlberg and collaborators (1974), who showed that the primer for the transcription of DNA from the RNA genome of Rous sarcoma virus is tRNA^{Trp}.

In our laboratories, a different experimental approach was employed to demonstrate that tRNA^{Trp} also serves as the primer for reverse transcription in avian myeloblastosis virus (Waters et al. 1975). The experimental approach included three steps: (1) a 60S–70S RNA complex of avian myeloblastosis virus was subjected to differential thermo-dissociation (at 60°C and then at 95°C) and the most tenaciously bound 4S RNA was identified as tRNA^{Trp} by aminoacylation and RPC-5 chromatography; (2) purified tRNA^{Trp} from cellular sources forms a complex with viral 30S–40S genomic RNA in in vitro hybridization reactions; and (3) the viral RNA in complex with tRNA^{Trp} is active in reverse transcription with viral polymerase. Actually, in the second step, a mixture of tRNAs, rather than a purified tRNA preparation, can be used in the hybridization reaction to show the selective association of tRNA^{Trp} and the genomic RNA of avian myeloblastosis virus. Thus, a selective binding of primer tRNA generally occurs when the genomic RNA of a retrovirus is incubated with a mixture of cellular tRNAs under the hybridization condition (Waters et al. 1974), implying the mutual recognition of complementary nucleotide sequences between viral genomic RNA and primer tRNA. This is now known to involve a 15- to 20-base sequence extending from the 3' end to the $\psi\psi$C loop region of the primer tRNA (Eiden et al. 1976; Dahlberg 1977). The site of primer tRNA binding was found to be near the 5' end of the retroviral RNA (Taylor and Illmensee 1975).

Although tRNA^{Trp} has been demonstrated to be the primer for DNA transcription in the two avian retroviruses mentioned, it does not serve the same role in murine retroviruses. Among various tRNA species associated with genomic RNA of leukemia virus of AKR mouse, the last dissociated from the 60S–70S RNA complex by heat treatment was found to be tRNA^{Pro} by aminoacylation measurement (Waters 1975). Subsequently, it was demonstrated that purified tRNA^{Pro} was able to form a complex with the 30S–40S genomic RNA of AKR murine leukemia virus and to function as its primer in in vitro reverse transcription reactions (Yang 1977). Independently, studies with Moloney murine leukemia virus have established that tRNA^{Pro} is also the primer for reverse transcription in this murine virus (Peters et al. 1977). These results indicate that different retroviruses may have different primer tRNAs and also that the differental heat dissociation/aminoacylation procedure (Canaani and Duesberg 1972; Waters 1975) is rather dependable to implicate a tRNA species as the candidate primer in retroviruses. Hence, this procedure has been extensively employed by Waters in all available retrovirus samples. The candidate primer tRNAs thus revealed are tRNA^{Pro} in two lines of feline leukemia virus; tRNA^{Pro} in woolly monkey leukemia/sarcoma viruses; tRNA^{Asp}, tRNA^{Gly}, and/or tRNA^{Lys} in RD114 virus; and tRNA^{Lys} in mouse mammary virus (Waters and Mullin 1977; Waters 1978). Since feline leukemia virus and simian sarcoma/leukemia viruses originate genetically from a rat ancestor species and from Asian mouse species, respectively, whereas RD114 virus is from a primate species and has become endogenous in domestic cats (Todaro 1975), the differences in primer tRNA utilization may be determined by the genetic and evolutional origin of retroviruses rather than the infected host species. However, this implication remains to be tested with a survey of primer tRNAs in all known retroviruses.

SELECTIVE tRNAs BOUND TO MOUSE 28S AND 18S rRNAs

The enormous effort required to prepare the virus materials for the studies of primer tRNAs in AKR murine leukemia virus and RD114 virus (Waters 1975; Yang 1977; Waters and Mullin 1977) made us realize that similar investigations with many other poorly reproducing retroviruses would be extremely difficult, since the differential thermo-dissociation experiment generally requires 50 μg or more of intact 70S RNA sample. Hence, our attention has been drawn toward intracellular virus-specific RNAs, which are known to be present in relatively large quantity in retrovirus-infected and/or -transformed cells (e.g., Tsuchida et al. 1972; Leong et al. 1972; Fan and Baltimore 1973). The intracellular virus-specific RNAs may therefore serve as substitutes for virion 60S–70S RNA; in the case of virus-transformed nonproducer cells, they are the only material available for the primer tRNA study. To test the feasibility of this idea, we prepared RNA from NIH/3T3 mouse fibroblasts, which were actively producing N-tropic retroviruses, and separated poly(A)-containing RNAs by oligo(dT)-cellulose chromatography. When poly(A)-RNA of 30S and larger sizes in milligram quantities were heated, no release of tRNA^{Pro} was detected. This indicates that no primer tRNA is associated with intracellular virus-specific RNAs and supports the view that the primer tRNA forms a complex with retrovirus genomic RNA extracellularly after the virus-budding process. Next, these poly(A) RNA preparations were tested for their capability to selectively bind tRNA. This was performed by using radiolabeled tRNA mixture in an in vitro hybridization, sucrose gradient centrifugation to separate the bound tRNAs, and reversed-phase RPC-5 chromatography to demonstrate the specificity of the binding, as previously performed with retrovirus genomic RNAs (Waters et al. 1974). The poly(A) RNAs of cells infected with AKR virus appeared to show selective binding of tRNA^{Pro}, although the results were complicated by a low recovery of the bound tRNAs due to an extensive aggregation of poly(A) RNAs during hybridization.

As a control in these experiments, a 28S-size RNA isolated by sucrose gradient centrifugation from the non-poly(A) RNA fraction of the same cells was used. Under the same hybridization condition, this RNA preparation showed a remarkable binding of tRNA radioactivity, which comigrated to the 28S region after sucrose gradient centrifugation. The 28S-associated radioactivity, upon heat treatment, was recovered as a 4S peak in another sucrose gradient centrifugation. Analysis by co-chromatography of this sample with total tRNA mixture in an RPC-5 column revealed a profile containing fewer peaks than the complex profile of the total tRNA (Fig. 1), suggesting that selective tRNA species were

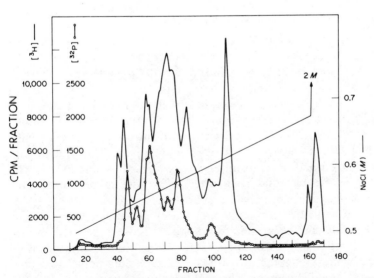

Figure 1 RPC-5 cochromatography of mouse cell tRNA(——) and tRNA species capable of binding to mouse 28S RNA (O—O). (Reprinted, with permission, from Yang et al. 1978.)

bound by the 28S non-poly(A) RNA preparation in the hybridization reaction.

To determine the identity of the selectively bound tRNA species, we employed a method using combinations of enzymatic aminoacylation and an amino acid analyzer (Davey and Howells 1974; Waters 1975). The 4S RNA released from the 28S RNA complex was incubated in a reaction mixture containing aminoacyl-tRNA synthetases and 17 ^{3}H-labeled amino acids (without asparagine, cysteine, and glutamine). Accepting activities for asparagine, cysteine, and glutamine were determined individually by the filter paper disc method (Yang and Novelli 1971). The labeled aminoacylated tRNAs were isolated by DEAE-cellulose chromatography. The esterified amino acids were discharged from the tRNAs by mild alkaline hydrolysis and subsequently identified by chromatography in the amino acid analyzer. The results are shown in Figure 2. Whereas a

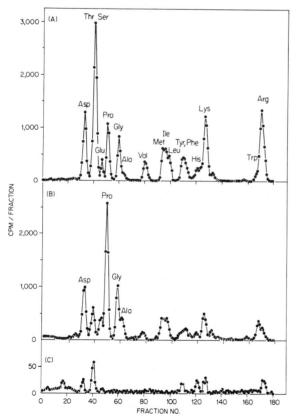

Figure 2 Amino acid analyzer profiles of amino acids accepted by total liver tRNA (*A*), tRNA bound to mouse 28S RNA in in vitro hybridization (*B*), and 4S RNA fraction from heat-treated mouse 28S RNA (*C*). (Reprinted, with permission, from Yang et al. 1978.)

total mouse tRNA preparation, subjected to the hybridization reaction in the absence of 28S RNA, accepted all 17 amino acids (Fig. 2A), the 28S-RNA-associated 4S RNA derived from this tRNA preparation accepted predominantly proline and small-to-negligible amounts of other amino acids (Fig. 2B). Repeated experiments were carried out with 28S non-poly(A) RNA preparations from embryos and young adult livers of normal mice; the results consistently demonstrated the selective binding of tRNA^{Pro}, as well as small but significant amounts of tRNA^{Gly}, tRNA^{Asp}, and tRNA^{Ala}. No binding of tRNA^{Asn}, tRNA^{Cys}, and tRNA^{Gln} was apparent by the individual aminoacylation reactions. To determine whether or not the tRNA·28S RNA complex might occur in vivo, we heated the isolated 28S RNA directly and detected no release of 4S RNA with amino acid accepting activities (Fig. 2C).

At this time we became aware of the results of Dahlberg and colleagues (1978), who observed selective binding of tRNA^{fMet} to *Escherichia coli* 23S rRNA in a similar in vitro hybridization reaction. Because of a possible implication of this observation in the initiation mechanism of protein biosynthesis, it was interesting to examine whether or not a similar interaction could be demonstrated with mammalian 28S rRNA and initiator tRNA^{Met}. Despite careful examinations, no significant methionine-accepting activity could be detected with our in-vitro-prepared tRNA·28S RNA complexes. Apparently, initiator tRNA^{Met} is not selectively bound by mouse 28S rRNA under our hybridization condition.

The 18S RNA, isolated from the non-poly(A) RNA fraction of mouse cells and mixed with total tRNA in the hybridization reaction, was also found to bind tRNAs selectively. However, rather than tRNA^{Pro}, three other kinds of amino acid tRNAs (tRNA^{Lys}, tRNA^{Glu}, tRNA^{Gly}) were the major species identified by the aminoacylation assay. Various 18S RNA preparations from mouse cells, either with or without infection with murine retroviruses, showed essentially the same selective binding of these three tRNAs.

In these experiments, it was found that most tRNAs bound with 18S or 28S RNA are unable to accept amino acids unless they are first released from the complex by heat treatment (95°C for 3 min, quickly chilled) and subsequently incubated in 10 mM MgCl$_2$ solution at 4°C. These included tRNA^{Pro}, tRNA^{Gly}, and tRNA^{Asp} bound to 28S RNA, as well as tRNA^{Glu} and tRNA^{Gly} bound to 18S RNA, suggesting that the tRNAs are structurally altered by the binding and no longer recognized by the respective synthetases for aminoacylation. An apparent exception was 18S-RNA-bound tRNA^{Lys}, which gave the same lysine-accepting activity with and without heat treatment of the complex. It is not known whether tRNA^{Lys} in complex with 18S RNA still retains the structural feature recognized by the lysyl-tRNA synthetase, or the synthetase preparation contains some factor that releases this tRNA from the complex during the aminoacylation reaction.

The question was then raised whether or not the observed selective tRNA binding properties are due to the rRNAs, which constituted most, if not all, of our 18S and 28S non-poly(A) RNA preparations. Calculations on the basis of tRNA-specific radioactivity as well as specific amino acid accepting activities revealed molar ratios of 0.2–0.6 mole of tRNA^{Pro} bound per mole of 28S RNA (1,750,000 daltons) and 0.1–0.2 mole each of tRNA^{Lys}, tRNA^{Glu}, and tRNA^{Gly} per mole of 18S RNA (700,000 daltons). These ratios appeared too high to attribute the binding to non-rRNAs in the non-poly(A) RNA preparations. Nevertheless, 40S and 60S ribosome subunits were isolated from mouse livers and used for preparing 18S and 28S rRNAs by phenol extraction and sucrose gradient centrifugation. These rRNA preparations showed essentially the same selective tRNA binding activities. Since 60S ribosome subunits prepared by sucrose gradient centrifugation might still contain contaminant RNAs of large size and since phenol extraction could possibly cause partial poly(A) breakage, the binding of tRNA^{Pro} could have been due to RNAs derived from endogenous murine retroviruses. To test this possibility, poly(A) RNAs of 28S and larger size were isolated from mouse livers and incubated with tRNAs in the hybridization reactions. No apparent binding of tRNA^{Pro} or other tRNA was observed. Further, when mixed with 28S rRNA, the poly(A) RNAs did not increase the selective tRNA binding capacities. Based on these results, it was concluded that rRNAs are responsible for the observed selective binding of tRNAs in vitro.

CHARACTERISTICS OF MOUSE tRNA·rRNA COMPLEXES

The hybridization reaction condition adopted in our studies presumed that any binding would reflect specific interaction between complementary nucleotide sequences of the input RNA molecules. With the observation of selective tRNAs bound by the mouse rRNAs, it was necessary for us to validate the presumption. This included determination of melting temperature, analysis of isoaccepting species of the bound tRNAs, binding specificity of rRNA fragments, primer-template activity of tRNA·rRNA complex in reverse-transcriptase-catalyzed reaction, and elucidation of the exact nucleotide sequence involved in the binding. Because of the functional role played by tRNA^{Pro} as the primer in murine retroviruses, emphasis of the characterization studies has been placed on the tRNA^{Pro}·28S rRNA complex.

Melting temperatures of tRNA·rRNA complexes were determined by two kinds of experiments. With the use of radiolabeled tRNA preparations in the initial hybridization, dissociation of tRNA·rRNA complexes at various temperatures was followed by measuring the release of tRNA radioactivity. Alternatively, since tRNAs (except tRNA^{Lys}) complexed with rRNAs are not aminoacylated by the synthetases, the release of tRNA can be detected by the appearance of specific amino acid accepting activity. Figure 3 shows that, in 0.1 M NaCl, the radioactive tRNAs were released from 18S rRNA and 28S rRNA precipitously at 55–65°C and 50–75°C, respectively. Aminoacylation showed that tRNA^{Gly}·28S rRNA dissociated at 50–60°C, tRNA^{Pro}·28S rRNA at 60–65°C, tRNA^{Gly}·18S rRNA at 55–65°C, and tRNA^{Glu}·18S rRNA at 65–75°C. The observed melting temperatures of tRNA·rRNA complexes generally increased when the NaCl concentration of the RNA solution was raised—a known characteristic of polynucleotide duplexes. The magnitude of the melting temperatures also suggested that the binding to rRNAs involves a portion rather than all of the tRNA sequence.

To investigate whether or not all isoaccepting species of a tRNA were involved in the selective binding to rRNAs, we employed reversed-phase RPC-5 chromatography, which separates mouse cell tRNA preparations into three peaks of tRNA^{Pro}, three peaks of tRNA^{Lys}, three peaks of tRNA^{Gly}, and four peaks of tRNA^{Glu}. The three peaks of tRNA^{Pro} were found equally capable of binding to mouse 28S as well as the three tRNAs^{Lys} to 18S rRNA; whereas two of the three isoaccepting tRNAs^{Gly} showed preferential binding to 18S rRNA and only two of the four tRNAs^{Glu} are capable of binding to 18S rRNA (L. R. Shugart et al., unpubl.). These results explain why tRNAs, bound to rRNA in vitro, generally yield more peaks by RPC-5 chromatography (e.g., Fig. 1) than

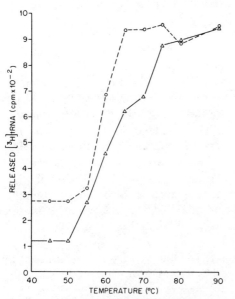

Figure 3 [^{3}H]Uridine-labeled tRNA released from the tRNA·[28S rRNA] complex (○) and the tRNA·[18S rRNA] complex (△) at various temperatures in a solution of 0.1 M NaCl, 10 mM Tris HCl (pH 7.6), and 1 mM EDTA.

the number of accepted amino acids (e.g., Fig. 2B). Obviously, the nucleotide sequences responsible for binding to rRNAs are present in all three isoaccepting tRNAs^{Pro} and in the three tRNAs^{Lys}, in two of the three tRNAs^{Gly}, and in two of the four tRNAs^{Glu}.

A brief heating at 85–90°C with subsequent prolonged incubation at 60–65°C (Table 1) apparently caused some fragmentation of 28S and 18S rRNA. After hybridization under this condition and removal of most of the free tRNAs, rRNAs with the bound tRNAs could be separated by sucrose gradient centrifugation into fractions corresponding to full-sized rRNA, large rRNA fragments, and small rRNA fragments. The bound tRNA could be detected by the appearance of amino acid accepting activities upon heat treatment of the fractions. Although the full-sized 28S rRNA binds three to four times more tRNA^{Pro} than tRNA^{Gly}, its large fragments were found to bind more tRNA^{Gly} than tRNA^{Pro}; the small fragments contained predominantly tRNA^{Pro}. Bound tRNA^{Glu} was found mainly in the small fragments of 18S rRNA, whereas tRNA^{Gly} was found mainly with the large fragments. Thus, different tRNAs appeared to be bound at different sites on the rRNA molecules.

The observation that significant rRNA fragmentation may occur during an in vitro hybridization reaction has led us to consider its possible effect on the measurement of tRNA binding. Table 1 lists the three reaction

Table 1 Reaction conditions of in vitro hybridization for demonstrating the binding of selective tRNAs to rRNAs

	Reaction conditions		
	1	2	3
tRNA/rRNA, A$_{260}$/ml	20/5	15/4	10/5
Pretreatment	85°C, 5 min	85–90°C, 5 min	65°C, 5 min
Salt concentration	0.4 M NaCl 1 mM EDTA	0.6 M NaCl 20 mM EDTA	0.6 M NaCl 60% formamide 1 mM EDTA
Incubation	60–65°C, 16 hr	65–67°C, 1 hr	50°C, 3 hr
tRNA binding to mouse 18S rRNA	Glu (0.04)[a] Gly (0.03) Lys (0.05)	n.d.[b]	Glu (0.26) Gly (0.10) Lys (0.12)
tRNA binding to mouse 28S rRNA	Pro (0.21) Gly (0.08) Asp (0.03)	Pro (0.20) Gly (0.05) Asp (0.10)	Pro (0.68) Gly (0.02) Asp (0.02)

[a]Molar ratio of tRNA (expressed as amino acid accepted) per rRNA molecule. The quantity of rRNA was estimated by assuming the m.w. of 1,750,000 for 28S rRNA and 700,000 for 18S rRNA.
[b]Not determined.

conditions that have been used by us to study tRNA·rRNA interaction. The first was used in most of our initial experiments. The second, following the procedure of Dahlberg and colleagues (1978), was used in vain to demonstrate binding of initiator tRNA^{Met} to mouse 28S rRNA. The third utilizes formamide in combination with relatively low temperatures to minimize rRNA degradation and also to add stringency to the hybridization. Table 1 also shows the results obtained by application of the three conditions to the same preparations of 18S and 28S rRNAs from mouse livers. The first and second conditions of hybridization produced very similar data, but the third (formamide) condition evidently improved the selective tRNA binding. For example, it promoted the molar ratio of tRNA^{Pro} bound per 28S rRNA from 0.2 to 0.6, apparently by causing less rRNA fragmentation; and the binding of the other tRNAs to 28S rRNA were decreased, presumably because of the added stringency. The binding activities of tRNA^{Glu}-tRNA^{Gly}-tRNA^{Lys} to mouse 18S rRNA were also enhanced by the application of the formamide condition. Thus, the observed modification of the results with the changes of hybridization condition further support the conclusion that selective tRNA binding to rRNAs involves recognition of complementary nucleotide sequences.

The formamide hybridization condition has made it possible to prepare a complex containing tRNA^{Pro} as the sole tRNA bound to mouse 28S rRNA. When the mouse tRNA^{Pro}·28S RNA complex is mixed with avian myeloblastosis virus polymerase in vitro, DNA transcription reaction can be observed. Heat dissociation of the complex eliminates the DNA transcription activity. Further, in experiments in which the complex was heated at various temperatures and then tested with the viral polymerase, the disappearance of DNA transcription activity followed the release of tRNA^{Pro}. Thus, the bound tRNA^{Pro} can apparently serve as the primer for reverse transcription of mouse 28S rRNA. Isolated DNA products from the reverse transcription were analyzed by alkaline sucrose gradient centrifugation and found to be of 5S or smaller in size.

The portion of the tRNA^{Pro} nucleotide sequence involved in the binding to mouse 28S rRNA has been elucidated by J. E. Dahlberg in collaboration with us (J. E. Dahlberg et al., in prep.). In this study, ^{32}P-labeled tRNA was chromatographed on Sepharose 4B and RPC-5 columns to obtain a fraction containing tRNA$^{Pro}_3$ at 830 pmoles/A$_{260}$ specific activity. The tRNA fraction was incubated with non-labeled mouse 28S rRNA, under the formamide hybridization condition. The [^{32}P]tRNA^{Pro}·[28S rRNA] complex was subjected to an extensive digestion with ribonucleases. Subsequent electrophoresis revealed that two fragments of tRNA^{Pro} were protected from the digestion. By oligonucleotide analyses, the two fragments were identified as GCGAGAGm^{7}GDm^5-Cm^{5}CCGGGψ and CGGACGAGCCCCC(A). Thus, the binding involves the 3′ half of the tRNA molecule extending from the anticodon stem to the 3′ CC(A) end, without the ψψC loop region or the terminal A (Fig. 4). A comparison with the results of a similar study done with the genomic RNA of Moloney murine leukemia virus (Dahlberg 1977) clearly indicates that tRNA^{Pro} binding to mouse 28S rRNA is distinctly different. However, in both cases, the 3′-end sequence of tRNA^{Pro} is bound and hence able to provide the primer function for DNA transcription.

DIFFERENT tRNA BINDING TO rRNAs OF DIFFERENT VERTEBRATES

The next obvious question was whether or not the same observation made with mouse rRNAs, i.e., binding of tRNA^{Pro} to 28S rRNA and that of tRNA^{Glu}-tRNA^{Gly}-tRNA^{Lys} to 18S rRNA, could be made with rRNA preparations from another animal species. Therefore, a survey of in vitro binding of tRNAs to rRNAs of various vertebrates was carried out (D. Hwang et al., in prep.). The survey used RNA preparations from adult livers, except that cultured cells were used in the cases of man, rhesus monkey, and Chinese hamster. The 18S and 28S rRNAs were isolated from the tRNA-depleted non-poly(A) RNA fraction by sucrose gradient centrifugation, as described above; tRNA was prepared from an RNA fraction that is 1 M NaCl soluble by DEAE-cellulose chromatography (Yang and Novelli 1971). Preliminary experiments indicated that tRNAs and rRNAs from the same cell source gave quantitatively more reproducible results of in vitro hybridization than RNAs from heterologous sources. Three aminoacyl-tRNA synthetase preparations from mouse livers, chicken livers, and human TE32 cells, respectively, gave the same results in the aminoacylation of homologous or heterologous tRNAs and hence were used simultaneously or alternatively for all tRNA identification. The formamide procedure (Table 1, condition 3) was uniformly used to minimize some less stringent hybridization such as binding of tRNA^{Gly} to all rRNA samples. The bound tRNAs were identified by amino acid accepting activities in combination with the amino acid analyzer and/or individual aminoacylation reactions.

In the survey of tRNA-rRNA interaction using this experimental approach, results were obtained for 14 vertebrates (trout, frog, chicken, mouse, rat, Chinese hamster, rabbit, calf, pig, dog, cat, baboon, rhesus monkey, and human). The major tRNAs bound by the 18S and 28S rRNAs of these animals are listed in Table 2. From the study, several points can be made.

1. In all vertebrate species examined, rRNA preparations show consistently the characteristic of selective tRNA binding in the in vitro hybridization reaction.
2. rRNAs of different vertebrates may bind different tRNA. Other than

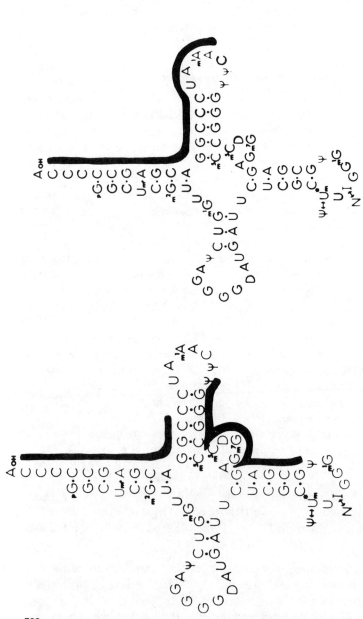

Figure 4 Sequence portions of tRNA^Pro responsible for specific binding to mouse 28S rRNA and genomic RNA of Moloney murine leukemia virus, as indicated by tRNA fragments protected from ribonuclease digestion of the binding RNA complexes (J. E. Dahlberg et al., in prep.).

Table 2 Predominant tRNAs bound to cellular 18S and 28S tRNAs (in vitro) and to retroviral RNAs (in vivo)

Animal species	Cellular RNA (in vitro) 18S	Cellular RNA (in vitro) 28S	Viral RNA (in vivo) 60S–70S	Viral RNA (in vivo) virus
Fish (trout)	Asp	Asp	—	—
Amphibian (frog)	Gly	Asp (Leu)[a]	—	—
Bird (chicken)	Trp (Gly)	Trp (Gly)	Trp	RSV[b]
			Trp	AMV[c]
Mouse	Glu, Gly, Lys	Pro (Gly)	Pro	MuLVs[d]
			Pro	(SiSV)[e]
			[Lys]	MMTV[f]
Rat	Glu (Gly)	Pro	[Pro]	(FeLV)[e]
Chinese hamster[g]	Gly	Gly	—	—
Rabbit	Gly	Gly	—	—
Pig	Glu	Glu	—	—
Dog	Glu	Ser	—	—
Cat	Glu	Ser (Lys, Glu)	[Gly, Asp]	RD114
Baboon	Glu (Gly)	Ser (Gly)	[Gly, Asp, Pro]	BaEV[h]
Rhesus monkey[g]	Glu (Gly)	Glu (Ser)	—	—
Human[g]	Glu	Glu	—	—

[a]Minor but distinctly bound tRNA species.
[b]Rous sarcoma virus.
[c]Avian myeloblastosis virus.
[d]Murine leukemia virus.
[e]Evolutionarily related endogenous retrovirus. SiSV is simian sarcoma/leukemia virus; FeLV is feline leukemia virus.
[g]Cultured cells, rather than liver tissue, were used for rRNA preparations.
[h]Baboon endogenous virus.

tRNA^{Pro}, mentioned above, the predominantly bound include also tRNA^{Trp} (e.g., to chicken rRNAs), tRNA^{Glu} (e.g., to human rRNAs), tRNA^{Asp} (e.g., to trout rRNAs), tRNA^{Gly} (e.g., to hamster and rabbit rRNAs), and tRNA^{Ser} (e.g., to 28S rRNA of dog, cat, and baboon).

3. Some vertebrate rRNA may reproducibly bind additional tRNAs, although to a lesser extent. Like the predominantly bound tRNAs, the subordinately bound tRNAs may differ from animal to animal (e.g., tRNA^{Phe} to frog 18S rRNA and tRNA^{Leu} to dog rRNAs).

4. 18S rRNA and 28S rRNA of the same animal may bind the same or different tRNAs.

5. Phylogenetically related vertebrates generally show more similar tRNA–rRNA interaction than phylogenetically distant vertebrates.

6. The predominant tRNA bound by the 28S rRNA of a vertebrate species appears to be the one serving as a primer for the reverse transcription of genomic RNA of type-C retroviruses that are endo-

genously derived from that vertebrate species (Table 2). This is true at least for tRNA^{Trp} in the avian system and tRNA^{Pro} in the murine system.
7. The tRNA · 28S rRNA complexes are usually active template primers for DNA transcription in reactions catalyzed by a retroviral polymerase. We have demonstrated this with preparations of tRNA · 28S rRNA complexes from human, cat, rabbit, rat, mouse, chicken, and trout sources.

DISCUSSION

After we made the initial observation of selective binding of tRNA^{Pro} to mouse 28S rRNA (Yang et al. 1978), the immediate question asked was whether the binding represents a recognition of specific complementary nucleotide sequences or is merely an artifact. Experimental results subsequently obtained have indicated that mouse 18S rRNA binds tRNAs other than tRNA^{Pro} under the same conditions of in vitro hybridization, that the extent of the selective tRNA–rRNA interaction is affected by the input concentration of the specific RNA, and that tRNA · rRNA complexes have distinct melting temperatures. These results have virtually ruled out the possibility of an artifact. It was then asked whether the binding was due to rRNAs or non-rRNA components of the preparation. This was answered by experiments using the ribosome-subunit-derived rRNA preparations, which showed the same selective tRNA binding, and using the formamide method of hybridization, which decreases rRNA degradation while improving the molar ratio of selective tRNA bound per rRNA molecule. Most important is the elucidation, by the collaboration of J. E. Dahlberg and us, of the sequence of tRNA^{Pro} that binds to mouse 28S rRNA; this not only establishes the distinctive feature of the binding but also makes it difficult for one to explain the binding as due to purely adventitious base-pair arrangements in the interacting tRNAs and rRNAs. Another point of interest is from the survey of various vertebrates, which has revealed the selective tRNA–rRNA interaction to be common rather than rare. The phylogenetic differences of this selective interaction add further curiosity to the phenomenon, especially from the biological point of view.

Whether or not the selective tRNA–rRNA interaction has a biological function remains to be determined. At present, only speculative considerations can be made mainly from three aspects.

The first regards its relationship to the protein synthesis function of ribosomes. Although rRNAs constitute about half the mass of ribosomes, the function (especially of 28S rRNA) is still unknown. The nucleotide sequence characteristics of these RNAs presumably plays an important role, such as in the cases of tRNA–5S rRNA interaction (Richter et al. 1973) and complementary sequences between the 3' end of 16S/18S rRNA and the 5' end of mRNA (Shine and Dalgarno 1974; Steitz and Jakes 1975). In this regard, the present finding of selective tRNA–rRNA interaction could be related to the protein synthesis function of rRNAs. However, there are three constraints to such considerations: (1) the binding sequences of 15–30 bases are presumably too long to allow flexibility of tRNA movement within and/or in and out of the ribosome; (2) the binding is selective not only among different tRNAs but also among some of the different isoacceptors, whereas the protein synthesis mechanism should be applicable to most, if not all, tRNAs; (3) the tRNA–rRNA interaction is heterogeneous among vertebrates, which is contrary to the expected conservation of protein synthesis machinery in the evolutionary process. Thus, even if the present finding of tRNA–rRNA interaction plays some role in the protein synthesis function, a very specific and unique mechanism has yet to be conceived.

The second regards the problem of tRNA gene location in the vertebrate rRNA genes. The presence of specific tRNA genes in the spacer region of prokaryotic rRNA genes is well established (Lund et al. 1976; Lund and Dahlberg 1977; Morgan et al. 1978). Whether a similar phenomenon occurs in eukaroytic rRNA genes remains to be explored. Since the binding of tRNA^{Pro} to mouse 28S rRNA involves the major portion of the 3' half of this tRNA, this implies that at least a gene sequence capable of coding for this portion of tRNA^{Pro} is present in the opposite strand of the DNA sequence coding for mouse rRNA. Recent studies have shown that in yeast some tRNA genes may contain intervening sequences inserted at the anticodon region, thus dividing the tRNA coding sequences into two noncolinear halves (Goodman et al. 1977; Knapp et al. 1978). However, the intervening sequences in yeast tRNA genes are relatively short. The absence of mouse 28S rRNA binding with the 5' half of tRNA^{Pro} would thus argue against the presence of a complete tRNA^{Pro} gene in mouse rRNA gene, unless the binding sequence for the 5' half of tRNA^{Pro} had been removed during processing of rRNA. Further studies are therefore needed to examine the possibility that selective tRNA–rRNA interaction represents a selective presence of tRNA genes in different vertebrate rRNA genes.

The third regards the cellular origin of primer tRNA utilization by endogenous retroviruses. It is now known that the reverse transcription, the most essential genetic process in the infection cycle of neoplasia-related retroviruses, is initiated from a primer identified as a selective tRNA molecule of host cell origin (Dahlberg et al. 1974; Waters et al. 1975; Harada et al. 1975; Faras et al. 1975; Waters 1975; Peters et al. 1977; Yang 1977). This is evidently dependent on the presence in retroviral genomic RNA of a nucleotide sequence complementary to the 3' portion of the tRNA molecule and hence capable of selective tRNA binding. Since retroviruses of different phylogenetic origin appear to utilize different tRNAs as the primer, the primer tRNA binding sequence is obviously subject to evolutional changes. In this regard, it is remarkable that chicken rRNAs and genomic RNA of avian endogenous retroviruses both contain the binding sequences that recognize tRNA^{Trp}; mouse 28S rRNA and genomic RNA of murine endogenous retroviruses both contain sequences that recognize tRNA^{Pro} (Table 2). Since Moloney murine leukemia virus was isolated through an extensive selection procedure, it should be determined whether the primer tRNA^{Pro} binding sequence in the genomic RNA of this virus (Fig. 4) remains the same as those of less evolved retroviruses of murine origin. Thus, it is important to find out whether tRNA binding sequences of rRNAs are genetically related to the tRNA binding sequences on the genomic RNA of retroviruses.

To determine the biological significance of selective tRNA·rRNA interaction, many problems remain to be explored. For example, the exact location of the tRNA binding site on the rRNA molecule is still unknown. Also, whether or not the binding nucleotide sequences of tRNA^{Trp}, tRNA^{Glu}, tRNA^{Gly}, and tRNA^{Asp}, are on the 3'-end portion of the molecule should be determined. Our earlier results of low tRNA : rRNA molar ratios and of apparent different binding capacities of rRNAs from embryos and adult livers gave the impression that there might be tRNA-binding and non-binding subpopulations of rRNAs; whether or not this was due to use of an inappropriate hybridization condition remains uncertain. For studying the phylogenetic relationship, the sequence structures of the DNA transcripts derived from different tRNA-primed rRNA will be needed.

However, for practical purposes, the present findings (Table 2) have apparently provided a new avenue for our research on retroviral genetic expression, especially in relation to neoplastic transformation of the cell. It is now well recognized that retrovirus isolates capable of causing neoplastic cell transformations are generally replication-defective. Actually, animal model studies have demonstrated that neoplastic transformation requires only partial (oncogene) expression rather than complete (virogene) expression of the oncogenic retroviruses. Furthermore, these viruses are often heterogeneous in genetic structure and their oncogene component may be completely different. Thus, the study of carcinogenesis by the experimental approach of virus isolation, particularly in the human system, has not been successful. In view of this, we have taken a different experimental approach, namely, to search for retrovirus-specific mRNA molecules in the cell. Recently, it has been found that 5'-terminal sequences of retroviral genomic RNA, which contains more than 100 nucleotides, is transposed to the 5' terminal of subgenomic-sized retrovirus-specific mRNAs (Weiss et al. 1977; Mellon and Duesberg 1977; Rothenberg et al. 1978). Our preliminary results indicate that subgenomic-sized poly(A)$^+$ RNA from cells infected with murine retrovirus contain an increased binding capacity for tRNA^{Pro}; this suggests that primer tRNA binding sequence is included in the transposed 5' terminals of the retrovirus-specific mRNAs. Work is now in progress in our laboratory to examine human cancer cells for the presence of poly(A)$^+$ RNA molecules with tRNA binding capacities, especially for those tRNAs that have been shown to interact selectively with rRNAs in vitro.

CONCLUDING REMARKS

The experimental results of selective tRNA–rRNA interaction described in this paper are all obtained from in vitro studies, in which isolated rRNA and tRNA molecules are first heat-denatured and then mixed together in reaction conditions that favor the hybridization. The results indicate the presence in rRNA structure of certain nucleotide sequences, which are heterologous among various vertebrates. The specific tRNA·rRNA complexes, demonstrated in vitro, have not been found in the cell and thus the biological significance of these observations is still in question.

Similar results can be obtained with the genomic RNAs of retroviruses; when first denatured and then mixed with total cellular tRNAs, these poly(A)$^+$ RNA molecules selectively bind the respective primer tRNAs. The retrovirus genomic RNA·primer tRNA complex, however, is present within the virion. Recent evidence indicated that the complex formation requires the presence of active viral reverse transcriptase (G. G. Peters et al., pers. comm.). In vitro experiments showed that intracellular retrovirus-specific mRNA may also contain the primer tRNA binding nucleotide sequence, although no such complexes are detectable in the cell.

It appears that rRNAs and retrovirus genome RNAs are similar not only in the selective tRNA binding capability but also in the particular tRNA molecule they recognize. tRNA^{Trp} is recognized by chicken rRNAs as well as by genomic RNAs of avian endogenous type-C retroviruses, whereas tRNA^{Pro} is recognized by mouse 28S rRNA as well as genomic RNAs of murine endogenous type-C retroviruses. Thus, our observation of selective binding of tRNA^{Glu} to human rRNAs should be assessed for a possible value of application in human cancer research.

ACKNOWLEDGMENTS

Our thanks go to D. Price, D. Moore, C. Stringer, J. Kiggans, and M. Yang for technical assistance at various phases of the study; to J. E. Dahlberg for the collaboration on mouse 28S rRNA binding sequence of tRNA^{Pro}; to F. C. Hartman for the use of the amino acid analyzer; and to W. E. Cohn and L. C. Waters for reading the manuscript. D. L. R. H. was a postdoctoral investigator supported by subcontract 3332 from the Biology Division of the Oak Ridge National Laboratory to the University of Tennessee. This research was sponsored jointly by the National Cancer Institute (YO1-CP9-0503) and the Division of Biomedical and Environmental Research, U.S. Department of Energy, under contract W-7405-eng-26 with the Union Carbide Corporation.

REFERENCES

Baltimore, D. and D. Smoler. 1971. Primer requirement and template specificity of the DNA polymerase of RNA tumor viruses. *Proc. Natl. Acad. Sci.* **68**:1507.

Beaudreau, G. S., L. Sverak, R. Zischka, and J. W. Beard. 1964. Attachment of C^{14}-amino acids to BAI strain A (myeloblastosis) avian tumor virus RNA. *Natl. Cancer Inst. Monogr.* **17**:791.

Bishop, J. M., W. E. Levinson, N. Quintrell, D. Sullivan, L. Fanshier, and J. Jackson. 1970. The low molecular weight RNAs of Rous sarcoma virus. I. The 4S RNA. *Virology* **42**:182.

Bonar, R. A., L. Sverak, D. P. Bolognesi, A. J. Langlois, D. Beard, and J. W. Beard. 1967. Ribonucleic acid components of BA1 strain A (myeloblastosis) avian tumor virus. *Cancer Res.* **27**:1138.

Canaani, E. and P. Duesburg. 1972. Role of subunits of 60 to 70S avian tumor virus ribonucleic acid in its template activity for the viral deoxyribonucleic acid polymerase. *J. Virol.* **10**:23.

Carnegie, J. W., A. O'C. Deeney, K. C. Olson, and G. S. Beaudreau. 1969. An RNA fraction from myeloblastosis virus having properties similar to transfer RNA. *Biochim. Biophys. Acta* **190**:274.

Dahlberg, J. E. 1977. RNA primer for the reverse transcriptases of RNA tumor viruses. In *Nucleic acid-protein recognition* (ed. H. J. Vogel), p. 345. Academic Press, New York.

Dahlberg, J. E., C. Kintner, and E. Lund. 1978. Specific binding of $tRNA_f^{Met}$ to 23S rRNA of *Escherichia coli*. *Proc. Natl. Acad. Sci.* **75**:1071.

Dahlberg, J. E., R. C. Sawyer, J. M. Taylor, A. J. Faras, W. E. Levinson, H. M. Goodman, and J. M. Bishop. 1974. Transcription of DNA from the 70S RNA of Rous sarcoma virus. I. Identification of a specific 4S RNA which serves as primer. *J. Virol.* **13**:1126.

Davey, R. A. and A. J. Howells. 1974. A method for the analysis of tRNA patterns in *Drosophila melanogaster* using an amino acid analyzer. *Anal. Biochem.* **60**:469.

Eiden, J. J., K. Quade, and J. L. Nichols. 1976. Interaction of tryptophan transfer RNA with Rous sarcoma virus 35S RNA. *Nature* **259**:245.

Elder, K. J. and A. E. Smith. 1973. Methionine transfer ribonucleic acids of avian myeloblastosis virus. *Proc. Natl. Acad. Sci.* **70**:2823.

Erikson, E. and R. L. Erikson. 1970. Isolation of amion acid acceptor RNA from purified avian myeloblastosis virus. *J. Mol. Biol.* **52**:387.

―――. 1971. Association of 4S ribonucleic acid with oncornavirus ribonucleic acids. *J. Virol.* **8**:254.

Fan, H. and D. Baltimore. 1973. RNA metabolism of murine leukemia virus:

Detection of virus-specific RNA sequences in infected and uninfected cells and identification of virus-specific messenger RNA. *J. Mol. Biol.* **80**:93.

Faras, A. J., J. E. Dahlberg, R. C. Sawyer, F. Harada, J. M. Taylor, W. E. Levinson, J. M. Bishop, and H. M. Goodman. 1974. Transcription of DNA from the 70S RNA of Rous sarcoma virus. II. Structure of a 4S RNA primer. *J. Virol.* **13**:1134.

Flügel, R. M. and R. D. Wells. 1972. Nucleotides at the RNA-DNA covalent bonds formed in the endogenous reaction by the avian myeloblastosis virus DNA polymerase. *Virology* **48**:394.

Gallagher, R. E. and R. C. Gallo. 1973. Chromatographic analyses of isoaccepting tRNAs from avian myeloblastosis virus. *J. Virol.* **12**:449.

Goodman, H. M., M. V. Olson, and B. D. Hall. 1977. Nucleotide sequence of a mutant eukaryotic gene: The yeast tyrosine-inserting ochre suppressor SUP4-O. *Proc. Natl. Acad. Sci.* **74**:5453.

Harada, F., R. C. Sawyer, and J. E. Dahlberg. 1975. A primer ribonucleic acid for initiation of in vitro Rous sarcoma virus deoxyribonucleic acid synthesis. Nucleotide sequence and amino acid acceptor activity. *J. Biol. Chem.* **250**:3487.

Knapp, G., J. S. Beckmann, P. F. Johnson, S. A. Fuhrman, and J. Abelson. 1978. Transcription and processing of intervening sequences in yeast tRNA genes. *Cell* **14**:221.

Larsen, C. J., R. Emanoil-Ravicovitch, A. Samso, J. Robin, and M. Boiron. 1973. Studies of the low molecular weight RNA components of the mouse leukemia-sarcoma virus complex (MSV-MLV). *Virology* **54**:552.

Leong, J. A., A. C. Garapin, N. Jackson, L. Fanshier, W. E. Levinson, and J. M. Bishop. 1972. Virus-specific ribonucleic acid in cells producing Rous sarcoma virus: Detection and characterization. *J. Virol.* **9**:891.

Lund, E. and J. E. Dahlberg. 1977. Spacer transfer RNAs in ribosomal RNA transcripts of *E. coli*: Processing of 30S ribosomal RNA in vitro. *Cell* **11**:247.

Lund, E., J. E. Dahlberg, L. Lindahl, S. R. Jaskunas, P. P. Denis, and M. Nomura. 1976. Transfer RNA genes between 16S and 23S rRNA genes in rRNA transcription units of *E. coli*. *Cell* **7**:165.

Mellon, P. and P. H. Duesberg. 1977. Subgenomic, cellular Rous sarcoma virus RNAs contain oligonucleotides from the 3' half and the 5' terminus of virion RNA. *Nature* **270**:631.

Morgan, E. A., T. Ikemura, L. Lindahl, A. M. Fallon, and M. Nomura. 1978. Some rRNA operons in *E. coli* have tRNA genes at their distant ends. *Cell* **13**:335.

Peters, G., F. Harada, J. E. Dahlberg, A. Panet, W. A. Haseltine, and D. Baltimore. 1977. Low-molecular-weight RNA's of Moloney murine leukemia virus: Identification of the primer for RNA-directed DNA synthesis. *J. Virology* **21**:1031.

Richter, D., V. A. Erdman, and M. Sprinzl. 1973. Specific recognition of GTψC loop (loop IV) of tRNA by 50S ribosomal subunits from *E. coli*. *Nat. New Biol.* **246**:132.

Rosenthal, L. J. and P. C. Zamecnik. 1973a. Amino acid acceptor activity of the "70S-associated" 4S RNA from avian myeloblastosis virus. *Proc. Natl. Acad. Sci.* **70**:1184.

———. 1973b. Minor base composition of "70S-associated" 4S RNA from avian myeloblastosis virus. *Proc. Natl. Acad. Sci.* **70**:865.

Rothenberg, E., D. J. Donoghue, and D. Baltimore. 1978. Analysis of a 5' leader sequence on murine leukemia virus 21S RNA: Heteroduplex mapping with long reverse transcriptase product. *Cell* **13**:435.
Sawyer, R. C. and J. E. Dahlberg. 1973. Small RNAs of Rous sarcoma virus: Characterization of two dimensional polyacrylamide gel electrophoresis and fingerprint analysis. *J. Virol.* **12**:1126.
Shine, J. and L. Dalgarno. 1974. The 3'-terminal sequence of *Escherichia coli* 16S ribosomal RNA: Complementary to nonsense triplets and ribosome binding sites. *Proc. Natl. Acad. Sci.* **71**:1342.
Sompayrac, L. and O. Maaløe. 1973. Transfer RNA activities of Rous sarcoma and Rous-associated viruses. *Nat. New Biol.* **242**:133.
Spiegelman, S., A. Burny, M. R. Das, J. Keydar, J. Schlom, M. Trávnicek, and K. Watson. 1970. Synthetic DNA-RNA hybrids and RNA-DNA duplexes as templates for the polymerases of the oncogenic RNA viruses. *Nature* **228**:430.
Steitz, J. A. and K. Jakes. 1975. How ribosomes select initiator regions in mRNA: Base pair formation between the 3' terminus of 16S rRNA and the mRNA during initiation of protein synthesis in *Escherichia coli. Proc. Natl. Acad. Sci.* **72**:4734.
Taylor, J. M. and R. Illmensee. 1975. Sites on the RNA of an avian sarcoma virus at which primer is bound. *J. Virol.* **16**:553.
Todaro, G. J. 1975. Evolution and modes of transmission of RNA tumor viruses. *Am. J. Pathol.* **81**:590.
Trávnicek, M. 1968. RNA with amino acid-acceptor activity isolated from an oncogenic virus. *Biochim. Biophys. Acta* **166**:757.
Tsuchida, N., M. S. Robin, and M. Green. 1972. Viral RNA subunits in cells transformed by RNA tumor viruses. *Science* **176**:1418.
Verma, I. M., N. L. Meuth, E. Bromfeld, K. F. Manly, and D. Baltimore. 1971. Covalently linked RNA-DNA molecule as initial product of RNA tumor virus DNA polymerase. *Nat. New Biol.* **233**:131.
Waters, L. C. 1975. Transfer RNAs associated with the 70S RNA of AKR murine leukemia virus. *Biochem. Biophys. Res. Commun.* **65**:1130.
———. 1978. Lysine tRNA is the predominant tRNA in murine mammary tumor virus. *Biochem. Biophys. Res. Commun.* **81**:822.
Waters, L. C. and B. C. Mullin. 1977. Transfer RNA in RNA tumor viruses. *Prog. Nucleic Acid Res. Mol. Biol.* **20**:131.
Waters, L. C., B. C. Mullin, E. G. Bailiff, and R. A. Popp. 1975a. tRNA's associated with the 70S RNA of avian myeloblastosis virus. *J. Virol.* **16**:1608.
Waters, L. C., B. C. Mullin, T. Ho, and W. K. Yang. 1974. *In vitro* association of unique species of cellular 4S RNA with 35S RNA of RNA tumor viruses. *Biochem. Biophys. Res. Commun.* **60**:489.
———. 1975b. Ability of tryptophan tRNA to hybridize with 35S RNA of avian myeloblastosis virus and to prime reverse transcription *in vitro. Proc. Natl. Acad. Sci.* **72**:2155.
Weiss, S. R., H. E. Varmus, and J. M. Bishop. 1977. The size and genetic composition of virus-specific RNAs in the cytoplasm of cells producing avian sarcoma-leukosis viruses. *Cell* **12**:983.
Yang, W. K. 1977. Primer function of proline tRNAs in the reverse transcription of murine leukemia virus 35S RNA. *Fed. Proc.* **36**:847.

Yang, W. K. and G. D. Novelli. 1971. Analysis of isoaccepting tRNAs in mammalian tissues and cells. *Methods Enzymol.* **20**:44.

Yang, W. K., D. L. R. Hwang, J. O. Kiggans, Jr., D. M. Yang, C. D. Stringer, D. J. Moore, and F. C. Hartman. 1978. *In vitro* association of selective tRNA species with 28S RNA of mouse cells. *Biochem. Biophys. Res. Commun.* **80**:443.

tRNA-like Structures in Viral RNA Genomes

Anne-Lise Haenni and François Chapeville
Laboratoire de Biochimie du Développement
Institut de Recherche en Biologie Moléculaire
du CNRS et de l'Université Paris VII
75221 Paris Cedex 05, France

Most of the eukaryotic mRNAs and a great number of genomes from plus-stranded RNA viruses contain at their 3' ends a poly(A) sequence of variable length. All known prokaryotic mRNAs and bacteriophage RNAs as well as a few eukaryotic mRNAs are devoid of such a sequence. Among plant RNA viruses, the genomes not possessing this poly(A) stretch are fairly numerous. Of the 16 or so groups of plant RNA viruses (Wildy 1971), in only one group does the RNA terminate by poly(A).

In all known cases of mRNAs or of viral RNAs, whether or not they contain poly(A), the coding regions are followed by an untranslated heteropolymeric sequence. In human and rabbit β-globin mRNAs, the poly(A) stretch is preceded by a sequence of 134 and 95 untranslated nucleotides, respectively (Proudfoot 1977). The sea urchin mRNA for histone H4, which does not possess poly(A), presumably has an untranslated region at its 3' end of about 30 nucleotides (Grunstein and Grunstein 1978). In the RNA of the bacteriophage MS2, the length of the 3' untranslated sequence that follows the replicase gene is 174 nucleotides long (Fiers et al. 1976); this RNA, as well as the genomes of several other RNA bacteriophages, terminates with the sequence CCA identical to that of tRNAs.

The physiological function of these nontranslated regions at the 3' end of mRNAs has not been completely elucidated. It has been reported that the poly(A) stretch protects the mRNAs, probably against $3' \rightarrow 5'$ exonucleases and thus confers greater stability to these RNAs (Huez et al. 1974). The function of the untranslated heteropolymeric sequences that follow the termination codon, whether or not these sequences are followed by poly(A), is not understood. However, this function must be very important, since these untranslated regions are highly conserved within a given group, and since when removed, as in bacteriophages (Kamen 1969; Rensing and August 1969) or in viruses (Steinschneider and Fraenkel-Conrat 1966), infectivity is abolished.

In plant RNA viruses, the length of the 3' noncoding region of genomes not terminating with poly(A) exceeds 100 nucleotides. In 1970, we showed that turnip yellow mosaic virus (TYMV) RNA incubated in the presence of a cell extract, ATP, and free amino acids is esterified at its 3' end with valine (Pinck et al. 1970; Yot et al. 1970). Subsequently, it was reported that several other plant viral RNAs are acceptors of a specific amino acid in conditions analogous to those used for the charging of tRNAs.

When TYMV RNA (m.w. 2×10^6) is fragmented, the 3'-end fragment of about 100 nucleotides is also an efficient acceptor of the amino acid (Prochiantz and Haenni 1973). This observation, plus the fact that viral RNAs that can be aminoacylated are also recognized by several tRNA-specific enzymes, led to the conclusion that they contain a tRNA-like structure at their 3' end.

In this paper, we shall review the basic experiments and results concerning these tRNA-like structures in viral RNA genomes and discuss their possible physiological role.

AMINOACYLATION OF VIRAL GENOMES

The best documented case of aminoacylation of a viral RNA is that of TYMV RNA. Rigorous analyses were necessary to demonstrate that charging with the amino acid was not an artifact due to a strong noncovalent binding of contaminant cellular tRNAs to the high-molecular-weight viral RNA. The existence of a possible covalent linkage between such a tRNA and the viral RNA also had to be investigated.

The methods used in these experiments and the results obtained were later extended to other RNA viruses.

Aminoacylation of TYMV RNA

In 1970, Pinck et al. and Yot et al. (1970) reported that after incubation of purified TYMV RNA in the presence of ATP, a mixture of ^{14}C-labeled amino acids and a bacterial extract devoid of tRNAs, the cold, acid-insoluble material contained ^{14}C that was alkali labile. The use of individual amino acids showed that only valine could be associated with this material. Sucrose gradient analyses indicated that valine was recovered essentially with high-molecular-weight RNA (Fig. 1).

About 1 mole of valine was retained per mole of acid-precipitable RNA. More recently, it has been reported that this stoichiometry might be slightly above 1 (Giégé et al. 1978). This is probably due to the presence of a population of smaller RNA molecules also encapsidated in the virion and that contain the coat-protein gene followed by the 3' terminal of the viral RNA.

To determine the nature of the linkage between valine and the high-molecular-weight RNA, the valine-containing RNA was digested by either pancreatic RNase or by snake venom phosphodiesterase. The

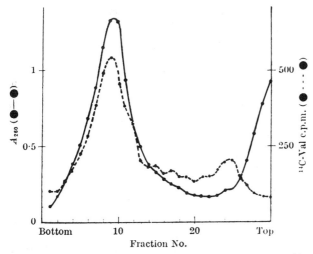

Figure 1 Sucrose gradient sedimentation pattern of ^{14}C-labeled valyl-RNA of TYMV. It was later shown that the radioactivity sedimenting with RNA of sedimentation constant lower than 23–25S most likely corresponds to degradation products of the viral RNA and the small radioactive peak close to the top of the gradient to a fragment pertaining from the 3' end of TYMV RNA and encompassing the coat-protein gene, and present in certain virus particles. (Reprinted, with permission, from Pinck et al. 1970.)

analysis of the valine-containing molecules yielded valyl-A or valyl-AMP, respectively (Pinck et al. 1970). Chemical acetylation of the RNA-bound valine led to N-acetylvalyl-RNA, showing that the amino group of valine in valyl-RNA is free, and thereby confirming that the carboxyl group of the amino acid must be esterified to the 3'-terminal A of the RNA (Yot et al. 1970).

As to the nature of the RNA acceptor of valine, these preliminary experiments could not entirely exclude the possibility that host tRNA^{Val} entrapped within the virion might somehow be strongly bound to the viral RNA. Various types of linkages could be responsible for such an interaction, for example, protein-mediated linkages, ionic bonds, and hydrogen bonds. However, all attempts to dissociate such a putative complex did not reveal the existence of a chargeable tRNA species in the TYMV RNA preparations. For instance, treatment of the charged RNA preparations with SDS, EDTA, 8 M urea, or DMSO followed by formaldehyde, did not liberate a valyl-tRNA (Yot et al. 1970).

The best demonstration that TYMV RNA was not contaminated by a host tRNA^{Val} molecule was brought about by the analysis of the RNase T1 digests of the host valyl-tRNA^{Val} and that of TYMV valyl-RNA. The host plant (Chinese cabbage) tRNAs gave rise to two main valyl-oligonucleotides deriving from at least two tRNAs isoacceptors of valine.

They differed from the valyl-oligonucleotide obtained by the same procedure from TYMV valyl-RNA (Fig. 2). These results excluded, therefore, also the possibility of a hypothetical ligation between a cellular tRNA^{Val} and the 3' end of the viral genome (Yot et al. 1970).

The nature of the enzyme responsible for the charging reaction was investigated using purified *E. coli* valyl-tRNA synthetase. This enzyme was able to esterify TYMV RNA with valine, provided that the tRNA nucleotidyl transferase was included in the incubation mixture (Yot et al. 1970). As will be shown later, this latter enzyme catalyzes the addition of 1 molecule of AMP to TYMV RNA. Valyl-tRNA synthetase of yeast, plant, and animal origin are also active in the presence of the tRNA nucleotidyl transferase (Yot 1970).

Using purified yeast valyl-tRNA synthetase, Giégé et al. (1978) have shown that the kinetic parameters (K_m and rate constant) for valylation of TYMV RNA do not differ significantly from those observed with the same enzyme for yeast tRNA^{Val}. The studies of these systems led to the conclusion that TYMV RNA differs from incorrect valylations observed, for instance, with yeast tRNA^{Phe} or yeast tRNA^{Met} by the yeast valyl-tRNA synthetase. Consequently, it is not very likely that aminoacylation of TYMV RNA occurs fortuitously.

When TYMV RNA is incubated in the presence of a partially purified *E. coli* extract containing RNase P, which is involved in the processing of tRNA precursors (Robertson et al. 1972), an RNA fragment of about 40,000 daltons (4.5S) pertaining from the 3' end of the viral genome can be isolated. This fragment retains the same charging capacity for valine as the high-molecular-weight RNA. Therefore, the features in the RNA

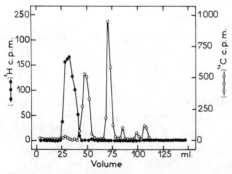

Figure 2 DEAE-cellulose chromatography of RNase T1 digests of TYMV ^{3}H-labeled valyl-RNA and cabbage ^{14}C-labeled valyl-tRNA. The positions of oligonucleotides 5 and 13 residues long that appeared in tubes 28–30 and tubes 94–96, respectively, were defined by treating *E. coli* ^{14}C-labeled valyl-tRNA with RNase T1 and analyzing the digests in similar conditions. (Reprinted, with permission, from Yot et al. 1970.)

recognized by the valyl-tRNA synthetase are not dispersed but localized in the 3' region of the genome (Prochiantz and Haenni 1973).

Very little is known about the aminoacylation of TYMV RNA in vivo. If such a reaction does occur, it must be transient, since in the virion the RNA is not aminoacylated; it even lacks the 3'-terminal A. Consequently, if in the infected leaves only a small portion of the viral RNA is transiently aminoacylated, the characterization of such valyl-RNAs will be difficult. The use of plant protoplasts will certainly facilitate such studies in vivo.

When TYMV RNA is injected into *Xenopus laevis* oocytes, TYMV valyl-RNA is recovered after incubation with valine (Joshi et al. 1978). This suggests that in the homologous system, the viral RNA might also be aminoacylated.

Aminoacylation of the Genome of Other Plant Viruses

The amino acid acceptor capacity of plant viral RNAs is not limited to the genome of TYMV. Over the last six years, it has been reported that the genomes of at least ten different viruses belonging to four different groups could be charged with a specific amino acid. The list of the main representatives of these viruses with the group to which they belong and the amino acid they can accept, is presented in Table 1.

Charging of the RNAs of various strains of the tobamovirus group with histidine was discovered by Öberg and Philipson (1972), and by Litvak et al. (1973b); that of the cowpea strain of tobacco mosaic virus (C_cTMV) with valine by Beachy et al. (1976). The work on the tymoviruses comes from the group of Duranton (Pinck et al. 1972, 1975).

Charging of the genome of bromoviruses and of cucumber mosaic virus (CMV) with tyrosine was reported by the groups of Kaesberg and Hall (Hall et al. 1972; Kohl and Hall 1974). These are multicomponent viruses whose genome contains four distinct RNA molecules of molecular weight ranging from 1.1 to 0.3×10^6. In every case, each of these molecules is an acceptor of tyrosine.

With the exception of the tobamoviruses in which the two representatives examined accept different amino acids, the RNAs of the viruses of a same group accept the same amino acid.

In those cases where such experiments were performed, it was observed that the aminoacyl-tRNA synthetase preparations from the host plant are active in the esterification of the viral RNA (Yot 1970; Kohl and Hall 1974). In some cases, the bacterial enzyme catalyzes the esterification of the viral RNA as, for instance, the genome of the tymoviruses that are charged with valine. In others, eukaryotic enzymes (yeast or animal) are needed (Litvak et al. 1973b; Kohl and Hall 1974). No case has been reported so far of a synthetase that would charge a viral RNA with a

Table 1 In vitro amino acid acceptor activity of viral RNA genomes

Source of viral RNA		Amino acid bound
Plant Viruses		
Bromovirus group	BMV	tyrosine
	BBMV[a]	tyrosine
	CCMV[b]	tyrosine
Cucumovirus group	CMV	tyrosine
Tobamovirus group	TMV	histidine
	C$_c$TMV	valine
Tymovirus group	TYMV	valine
	CYMV[c]	valine
	EMV	valine
	OMV	valine
	WCMV[d]	valine
Animal Viruses		
Picornavirus group	EMCV	serine
	mengovirus	histidine

The RNA of the following strains of TMV were shown to accept histidine: U1, U2, HRG, and Vulgare; the Dahlemense strain was a very poor acceptor of histidine.
[a]Broad bean mottle virus.
[b]Cowpea chlorotic mottle virus.
[c]Cacao yellow mosaic virus.
[d]Wild cucumber mosaic virus.

given amino acid, but would not charge with the same amino acid the tRNA of the infected plant. Moreover, no case of mischarging (whatever the synthetase or its origin) of a viral RNA has been observed.

Aminoacylation of the Genome of Animal Viruses

Contrary to plant viruses that can be obtained in large amounts, animal viruses are only available in very small quantities; this makes their aminoacylation studies difficult. Consequently, only a few animal viruses have been examined for their capacity to be esterified by an amino acid.

Salomon and Littauer (1974) reported that mengovirus RNA could be aminoacylated with histidine, and later, Lindley and Stebbing (1977) showed that encephalomyocarditis virus (EMCV) RNA can be aminoacylated with serine. These results are surprising, since in both cases the viral RNA normally contains a poly(A) stretch at its 3' end and it is highly unlikely that the amino acid esterifies the A of the poly(A) portion. Consequently, the acceptor of the amino acid must be either molecules deprived of their poly(A) stretch or fragments of the viral genome appearing as a result of degradation during the incubation with partially purified aminoacyl-tRNA synthetases. In the latter case, the tRNA-like structure would be located somewhere other than at the 3' end of the RNA molecule.

RECOGNITION OF VIRAL RNAs BY OTHER tRNA-SPECIFIC ENZYMES

Besides their reaction with aminoacyl-tRNA synthetases, the tRNA molecules are substrates of many other enzymes that, in general, are less specific than the synthetases; these enzymes often recognize all species of tRNA molecules, whereas the synthetases recognize only the isoacceptors. Therefore, it seemed interesting to examine to what extent other tRNA enzymes can also use the viral RNAs as substrates.

tRNA Nucleotidyl Transferase

In early experiments with TYMV RNA, when instead of using only partially purified extracts, a homogeneous valyl-tRNA synthetase was employed, it was found that valylation of the RNA did not occur. The addition to the incubation mixture of purified *E. coli* tRNA nucleotidyl transferase reestablished the reaction (Yot et al. 1970; Litvak et al. 1970). Table 2 shows the results of such an experiment. Similar results have been obtained with the 4.5S 3'-terminal RNA fragment as an acceptor of valine. It was thus established that viral RNA is essentially free of a 3'-terminal A and that it is a substrate of the tRNA nucleotidyl transferase.

These results led to further investigations that showed that when TYMV RNA is incubated with snake venom phosphodiesterase in conditions that remove the CCA terminal from tRNAs (Litvak et al. 1973a), it becomes an acceptor of 2 molecules of cytidylate and 1 molecule of adenylate. Consequently, it behaves the same way as all tRNA molecules.

Table 2 Acceptor activity of host tRNAs and of TYMV RNA for ^{14}C-labeled valine

Enzymes added	Cabbage tRNA	TYMV RNA
DEAE enzyme preparation	340	56
Valyl-tRNA synthetase	378	5
tRNA nucleotidyl transferase and valyl-tRNA synthetase	279	53

All enzymes were from *E. coli*; results are expressed as cpm/μg RNA (from Yot et al. 1970).

Contrary to TYMV RNA (Yot et al. 1970; Litvak et al. 1970) and to okra mosaic virus (OMV) RNA (Génevaux et al. 1976), the other plant viral RNAs that can be aminoacylated in vitro seem to contain a terminal A residue adjacent to the sequence CC at their 3' end. Such is the case, for example of brome mosaic virus (BMV) RNA and of tobacco mosaic virus (TMV) RNA. If the CCA terminal of these RNAs is removed by snake venom phosphodiesterase, the tRNA nucleotidyl transferase can restore the original sequence (Benicourt and Haenni 1974; Busto et al. 1976).

Peptidyl-tRNA Hydrolase

This enzyme catalyzes the hydrolysis of the ester linkage in peptidyl-tRNAs or generally in N-substituted aminoacyl-tRNAs, except for formylmethionyl-tRNA^{iMet}, which is a poor substrate (Cuzin et al. 1967; Kössel and RajBhandary 1968). When valyl-RNA of TYMV was chemically acetylated and submitted to an incubation with purified *E. coli* peptidyl-tRNA hydrolase, free acetylvaline was recovered (Yot et al. 1970).

Elongation Factors EF-1 (Wheat Germ) or EF-T (*E. coli*)

These factors involved in the elongation of polypeptide chains interact with aminoacyl-tRNAs in the presence of GTP. Using purified wheat germ EF-1, Litvak et al. (1973b) demonstrated that TYMV or TMV RNA when aminoacylated interact stoichiometrically with the binary complex EF-1·GTP, as determined by the Millipore filtration technique. The direct formation of a ternary complex was not investigated by these authors, but Bastin and Hall (1976) have reported that with BMV RNA a ternary complex cannot be isolated, probably because of its low stability.

The recognition of TYMV valyl-RNA by purified *E. coli* EF-T in the presence of GTP has also been reported (Haenni et al. 1974).

RNase P

As mentioned previously, in the presence of an extract containing this enzyme, TYMV RNA yields a 4.5S RNA fragment from its 3' end, of 112 nucleotides in length, that can be charged with valine. The main part of the genome is however also fragmented during the incubation with the extract, but it has not been determined whether this is due to the action of RNase P or to other nucleases contaminating the enzyme preparations (Prochiantz and Haenni 1973).

In the presence of host plant extracts, TYMV valyl-RNA also yields a valyl-RNA fragment of similar length; the same results were obtained when TYMV RNA was injected into oocytes (Joshi et al. 1978). All these results suggest that either an RNase-P-like activity is present in these two other systems, or that the enzymes are different from RNase P, but that the sequence of nucleotides around position 110 from the 3' end of the viral genome is particularly exposed and accessible to various endonucleases.

tRNA Methyl Transferases

An *E. coli* preparation of tRNA (uracil-5)methyl transferase, a tRNA-specific enzyme for forming T, quantitatively methylates the U residue in position 32 from the 3' end of TMV RNA. The U that is methylated is located in a site resembling the usual T position of tRNAs (Marcu and Dudock, 1975; Lesiewicz and Dudock 1978).

In the case of TYMV RNA, it has been reported that a tRNA (cytosine-5)methyl transferase of animal origin and specific for tRNAs transfers one or two methyl groups to the viral RNA (Haenni et al. 1975).

The physiological significance of these results, besides the fact that these tRNA methyl transferases do recognize the viral RNA, cannot be easily interpreted. Indeed, as will be described later, the tRNA-like structures are devoid of modified nucleotides.

Ribosomal System

In early experiments with TYMV valyl-RNA, we reported that, although poorly, valine residues could be donated for the synthesis of peptide chains using an *E. coli* ribosomal system (Haenni et al. 1973). The experiments suggesting this transfer were performed in the presence of all *E. coli* tRNAs, free amino acids, ATP, and in the presence of valinol-AMP as inhibitor of the valyl-tRNA synthetase that would otherwise catalyze the transfer of valine from TYMV RNA onto the cognate tRNA. In these experiments, the control showing the absence of such a direct transfer was not performed and it is likely that such a transfer could explain valine incorporation into polypeptides. Recently, experiments in our laboratory have been in favor of this conclusion (A.-L. Haenni and F. Chapeville, unpubl.).

In two other cases, BMV tyrosyl-RNA (Chen and Hall 1973) and eggplant mosaic virus (EMV) valyl-RNA (T. Hall, pers. comm.), no transfer of the amino acid into peptides could be observed.

In conclusion, even though several enzymes specific for tRNAs, including EF-1 and EF-T, efficiently recognize the viral RNA, the ribosomal system is probably not able to use it as an amino acid donor in polypeptide synthesis. It would be interesting to investigate if these tRNA-like structures can nevertheless compete with the tRNAs at the ribosomal level. Fragments of the viral RNA containing the tRNA-like structure should be used for such experiments. It is not unlikely that such fragments might possess a certain affinity for the aminoacyl site of the ribosome and lead to peptide chain termination and release by a puromycinlike reaction, or, at least in the case of TYMV RNA, to inhibition of elongation by blocking the aminoacyl site.

STRUCTURAL FEATURES OF THE AMINO ACID ACCEPTOR REGIONS OF VIRAL RNAs

The fact that several enzymes highly specific for tRNAs efficiently recognize the 3'-terminal part of certain viral genomes strongly suggested that these RNAs must have structures similar to those of tRNAs. The results obtained on the nucleotide sequence of the amino acid acceptor 3'-terminal part of three viral RNAs, surprisingly, did not show much similarity to tRNAs. These observations are of great interest because they stress how great is our ignorance of the mechanisms of enzymatic recognition of nucleic acids.

The three different primary structures that have been established and are reproduced below have been drawn to bring out as much as possible their resemblances to the secondary structure of tRNAs. Only very little similarity can be found and certainly it is not sufficient to account for the efficient reactivity of these structures with many tRNA-specific enzymes. Therefore, there is no doubt that structures of a higher order somehow determine the features required for the recognition of these molecules. It would be of great interest for a better understanding of nucleic acid–protein interactions, to gain more information on the tertiary folding of these viral tRNA-like structures.

None of the tRNA-like structures whose sequences have been established contain modified nucleotides. The fact that they are aminoacylated demonstrated for the first time that base modification is not a prerequisite for recognition by aminoacyl-tRNA synthetases. The same holds true for the other tRNA-specific enzymes. Moreover, since these tRNA-like structures, when they are part of the genome, are efficiently recognized by several tRNA enzymes, they must be free of strong associations with other parts of the RNA molecule.

TYMV RNA

The 3'-terminal fragment (112 nucleotides long) of TYMV RNA acceptor of valine, a fragment obtained in the presence of an RNase P preparation, was sequenced (Silberklang et al. 1977) and is presented in Figure 3. Briand et al. (1977) simultaneously sequenced the same region of the RNA, extending sequence analysis into the coat-protein gene, which is adjacent to the tRNA-like structure.

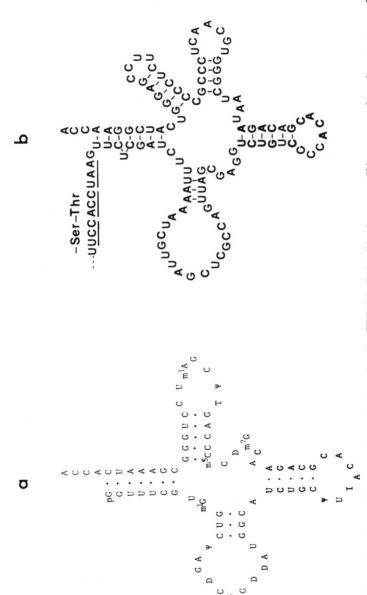

Figure 3 A possible secondary structure for the TYMV tRNA-like fragment (*b*), compared to the structure of yeast tRNA$_1^{Val}$ (*a*). Dark areas correspond to regions of significant sequence homologies. The last two codons of the coat-protein gene corresponding to serine and threonine and the adjacent termination codon are underlined. (Reprinted, with permission, from Silberklang et al. 1977.)

It is interesting to note that the termination codon UAA of the coat-protein gene precedes the tRNA-like structure that contains four other termination codons in phase, but none out of phase. This indicates that the tRNA-like structure is a noncoding region. As far as its resemblances to tRNAs, other than the ACC(A) end, the only significant sequence homologies reside in the anticodon stem and loop regions. The sequence GUGCA in the tRNA-like structure occurs in the TψCG loop of certain tRNAs.

BMV RNA

The sequence of 161 nucleotides pertaining from the 3' end of the four RNAs that make up the BMV genome (Figure 4) was reported by Dasgupta and Kaesberg (1977). This sequence has been highly preserved during evolution, since in the four RNAs of the virus, there are only one or two base substitutions in the fragment.

The only significant sequences found in this fragment and in all tRNA^{Tyr} examined, comprise 9 nucleotides in the CCA stem, and 11 nucleotides in the anticodon stem and loop of the tRNA. In the BMV

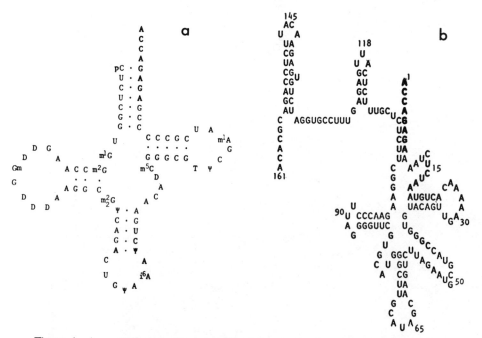

Figure 4 A possible secondary structure of the 161 nucleotides at the 3' end of RNAs 3 and 4 of BMV (*b*) and its comparison to tRNA^{Tyr} (*a*) of *Torulopsis*. Dark areas correspond to sequence homologies of more than 4 bases. (Reprinted, with permission, from Dasgupta and Kaesberg 1977.)

RNA fragment, however, these two sequences are separated by only 5 nucleotides.

Gould and Symons (1978) using complementary DNA probes of CMV RNA, a multicomponent virus whose RNA also accepts tyrosine, have demonstrated that here again, all four RNAs have a common stretch corresponding to about 200 nucleotides at their 3' ends.

TMV RNA

A fragment of 48 nucleotides from the 3' end of TMV RNA has been sequenced (Figure 5) by Guilley et al. (1975). Here again, sequence

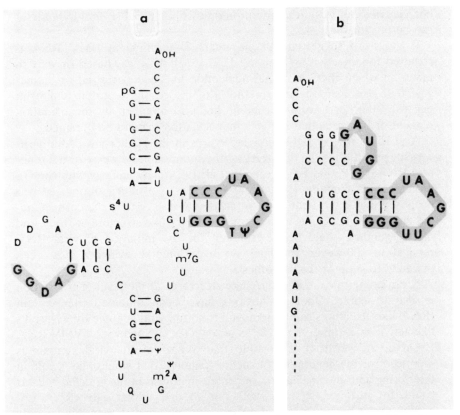

Figure 5 Sequence of the 48 nucleotides at the 3' end of TMV RNA (*b*) and its comparison to *E. coli* tRNA^{His} (*a*). Dark areas correspond to sequence homologies of more than 4 bases (from Lamy et al. 1975). Recent experiments have shown that the C4 tract in position 13–16 as well as in an extra C residue should be included in the C5 tract in position 21–25; moreover, the primary sequence of the next 22 nucleotides has now been established (H. Guilley, pers. comm.). (Reprinted, with permission, from Guilley et al. 1975.)

homologies with tRNA^{His} are found, but they also occupy very different positions in the two types of molecules. This is also the case of the 3' end of the RNA of the green tomato atypical mosaic virus strain, a member of the tobamovirus group acceptor of histidine (Lamy et al. 1975).

POSSIBLE PHYSIOLOGICAL ROLE OF tRNA-LIKE STRUCTURES

Very little is known about the physiological role of these structures in viral RNA genomes. The fact that they are efficiently recognized by tRNA-specific enzymes and are highly conserved, probably more so than the genes themselves, indicates that they must play an important physiological role. Up to the present time, only very few experiments have been carried out to gain information on this point, but several hypotheses can be put forward.

It has been suggested that an intact 3' terminal in BMV RNA is required for infectivity (Kohl and Hall 1977). This was based on experiments in which the 3'-terminal nucleotide had been oxidized or aminoacylated. However, the interpretation of the results reported is difficult because other parts of the genome are also modified by the oxidation reagent or because quantitative aminoacylations cannot be obtained.

A possible role of tRNA-like structures in the replication of the virus was suggested after the discovery of the interaction between viral aminoacyl-RNA, GTP, and EF-1 (Litvak et al. 1973b). By analogy with the Qβ replicase that contains EF-Tu and EF-Ts as two of its subunits, it was proposed that if the viral RNA replicase were to contain the corresponding eukaryotic factor, its affinity for the tRNA-like structure at the 3' end of the genome could be determinant for the initiation of viral RNA replication. However to date, no experimental evidence has been obtained to support this hypothesis.

A physiological role of tRNA-like structures at the level of translation is also probable. Good indications have been obtained that certain tRNA-like structures can be removed from the viral genome by RNase-P-like activities, as observed in *X. laevis* oocytes injected with TYMV RNA (Joshi et al. 1978). It is possible that a part of the viral genome is sacrificed for the generation of such a fragment that would play a role in interfering with the mechanism of protein synthesis in the infected cells. It could, for instance, compete with one of the cellular valyl-tRNA isoacceptors not needed for viral protein synthesis and thereby hinder only cellular peptide chain elongation.

A possible role of aminoacylation in virus morphogenesis cannot be excluded. However, the possibility of such a role is weak for the following reasons: (1) in the virion, the RNA is not aminoacylated, and (2) in some cases, e.g. in TYMV RNA, even the 3' terminal A is absent from the RNA.

The involvement of tRNA-like structures in some unpredictable regulation step must also be envisaged if one considers the diversity of functions of tRNAs in the economy of the cell. Indeed, several examples are known in which small modifications in the structure of a given tRNA lead to a new function of that molecule. This is the case, for example, of the staphylococcal $tRNA_1^{Gly}$ species involved in bacterial cell wall synthesis (Stewart et al. 1971).

POSSIBLE ROLE OF ANCESTRAL tRNAs IN VIRAL RNA GENOME FORMATION

Unquestionably, the elucidation of the significance of viral tRNA-like structures would lead to new approaches in the study of evolution of viruses. For the present time, only speculations are possible.

In early evolution, structures resembling modern tRNA molecules were probably of paramount importance in the formation of the first organisms and in their diversification.

Some of these primitive molecules evolved into tRNAs that are used either as fundamental elements for the translation of genes into proteins or for other very different functions. The use of certain specific tRNAs, such as the tRNAs specific for tryptophan or for proline as primers for the reverse transcriptase (see Dahlberg, this volume) and that of a specifically evolved $tRNA^{Gly}$ for cell-wall peptide synthesis (Stewart et al. 1971), illustrate just how important this diversification is. With our knowledge, nothing can be said concerning the reasons why these structures were chosen to perform so many disparate functions.

Other ancestral tRNA molecules have played an important role in the formation of various RNA genomes in which they are now found and where they probably operate as signals in the regulation of replication or of the expression of these genomes. The evolution of these ancestral molecules did not necessarily follow the same pathway as the one adopted by the molecules destined to become tRNAs. Since probably not all the structural elements present in modern tRNAs are necessary to perform these virus-specific regulatory functions, one can expect to find a great diversity in the resemblance existing between these regulatory structures and tRNAs. This diversity will depend on the necessary or permissive modifications of these structures for their efficiency as regulators and also on the size of the structure involved in the process of regulation. It is very likely that these structures evolved rather early, and since one finds no significant differences among them in multicomponent viruses, for instance, their conservation must be essential for the survival of the virus.

Comparative structural studies of a great number of tRNA-like structures in viral genomes and of the corresponding host tRNAs would probably enable us to deduce the structures of such tRNA ancestors.

ACKNOWLEDGMENTS

This work was financed, in part, by a grant from the Action Thématique Programmée: Phytopathologie (no. 3608) and, in part, by a grant from the Ecole Pratique des Hautes Etudes.

REFERENCES

Bastin, M. and T. C. Hall. 1976. Interaction of elongation factor 1 with aminoacylated brome mosaic virus and tRNAs. *J. Virol.* **20:** 117.

Beachy, R. N., M. Zaitlin, G. Bruening, and H. W. Israel. 1976. A genetic map for the cowpea strain of TMV. *Virology* **73:** 498.

Benicourt, C. and A. L. Haenni. 1974. Recognition of TMV RNA by the tRNA nucleotidyltransferase. *FEBS Lett.* **45:** 228.

Briand, J. P., G. Jonard, H. Guilley, K. Richards, and L. Hirth. 1977. Nucleotide sequence ($n = 159$) of the amino-acid accepting 3'-OH extremity of turnip yellow mosaic virus RNA and the last portion of its coat protein cistron. *Eur. J. Biochem.* **72:** 453.

Busto, P., E. Carriquiri, L. Tarrago-Litvak, M. Castroviejo, and S. Litvak. 1976. Interactions of plant viral RNAs and tRNA nucleotidyltransferase. *Ann. Microbiol.* **127A:** 39.

Chen, J. M. and T. C. Hall. 1973. Comparison of tyrosyl transfer ribonucleic acid and brome mosaic virus tyrosyl ribonucleic acid as amino acid donors in protein synthesis. *Biochemistry* **12:** 4570.

Cuzin, F., N. Kretchmer, R. E. Greenberg, R. Hurwitz, and F. Chapeville. 1967. Enzymatic hydrolysis of N-substituted aminoacyl-tRNA. *Proc. Natl. Acad. Sci.* **58:** 2079.

Dasgupta, R. and P. Kaesberg. 1977. Sequence of an oligonucleotide derived from the 3' end of each of the four brome mosaic viral RNAs. *Proc. Natl. Acad. Sci.* **74:** 4900.

Fiers, W., R. Contreras, F. Duerinck, G. Haegeman, D. Iserentant, D. J. Merregaert, W. Min Jou, F. Molemans, A. Raeymaekers, A. Van den Berghe, G. Volckaert, and M. Ysebaert. 1976. Complete nucleotide sequence of bacteriophage MS2 RNA: Primary and secondary structure of the replicase gene. *Nature* **260:** 500.

Génevaux, M., M. Pinck, and H. M. Duranton. 1976. Amino acid accepting structures in tymovirus RNAs. *Ann. Microbiol.* **127A:** 47.

Giégé, R., J. P. Briand, R. Mengual, J. P. Ebel, and L. Hirth. 1978. Valylation of the two RNA components of turnip yellow mosaic virus and specificity of the tRNA aminoacylation reaction. *Eur. J. Biochem.* **84:** 251.

Gould, A. R. and R. H. Symons. 1978. Alfalfa mosaic virus. Determination of the sequence homology between the four RNA species and a comparison with the four RNA species of cucumber mosaic virus. *Eur. J. Biochem.* **91:** 269.

Grunstein, M. and J. E. Grunstein. 1978. The histone H4 gene of *Strongylocentrotus purpuratus*: DNA and mRNA sequence at the 5' end. *Cold Spring Harbor Symp. Quant. Biol.* **42:** 1083.

Guilley, H., G. Jonard, and L. Hirth. 1975. Sequence of 71 nucleotides at the 3'-end of tobacco mosaic virus RNA. *Proc. Natl. Acad. Sci.* **72:** 864.

Haenni, A. L., A. Prochiantz, and P. Yot. 1974. tRNA structures in viral RNA genomes. In *Lipmann Symposium*: *Energy, regulation, and biosynthesis in molecular biology* (ed. D. Richter), p. 264. de Gruyter, Berlin.

Haenni, A. L., A. Prochiantz, O. Bernard, and F. Chapeville. 1973. TYMV valyl-RNA as an amino-acid donor in protein synthesis. *Nat. New Biol.* **241**:166.

Haenni, A. L., C. Benicourt, S. Teixeira, A. Prochiantz, and F. Chapeville. 1975. The specific role of tRNAs and tRNA-like structures in viral RNA genome replication and translation. *FEBS Proc. Meet.* **39**:121.

Hall, T. C., D. S. Shih, and P. Kaesberg. 1972. Enzyme-mediated binding of tyrosine to brome mosaic virus ribonucleic acid. *Biochem. J.* **129**:969.

Huez, G., G. Marbaix, E. Hubert, M. Leclercq, U. Nudel, H. Sorecq, R. Salomon, B. Lebleu, M. Revel, and U. Z. Littauer. 1974. Role of the polyadenylated segment in the translation of globin messenger RNA in *Xenopus* oocytes. *Proc. Natl. Acad. Sci.* **71**:3143.

Joshi, S., A. L. Haenni, E. Hubert, G. Huez, and G. Marbaix. 1978. In vivo aminoacylation and "processing" of turnip yellow mosaic virus RNA in *X. laevis* oocytes. *Nature* **275**:339.

Kamen, K. 1969. Infectivity of bacteriophage R17 RNA and sequential removal of 3′ terminal nucleotides. *Nature* **221**:321.

Kohl, R. J. and T. C. Hall. 1974. Aminoacylation of RNA from several viruses: Amino acid specificity and differential activity of plant, yeast, and bacterial synthetases. *J. Gen. Virol.* **25**:257.

———. 1977. Loss of infectivity of brome mosaic virus RNA after chemical modification of the 3′- or 5′-terminus. *Proc. Natl. Acad. Sci.* **74**:2682.

Kössel, H. and U. L. RajBhandary. 1968. Studies on polynucleotides. LXXXVI. Enzymatic hydrolysis of N-acylaminoacyl-transfer RNA. *J. Mol. Biol.* **35**:539.

Lamy, D., G. Jonard, H. Guilley, and L. Hirth. 1975. Comparison between the 3′ OH end RNA sequence of two strains of tobacco mosaic virus (TMV) which may be aminoacylated. *FEBS Lett.* **60**:202.

Lesiewicz, J. and B. Dudock. 1978. In vitro methylation of tobacco mosaic virus RNA with ribothymidine-forming tRNA methyltransferase of *Escherichia coli*. Characterization and specificity of the reaction. *Biochim. Biophys. Acta* **520**:411.

Lindley, I. J. D. and N. Stebbing. 1977. Aminoacylation of encephalomyocarditis virus RNA. *J. Gen. Virol.* **34**:177.

Litvak, S., D. Carré, and F. Chapeville. 1970. TYMV RNA as a substrate of the tRNA nucleotidyltransferase. *FEBS Lett.* **11**:316.

Litvak, S., L. Tarrago-Litvak, and F. Chapeville. 1973a. Turnip yellow mosaic virus RNA as a substrate of the transfer RNA nucleotidyl transferase. II. Incorporation of a cytidine 5′-monophosphate and determination of a short nucleotide sequence at the 3′ end of the RNA. *J. Virol.* **11**:238.

Litvak, S., A. Tarrago, L. Tarrago-Litvak, and J. E. Allende. 1973b. Elongation factor-viral genome interaction dependent on the aminoacylation of TYMV and TMV RNAs. *Nat. New Biol.* **241**:88.

Marcu, K. and B. Dudock. 1975. Methylation of TMV RNA. *Biochem. Biophys. Res. Commun.* **62**:798.

Öberg, B. and L. Philipson. 1972. Binding of histidine to tobacco mosaic virus RNA. *Biochem. Biophys. Res. Commun.* **48**:927.

Pinck, L., M. Genevaux, J. P. Bouley and M. Pinck. 1975. Amino acid acceptor activity of replicative form from some tymovirus RNAs. *Virology* **63**:589.

Pinck, M., P. Yot, F. Chapeville, and H. Duranton. 1970. Enzymatic binding of valine to the 3'-end of TYMV RNA. *Nature* **226**:954.

Pinck, M., S. K. Chan, M. Genevaux, L. Hirth, and H. Duranton. 1972. Valine specific tRNA-like structure in RNAs of two viruses of turnip yellow mosaic virus group. *Biochimie* **54**:1093.

Prochiantz, A. and A. L. Haenni. 1973. TYMV RNA as a substrate of the tRNA maturation endonuclease. *Nat. New Biol.* **241**:168.

Proudfoot, N. J. 1977. Complete 3' noncoding region sequence of rabbit and human β-globin messenger RNAs. *Cell* **10**:559.

Rensing, U. and J. T. August. 1969. The 3'-terminus and replication of phage RNA. *Nature* **224**:853.

Robertson, H. D., S. Altman, and J. D. Smith. 1972. Purification and properties of a specific *Escherichia coli* ribonuclease which cleaves a tyrosine transfer ribonucleic acid precursor. *J. Biol. Chem.* **247**:5243.

Salomon, R. and U. Z. Littauer. 1974. Enzymatic acylation of histidine to mengovirus RNA. *Nature* **249**:32.

Steinschneider, A. and H. Fraenkel-Conrat. 1966. Studies of nucleotide sequences in tobacco mosaic virus ribonucleic acid. IV. Use of aniline in stepwise degradation. *Biochemistry* **5**:2735.

Silberklang, M., A. Prochiantz, A. L. Haenni, and U. L. RajBhandary. 1977. Studies on the sequence of the 3'-terminal region of turnip yellow mosaic virus RNA. *Eur. J. Biochem.* **72**:465.

Stewart, T. S., R. J. Roberts, and J. L. Strominger. 1971. Novel species of tRNA. *Nature* **230**:36.

Wildy, P. 1971. Classification and nomenclature of viruses. *Monogr. Virol.* **5**:1.

Yot, P. 1970. "Esterification enzymatique du RNA du virus de la mosaïque jaune du navet." Ph.D. thesis, University of Paris.

Yot, P., M. Pinck, A. L. Haenni, H. M. Duranton, and F. Chapeville. 1970. Valine specific tRNA-like structure in turnip yellow mosaic virus RNA. *Proc. Natl. Acad. Sci.* **67**:1345.

Appendices

APPENDIX I
Localization of tRNA Genes on *Drosophila melanogaster* Polytene Salivary Gland Chromosome

Eric Kubli
Zoologisches Institut der Universität Zürich
CH-8057 Zürich, Switzerland

The natures of the tRNA genes at the arrows without amino acid symbols have not been identified. The bars delineate *Minute* deficiencies (Zimm and Lindsley. 1976. In *Handbook of biochemistry and molecular biology, 3rd edition* [ed. G. D. Fasman], vol. II, p. 848. CRC Press, Cleveland, Ohio). Hybridization data are taken from Elder et al., Kubli et al., and Tener et al. (all this volume). (Reprinted, with permission, from Kubli, E. 1980. *Adv. Genet.* [in press].)

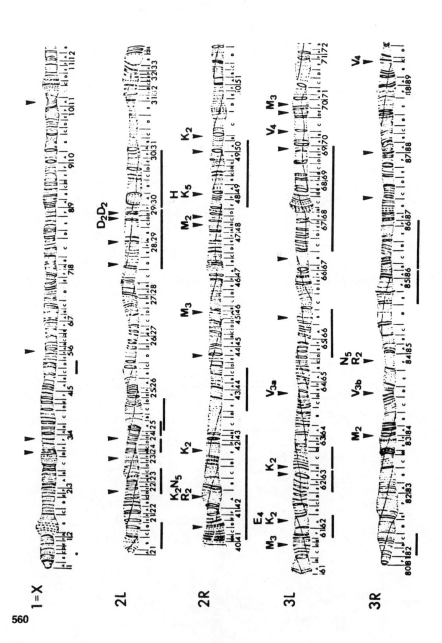

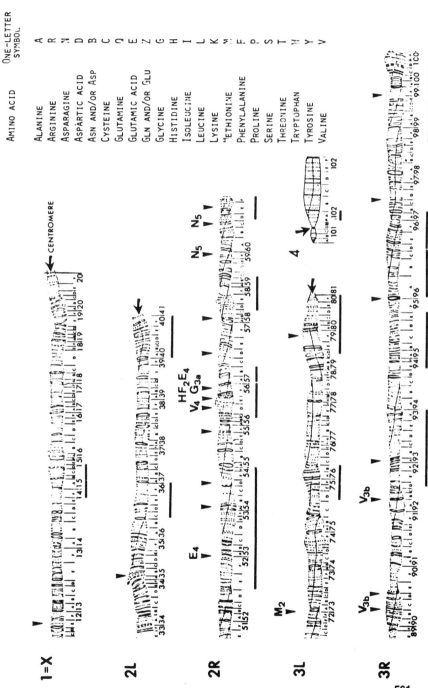

APPENDIX II
Known Locations of tRNA Genes on the *Escherichia coli* Map

Alice Y. P. Cheung
Department of Biophysics and Biochemistry
Yale University
New Haven, Connecticut 06520

The various isoacceptors are indicated, followed by the number of gene copies written in brackets or parentheses. Brackets indicate that this number was determined by nucleotide sequencing or other physical means; parentheses indicate the use of genetic techniques for this purpose. The letters C, E, K, M, P, X, and Y are unidentified genes for tRNA. The numbers inside the circle correspond to the minutes on the *Escherichia coli* map (Bachmann, B. J., K. B. Low, and A. L. Taylor. 1976. *Bacteriol. Rev.* **40:**116). Data are compiled from the following: Hill, C. W., C. Squires, and J. Carbon. 1970. *J. Mol. Biol.* **52:**557; Squires, C. and J. Carbon. 1971. *Nat. New Biol.* **233:**274; Smith, J. D. 1972. *Ann. Rev. Genet.* **6:**235; Steege, D. A. and B. Low. 1975. *J. Bacteriol.* **122:**120; Hill, C. W. 1976. *Cell* **6:**419; Lund, E., J. E. Dahlberg, L. Lindahl, S. R. Jaskunas, R. P. Dennis, and M. Nomura. 1976. *Cell* **7:**165; Ikemura, T. and H. Ozeki. 1977. *J. Mol. Biol.* **117:**419; Morgan, E. A., T. Ikemura, and M. Nomura. 1977. *Proc. Natl. Acad. Sci.* **74:**2710; Morgan E. A., T. Ikemura, L. Lindahl, A. M. Fallon, and M. Nomura. 1978. *Cell* **13:**335; Rossi, J. J. and A. Landy. 1979. *Cell* **16:**523; Yarus, M., S. P. Eisenberg, and L. Soll. 1979. *Mol. Gen. Genet.* **170:**299; Young, R., R. Macklis, and J. Steitz. 1979. *J. Biol. Chem.* **254:**3264; Morgan et al., this volume; Ozeki et al., this volume.

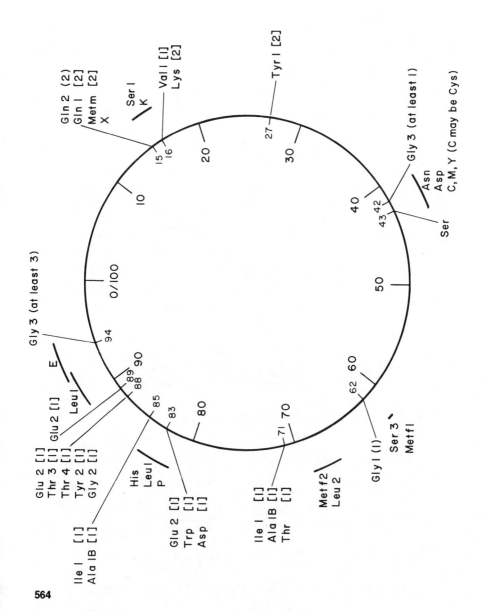

APPENDIX III
Codon Usage
in Several Organisms

Henri Grosjean
Laboratoire de Chimie Biologiqué
Université Libre de Bruxelles
B-1640, Rhode-St. Genese, Belgium

The genetic code is presented in the conventional manner with its corresponding amino acid assignment. The numbers correspond to the frequency with which a triplet is used in the coding regions of different genes or in the mRNA. The initiator (second number in the case of AUG and GUG) and terminator codon signals are also indicated. Further comments and analysis on codon usage are found in research by Fiers and Grosjean (1979. *Nature* **277**: 328), Grosjean et al. (1978. *J. Mol. Evol.* **12:** 113), Grantham (1978. *FEBS Lett.* **95:** 1), and Chavancy et al. (1979. *Biochimie* **61:** 71). The sources of original data on codon usage in the different mRNA or genes are listed at the end of the Appendix, in the same order of appearance as for genes or mRNA in the notes of each table.

Table 1 Codon usage in *Escherichia coli*

	U/T		C		A		G	
U/T	Phe	11	Ser	20	Tyr	6	Cys	2	U/T
U/T	Phe	15	Ser	19	Tyr	11	Cys	4	C
U/T	Leu	6	Ser	5	ocher	5	opal	2	A
U/T	Leu	9	Ser	6	amber	0	Trp	3	G
C	Leu	4	Pro	0	His	3	Arg	21	U/T
C	Leu	5	Pro	8	His	8	Arg	29	C
C	Leu	3	Pro	5	Gln	22	Arg	4	A
C	Leu	79	Pro	30	Gln	34	Arg	2	G
A	Ile	27	Thr	23	Asn	10	Ser	6	U/T
A	Ile	34	Thr	29	Asn	27	Ser	13	C
A	Ile	1	Thr	4	Lys	75	Arg	2	A
A	Met	32+12	Thr	9	Lys	12	Arg	0	G
G	Val	53	Ala	63	Asp	24	Gly	34	U/T
G	Val	17	Ala	25	Asp	32	Gly	37	C
G	Val	33	Ala	41	Glu	59	Gly	1	A
G	Val	21+1	Ala	41	Glu	15	Gly	4	G

Codon usage in *E. coli* in *lacI* gene (360 codons assigned); in ribosomal protein genes *L11* (142), *L1* (234), *L10* (131), and *L7/L12* (124); in fragments of β gene (33), *S12* gene (21), *L14* gene (20), and *S17* gene (25); in fragments of mRNA recognized by *E. coli* ribosomes for *trp* leader (8), *trpBtrpA* (68), *trpF* (8), *araB* (5), *galE* (6), *galT* (6), and *lacZ* (7). The total number of codons assigned was 1198.

Table 2 Codon usage in RNA coliphages

		U		C		A		G	
U	Phe	26	Ser	26	Tyr	11	Cys	9	U
U	Phe	35	Ser	24	Tyr	38	Cys	6	C
U	Leu	24	Ser	18	ocher	4	opal	0	A
U	Leu	15	Ser	30	amber	2	Trp	28	G
C	Leu	23	Pro	21	His	9	Arg	29	U
C	Leu	31	Pro	11	His	9	Arg	28	C
C	Leu	22	Pro	13	Gln	25	Arg	15	A
C	Leu	14	Pro	18	Gln	33	Arg	11	G
A	Ile	19	Thr	29	Asn	24	Ser	10	U
A	Ile	32	Thr	29	Asn	38	Ser	20	C
A	Ile	19	Thr	21	Lys	26	Arg	10	A
A	Met	19 + 10	Thr	18	Lys	31	Arg	7	G
G	Val	25	Ala	33	Asp	35	Gly	46	U
G	Val	23	Ala	25	Asp	24	Gly	18	C
G	Val	22	Ala	32	Glu	21	Gly	13	A
G	Val	23 + 1	Ala	26	Glu	32	Gly	16	G

Codon usage in RNA coliphages in phage MS2 for coat protein (130 codons assigned), A protein (394), replicase (545), and lysis protein (75); in fragments of phage Qβ for coat protein (6), A2 protein (89), and replicase (22); in fragments of phage R17 for coat protein (75), A protein (9), and replicase (6); and in a fragment of phage f2 for coat protein (34). The total number of codons assigned was 1385.

Table 3 Codon usage in DNA coliphages

		U/T		C		A		G	
U/T	Phe	189	Ser	277	Tyr	187	Cys	42	U/T
U/T	Phe	154	Ser	87	Tyr	76	Cys	37	C
U/T	Leu	151	Ser	107	ocher	17	opal	22	A
U/T	Leu	111	Ser	61	amber	1	Trp	93	G
C	Leu	197	Pro	145	His	79	Arg	141	U/T
C	Leu	95	Pro	42	His	41	Arg	118	C
C	Leu	37	Pro	54	Gln	165	Arg	39	A
C	Leu	141	Pro	69	Gln	158	Arg	21	G
A	Ile	220	Thr	192	Asn	236	Ser	50	U/T
A	Ile	115	Thr	118	Asn	162	Ser	39	C
A	Ile	44	Thr	88	Lys	321	Arg	44	A
A	Met	179 + 42	Thr	68	Lys	163	Arg	17	G
G	Val	265	Ala	310	Asp	220	Gly	249	U/T
G	Val	85	Ala	107	Asp	195	Gly	157	C
G	Val	75	Ala	108	Glu	176	Gly	60	A
G	Val	68 + 1	Ala	81	Glu	158	Gly	28	G

Codon usage in DNA coliphages in phage φX174 for genes A (513 codons assigned), A* (341), B (120), K (56), C (84), D (152), E (91), J (38), F (421), G (175), and H (328); in phage G4 for genes A (554), A* (341), B (120), K (56), C (86), D (152), E (96), J (25), F (427), G (177), and H (337); in phage fd for genes II (404), V (88), VII (33), VIII (75), III (424), VI (112), I (350), and IV (426); (fragments of G3 mRNA coat protein correspond to gene VIII of phage fd); in phage M13 for genes VII (33) and IX (33) and for fragments of genes VIII (32) and V (14); in phage λ for genes cro (66), cII (97), the replicative origin (299), and the cI repressor (237); in phage 434 for genes cro (72) and cII (98); in phage T3 at the initiation sequences recognized by *Escherichia coli* ribosomes (6); and in phage T7 for gene 0.3 at site a (6). The total number of codons assigned was 7595.

Table 4 Codon usage in mammalian oncogenic DNA virus

	T		C		A		G	
T	Phe	82	Ser	34	Tyr	38	Cys	18	T
T	Phe	8	Ser	14	Tyr	30	Cys	32	C
T	Leu	51	Ser	25	ocher	6	opal	1	A
T	Leu	44	Ser	2	amber	11	Trp	39	G
C	Leu	30	Pro	52	His	26	Arg	3	T
C	Leu	6	Pro	21	His	15	Arg	6	C
C	Leu	24	Pro	33	Gln	52	Arg	3	A
C	Leu	33	Pro	1	Gln	40	Arg	5	G
A	Ile	55	Thr	49	Asn	52	Ser	44	T
A	Ile	0	Thr	31	Asn	28	Ser	17	C
A	Ile	25	Thr	34	Lys	101	Arg	51	A
A	Met	50+7	Thr	1	Lys	48	Arg	32	G
G	Val	50	Ala	88	Asp	76	Gly	24	T
G	Val	4	Ala	20	Asp	50	Gly	19	C
G	Val	32	Ala	25	Glu	83	Gly	43	A
G	Val	37	Ala	0	Glu	53	Gly	33	G

Codon usage in mammalian oncogenic DNA virus in SV40 for genes VP1 (361 codons assigned), VP2 (352), small T antigen (173), VP2 leader X (62), and large T gene (626); in BK virus for gene of small T antigen (172); in BK virus (MM species) for fragments of the genes for small T (30), leader X (66), VP2 (25), and T2 (61); and in polyoma virus for the gene corresponding to small T antigen (124). The total number of codons assigned was 2047.

Table 5 Codon usage in genes or mRNA from eukaryotes (vertebrates)

	U/T		C		A		G	
U/T	Phe	46	Ser	48	Tyr	30	Cys	20	U/T
U/T	Phe	80	Ser	53	Tyr	62	Cys	20	C
U/T	Leu	9	Ser	35	ocher	7	opal	5	A
U/T	Leu	22	Ser	13	amber	7	Trp	24	G
C	Leu	34	Pro	44	His	34	Arg	39	U/T
C	Leu	84	Pro	45	His	55	Arg	54	C
C	Leu	24	Pro	35	Gln	35	Arg	17	A
C	Leu	145	Pro	14	Gln	87	Arg	17	G
A	Ile	35	Thr	53	Asn	34	Ser	35	U/T
A	Ile	94	Thr	85	Asn	78	Ser	52	C
A	Ile	14	Thr	44	Lys	58	Arg	25	A
A	Met	56+19	Thr	17	Lys	185	Arg	33	G
G	Val	40	Ala	92	Asp	50	Gly	59	U/T
G	Val	52	Ala	136	Asp	79	Gly	95	C
G	Val	15	Ala	56	Glu	81	Gly	47	A
G	Val	99	Ala	20	Glu	126	Gly	32	G

Codon usage in genes or mRNA from eukaryotes (vertebrates) in nine genes or fragments of genes for histones in sea urchin (370 codons assigned for H3, H2A, and H2B, in *Strongylocentrotus purpuratus* and 539 codons assigned for H2B, H2A, H3, H4, and H1 in *Psammechinus miliaris*); in five globin mRNAs for rabbit hemoglobin α chain (141) and β chain (146), human β chain (142), mouse β chain (147), and partial sequence from human γ chain (31); in ovalbumin mRNA (387); in mRNAs for the human chorionic somatomammotropin hormone (167) and adrenocorticotrophic hormone β-LPH precursor protein (257); in genes of rat pregrowth hormone (216), rat preproinsulin I (100), and rat preprolactin (160); and in three immunoglobulin genes from mouse embryonic $V_{\lambda II}$ (160) and $V_{\lambda 1}$ (173) light chains and MOPC-21 light constant chain (108). The total number of codons assigned was 3240.

Table 6 Codon usage in viral plant RNA

	U			C			A			G		
U	Phe	4	Ser	10	Tyr	2	Cys	1	U
U	Phe	9	Ser	5	Tyr	5	Cys	4	C
U	Leu	7	Ser	7	ocher	1	opal	1	A
U	Leu	4	Ser	4	amber	0	Trp	5	G
C	Leu	3	Pro	5	His	2	Arg	2	U
C	Leu	11	Pro	11	His	1	Arg	1	C
C	Leu	4	Pro	8	Gln	11	Arg	1	A
C	Leu	1	Pro	4	Gln	6	Arg	2	G
A	Ile	3	Thr	16	Asn	10	Ser	3	U
A	Ile	14	Thr	17	Asn	4	Ser	4	C
A	Ile	8	Thr	3	Lys	6	Arg	5	A
A	Met	3+2	Thr	6	Lys	3	Arg	3	G
G	Val	7	Ala	7	Asp	3	Gly	5	U
G	Val	11	Ala	12	Asp	12	Gly	4	C
G	Val	5	Ala	4	Glu	9	Gly	5	A
G	Val	5	Ala	6	Glu	4	Gly	1	G

Codon usage in viral plant RNA in coat protein gene of tobacco mosaic virus (166 codons assigned) and of turnip yellow mosaic virus (186 codons assigned). The total number of codons assigned was 352.

Sources

Table 1
Farabaugh, P. 1978. *Nature* **274**:765; Post, L. E. et al. 1979. *Proc. Natl. Acad. Sci.* **76**:1697; Post, L. E. et al. 1978. *Cell* **15**:215; Platt, T. and C. Yanofsky. 1975. *Proc. Natl. Acad. Sci.* **72**:2399; Steitz, J. A. 1977. In *Biological regulation and developments* (ed. R. F. Goldberger), p. 349. Plenum Press, New York.

Table 2
Fiers, W. et al. 1976. *Nature* **260**:500 (and references therein); Atkins, J. F. (pers. comm.); Barrell, B. G. and B. F. C. Clark. 1974. *Handbook of nucleic acid sequences.* Joynson-Bruvvers, Oxford.

Table 3
Sanger, F. et al. 1977. *Nature* **265**:687; Godson, G. et al. 1978. *Nature* **276**:236; Shaw, D. C. et al. 1978. *Nature* **272**:510; Sugimoto, K. et al. 1978. *Nucleic Acids Res.* **5**:4495; Hulsebos, T. and J. G. Schoenmakers. 1978. *Nucleic Acids Res.* **5**:4677; Schwaz, E. et al. 1978. *Nature* **272**:410; Scherer, G. 1978. *Nucleic Acids Res.* **9**:3141; Sauer, R. 1978. *Nature* **276**:301; Grosschedl, R. and E. Schwarz. 1979. *Nucleic Acids Res.* **6**:867; Steitz, J. A. 1977. In *Biological regulation and developments* (ed. R. F. Goldberger), p. 349. Plenum Press, New York.

Table 4
Fiers, W. et al. 1978. *Nature* **273**:113; Reddy, V. B. et al. 1978. *Science* **200**:494; Dhar, R., I. Seif, and G. Khoury. 1979. *Proc. Natl. Acad. Sci.* **76**:565; Yang, R. C. and

R. Wu. 1979. *Proc. Natl. Acad. Sci.* **76:**1179; Soeda, E. and B. E. Griffin. 1978. *Nature* **276:**296.

Table 5

Sures, I., J. Lowry, and L. H. Kedes. 1978. *Cell* **15:**1033; Schaffner, W. et al. 1978. *Cell* **14:**655; Heindell, P. H. et al. 1978. *Cell* **15:**43; Efstratiadis, A., F. C. Kafatos, and T. Maniatis. 1977. *Cell* **10:**571; Marotta, C. A. et al. 1977. *J. Biol. Chem.* **252:**5040; Konkel, D. A., S. M. Tilghman, and P. Leder. 1978. *Cell* **15:**1125; Poon, R. et al. 1978. *Nature* **273:**723; Shine, J. et al. 1977. *Nature* **270:**494; Nakanishi, S. et al. 1979. *Nature* **278:**423; Seeburg, P. H. et al. 1977. *Nature* **270:**486; Ullrich, A. et al. 1977. *Science* **196:**1313; Gubbins, E. J. et al. 1979. *Nucleic Acids Res.* **6:**915; Tonegawa, S. et al. 1978. *Proc. Natl. Acad. Sci.* **75:**1485; Bernard, O., N. Hozumi, and S. Tonegawa. 1978. *Cell* **15:**1133; Hamlyn, P. H. et al. 1978. *Cell* **15:**1067.

Table 6

Guilley, H. et al. 1979. *Nucleic Acids Res.* **6:**1287; Guilley, H. and Briand, J. 1978. *Cell* **15:**113.

Author Index

Abelson, J.N., 173, 211
Altman, S., 71
Altruda, F., 407
Apirion, D., 139
Atchison, R., 83
Atkins, J.F., 439

Beckmann, J.S., 173
Belagaje, R., 245
Benditt, J., 267
Berman, M.L., 221
Bock, R.M., 395
Bohnert, H.-J., 281
Bowman, E.J., 71
Bram, R.J., 99
Brown, E.L., 245
Buckingham, R.H., 421
Burkard, G., 281

Chapeville, F., 539
Cheung, A.Y.P., 563
Cordell, B., 191
Cortese, R., 287
Crouse, E.J., 281

Dahlberg, J.E., 123, 507
Daniel, V., 29
DeFranco, D., 325
Delaney, A., 295
Deutscher, M.P., 59
Driesel, A.J., 281
Dunn, R., 295

Egan, J., 221
Egg, A.H., 309
Eisenberg, S.P., 469
Elder, R.T., 317

Fritz, H-J., 245
Fuhrman, S.A., 173

Garber, R.L., 71
Gegenheimer, P., 139
Ghora, B.K., 139
Ghosh, R.K., 59
Gillam, I.C., 295
Goldfarb, A., 29
Goldman, E., 427
Goodman, H.M., 191
Gordon, K., 281
Gorman, J., 395
Grigliatti, T.A., 295
Grosjean, H., 565
Guthrie, C., 83, 123

Haenni, A-L., 539
Hall, B.D., 267
Hatfield, G.W., 427
Herrmann, R.G., 281
Hayashi, S., 295
Hovemann, B., 325
Hwang, D.L.R., 517

Ikemura, T., 259, 341
Inokuchi, H., 341

Johnson, P.F., 173

Kang, H.S., 173
Kaufman, T.C., 295
Keller, M., 281
Khorana, H.G., 245
Knapp, G., 173
Kodaira, M., 341
Kohli, J., 407

Kole, R., 71
Koski, R.A., 71
Kubli, E., 309, 559
Kubokawa, S., 43
Kurjan, J., 267
Kurland, C.G., 421
Kwong, T., 407

Landy, A., 221
Laten, H., 395
Leupold, U., 407
Loughney, K., 267
Lund, E., 123

Mao, J., 325
Maynard, T., 191
Mazzara, G.P., 3
McClain, W.H., 3, 107
Melton, D., 287
Meyhack, B., 155
Misra, T.K., 139
Morgan, E.A., 259
Mubumbila, M., 281

Nagawa, F., 43
Nomura, M., 259

Oeschger, M.P., 363
O'Farrell, P.Z., 191
Ogden, R.C., 173
Olson, M.V., 267
Oxender, D.L., 481
Ozeki, H., 43, 341

Pace, B., 155
Pace, N.R., 155
Page, G.S., 267

571

Author Index

Peebles, C.L., 173
Pelle, E.G., 107
Piper, P.W., 379
Plautz, G., 139
Post, L.E., 259

Quay, S.C., 481

Rafalski, A., 407
Robertson, H.D., 107
Rossi, J., 221
Rutter, W.J., 191
Ryan, M.J., 245

Sakano, H., 43, 341
Schmidt, O., 325

Schmidt, T., 309
Sentenac, A., 267
Shimura, Y., 43, 341
Silverman, S., 325
Smith, J.D., 287
Soffer, R.L., 493
Sogin, M.L., 155
Söll, D., 325, 407
Soll, L., 469
Stark, B.C., 71
Steinmetz, A., 281
Steitz, J.A., 99
Suzuki, D.T., 295
Szabo, P., 317

Tener, G.M., 295
Tranquilla, T., 287

Uhlenbeck, O.C., 317
Umbarger, H.E., 453

Valenzuela, P., 191

Wahl, G., 407
Weil, J.H., 281
Wetzel, R., 407

Yamada, H., 325
Yamao, F., 341
Yang, W.K., 517
Yarus, M., 469
Young, R.A., 99

Zeevi, M., 29

Subject Index

ABL 1, 108–120
A49, 37, 46, 108–120
Allosuppressors, 410–411
Amber mutations, 366–369, 373–374
  in DNA metabolism, 367
  in membrane proteins, 367–368
  in ribosomes, 369
  in RNA polymerase, 368–369
Amber suppressors. *See* Suppressor tRNAs
Amino acid biosynthesis
  role of aminoacyl-tRNA in, 453–463
  Lee and Yanofsky model, 458–459
  regulation of, 453–463
  tRNA-mediated control of, 457–459
Amino acid transport, 461, 481–489
Aminoacylation
  animal viral genome, 544–545
  plant viral genome, 540–544
Aminoacyl-tRNA, regulation of amino acid biosynthesis, role in, 453–463
Aminoacyl-tRNA-protein transferases, 493–503
  arginyl-tRNA-protein transferases, 494–497
  leucyl-tRNA-protein transferases, 497–502
  transferase-less mutants, 502
Aminoacyl-tRNA synthetases
  derepression of, 461–463
  genes encoding
    location on *E. coli* chromosome, 358
    regulation of, 461–463
  metabolic regulation of, 463
  mutations in Chinese hamster ovary cells of, 488
  role in regulation of amino acid transport, 481–485

Antisuppressors
  in *S. cerevisiae*, 381, 395–404
  in *S. pombe*, 410–411
Arginyl-tRNA-protein transferases, 494–497
Attenuation, 469–477
  effect of mutations on, 472–476
  su^+7, 475–476
  temperature sensitive tRNA^{Trp}, 474–475
  model for, 458–461, 470–472
Attenuators, 458–461, 470–472
  hairpin loops in, 470–471
  leader region of, 470

Bacillus subtilis
  biosynthesis of 5S rRNA of, 155–170
  5S rRNA genes of, 155–156
Bacteriophage ϕ80, suppressor derivatives of, 7, 29–39
Bacteriophage T4
  biosynthesis of tRNAs of, 3–6, 8–11, 17–19, 33–39, 213–214
  precursor tRNA structures in, 5–11, 19, 84–95, 213–215
  suppressor tRNAs of, 10, 33–38
  tRNA genes of , 5, 6, 33–39, 213–215
BN exonuclease
  precursor tRNA processing by, 10, 19
  sequence specificity of, 20
Bombyx mori
  biosynthesis of tRNAs of, 11, 18
  RNase P-like activity in, 80
  tRNA genes of, 11

Caenorhabditis elegans. *See* Nematode

573

CCA enzyme. *See* tRNA nucleotidyl transferase
cca mutants, 21–22
CCA 3′-terminal sequence, 59–67
  biosynthesis of, 12–13, 19–22, 55
  evolutionary implications of, 21–22
  synthetic addition of, 60–61
Chemical synthesis of tRNA^{Tyr} gene, 245–257
Chloroplast tRNA genes, from *S. oleracea*, 281–285
  fractionation of, 282–283
  genetic map of, 283–284
Cloning
  D. melanogaster tRNA genes, 326–327
  modified synthetic tRNA^{Tyr} gene, 252–254
  natural suppressor tRNA^{Tyr} gene, 254
  nematode tRNA genes, 288
  synthetic suppressor tRNA^{Tyr} gene, 247–250
Codon-anticodon interactions, 421
Codon recognition
  third-code-letter discrimination in, 435
  by tRNA isoaccepting species, 427–436
  by tRNA^{Leu} isoaccepting species, 427–436
  two-out-of-three mechanism of, 435
Codon specificities, 421
Codon usage frequencies, 565–568
Cytokinin activity, in i^{6}A-deficient tRNA, 402

Derepression
  of amino acid transport, 482–483
  of aminoacyl-tRNA synthetases, 461–463
dimeric tRNA precursors, 6–10, 51–53, 84–93, 103–104
Distal tRNA genes. *See also* Risosomal RNA
  arrangement, 263
  function, 264
Drosophila melanogaster
  polytene chromosomes of, 295, 301–302
  precursor tRNA of, 323–334
  tRNA genes of, 295–305, 309–313, 317–323, 325–336
    arrangment of, 326–328
    cloning of, 326–327
    localization of, 299–305, 309–313, 559
    number of, 215, 295–299, 320
    sequence analysis of, 328–333
    transcription of, 325–326, 332–336
  tRNA nucleotide sequence of, 331

Endoribonucleases of RNA processing
  RNase E, 139–152
  RNase F, 139, 143–152
  RNase M5, 146, 148–150, 155–170
  RNase M16, 102–103, 146–150
  RNase M23, 147–150
  RNase O, 8, 16–17, 33, 52–54, 109
  RNase P, 14–18, 37, 43–55, 71–80, 83–95, 107–112, 126–133, 139–152, 212, 542–547
  RNase P2, 8, 16–17, 33, 39, 109, 112, 120, 150
  RNase P4, 18
  RNase III, 17–18, 99–104, 107–115, 123–134, 139–152, 262–264
  splicing activity, 173–189, 191–206
Escherichia coli. *See also* individual *tRNAs*; Distal tRNA genes; Spacer tRNA genes; tRNA gene clusters
  5S rRNA nucleotide sequence of, 168
  precursor tRNA structures of, 7, 47–52
  rRNA gene clusters of, 4–5, 99–105, 123, 139–152
  tRNA genes of, 212, 563
Excision-ligase activity. *See* Splicing activity
Exoribonucleases of RNA processing
  BN exonuclease, 19–20
  RNase D, 19–20, 61–67, 146
  RNase P3, 18, 53–54, 60, 67, 146
  RNase Q, 18, 53–55, 60, 67
  RNase Y, 60, 67
  RNase II, 18, 53–54, 59–67

5S rRNA precursor, 158–170
  nuclease recognition elements in, 158–170
  nucleotide sequence of, 156
  synthetic, 160–170
Frameshift suppressors, 439–443
Frog oocyte injection, 216, 287–291

his T, 442, 455, 457, 485–488

In situ hybridization, 216, 295, 299–305, 309–313, 317–323, 559
Intervening sequences, 12–13, 173–189, 191–206, 276–277
In vitro suppression, 411–412
In vitro transcription. *See* Transcription
Iodinated tRNA, 296–302, 310–313, 317–323
i^{6}A, 380, 395–404, 410
i^{6}A-deficient tRNAs, 400–403

Lee and Yanofsky model, 458–461, 470–472
Leucyl-tRNA^{Leu}, 427–436
Luecyl-tRNA-protein transferases, 497–502

Minute loci in *D. melanogaster,* 300, 312, 323
Mischarging tRNAs, 350–354
Mitochondrial tRNA genes, 217
Modification enzymes, 22–23
Modified nucleosides, 22–23, 54, 194–196, 333
  function of, 23, 54

Subject Index   575

i⁶A, 380, 395–404
ms²i⁶A, 404, 472
ms²i⁶A, 404, 472
M2 RNA, 75
Mutations in tRNA genes. *See also individual organisms and genes*
  mischarging, 350–354
  temperature sensitive, 350–354

Nematode *C. elegans* tRNA genes
  cloning of, 288
  modification of, 291
  transcription of, 289–291
Nonsense suppressor tRNA. *See also individual suppressor tRNAs*
  in *E. coli*, 339–359
  in *S. cerevisiae*, 379–392
  in *S. pombe*, 407–417
Nontriplet tRNA-mRNA interactions, 439–447
  overlapping reading in, 445
  quadruplet base paring in, 441
  rel A mistranslation and, 443
  ribosomal mutants and, 443

Ocher suppressors. *See* Suppressor tRNAs
Opal suppressors. *See* Suppressor tRNAs

Peptide mapping, 430
Plant viruses, 539–553
Polycistronic tRNA precursors, 30–39, 49–50. *See also* Dimeric tRNA precursors
Polytene chromosomes, 295, 317–323, 559
Precursor rRNAs
  5S
    nucleotide sequence of, 156
    processing of, 155–170
    synthetic, 160–170
  30S
    mutants accumulating, 107–121, 123–135, 139–152
    processing of, 103–105, 123–135, 139–152
Precursor tRNAs
  aminoacylation of, 13, 198–200
  dimeric forms, 6–10, 51–53, 84–93, 103–104
  intervening-sequence-containing
    aminoacylation of, 198–200
    consensus sequence of, 179
    mutants accumulating. *See ts136*
    secondary structure of, 178–179, 196–198
    splicing of, 173–189, 200–202
    tertiary structure of, 187–189, 196–198
  monomeric forms of, 8, 48–49
  mutants accumulating, 7–8, 13, 37, 107–121, 139–152, 173–177, 192–196. *See also individual mutants*

  mutations in, 20, 72
  polycistronic, 30–39, 49–50
  processing of, 3, 8–23, 35–36, 52–54, 59–67, 83–95, 177–185, 200–202
    from rRNA spacer regions, 103–105, 123–135
    sequential processing models for, 8–10, 53–55, 66–67
  structures of
    in Bacteriophage T4, 8–11, 213–215
    in *D. melanogaster*, 332–333
    in *E. coli*, 47–52, 78, 212
    in eukaryoutes, 11–13
    in nematodes, 291
    in yeast, 177–180
  synthetic, 60–67
  three-dimensional structure of, 11, 187–189, 196–198
Primer tRNAs, 507–512, 517–533
  reverse transcriptase interaction with, 511–512
  rRNA binding by, 519–530
  structure of, 508–509
  template interactions with, 508–511
  tRNA^Pro, 508–509, 519–533
  tRNA^Trp, 508–512, 518–533
Processing enzymes. *See also individual enzymes*
  precursor maturation by, 14–23, 123, 146–150
  purification of, 61
  sequence of utilization of, 8–10, 53–55, 66–67, 146–152
  signals for, 120, 151–152, 158–159
  three-dimensional specificity of, 10, 187–189, 196–198
Proline dehydrogenase, 500–501
Proline oxidase, 500–501
Protein synthesis, in vitro, 427–436
  mistranslation in, 428, 434
  MS2 RNA and, 428–433, 444–447
  tRNA as tracers, 430–433
  tRNA-dependence of, 428–430
psi⁺ factor, 391

Q base, 312

Repeated sequences, in tRNA^Tyr genes, 226–227, 237–239
Retroviruses, 517–519
Reverse transcriptase, 507
  primer tRNA interaction with, 511–512
  structure of, 511
Reverse transcription, 531
Rho-dependent transcription termination, 227
Ribonuclease D
  cation requirement of, 62

Ribonuclease D (*continued*)
  purification of, 61, 62
  specificity of, 19-20, 63-66
  tRNA biosynthesis, role in, 19, 61-67
  tRNA degradation by, 20, 64-65
Ribonuclease E, 139-152
Ribonuclease F, 143-152
Ribonuclease M5, 146, 148-150, 155-170
  cleavage sites of, 149-150, 155-170
  substrate specificity of
    hairpin loops, 158-159, 164
    symmetry features of, 158-164
    synthetic substrates and, 160-170
  subunit structure of, 157, 163-164
Ribonuclease M16, 103, 146-150
Ribonuclease M23, 147-150
Ribonuclease O, 8, 16-17, 33, 52-54, 109
Ribonuclease P, 44-55, 71-80, 83-95, 126-133, 139-152, 542-547
  cleavage sites of, 14-15, 73, 105, 126-133, 139-152
  eukaryotic forms of, 18, 79-80
  mutations affecting, 7, 14-17, 44-55, 73, 107-112, 139-152. *See also individual mutations*
  protein components of, 89-94
  purification of, 74-77, 85-87, 124
  as ribonucleoprotein
    buoyant density of, 75-76
    nuclease sensitivity of, 76-77
    reconstitution of, 77-78, 89-93
  RNA component of, 16, 75-79, 87-95
  substrate recognition by, 15-18, 79, 542-547
  tRNA biosynthesis, role in, 44-55, 71-80
Ribonuclease P2, 8, 16-17, 33, 39, 109, 112, 120, 150
Ribonuclease P3, 18, 53-54, 60, 67
Ribonuclease P4, 18
Ribonuclease Q, 18, 53-55, 60, 67
Ribonuclease Y, 60, 67
Ribonuclease II, 18, 53-54, 59-67
Ribonuclease III
  mutations in, 17-18, 107-115, 139-152
  30S rRNA processing by, 11, 17, 99-105, 109-120, 123-134, 139-152
    cleavage sites of, 101-104, 115-120, 126-134, 262, 264
    recognition sites of, 101-103, 113-115, 119-120
Ribonucleoprotein, RNase P as, 84-95
Ribosomal RNA (rRNA)
  gene clusters encoding, 139-152, 259-264
    primary transcript of, 17-18, 99-105, 139-152
  tRNA genes within, 5, 99-105, 139-140
  precursors of
    5S, 155-170
    30S, 84, 101-105, 139-152
  primer tRNA binding of, 519-530

rna-1 mutants, 12-13, 173-177
RNA polymerase III, 289-290
  initiation sites of, 334
  termination sequences of, 329
RNases. *See individual ribonucleases*
RNA tumor viruses, 507-513
rnc-105, 108-109, 139-152
rne, 139-152
rnp A, 45-52, 83
rnp B, 45-48, 83
rrn D, 99-105, 127-132
rrn E, 127-132
rrn-49, 108-109, 139-152
rrn X, 99-103, 127

Saccharomyces cerevisiae
  antisuppressors of, 395-404
  biosynthesis of tRNAs of, 11-13, 173-189, 191-206
  nonsense suppressor tRNAs of, 267, 379-392
  tRNA genes of, 6, 11, 13, 173-176, 191-196, 267-278
    mitochondrial, 217
    organization of, 215-216
  suppressor tRNA genes of, 267-278
    cloning of, 270
    linkage analysis of, 271-272
    sequence analysis of, 273-275
    SUP4, 271-276
Schizosaccharomyces pombe
  allosuppressors of, 410-411
  antisuppressors of, 410-411
  nonsense suppressor tRNAs of, 407-417
    amino acid specificity of, 412-414
    fine structure gene map of, 309-410
    genetic characterization of, 408-409
    in vitro characterization of, 411-412
    nucleotide sequences of, 414-415
    opal, 415
    UGA, 415
Second-site mutations in tRNA genes, 275-276
Spacer tRNA genes, 5, 103-105, 123-134, 143-149, 259-264, 281-285
Spinacia oleracea chloroplast tRNA genes, 281-285
Splicing activity, 173-189, 191-206
  endonuclease activity of
    cleavage sites of, 179-186, 195
    conformational requirements for, 187-189
    excision products of, 182-184
    intracellular location of, 203-205
  ligase activity of
    ATP requirement, 177-181
    substrate specificity, 179-182, 205
    purification of, 201-205
    substrate recognition by, 201-205

Subject Index

SupD, 364–366
SUP genes. *See Saccharomyces cerevisiae*, suppressor tRNA genes of Suppressor tRNAs. *See also* individual tRNAs
amber
  su^+1 (SupD), 347–348, 364–366
  su^+2, 345, 350, 355
  su^+3, 245–257, 343–344, 351–353
  su^+6, 348
  su^+7, 263, 344–345, 351–353
enhancing mutations, 369–373
frameshift, 439–443
genes
  in *S. cerevisiae*, 267–278
  in *S. pombe, 409–410*
  tRNA^{Tyr}, *E. coli*, 29–39, 245–257, 342
mapping tRNA genes with, 4, 5, 10
mischarging mutants, 350–354
nonsense. *See* Nonsense suppressor tRNAs
ocher
  su^+B, 346, 349
  $su\beta$, 347
  su^+4, 343–344
  su^+5, 347
  su^+8, 344–345
opal
  su^+9, 345
in *S. cerevisiae*
  serine inserting, 267
  tyrosine inserting, 267
temperature sensitive, 349–350, 363–374
UGA, 415, 421–424, 472

Temperature-sensitive tRNAs, 363–374
30S rRNA
primary transcript processing, 99–105, 109–120, 123–134, 139–152
Transcription
attenuation of, 469–477
initiation of, 334
in vitro, 30–39, 325–326, 332–336
termination of
  in *E. coli* tRNA^{Tyr} gene clusters, 236–237
  rho-dependent, 227
  sequence, 329
of tRNA genes
  in Bacteriophage T4, 6, 33–35, 39
  in *D. melanogaster*, 325–326, 332–336
  in *E. coli*, 30–33, 37–39, 240
  in Nematode, 289–291
of suppressor tRNA genes, 255–257
of synthetic suppressor tRNA^{Tyr} gene, 250–252
in *X. laevis* germinal vesicles, 13, 325–326, 332–336
Transcriptional unit, 240
Transducing phages carrying tRNA genes, 222, 259–263, 342

Transport
of branched-chain amino acids, 481–489
effect of *hisT*, 485–488
in mammalian cells, 487–488
model, 489
rho factors, 487
tRNA$^{Ala}_{1B}$, *E. coli*
biosynthesis of, 126–134
rRNA spacer location of, 5, 100, 103, 123, 259–264
tRNA$^{Arg}_2$, *D. melanogaster*
gene, 327, 330
transcript of, 332–336
tRNA^{Asp}, *D. melanogaster*
gene, 312–313
tRNA gene clusters
in *D. melanogaster*, 326–333
in *E. coli*, 222–235, 240–241, 356–359
tRNA gene duplication, 354–359
tRNA genes
chemical synthesis of, 245–257
in *D. melanogaster*, 295–305, 309–313, 317–323, 325–336, 559
  arrangement of, 326–328
  cloning of, 326–327
  localization of, 299–305, 309–313, 559
  mutations of, 300–305, 310–312, 323
  number of, 215, 295–299, 320
  sequence analysis, 328–333
  transcription of, 325–326, 332–336
in *E. coli*, 4–5, 30–33, 212
  location of chromosome, 212, 358, 563
  in rRNA operons. *See* Distal tRNA genes and Spacer tRNA genes.
  suppressor tRNA genes, 212. *See also* Suppressor tRNA genes.
  suppressor tRNA^{Tyr} genes, 245–257
in eukaryotes, 215–216
intervening sequences in, 173–189, 191–206, 276–277
number per organism, 6
in organelles, 217, 281–285
organization of, 211–217
  in Bacteriophage T4, 5, 33–39, 213–215
  in *B. mori*, 11
  in *D. melanogaster*, 295–305, 326–332
  in *E. coli*, 4, 5, 30–33, 212, 221–242
  in *S. cerevisiae*, 173–176, 191–196, 216
  in *X. laevis*, 216
plasmids carrying, 259–263
second-site mutations in, 275–276
transcription of. *See* Transcription.
transcriptional unit of, 211–212
transducing phages carrying, 222, 259–263, 342
tRNA^{Gln}, Bacteriophage T4
biosynthesis of, 17
dimeric precursor containing, 84–95
tRNA$^{Glu}_4$ gene, *D. melanogaster*, 309–312

578 Subject Index

tRNA$_2^{Glu}$, *E. coli*
  biosyntheis of, 126–134
  genes, rRNA spacer location of, 5, 123, 259–265
tRNA$_1^{Ile}$, *E. coli*
  biosynthesis of, 126–134
  genes, rRNA spacer location of, 5, 100, 103, 123, 259–265
tRNA isoacceptors, 283, 285, 312
tRNA^{Leu}, Bacteriophage T4, 84–95
tRNA$_3^{Leu}$, *S. cerevisiae*
  precursor nucleotide sequence of, 178
  precursor, splicing in, 176–182
tRNA-like structure of viral genomes, 539–553
  physiological role of, 552–553
  recognition of
    by elongation factors, 547
    by peptidyl transferase, 547
    by RNase P, 542, 546–547
    by ribosomal complex, 547–553
    by tRNA methyl transferase, 547
    by tRNA nucleotidyl transferase, 545
tRNA$_2^{Lys}$ gene. *D. melanogaster*, 296–305, 327–330
  transcription of, 332–336
  transcription initiation in, 334
tRNA$_2^{Met}$, *D. melanogaster*, 320–321
tRNA-mRNA interactions, nontriplet, 439–447
tRNA nucleotidyl transferase
  mutations affecting level of, 21–22
  recognition of tRNA-like structures, 545
  repair function of, 22
  role in tRNA biosynthesis, 9–10, 19–21, 59–67
tRNA^{Phe}, *E. coli*, 12
tRNA^{Phe}, *S. cerevisiae*
  gene, 191
  precursor
    nucleotide sequence of, 178
    splicing of, 177–180
tRNA precursors. *See* Precursor tRNA
tRNA^{Pro}, Bacteriophage T4, 8–11, 19
tRNA^{Pro}, role as primer tRNA. 508–512
tRNA-rRNA complexes, 523–527
tRNA^{Ser}, Bacteriophage T4
  biosynthetic pathway of, 8–11, 19–21
  suppressor derivative of, 10
tRNA$_{UCG}^{Ser}$, *S. cerevisiae*
  precursor, nucleotide sequence of, 179
  precursor, splicing of, 176–182
tRNA$_4^{Thr}$ gene, *E. coli*, 222, 223–235, 241
tRNA^{Trp}
  precursor
    nucleotide sequence of, 178
    splicing of, 176–182
  role as primer tRNA, 508–512
  su$^+$7, effect of attenuation, 475–476

temperature sensitive, 474–475
UGA suppressor, 421–424, 472
tRNA$_1^{Tyr}$ gene, *E. coli*
  promoter of, 223–225
  promoter mutations of, 223–225, 239–240
  repeated sequences in, 226–227, 237–239
  rho-dependent transcription termination site, 227
  structure of, 222–228
  3'-flanking region, 226–228
tRNA$_2^{Tyr}$ gene, *E. coli*, 229–233
tRNA$_3^{Tyr}$, *E. coli*
  biosynthesis of, 32–33, 111–113, 221
  gene, 221–241
    chemical synthesis of, 245–257
  precursor nucleotide sequence of, 74
  suppressor derivatives of, 7, 29–32, 37, 43, 72–74, 245–257
tRNA^{Tyr}, *S.cerevisiae*
  biosynthesis of, 11–13, 193–196
  genes, 191, 194–195
  precursor
    confirmation of, 192, 196–198
    isolation of, 192–193
    nucleotide sequence of, 178, 194–195
    splicing of, 181–185, 200–202
  suppressor genes of, 267
tRNA$_4^{Val}$ gene, *D. melanogaster*, 299–305
trp operon, 469–477
trpX, 472–474
Trytophanyl-tRNA^{Trp}, 469–477
ts136, 12–13, 173–177, 192–196
two-base recognition, 435, 444
TYMV RNA, 540–543
  aminoacylation of, 540–543
  structure of, 548–550

UGA suppressor tRNA, 421–424
  phosphorolysis of, 424
  photocrosslinking of, 424

Viral RNAs, 512, 517, 519

Wobble hypothesis, 431, 435

Xenopus laevis
  germinal vesicle extract of, 325–326, 332–336
  oocyte injection of, 216, 287–291
  tRNA genes, 6, 215

Yeast. *See Saccharomyces cerevisiae* and *Schizosaccharomyces pombe*